STATISTICAL MECHANICS

McGRAW-HILL SERIES IN ADVANCED CHEMISTRY

Senior Advisory Board

W. Conard Fernelius   Louis P. Hammett

Editorial Board

David N. Hume   Gilbert Stork
Edward L. King   Harold H. Williams
John A. Pople   Dudley R. Herschbach

BAIR   Introduction to Chemical Instrumentation
BALLHAUSEN   Introduction to Ligand Field Theory
BENSON   The Foundations of Chemical Kinetics
BIEMANN   Mass Spectrometry
DAVIDSON   Statistical Mechanics
DAVYDOV   (*Trans*. Kasha and Oppenheimer)   Theory of Molecular Excitons
DEAN   Flame Photometry
DJERASSI   Optical Rotatory Dispersion
ELIEL   Stereochemistry of Carbon Compounds
FITTS   Nonequilibrium Thermodynamics
HELFFERICH   Ion Exchange
HILL   Statistical Mechanics
HINE   Physical Organic Chemistry
KIRKWOOD AND OPPENHEIM   Chemical Thermodynamics
KOSOWER   Molecular Biochemistry
LAITINEN   Chemical Analysis
McDOWELL   Mass Spectrometry
PITZER AND BREWER   (*Revision of* Lewis and Randall)   Thermodynamics
POPLE, SCHNEIDER, AND BERNSTEIN   High-resolution Nuclear Magnetic Resonance
PRYOR   Mechanisms of Sulfur Reactions
ROBERTS   Nuclear Magnetic Resonance
ROSSOTTI AND ROSSOTTI   The Determination of Stability Constants
STREITWIESER   Solvolytic Displacement Reactions
WIBERG   Laboratory Technique in Organic Chemistry

STATISTICAL MECHANICS

NORMAN DAVIDSON

Professor of Chemistry
California Institute of Technology

McGRAW-HILL BOOK COMPANY, INC.   1962
New York   San Francisco   Toronto   London

STATISTICAL MECHANICS

Copyright © 1962 by the McGraw-Hill Book Company, Inc. Printed in the United States of America. All rights reserved. This book, or parts thereof, may not be reproduced in any form without permission of the publishers. *Library of Congress Catalog Card Number* 61-12946

THE MAPLE PRESS COMPANY, YORK, PA.

II

15454 -

Dedicated to the memory of
William Moffitt

Preface

The statistical mechanics of dilute systems of independent particles at equilibrium is a subject which is essentially fully developed. The practicing chemist should be able to apply this theory with assurance and accuracy to calculate the thermodynamic properties of substances in the ideal-gas state from molecular structure data.

My first objectives have been to develop this theory in a simple, logical, and understandable way, and to describe the applications in sufficient detail, with illustrative problems, so that the reader can acquire the desired skill in this elementary part of the subject.

In the latter part of the book, I have tried to give a simple but rigorous account of the theories of the canonical ensemble and of the grand ensemble for treating systems of interacting particles. Several rather elementary but important examples are then studied, to illustrate the applications of these general theories. The most prominent and important manifestation of fluctuation phenomena for the experimental scientist is noise. I have therefore included a discussion of this topic in the chapter on fluctuations. However, I have not included a serious discussion of practical theories of liquids and dense gases. This is an important topic in statistical mechanics but it is difficult and complicated. The theoretical calculations do not as yet give good agreement with experiment. There are several excellent treatises by experts in the field, and I prefer to refer the interested student to this literature. I hope that I have carried the theory of systems of interacting particles far enough to provide a good foundation for such further study.

In addition, the latter half of the book presents discussions of some topics—paramagnetism, dielectrics, and ionic solutions—essentially from the point of view of the statistical mechanics of independent particles—but with at least an indication of how the theories of the canonical ensemble and the grand ensemble can be applied when interparticle interactions are important.

This text has developed from my lecture notes for a one-year graduate course at the California Institute. Many of the students are first-year graduate students (with an occasional bright senior) who have not had a serious course in quantum mechanics, although they are usually taking such a course simultaneously. I have therefore included a chapter which is an introductory discussion of the relevant parts of quantum mechanics. Fortunately, the principles of equilibrium statistical mechanics can be developed very satisfactorily on the basis of an elementary and unsophisticated formulation of quantum mechanics.

I hope that the problems are instructive and/or interesting. They do not however cover the principles and their application so well that the ability to do them demonstrates a mastery of statistical mechanics as expounded in the text. Therefore, in examinations and homework assignments, in addition to problems I often ask for derivations which are given in the text. Some problems are inserted in the body of the text because I consider it desirable that they be done before proceeding further.

At the end of the course, I usually ask for a short paper on a subject of the student's own choosing. There is a tendency for the student to bite off more than he can chew; nevertheless, the results are generally healthy. Occasionally, I ask my students to make up some new problems and, if possible, to solve them. The results of this assignment are interesting and informative. Several of the problems in the text were obtained in this way.

I have solicited and received help and advice from so many colleagues that it would be unwise to attempt individual acknowledgments. The students in my course have contributed much to this book by their conscious criticisms and by my observations of their natural reactions (including some yawning and sleeping). The cheerful cooperation and painstaking care of Mrs. Ruth Hanson, Della Brown, and Allene Luke of the departmental secretarial staff are deeply appreciated.

I do have a special debt to Dr. Verner Schomaker, Dr. Robert Mazo, and the late Dr. William Moffitt. Each, in his own way, has added greatly to my understanding by answering many questions and discussing many problems with me.

I am dedicating this book to William Moffitt as an expression of my admiration and affection. It was from my conversations with him during my one-year stay at Harvard that I first gained confidence in the validity of my approach to statistical mechanics. Without this confidence, I would not have had the courage to write this book. In his own work, Bill Moffitt was a theorist with a passion for elegance and generality; but he insisted that the function of the theorist was to be useful. I hope that, were he still alive, he would think this book useful.

Norman Davidson

Contents

PREFACE vii

1. Introduction 1
2. Classical Mechanics 4
3. Quantum Mechanics 22
4. Thermodynamics 43
5. Mathematics 65
6. The Statistical Mechanics of a System of Independent Particles 75
7. Statistical Mechanics and Chemical Equilibrium 97
8. Distribution Laws, Partition Functions, and Thermodynamic Functions for Atoms and Diatomic Molecules 105
9. Nuclear Spin Statistics. Isotope Effects. 130
10. Classical Statistical Mechanics. The Kinetic Theory of Gases. 146
11. Polyatomic Molecules 169
12. Black-body Radiation 211
13. Canonical and Grand Canonical Ensembles 240
14. Fluctuations, Noise, Brownian Motion 265
15. Real Gases 310
16. Lattice Vibrations and the Thermodynamic Properties of Crystals. Order-Disorder Phenomena 351
17. A Digression on Electricity, Magnetism, and Units. . . . 394
18. Dielectric Phenomena 402
19. Magnetic Phenomena 428
20. Distribution Functions and the Theory of Dense Fluids . . 468
21. Solutions of Electrolytes. The Debye-Hückel Theory . . 485

GENERAL REFERENCES 528
Appendix 1. Physical Constants and Conversion Factors. . 530
Appendix 2. Thermodynamic Functions for the Harmonic Oscillator 532

INDEX 535

1
Introduction

A macroscopic system at equilibrium has certain properties, for example, the energy, heat capacity, entropy, volume, pressure, and coefficient of expansion, which are of particular interest in thermodynamics. From the standpoint of thermodynamics, some of these quantities must be determined by experiment; the laws of thermodynamics provide relations by which it is then possible to calculate other quantities. Thus, if we know from experiment the equation of state of a substance in the form $V = V(T,P)$ and we know the entropy at one pressure, P, we can calculate the entropy at any other pressure and the same temperature from the thermodynamic relation $(\partial S/\partial P)_T = -(\partial V/\partial T)_P$.

Our intuition tells us that it should not be necessary to measure the macroscopic properties of a system but that it should be possible to calculate them if the properties of the constituent molecules and the laws of force (the intermolecular interactions) between the molecules are known.

Statistical mechanics is a method, and in practice *the* method, for calculating the properties of macroscopic systems from the properties of the constituent molecules. Quantum mechanics provides the fundamental laws for calculating the properties of individual molecules and their intermolecular interactions. Statistical mechanics starts with these results and introduces a statistical hypothesis about the behavior of systems containing a large number of molecules. (The hypothesis that we shall use is actually that of equal a priori probabilities of individual quantum states; but its exact nature does not concern us right now.) It is then possible to predict many of the important properties of macroscopic systems.

This is not the only conceivable method for making predictions about macroscopic systems. One could, in principle, resort to a straightforward mechanical calculation. Consider, for example, a gas containing 10^{24} atoms. Suppose that, in this case, classical mechanics is a satisfactory approximation and that we need not use quantum mechanics. It is necessary to know, at some initial time t_0, the 3×10^{24} position coordinates and the 3×10^{24} velocity coordinates of all the particles. As we shall see in the next chapter, one can then, in principle, solve the equations of motion and calculate the positions and velocities of all the particles at all future times.

There are two difficulties with this direct approach. In the first place, the calculation is far too complex and cannot actually be performed. But suppose that, with a fantastically effective computer, it were possible to calculate the trajectories of all the particles. The results might be a gigantic data sheet giving the 3×10^{24} position coordinates of the atoms every 10^{-11} sec. (This interval of time is chosen as reasonable because a typical atom at room temperature and atmospheric pressure undergoes a collision about every 10^{-10} sec.) The history of the system for 1 sec would require 3×10^{35} entries. Such an enumeration of the data would be quite indigestible. We would look for a statistical summary of the data, and we would calculate certain statistical functions: the number of atoms with velocities in a certain interval, the number of collisions between atoms per second, the average number of atoms that are within a given distance of another atom at any particular time, the momentum exchange with the walls in any time interval, etc.

We shall see that by the methods of statistical mechanics it is possible to calculate these functions directly, without first calculating the detailed behavior of the system. A knowledge of these statistical functions is usually sufficient for the calculation of the macroscopic properties of a system. The statistical mechanical calculation is not just a cowardly expedient that we resort to because of our inability to make a complete calculation (although it is that, too); for most problems it contains all the information that we want about the system without going into unnecessary detail.*

Equilibrium statistical mechanics treats the properties of systems at equilibrium. The calculation of the time-varying properties of a system which is not at equilibrium is more difficult. We are then interested in such properties as viscosity, heat conductivity, paramagnetic relaxation times, and chemical reaction rates. This is, in general, nonequilibrium statistical mechanics. It is not yet nearly so well developed a subject as equilibrium statistical mechanics. We shall be principally, but not exclusively, concerned with topics in equilibrium statistical mechanics.

Statistical mechanics is firmly based on quantum mechanics. The

---

* It is interesting to note, however, that some difficult statistical mechanical problems are now being studied by detailed calculations of the mechanical behavior of small prototype systems. In one such calculation, the behavior of a system of hard spheres containing 32 particles was computed through 7,000 total collisions in an hour with an IBM-704 computer [B. J. Alder and T. E. Wainwright, *J. Chem. Phys.*, **27**: 1208 (1957)].

usual presentations of quantum mechanics presume a prior knowledge of classical mechanics; furthermore, classical mechanics is directly useful for many problems in statistical mechanics. We shall see that thermodynamics is closely related to statistical mechanics. Therefore, in the next three chapters, we review classical mechanics, quantum mechanics, and thermodynamics.

2

Classical Mechanics*

2-1. Introduction. We begin with a brief review of classical mechanics. We shall derive the equations of motion in Hamiltonian form from the more familiar Newtonian equations and shall introduce the concept of phase space.

2-2. Mathematical Prelude. Our object in this section is to illustrate, for a simple case, some of the mathematical operations needed in transforming the equations of motion from one system of coordinates to another. We shall treat the same problem for the general case in the next section.

Consider a single particle constrained to move in the xy plane. Newton's equations of motion are

$$m\frac{d^2x}{dt^2} = m\ddot{x} = F_x \qquad m\frac{d^2y}{dt^2} = m\ddot{y} = F_y \qquad (2\text{-}1)$$

where F_x and F_y are the forces in the x and y directions on the particle. We assume that the forces are derivable from a potential energy $U(x,y)$,

$$F_x = -\frac{\partial U(x,y)}{\partial x} \qquad F_y = -\frac{\partial U(x,y)}{\partial y} \qquad (2\text{-}2)$$

and that the potential energy is a function of the position coordinates of the particle, but not an explicit function of time or of the velocities $\dot{x}$ and $\dot{y}$. If these conditions are satisfied, the system is said to be conservative.

Suppose, for example, that the potential function is

$$U(x,y) = \tfrac{1}{2}ax^2 + bxy + \tfrac{1}{2}cy^2 \qquad (2\text{-}3)$$

The equations of motion then are

$$\begin{aligned} m\ddot{x} &= -ax - by \\ m\ddot{y} &= -bx - cy \end{aligned} \qquad (2\text{-}4)$$

---

* A word of apology is in order. As regards mathematical content, this beginning chapter is one of the more difficult ones in the text. This may be bad pedagogy, but it is advantageous to base our further studies on a more general formulation of classical mechanics than Newton's laws of motion as presented in elementary physics classes.

Sec. 2-2]   CLASSICAL MECHANICS

The equations of motion (2-4) can be integrated to give a solution

$$x = x(t) \qquad y = y(t) \qquad (2\text{-}5)$$

which we call a trajectory. The particular trajectory depends upon the initial conditions, for example, the positions and velocities at $t = 0$, and there is a family of solutions for different initial conditions.

The kinetic energy K is

$$K(\dot{x},\dot{y}) = \tfrac{1}{2}m\left(\frac{dx}{dt}\right)^2 + \tfrac{1}{2}m\left(\frac{dy}{dt}\right)^2 = \tfrac{1}{2}m\dot{x}^2 + \tfrac{1}{2}m\dot{y}^2 \qquad (2\text{-}6)$$

The Lagrangian function L is defined as

$$L(x,y,\dot{x},\dot{y}) = K - U = \tfrac{1}{2}m\dot{x}^2 + \tfrac{1}{2}m\dot{y}^2 - \tfrac{1}{2}ax^2 - bxy - \tfrac{1}{2}cy^2 \qquad (2\text{-}7)$$

If we transform to polar coordinates,

$$\begin{aligned} x &= r \cos \phi \\ y &= r \sin \phi \end{aligned} \qquad (2\text{-}8a)$$

then

$$\begin{aligned} \dot{x} &= \dot{r} \cos \phi - r \sin \phi \, \dot{\phi} \\ \dot{y} &= \dot{r} \sin \phi + r \cos \phi \, \dot{\phi} \end{aligned} \qquad (2\text{-}8b)$$

By substitution in (2-6) and (2-3), we find

$$K = \tfrac{1}{2}m\dot{r}^2 + \tfrac{1}{2}mr^2\dot{\phi}^2 \qquad (2\text{-}9a)$$
$$U = \tfrac{1}{2}ar^2 \cos^2 \phi + br^2 \sin \phi \cos \phi + \tfrac{1}{2}cr^2 \sin^2 \phi \qquad (2\text{-}9b)$$

so that

$$L(r,\phi,\dot{r},\dot{\phi}) = \tfrac{1}{2}m\dot{r}^2 + \tfrac{1}{2}mr^2\dot{\phi}^2 - \tfrac{1}{2}ar^2 \cos^2 \phi \\ - br^2 \sin \phi \cos \phi - \tfrac{1}{2}cr^2 \sin^2 \phi \qquad (2\text{-}9c)$$

We notice that the position variable r enters into the expression for K and that the expression is quadratic in the velocities.

In Eq. (2-7), we can regard the Lagrangian L as a function of the independent variables x, y, $\dot{x}$, and $\dot{y}$. We can then write

$$\frac{\partial L(x,y,\dot{x},\dot{y})}{\partial \dot{x}} = m\dot{x} \qquad \frac{\partial L}{\partial \dot{y}} = m\dot{y}$$

$$\frac{\partial L(x,y,\dot{x},\dot{y})}{\partial x} = -ax - by \qquad \frac{\partial L}{\partial y} = -bx - cy$$

For L as a function of polar coordinates r, ϕ, $\dot{r}$, $\dot{\phi}$,

$$\frac{\partial L(r,\phi,\dot{r},\dot{\phi})}{\partial \dot{\phi}} = mr^2\dot{\phi}$$

$$\frac{\partial L(r,\phi,\dot{r},\dot{\phi})}{\partial r} = mr\dot{\phi}^2 - ar \cos^2 \phi - 2br \sin \phi \cos \phi - cr \sin^2 \phi$$

and similar equations for $\partial L/\partial \dot{r}$, $\partial L/\partial \phi$.

Sometimes the variables which are held fixed during partial differentiation will be indicated by the notation

$$\left(\frac{\partial L}{\partial \phi}\right)_{r,\phi,\dot{r}}$$

This can also be indicated by writing

$$\frac{\partial L(r,\phi,\dot{r},\phi)}{\partial \phi}$$

Where the context makes the meaning clear, we shall often write simply

$$\frac{\partial L}{\partial \phi}$$

[Incidentally, note the difference between the ways in which a natural scientist and a mathematician regard function notation. When we write $L(x,y,\dot{x},\dot{y})$ we mean the physical quantity, the Lagrangian function, $K - U$, expressed in cartesian coordinates, as in Eq. (2-7), and $L(r,\phi,\dot{r},\phi)$ means the same physical quantity, $K - U$, expressed in polar coordinates. To a mathematician, the function $L(x,y,\dot{x},\dot{y})$ means the functional form

$$L(x,y,\dot{x},\dot{y}) = \frac{m}{2}\dot{x}^2 + \frac{m}{2}\dot{y}^2 - \tfrac{1}{2}ax^2 - bxy - \tfrac{1}{2}cy^2$$

so that $L(r,\phi,\dot{r},\phi)$ would be the same form with r in place of x, ϕ in place of y, etc.; i.e.,

$$L(r,\phi,\dot{r},\phi) = \frac{m}{2}\dot{r}^2 + \frac{m}{2}\dot{\phi}^2 - \tfrac{1}{2}ar^2 - br\phi - \tfrac{1}{2}c\phi^2$$

For correct mathematical use of function notation, if $L(x,y,\dot{x},\dot{y})$ is defined by (2-7), then the transformation (2-8) would transform L to a new function $M(r,\phi,\dot{r},\phi)$, with

$$L[x(r,\phi),y(r,\phi),\dot{x}(r,\phi,\dot{r},\phi),\dot{y}(r,\phi,\dot{r},\phi)] = M(r,\phi,\dot{r},\phi)$$
$$M(r,\phi,\dot{r},\phi) = \tfrac{1}{2}m\dot{r}^2 + \tfrac{1}{2}mr^2\dot{\phi}^2 - \tfrac{1}{2}ar^2\cos^2\phi$$
$$- br^2 \sin\phi \cos\phi - \tfrac{1}{2}cr^2 \sin^2\phi$$

so that the function M is the function which, in physical language, we called $L(r,\phi,\dot{r},\phi)$ in (2-9c).]

For arbitrary variations in x, y, $\dot{x}$, and $\dot{y}$, the variation in L is given by

$$dL = \frac{\partial L}{\partial x}dx + \frac{\partial L}{\partial y}dy + \frac{\partial L}{\partial \dot{x}}d\dot{x} + \frac{\partial L}{\partial \dot{y}}d\dot{y}$$

For a particular trajectory, x, y, $\dot{x}$, and $\dot{y}$ are known functions of t. The variation of L with time can then be calculated from the equation

$$\frac{dL}{dt} = \frac{\partial L}{\partial x}\frac{dx}{dt} + \frac{\partial L}{\partial y}\frac{dy}{dt} + \frac{\partial L}{\partial \dot{x}}\frac{d\dot{x}}{dt} + \frac{\partial L}{\partial \dot{y}}\frac{d\dot{y}}{dt} = \frac{\partial L}{\partial x}\dot{x} + \frac{\partial L}{\partial y}\dot{y} + \frac{\partial L}{\partial \dot{x}}\ddot{x} + \frac{\partial L}{\partial \dot{y}}\ddot{y}$$

Problem 2-1. For a system of two particles moving in two dimensions, with masses m_1 and m_2 and cartesian coordinates x_1, y_1, x_2, y_2, the kinetic energy is

$$K = \tfrac{1}{2} m_1 (\dot{x}_1^2 + \dot{y}_1^2) + \tfrac{1}{2} m_2 (\dot{x}_2^2 + \dot{y}_2^2)$$

We now replace x_1, y_1, x_2, y_2 by four new variables, X, Y, x_{12}, y_{12}, where

$$X = \frac{m_1 x_1 + m_2 x_2}{m_1 + m_2} \qquad Y = \frac{m_1 y_1 + m_2 y_2}{m_1 + m_2}$$
$$x_{12} = x_2 - x_1 \qquad\qquad y_{12} = y_2 - y_1$$

X and Y are the coordinates of the center of gravity; x_{12} and y_{12} are the relative coordinates, which give the position of the second particle with respect to the first. Express K in terms of the velocities $\dot{X}$, $\dot{Y}$, $\dot{x}_{12}$, $\dot{y}_{12}$ in the new system of variables. Explain the significance of this calculation.

By analogy, you can now write the corresponding expressions for the kinetic energy for a system of two particles in three dimensions.

Problem 2-2. The transformation between spherical polar coordinates and cartesian coordinates, as illustrated in Fig. 2-1, is

$$x = r \sin \theta \cos \phi$$
$$y = r \sin \theta \sin \phi$$
$$z = r \cos \theta$$

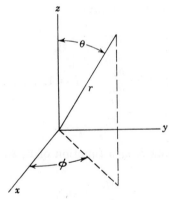

Fig. 2-1. Relation between cartesian coordinates and spherical polar coordinates.

Express the kinetic energy of a single particle in terms of its spherical coordinates r, θ, ϕ, and $\dot{r}$, $\dot{\theta}$, $\dot{\phi}$.

2-3. The Lagrangian Equations of Motion.* There are several formulations of the laws of mechanics that are more general than Newton's

---

* The subject matter of the next few sections is discussed in innumerable texts on classical mechanics. There are also clear expositions in Pauling and Wilson, "Introduction to Quantum Mechanics" [21], and Eyring, Walter, and Kimball, "Quantum Chemistry" [20].

equations. The two that we shall consider are the equations of motion in the Lagrangian form and in the Hamiltonian form. These formulations are easier to apply than Newton's equations for a number of problems in mechanics—especially when the most suitable coordinate system is not cartesian coordinates. However, our principal purpose is to derive Hamilton's equations, because these equations are used in the formulation of quantum mechanics and because they play a central role in statistical mechanics. In our treatment, Hamilton's equations of motion will be derived from Lagrange's equations, which will be derived from Newton's equations.

Consider a system composed of n particles, with masses $m_1, \ldots, m_i, \ldots, m_n$, and cartesian coordinates $x_1, y_1, z_1, \ldots, x_n, y_n, z_n$.

The kinetic energy is given by

$$K = \frac{1}{2} \sum_{i=1}^{n} m_i(\dot{x}_i^2 + \dot{y}_i^2 + \dot{z}_i^2) \tag{2-10}$$

We assume that there is a potential, $U(x_1, \ldots, z_n)$, which is a function of the position coordinates only.

Newton's equations of motion are

$$\begin{aligned} m_i \ddot{x}_i &= -\frac{\partial U}{\partial x_i} \\ m_i \ddot{y}_i &= -\frac{\partial U}{\partial y_i} \quad i = 1, \ldots, n \\ m_i \ddot{z}_i &= -\frac{\partial U}{\partial z_i} \end{aligned} \tag{2-11}$$

From (2-10), regarding K as a function of $\dot{x}_i$, $\dot{y}_i$, and $\dot{z}_i$, we have

$$\frac{\partial K}{\partial \dot{x}_i} = m_i \dot{x}_i$$

The Lagrangian function L is defined by

$$L(x_1, \ldots, z_n; \dot{x}_1, \ldots, \dot{z}_n) = K(\dot{x}_1, \ldots, \dot{z}_n) - U(x_1, \ldots, z_n) \tag{2-12}$$

and we see that

$$\frac{\partial L}{\partial \dot{x}_i} = \frac{\partial K}{\partial \dot{x}_i} \qquad \frac{\partial L}{\partial x_i} = -\frac{\partial U}{\partial x_i} \tag{2-13}$$

so that Eqs. (2-11) can be rewritten as

$$\frac{d}{dt}\left(\frac{\partial L}{\partial \dot{x}_i}\right) = \frac{\partial L}{\partial x_i}$$
$$\frac{d}{dt}\left(\frac{\partial L}{\partial \dot{y}_i}\right) = \frac{\partial L}{\partial y_i} \qquad (2\text{-}14)$$
$$\frac{d}{dt}\left(\frac{\partial L}{\partial \dot{z}_i}\right) = \frac{\partial L}{\partial z_i}$$

These are the equations of motion in Lagrangian form for cartesian coordinates. Let us see what happens to these equations under a transformation of coordinates. Let there be $3n$ generalized coordinates $q_1, \ldots, q_{3n}$ (center of mass plus internal coordinates, spherical polar coordinates, elliptical coordinates, or any other suitable coordinates for the problem at hand), which are related to the cartesian coordinates of the individual particles by the $3n$ transformation equations

$$\begin{aligned} x_1 &= x_1(q_1, \ldots, q_{3n}) \\ y_1 &= y_1(q_1, \ldots, q_{3n}) \\ &\ldots\ldots\ldots\ldots\ldots \\ x_i &= x_i(q_1, \ldots, q_{3n}) \\ &\ldots\ldots\ldots\ldots\ldots \\ z_n &= z_n(q_1, \ldots, q_{3n}) \end{aligned} \qquad (2\text{-}15a)$$

The velocities are then given by

$$\begin{aligned} \dot{x}_1 &= \sum_{j=1}^{3n} \frac{\partial x_1}{\partial q_j} \dot{q}_j \\ \dot{y}_i &= \sum_{j=1}^{3n} \frac{\partial y_i}{\partial q_j} \dot{q}_j \qquad i = 1, \ldots, n \qquad (2\text{-}15b) \\ \dot{z}_n &= \sum_{j=1}^{3n} \frac{\partial z_n}{\partial q_j} \dot{q}_j \end{aligned}$$

We now want to regard $x_1, \ldots, z_n; \dot{x}_1, \ldots, \dot{z}_n$ as $6n$ independent variables for expressing L and $q_1, \ldots, q_{3n}; \dot{q}_1, \ldots, \dot{q}_{3n}$ as another set of $6n$ variables for expressing L. The relations between the two sets of variables are given by the $6n$ equations (2-15). According to Eq. (2-15a), the functional relationships between the position coordinates do not contain the velocities, so that

$$\frac{\partial x_i}{\partial \dot{q}_j} = \frac{\partial y_i}{\partial \dot{q}_j} = \frac{\partial z_i}{\partial \dot{q}_j} = 0 \qquad i = 1, \ldots, n; j = 1, \ldots, 3n \qquad (2\text{-}16)$$

Furthermore, according to (2-15b), the velocity $\dot{y}_i$ is a linear function of all the $\dot{q}_j$, and

$$\frac{\partial \dot{y}_i(q_1, \ldots, q_{3n}; \dot{q}_1, \ldots, \dot{q}_{3n})}{\partial \dot{q}_j} = \frac{\partial y_i}{\partial q_j} \qquad (2\text{-}17)$$

We can now express $\partial L/\partial q_k$ and $\partial L/\partial \dot{q}_k$ in terms of the derivatives $\partial L/\partial x_i$ and $\partial L/\partial \dot{x}_i$, using the transformation relations (2-15).

We have

$$\frac{\partial L}{\partial q_k} = \sum_{i=1}^{n} \left(\frac{\partial L}{\partial x_i} \frac{\partial x_i}{\partial q_k} + \frac{\partial L}{\partial y_i} \frac{\partial y_i}{\partial q_k} + \frac{\partial L}{\partial z_i} \frac{\partial z_i}{\partial q_k} \right)$$

$$+ \sum_{i=1}^{n} \left(\frac{\partial L}{\partial \dot{x}_i} \frac{\partial \dot{x}_i}{\partial q_k} + \frac{\partial L}{\partial \dot{y}_i} \frac{\partial \dot{y}_i}{\partial q_k} + \frac{\partial L}{\partial \dot{z}_i} \frac{\partial \dot{z}_i}{\partial q_k} \right) \qquad k = 1, \ldots, 3n \qquad (2\text{-}18)$$

$$\frac{\partial L}{\partial \dot{q}_k} = \sum_{i=1}^{n} \left(\frac{\partial L}{\partial x_i} \frac{\partial x_i}{\partial \dot{q}_k} + \frac{\partial L}{\partial y_i} \frac{\partial y_i}{\partial \dot{q}_k} + \frac{\partial L}{\partial z_i} \frac{\partial z_i}{\partial \dot{y}_k} \right)$$

$$+ \sum_{i=1}^{n} \left(\frac{\partial L}{\partial \dot{x}_i} \frac{\partial \dot{x}_i}{\partial \dot{q}_k} + \frac{\partial L}{\partial \dot{y}_i} \frac{\partial \dot{y}_i}{\partial \dot{q}_k} + \frac{\partial L}{\partial \dot{z}_i} \frac{\dot{z}_i}{\dot{q}_k} \right) \qquad (2\text{-}19)$$

The first sum of terms in (2-19) is zero because $\partial x_i/\partial \dot{q}_k = 0$ [Eq. (2-16)]. In the second sum, we substitute $\partial x_i/\partial q_k$ for $\partial \dot{x}_i/\partial \dot{q}_k$, which is justified by Eq. (2-17). Then take the total derivative of both sides of (2-19) with respect to time:

$$\frac{d}{dt}\left(\frac{\partial L}{\partial \dot{q}_k}\right) = \sum_{i=1}^{n} \frac{\partial x_i}{\partial q_k} \frac{d}{dt}\left(\frac{\partial L}{\partial \dot{x}_i}\right) + \frac{\partial y_i}{\partial q_k} \frac{d}{dt}\left(\frac{\partial L}{\partial \dot{y}_i}\right) + \frac{\partial z_i}{\partial q_k} \frac{d}{dt}\left(\frac{\partial L}{\partial \dot{z}_i}\right)$$

$$+ \sum_{i=1}^{n} \frac{\partial L}{\partial \dot{x}_i} \frac{d}{dt}\left(\frac{\partial x_i}{\partial q_k}\right) + \frac{\partial L}{\partial \dot{y}_i} \frac{d}{dt}\left(\frac{\partial y_i}{\partial q_k}\right) + \frac{\partial L}{\partial \dot{z}_i} \frac{d}{dt}\left(\frac{\partial z_i}{\partial q_k}\right) \qquad (2\text{-}20)$$

In the first sum of (2-20), we set

$$\frac{d}{dt}\left(\frac{\partial L}{\partial \dot{x}_i}\right) = \frac{\partial L}{\partial x_i}$$

by Lagrange's equation. For the second sum, we have

$$\frac{d}{dt}\left(\frac{\partial x_i}{\partial q_k}\right) = \frac{\partial \dot{x}_i}{\partial q_k}$$

since the order of differentiation is immaterial. With these substitutions, (2-20) becomes

$$\frac{d}{dt}\left(\frac{\partial L}{\partial \dot{q}_k}\right) = \sum_{i=1}^{n}\left(\frac{\partial L}{\partial x_i}\frac{\partial x_i}{\partial q_k} + \frac{\partial L}{\partial y_i}\frac{\partial y_i}{\partial q_k} + \frac{\partial L}{\partial z_i}\frac{\partial z_i}{\partial q_k}\right)$$

$$+ \sum_{i=1}^{n}\left(\frac{\partial L}{\partial \dot{x}_i}\frac{\partial \dot{x}_i}{\partial q_k} + \frac{\partial L}{\partial \dot{y}_i}\frac{\partial \dot{y}_i}{\partial q_k} + \frac{\partial L}{\partial \dot{z}_i}\frac{\partial \dot{z}_i}{\partial q_k}\right) \qquad k = 1, \ldots, 3n \quad (2\text{-}21)$$

We recognize from (2-18) that the right-hand side of (2-21) is just $\partial L/\partial q_k$. Therefore

$$\frac{d}{dt}\left(\frac{\partial L}{\partial \dot{q}_k}\right) = \frac{\partial L}{\partial q_k} \qquad k = 1, \ldots, 3n \quad (2\text{-}22)$$

We have thus proved the important theorem that Lagrange's equations of motion are the same in all coordinate systems. If by one means or another we can express the Lagrangian $L = K - U$ in terms of the coordinates q, $\dot{q}$, we can immediately write the equations of motion (2-22) in the q coordinate system.

Problem 2-3. For a particle moving in three dimensions under the influence of a spherically symmetrical potential $U = U(r)$, use the results of Prob. 2-2 to write out the Lagrangian and the equations of motion in spherical coordinates.

Problem 2-4. Given a system of two particles as in Prob. 2-1, with the only potential their potential of interaction, which is a function of the relative coordinates only,

$$U = U(x_2 - x_1, y_2 - y_1, z_2 - z_1) = U(x_{12}, y_{12}, z_{12})$$

Use Lagrange's equations to write the equations of motion for the system in terms of the coordinates X, Y, Z and x_{12}, y_{12}, z_{12}. Describe the significance of the results in words.

2-4. Hamilton's Equations. The Lagrangian function L is regarded as a function of the $6n$ coordinates $q_1, \ldots, q_{3n}; \dot{q}_1, \ldots, \dot{q}_{3n}$. We shall eliminate the $\dot{q}$'s as independent variables by introducing $3n$ new variables by the equations

$$p_k = \frac{\partial L(q_1, \ldots, q_{3n}; \dot{q}_1, \ldots, \dot{q}_{3n})}{\partial \dot{q}_k} \qquad k = 1, \ldots, 3n \quad (2\text{-}23)$$

By solving the $3n$ equations (2-23), we can express each $\dot{q}$ as a function of $q_1, \ldots, q_{3n}; p_1, \ldots, p_{3n}$:

$$\dot{q}_j = \dot{q}_j(q_1, \ldots, q_{3n}; p_1, \ldots, p_{3n}) \qquad j = 1, \ldots, 3n \quad (2\text{-}24)$$

The p variables are called the generalized momenta corresponding to the generalized coordinates q. The two variables q_i and p_i are spoken of as being conjugate variables.

We shall define Hamilton's function H as being the total energy $K + U$:

$$H(q_1, \ldots, q_{3n}; p_1, \ldots, p_{3n}) = K + U = 2K - L \quad (2\text{-}25)$$

In (2-25), we can think that K and L were initially given as functions of the q's and $\dot{q}$'s; by using Eq. (2-24), they can be expressed as functions of the q's and p's.

It is essential to realize that the kinetic energy is always a quadratic function of the velocities $\dot{q}_i$, that is,

$$K(q_1, \ldots, q_{3n}; \dot{q}_1, \ldots, \dot{q}_{3n}) = \sum_{i=1}^{3n} \sum_{j=1}^{3n} \dot{q}_i a_{ij} \dot{q}_j$$

$$= a_{11}\dot{q}_1^2 + 2a_{12}\dot{q}_1\dot{q}_2 + \cdots + 2a_{3n-1,3n}\dot{q}_{3n-1}\dot{q}_{3n} + a_{3n3n}\dot{q}_{3n}^2 \quad (2\text{-}26)$$

The coefficients a_{ij} in (2-26) are in general functions of the q's as in (2-9a), where the coefficient of $\dot{\phi}^2$ is $\frac{1}{2}mr^2$. We can always choose the coefficients $a_{ij} = a_{ji}$ in the double sum of (2-26), so that we write, for example, $2a_{12}\dot{q}_1\dot{q}_2$ instead of $a_{12}\dot{q}_1\dot{q}_2 + a_{21}\dot{q}_2\dot{q}_1$. Then, since U is not a function of the velocities,

$$p_k = \frac{\partial L}{\partial \dot{q}_k} = \frac{\partial (K-U)}{\partial \dot{q}_k} = \frac{\partial K(q_1, \ldots, q_{3n}; \dot{q}_1, \ldots, \dot{q}_{3n})}{\partial \dot{q}_k}$$

$$= 2 \sum_{j=1}^{3n} a_{kj}\dot{q}_j \quad (2\text{-}27)$$

[Readers who are not experienced with general summation notation can verify (2-27) by differentiating the expression for K as displayed in detail in the right-hand expression of (2-26).]

It follows from (2-27) that

$$\sum_{k=1}^{3n} \dot{q}_k p_k = \sum_{k=1}^{3n} \sum_{j=1}^{3n} 2\dot{q}_k a_{kj} \dot{q}_j = 2K \quad (2\text{-}28)$$

We can therefore rewrite (2-25) as

$$H(q_1, \ldots, q_{3n}; p_1, \ldots, p_{3n})$$
$$= \sum_{k=1}^{3n} p_k \dot{q}_k - L(q_1, \ldots, q_{3n}; \dot{q}_1, \ldots, \dot{q}_{3n}) \quad (2\text{-}29)$$

where it is understood that the $\dot{q}$'s can be expressed in terms of q's and p's by (2-24). [We may remark that Eq. (2-29) is commonly taken as the definition of the Hamiltonian function; it has the advantage of being applicable for the more general case of nonconservative systems. For our purposes, the definitions (2-29) and (2-25) are equivalent.]

For an arbitrary variation in the q's and p's, we have, from (2-29),

$$dH = \sum \dot{q}_k \, dp_k + \sum p_k \, d\dot{q}_k - \sum_{k=1}^{3n} \frac{\partial L(q_1, \ldots, q_{3n}; \dot{q}_1, \ldots, \dot{q}_{3n})}{\partial \dot{q}_k} d\dot{q}_k$$

$$- \sum_{k=1}^{3n} \frac{\partial L(q_1, \ldots, q_{3n}; \dot{q}_1, \ldots, \dot{q}_{3n})}{\partial q_k} dq_k \quad (2\text{-}30)$$

In view of the original definition of p_k as $\partial L/\partial \dot{q}_k$, the second and third terms of (2-30) add to zero. Furthermore for the coefficients $\partial L/\partial q_k$ in the last sum in (2-30), we can use Lagrange's equation

$$\frac{d}{dt}\left(\frac{\partial L}{\partial \dot{q}_k}\right) = \frac{\partial L}{\partial q_k}$$

plus the definition $p_k = \partial L/\partial \dot{q}_k$, to write

$$\dot{p}_k = \frac{\partial L}{\partial q_k} \tag{2-31}$$

and then (2-30) becomes

$$dH = \sum_{k=1}^{3n} \dot{q}_k\, dp_k - \sum_{k=1}^{3n} \dot{p}_k\, dq_k \tag{2-32}$$

Therefore

$$\frac{\partial H(q_1,\ldots,q_{3n};p_1,\ldots,p_{3n})}{\partial p_k} = \dot{q}_k \qquad k = 1,\ldots,3n \tag{2-33a}$$

$$\frac{\partial H(q_1,\ldots,q_{3n};p_1,\ldots,p_{3n})}{\partial q_k} = -\dot{p}_k \tag{2-33b}$$

Equations (2-33) are Hamilton's equations of motion. A mechanical system is characterized by its Hamiltonian function; the $6n$ first-order differential equations (2-33) can then be solved to give the p's and q's as a function of time.

By contrast, Lagrange's equations are $3n$ second-order differential equations for determining the q's as functions of time. The two sets of equations are, of course, equivalent.

From (2-32), we can calculate dH/dt along a trajectory:

$$\frac{dH}{dt} = \sum_k \dot{q}_k \frac{dp_k}{dt} - \sum_k \dot{p}_k \frac{dq_k}{dt} = \sum_k \dot{q}_k \dot{p}_k - \dot{p}_k \dot{q}_k = 0 \tag{2-34}$$

This is the mechanical theorem of the conservation of the total energy $K + U$, or H, for a conservative system.

2-5. Some Simple Examples. For a system of n particles and cartesian coordinates, as envisaged at the beginning of Sec. 2-3,

$$L(x_1,\ldots,z_n;\dot{x}_1,\ldots,\dot{z}_n) = K - U$$

$$= \frac{1}{2}\sum_{i=1}^{n} m_i(\dot{x}_i^2 + \dot{y}_i^2 + \dot{z}_i^2) - U(x_1,\ldots,z_n) \tag{2-35}$$

Then $\qquad p_{x_i} = \dfrac{\partial L}{\partial \dot{x}_i} = m_i \dot{x}_i \qquad p_{y_i} = \dfrac{\partial L}{\partial \dot{y}_i} = m_i \dot{y}_i \qquad$ etc. $\qquad$ (2-36)

We recognize that the conjugate momenta here are the ordinary linear momenta. By the substitution $\dot{x}_i = p_{x_i}/m_i$ in K we obtain

$$H = K + U = \frac{1}{2}\sum_{i=1}^{n} \frac{p_{x_i}^2 + p_{y_i}^2 + p_{z_i}^2}{2m_i} + U(x_1, \ldots, z_n) \quad (2\text{-}37)$$

and Hamilton's equations become

$$\dot{x}_i = \frac{\partial H}{\partial p_{x_i}} = \frac{p_{x_i}}{m} \qquad i = 1, \ldots, n, \text{ and for } x, y, z \quad (2\text{-}38a)$$

$$\dot{p}_{x_i} = -\frac{\partial H}{\partial x_i} = -\frac{\partial U}{\partial x_i} \quad (2\text{-}38b)$$

Equations (2-38a) repeat the relations (2-36). By substituting from (2-38a) into (2-38b), we obtain Newton's equation again:

$$m_i \ddot{x}_i = -\frac{\partial U}{\partial x_i} \qquad i = 1, \ldots, n, \text{ and for } x, y, z$$

Let us set up Hamilton's equations for the motion of a single particle in two dimensions, using polar coordinates, and with a potential function $U(r,\phi)$. The kinetic energy is

$$K = \tfrac{1}{2}m\dot{r}^2 + \tfrac{1}{2}mr^2\dot{\phi}^2$$

and the conjugate momenta are

$$p_r = \frac{\partial L}{\partial \dot{r}} = m\dot{r} \qquad p_\phi = mr^2\dot{\phi} \quad (2\text{-}39)$$

Then

$$H = \frac{p_r^2}{2m} + \frac{p_\phi^2}{2mr^2} + U(r,\phi)$$

and the equations of motion are

$$\dot{r} = \frac{\partial H}{\partial p_r} = \frac{p_r}{m} \qquad \dot{\phi} = \frac{\partial H}{\partial p_\phi} = \frac{p_\phi}{mr^2} \quad (2\text{-}40a)$$

and

$$\dot{p}_r = -\frac{\partial H}{\partial r} = \frac{p_\phi^2}{mr^3} - \frac{\partial U}{\partial r}$$

$$\dot{p}_\phi = -\frac{\partial H}{\partial \phi} = -\frac{\partial U}{\partial \phi} \quad (2\text{-}40b)$$

We again note that the equations for the $\dot{q}$ [(2-40a)] essentially repeat the definitions of the momenta [(2-39)].

Another important simple example is assigned as Prob. 2-5.

Problem 2-5. Write the equations of motion in Hamiltonian form, using spherical polar coordinates, for a single particle in a spherically symmetrical potential, as in Prob. 2-3.

2-6. Two Interacting Particles. An important example for us is the system of two particles interacting according to a potential which is a function only of their distance apart, $r_{12}^2 = (x_2 - x_1)^2 + (y_2 - y_1)^2 + (z_2 - z_1)^2$. We summarize the results which, it is hoped, the reader has already derived as the solutions to Probs. 2-1 to 2-5.

We transform to center-of-gravity coordinates X, Y, Z and internal, or relative, coordinates x, y, z (which we have called x_{12}, y_{12}, z_{12} in Probs. 2-1 and 2-4):

$$X = \frac{m_1 x_1 + m_2 x_2}{m_1 + m_2} \qquad Y = \frac{m_1 y_1 + m_2 y_2}{m_1 + m_2} \qquad Z = \frac{m_1 z_1 + m_2 z_2}{m_1 + m_2} \qquad (2\text{-}41)$$
$$x = x_2 - x_1 \qquad\qquad y = y_2 - y_1 \qquad\qquad z = z_2 - z_1$$

The kinetic energy then is

$$K = \tfrac{1}{2}(m_1 + m_2)(\dot X^2 + \dot Y^2 + \dot Z^2) + \frac{1}{2} \frac{m_1 m_2}{m_1 + m_2}(\dot x^2 + \dot y^2 + \dot z^2) \qquad (2\text{-}42)$$

Let $\mu = m_1 m_2/(m_1 + m_2)$; μ is called the reduced mass. The potential energy is a function only of the internal coordinates x, y, z, so that

$$L(X,Y,Z,x,y,z,\dot X,\dot Y,\dot Z,\dot x,\dot y,\dot z) = \tfrac{1}{2}(m_1 + m_2)(\dot X^2 + \dot Y^2 + \dot Z^2)$$
$$+ \frac{\mu}{2}(\dot x^2 + \dot y^2 + \dot z^2) - U(x,y,z) \qquad (2\text{-}43)$$

The equations of motion therefore are

$$(m_1 + m_2)\ddot X = (m_1 + m_2)\ddot Y = (m_1 + m_2)\ddot Z = 0 \qquad (2\text{-}44)$$

$$\mu\ddot x = -\frac{\partial U}{\partial x} \qquad \mu\ddot y = -\frac{\partial U}{\partial y} \qquad \mu\ddot z = -\frac{\partial U}{\partial z} \qquad (2\text{-}45)$$

According to (2-44), the motion of the coordinates of the center of mass is that of a free particle of mass equal to the total mass and subject to no forces. Integration of (2-44) gives constant velocities $\dot X, \dot Y, \dot Z$ and a constant kinetic energy in these degrees of freedom. According to (2-45), the internal motion is that of a particle with the reduced mass μ moving in the potential $U(x,y,z)$. The total energy is the sum of the "internal" energy plus the translational energy of the center of mass.

We now transform the internal coordinates to spherical polar coordinates.

$$z = r \cos \theta$$
$$x = r \sin \theta \cos \phi \qquad (2\text{-}46)$$
$$y = r \sin \theta \sin \phi$$

The internal kinetic energy is

$$K = \tfrac{1}{2}\mu \dot r^2 + \tfrac{1}{2}\mu r^2 \dot\theta^2 + \tfrac{1}{2}\mu r^2 \sin^2 \theta \dot\phi^2 \qquad (2\text{-}47)$$

By assumption, the potential is $U(r)$. The Lagrangian function is

$$L = \tfrac{1}{2}\mu \dot r^2 + \tfrac{1}{2}\mu r^2 \dot\theta^2 + \tfrac{1}{2}\mu r^2 \sin^2 \theta \dot\phi^2 - U(r) \qquad (2\text{-}48)$$

The conjugate momenta are therefore

$$p_r = \frac{\partial L}{\partial \dot{r}} = \mu \dot{r} \qquad p_\theta = \mu r^2 \dot{\theta} \qquad p_\phi = \mu r^2 \sin^2\theta \, \dot{\phi} \qquad (2\text{-}49)$$

By substitution for $\dot{r}$, $\dot{\theta}$, and $\dot{\phi}$, we obtain for the Hamiltonian function

$$H(r,\theta,\phi,p_r,p_\theta,p_\phi) = \frac{p_r^2}{2\mu} + \frac{p_\theta^2}{2\mu r^2} + \frac{p_\phi^2}{2\mu r^2 \sin^2\theta} + U(r) \qquad (2\text{-}50)$$

and the equations of motion are

$$\dot{r} = \frac{p_r}{\mu} \qquad \dot{\theta} = \frac{p_\theta}{\mu r^2} \qquad \dot{\phi} = \frac{p_\phi}{\mu r^2 \sin^2\theta} \qquad (2\text{-}51)$$

which are identical with (2-49), and

$$\dot{p}_r = -\frac{\partial U}{\partial r} + \frac{p_\theta^2}{\mu r^3} + \frac{p_\phi^2}{\mu r^3 \sin^2\theta} \qquad (2\text{-}52a)$$

$$\dot{p}_\theta = \frac{p_\phi^2 \cos\theta}{\mu r^2 \sin^3\theta} \qquad (2\text{-}52b)$$

$$\dot{p}_\phi = 0 \qquad (2\text{-}52c)$$

The integral of (2-52c) is

$$p_\phi = \text{constant} \qquad (2\text{-}53)$$

In cartesian coordinates, the angular-momentum components M_x, M_y, M_z for a particle are given in terms of the linear momenta p_x, p_y, p_z by

$$M_x = yp_z - zp_y \qquad M_y = zp_x - xp_z \qquad M_z = xp_y - yp_x \qquad (2\text{-}54)$$

The total square angular momentum is $M^2 = M_x^2 + M_y^2 + M_z^2$. By straightforward calculations, it can be shown that $M_z = p_\phi$. [The transformations between p_x, p_y, p_z and p_r, p_θ, p_ϕ are given as Eqs. (2-67) later.] The result p_ϕ = constant therefore states that the angular momentum around the z axis is constant during the motion. There is nothing unique about the z axis, and the component of angular momentum around any axis must also be a constant of the motion. By more elaborate calculations, the result can be explicitly proved, of course. The total squared angular momentum can be shown to be given by

$$M^2 = p_\theta^2 + \frac{p_\phi^2}{\sin^2\theta} \qquad (2\text{-}55)$$

We then see that

$$\frac{dM^2}{dt} = 2p_\theta \dot{p}_\theta + \frac{2p_\phi \dot{p}_\phi}{\sin^2\theta} - \frac{2p_\phi^2 \cos\theta \, \dot{\theta}}{\sin^3\theta} \qquad (2\text{-}56)$$

By using the expression for $\dot{\theta}$ [(2-51)] and for $\dot{p}_\theta$ [(2-52b)] we see that $dM^2/dt = 0$. Thus the total squared angular momentum is also a constant of the motion.

These theorems concerning the conservation of angular momentum can be shown to be true for any isolated system of particles—i.e., when the potential energy depends only on the relative positions of the particles.*

It is obvious that the linear momenta corresponding to the motion of the center of mass are $p_X = (m_1 + m_2)\dot{X}$, $p_Y = (m_1 + m_2)\dot{Y}$, $p_Z = (m_1 + m_2)\dot{Z}$, and the Hamiltonian function for the translational energy of the center of mass is

$$H_{cm} = \frac{p_X^2}{2(m_1 + m_2)} + \frac{p_Y^2}{2(m_1 + m_2)} + \frac{p_Z^2}{2(m_1 + m_2)}$$

The total H for the two-particle system is the sum of H_{cm} and H for the internal motion. There is no interaction whatsoever between the motion of the center of gravity and the internal motions for an isolated system.

2-7. Phase Space. For a single, structureless particle moving in three dimensions, there are six coordinates and conjugate momenta x, y, z, p_x, p_y, p_z. The equations of motion can, at least in principle, be solved to give the values of these six quantities as a function of time. The phase space for this mechanical system with this cartesian-coordinate system is defined as a six-dimensional orthogonal space, with six orthogonal axes, along the x, y, z, p_x, p_y, p_z "directions." A particular set of values for x, y, z, p_x, p_y, p_z is a point in this phase space. The trajectory of a particle in which x, y, z, p_x, p_y, p_z are given as functions of time can be described as the motion of a point in phase space. If spherical coordinates are used for the particle, there is a phase space with six axes at right angles to each other, along the r, θ, ϕ, p_r, p_θ, p_ϕ directions. It must be emphasized that, in phase space, θ is not an angle: it is a "distance" along a direction perpendicular to the r, ϕ, p_r, p_θ, and p_ϕ directions.

For a system of n particles with generalized (Hamiltonian) coordinates $q_1, \ldots, q_{3n}; p_1, \ldots, p_{3n}$, there is a $6n$-dimensional phase space, with orthogonal axes along the $6n$ directions, $q_1, \ldots, q_{3n}; p_1, \ldots, p_{3n}$. The motion of a system can be described as the motion of a point in this phase space.

Consider any set of variables $x_1, \ldots, x_n$ and a function $f(x_1, \ldots, x_n)$. We may be interested in the integral

$$I = \int \cdots \int_{R_x} f(x_1, \ldots, x_n)\, dx_1 \cdots dx_n \qquad (2\text{-}57)$$

where R_x is a region in $x_1, \ldots, x_n$ space. We transform to a new set of variables, $y_1, \ldots, y_n$, by the equations

$$\begin{aligned} x_1 &= x_1(y_1, \ldots, y_n) \\ &\cdots\cdots\cdots\cdots \\ x_n &= x_n(y_1, \ldots, y_n) \end{aligned} \qquad (2\text{-}58)$$

---

* Tolman, "The Principles of Statistical Mechanics" [2].

The question is: How do we evaluate the integral (2-57) using the y variables? Let $f(y_1, \ldots, y_n)$ represent the integrand function. We use function notation in the "physical" sense discussed previously. That is, by $f(y_1, \ldots, y_n)$ we mean the value of the quantity $f(x_1, \ldots, x_n)$ when the values of x_i are calculated from the values of y_i by (2-58). In rigorous mathematical notation, we mean

$$f[x_1(y_1, \ldots, y_n), x_2(y_1, \ldots, y_n), \ldots, x_n(y_1, \ldots, y_n)]$$

The volume element $dx_1 \cdots dx_n$ becomes

$$dx_1 \cdots dx_n = \begin{vmatrix} \dfrac{\partial x_1}{\partial y_1} & \cdots & \dfrac{\partial x_1}{\partial y_n} \\ \cdots & \cdots & \cdots \\ \dfrac{\partial x_n}{\partial y_1} & \cdots & \dfrac{\partial x_n}{\partial y_n} \end{vmatrix} dy_1 \cdots dy_n \quad (2\text{-}59a)$$

or
$$dx_1 \cdots dx_n = \frac{\partial(x_1, \ldots, x_n)}{\partial(y_1, \ldots, y_n)} dy_1 \cdots dy_n \quad (2\text{-}59b)$$

The Jacobian $\partial(x_1, \ldots, x_n)/\partial(y_1, \ldots, y_n)$ is the determinant in (2-59a). The integral I then becomes

$$I = \int \cdots \int_{R_y} f(y_1, \ldots, y_n) \frac{\partial(x_1, \ldots, x_n)}{\partial(y_1, \ldots, y_n)} dy_1 \cdots dy_n \quad (2\text{-}60)$$

where R_y is the region in y space corresponding to R_x of (2-57).

For example, consider the integral

$$I = \int_{-\infty}^{+\infty} \int_{-\infty}^{+\infty} e^{-\beta(x^2+y^2)} dx\, dy$$
$$= \left(\int_{-\infty}^{+\infty} e^{-\beta x^2} dx\right)\left(\int_{-\infty}^{+\infty} e^{-\beta y^2} dy\right) \quad (2\text{-}61a)$$

The integral $\int_{-\infty}^{+\infty} e^{-\beta x^2} dx$ cannot be directly evaluated by elementary means; however, its value is $(\pi/\beta)^{1/2}$. Hence

$$I = \frac{\pi}{\beta} \quad (2\text{-}61b)$$

If we use polar coordinates

$$x = r \cos \phi$$
$$y = r \sin \phi$$
$$\begin{vmatrix} \dfrac{\partial x}{\partial r} & \dfrac{\partial x}{\partial \phi} \\ \dfrac{\partial y}{\partial r} & \dfrac{\partial y}{\partial \phi} \end{vmatrix} = r \quad (2\text{-}62)$$

This gives the result with which we are already familiar:

$$dx\, dy = r\, dr\, d\theta \quad (2\text{-}63)$$

for an element of area in two dimensions expressed in polar coordinates. The integral (2-61) then becomes

$$I = \int_{r=0}^{r=\infty} \int_{\phi=0}^{\phi=2\pi} e^{-\beta r^2} r \, dr \, d\phi \tag{2-64}$$

Since $de^{-\beta r^2} = e^{-\beta r^2}(-2\beta r \, dr)$, the integration in (2-64) can be performed by elementary means and $I = \pi/\beta$. The comparison of (2-63) and (2-64) is actually a simple proof that $\int_{-\infty}^{+\infty} e^{-\beta x^2} \, dx = (\pi/\beta)^{1/2}$.

Suppose that, for a given mechanical system, there are two sets of coordinates and conjugate momenta:

$$q_1, \ldots, q_{3n}; p_1, \ldots, p_{3n}$$
and
$$Q_1, \ldots, Q_{3n}; P_1, \ldots, P_{3n}$$

There are equations of transformation between the (q,p) coordinates and the (Q,P) coordinates. One of the most important properties of phase space arises from the fact that the Jacobian of the transformation is unity:

$$\frac{\partial(q_1, \ldots, q_{3n}; p_1, \ldots, p_{3n})}{\partial(Q_1, \ldots, Q_{3n}; P_1, \ldots, P_{3n})} = 1 \tag{2-65}$$

Thus the "volume elements" in the two phase spaces are equal.

$$dq_1 \cdots dq_{3n} \, dp_1 \cdots dp_{3n} = dQ_1 \cdots dQ_{3n} \, dP_1 \cdots dP_{3n} \tag{2-66}$$

The equality (2-66) is the justification for thinking of phase space as consisting of mutually orthogonal axes along the $6n$ directions, $q_1, \ldots, q_{3n}; p_1, \ldots, p_{3n}$, since the volume element in such a cartesian space is given by (2-66). A general proof of Eq. (2-65) is beyond the scope of our treatment.*

For example, for a single particle in three dimensions, Eqs. (2-46) are the transformation between (x,y,z) and (r,θ,ϕ). By straightforward differentiation, and using the conjugate momenta, $p_x = m\dot{x}$, $p_y = m\dot{y}$, $p_z = m\dot{z}$, $p_r = m\dot{r}$, $p_\theta = mr^2\dot\theta$, $p_\phi = mr^2 \sin^2\theta \dot\phi$, one can arrive at the relations

$$p_x = \sin\theta\cos\phi \, p_r + \frac{\cos\theta\cos\phi}{r} p_\theta - \frac{\sin\phi}{r\sin\theta} p_\phi$$

$$p_y = \sin\theta\sin\phi \, p_r + \frac{\cos\theta\sin\phi}{r} p_\theta + \frac{\cos\phi}{r\sin\theta} p_\phi \tag{2-67}$$

$$p_z = \cos\theta \, p_r - \frac{\sin\theta \, p_\theta}{r}$$

By straightforward and extremely tedious calculations, one can then

---

* The proof is given by Wilson, "Thermodynamics and Statistical Mechanics," p. 119 [4]; an alternative proof based on Liouville's theorem is given by Tolman, pp. 49–57 [2].

show that

$$\frac{\partial(x,y,z,p_x,p_y,p_z)}{\partial(r,\theta,\phi,p_r,p_\theta,p_\phi)} = 1 \tag{2-68}$$

so that
$$dx\,dy\,dz\,dp_x\,dp_y\,dp_z = dr\,d\theta\,d\phi\,dp_r\,dp_\theta\,dp_\phi \tag{2-69}$$

It is to be noticed that the volume element in real space, $dx\,dy\,dz$, is in polar coordinates $r^2 \sin\theta\,dr\,d\theta\,d\phi$; however, the factor $r^2 \sin\theta$ does not appear for the volume element in the phase space, $dr\,d\theta\,d\phi\,dp_r\,dp_\theta\,dp_\phi$. The necessary factors to make the dimensions the same on the two sides of (2-69) arise from the fact that, while y and θ, for example, have different dimensions, the products yp_y and θp_θ both have dimensions of ml^2t^{-1}.

While it is tedious to prove the relation (2-68), it is rather easy to see that the use of this equality gives consistent results for certain integrals over phase space which can be evaluated in cartesian and polar coordinates (Prob. 2-6).

We could of course describe the motion of a system as the motion of a point in a space composed of position and velocity coordinates, $q_1, \ldots, q_{3n}; \dot{q}_1, \ldots, \dot{q}_{3n}$. The invariance of the volume element in a transformation of coordinates would not in general be true for such spaces, however. Furthermore, as we shall see, the volume in phase space has a fundamental meaning in quantum mechanics which makes it particularly appropriate for statistical mechanics.

Problem 2-6. Evaluate the phase integral

$$\xi = \iiint_V \int_{p_x=-\infty}^{+\infty} \int_{p_y=-\infty}^{+\infty} \int_{p_z=-\infty}^{+\infty} e^{-\beta H}\,dp_z\,dp_y\,dp_x\,dx\,dy\,dz$$

for a free particle $[U(x,y,z) = 0]$ with $H = (p_x^2 + p_y^2 + p_z^2)/2m$, and with the limits of integration for x, y, z being those of a container of volume V.

Now evaluate the same phase integral for phase space with spherical polar coordinates,

$$\xi = \iiint_V \int_{p_r=-\infty}^{\infty} \int_{p_\theta=-\infty}^{+\infty} \int_{p_\phi=-\infty}^{+\infty} e^{-\beta H}\,dp_\phi\,dp_\theta\,dp_r\,d\phi\,d\theta\,dr$$

using the appropriate expression for H. The limits of integration for the configuration coordinates r, θ, ϕ are those of a container of volume V, say, $0 \le \phi \le 2\pi$, $0 \le \theta \le \pi$, $0 \le R \le (3V/4\pi)^{1/3}$.

PROBLEMS

2-7. For a single particle constrained to move in the xy plane as discussed in Sec. 2-2, write the equations of motion in polar coordinates. For the potential function

$$U = \tfrac{1}{2}ax^2 + bxy + \tfrac{1}{2}cy^2$$

take $a = c$, $b = 0$, and write the equations of motion in polar coordinates. What can you say about the value of the angular momentum, $mr^2\dot{\theta}$, for such a motion?

2-8. For a free, rigid dumbbell rotor, with moment of inertia I, the Hamiltonian function is

$$H(\theta,\phi) = \frac{p_\theta^2}{2I} + \frac{p_\phi^2}{2I \sin^2 \theta}$$

[This is the kinetic-energy part of the Hamiltonian of Eq. (2-50), with $I = \mu r^2$ and with $p_r = 0$.]

Evaluate the phase integral

$$\xi = \int \cdots \int e^{-\beta H} \, d\theta \, d\phi \, dp_\theta \, dp_\phi$$

noting that the range of the variables is

$$0 \leq \phi \leq 2\pi \quad 0 \leq \theta \leq \pi \quad -\infty < p_\phi < +\infty \quad -\infty < p_\theta < \infty$$

3
Quantum Mechanics

3-1. Introduction. The fundamental laws for the behavior of matter at the atomic and molecular level are the laws of quantum mechanics. Modern statistical mechanics is firmly based on quantum mechanics. However, almost all the important results of statistical mechanics can be developed by using only certain very elementary concepts and results of quantum mechanics. We take from quantum mechanics the existence of discrete quantum states, each with a definite energy, plus the symmetry requirements of the wave functions for systems containing several identical particles; and we add a statistical assumption about an equal a priori probability for each quantum state. We can then derive, as in Chap. 6, the basic laws of statistical mechanics.

In order to apply the general laws to particular systems, it is necessary to know the energy-level structure for that system (although, as we shall see in due course, even this information is not needed for cases where the "classical" approximation is valid). The energy levels for a system may not be known, they may be known from experiment, and/or they may be obtained by quantum-mechanical calculations.

Only those features of quantum mechanics which are important for statistical mechanics are included in the concise and elementary review in this chapter. In the interests of brevity, our treatment is necessarily dogmatic. We shall not justify all our statements or distinguish between postulates and theorems. It is assumed that the reader has at least a qualitative familiarity with the basic physical phenomena—the existence of discrete energy levels and the wave-particle duality for both matter and radiation—that led to the formulation of quantum theory.*

3-2. Wave Functions. We begin with a mechanical system containing N particles (electrons and nuclei, for example). Classically, the particles would have coordinates $x_1, \ldots, z_N$, or, using generalized coordinates, $q_1, \ldots, q_{3N}$.

---

* There are a number of excellent treatises on quantum mechanics at all levels, including Landau and Lipschitz [19], Eyring, Walter, and Kimball [20], Pauling and Wilson [21], and Leighton [22]; the last of these is especially suitable for further study at the level needed here.

We shall use the vector notation **q** as an abbreviation for the $3N$-component vector $q_1, \ldots, q_{3N}$. We call the volume element in cartesian space for the N-particle system, $d\tau$.

$$d\tau = dx_1\, dy_1 \cdots dy_N\, dz_N$$
$$= \frac{\partial(x_1, \ldots, z_N)}{\partial(q_1, \ldots, q_{3N})}\, dq_1 \cdots dq_{3N}$$

In classical mechanics one can say that at a given time the system is at certain particular values of $q_1, \ldots, q_{3N}$, but in quantum mechanics it is in general not possible to locate a particle exactly. The state of a system is described by a wave function:

$$\Psi(q_1, \ldots, q_{3N}, t) = \Psi(\mathbf{q}, t)$$

The probability that the system will be found at time t to be at $q_1, \ldots, q_{3N}$ (or, speaking more precisely, to have coordinates in the range q_1 to $q_1 + dq_1, \ldots, q_{3N}$ to $q_{3N} + dq_{3N}$) is

$$\Psi^*(\mathbf{q}, t)\Psi(\mathbf{q}, t)\, d\tau \tag{3-1}$$

where Ψ^* is the complex conjugate of the function Ψ. The integral of this probability over all space must be unity:

$$\int \cdots \int \Psi^*(q_1, \ldots, q_{3N}, t)\Psi(q_1, \ldots, q_{3N}, t)\, d\tau = \langle \Psi | \Psi \rangle = 1 \tag{3-2}$$

The notation $\langle \Psi | \Psi \rangle$ for the integral as introduced in (3-2) is often convenient.

In addition to satisfying the integrability condition (3-2), a satisfactory wave function must be continuous.

3-3. Operators and Observable Quantities. The properties of a system can be calculated from the wave function according to the prescriptions of quantum mechanics. There is a class of observable quantities which are called dynamical variables: momentum, angular momentum, dipole moment, and energy are examples. Each dynamical variable is represented in quantum mechanics by an operator.

An operator is simply a prescription for a mathematical operation on a function. Thus the operator $\partial/\partial x_1$ operating on any function of x_1 and x_2, $f(x_1, x_2)$ means simply $\partial f(x_1, x_2)/\partial x_1$. The operator x_1^2 operating on $f(x_1, x_2)$ means the product function $x_1^2 f(x_1, x_2)$.

For each dynamical quantity B there is a classical expression in terms of the coordinates and conjugate momenta, $B(q_1, \ldots, q_{3N}; p_1, \ldots, p_{3N})$. There is a corresponding quantum-mechanical operator, $\hat{B}(q_1, \ldots, q_{3N}; \partial/\partial q_1, \ldots, \partial/\partial q_{3N})$, or $\hat{B}(\mathbf{q}, \partial/\partial \mathbf{q})$, which, as indicated, depends on the q_i's and the derivatives $\partial/\partial q_i$. The rules for constructing this operator will be stated shortly.

If a large number of observations are made on a set of identically prepared systems at time t, the observed values for the physical quantity B

will not necessarily all be the same. The system cannot be said to have a definite value of B; this is a fundamental uncertainty which is inherent in the nature of things. The average or expectation value of B, $\langle B \rangle$, as obtained by averaging a large number of such observations, is given by

$$\langle B \rangle = \int \cdots \int \Psi^*(\mathbf{q},t) \hat{B}\left(\mathbf{q}, \frac{\partial}{\partial \mathbf{q}}\right) \Psi(\mathbf{q},t) \, d\tau \qquad (3\text{-}3a)$$

$$\langle B \rangle = \langle \Psi | B | \Psi \rangle \qquad (3\text{-}3b)$$

where (3-3b) defines a convenient notation for the integral of (3-3a).

The operators in quantum mechanics are all linear; i.e.,

$$\hat{B}(a\Psi_1 + b\Psi_2) = a\hat{B}\Psi_1 + b\hat{B}\Psi_2 \qquad (3\text{-}4)$$

where a and b are constants. The operators are Hermitian; i.e.,

$$\int \Psi_2^* \hat{B}\Psi_1 \, d\tau = \int \Psi_1 \hat{B}^* \Psi_2^* \, d\tau$$
$$\langle \Psi_2 | \hat{B} | \Psi_1 \rangle = \langle \Psi_1^* | \hat{B}^* | \Psi_2^* \rangle \qquad (3\text{-}5)$$

where $\hat{B}^*$ is the operator which is the complex conjugate of B.

It is to be emphasized that, in Eqs. (3-4) and (3-5), Ψ_1 and Ψ_2 are arbitrary continuous functions.

The product of two operators $\hat{A}\hat{B}$ is the operator which means operate on the function Ψ with B and then operate on the result with A:

$$\hat{A}\hat{B}\Psi = \hat{A}[\hat{B}\Psi]$$

Operators do not necessarily satisfy the commutative rules. That is, $\hat{A}\hat{B}$ is not necessarily equal to $\hat{B}\hat{A}$. The most important example is just the product of the operators q and $\partial/\partial q$ and

$$q \frac{\partial}{\partial q}[\psi(q)] = q \frac{\partial \psi}{\partial q}$$

whereas
$$\frac{\partial}{\partial q} q[\psi(q)] = \frac{\partial}{\partial q}[q\psi(q)] = \psi(q) + q \frac{\partial \psi}{\partial q}$$

Operators for which $\hat{A}\hat{B} = \hat{B}\hat{A}$ are said to commute.

The rule for constructing a quantum-mechanical operator is the following: Take the classical expression for B in terms $q_1, \ldots, q_{3N}$; $p_1, \ldots, p_{3N}$, and replace each p_j by $(\hbar/i)\partial/\partial q_j$ (where $\hbar = h/2\pi$).

Some of the common and important quantum-mechanical operators therefore are

Position: $\qquad\qquad\qquad\qquad \hat{q}_j \qquad\qquad\qquad\qquad (3\text{-}6a)$

Conjugate momentum: $\qquad \hat{p}_j = \frac{\hbar}{i} \frac{\partial}{\partial q_j} \qquad\qquad (3\text{-}6b)$

Kinetic energy, in cartesian coordinates:

$$\hat{K} = \sum_j \frac{p_{x_j}^2 + p_{y_j}^2 + p_{z_j}^2}{2m_j} = -\hbar^2 \sum_j \frac{1}{2m_j}\left(\frac{\partial^2}{\partial x_j^2} + \frac{\partial^2}{\partial y_j^2} + \frac{\partial^2}{\partial z_j^2}\right) \qquad (3\text{-}6c)$$

(where the sum extends over all the particles).
Angular momentum around the z axis:

$$\hat{L}_z = \sum x_j p_{y_j} - y_j p_{x_j} = \frac{\hbar}{i} \sum x_j \frac{\partial}{\partial y_j} - y_j \frac{\partial}{\partial x_j}$$

(with similar expressions for $\hat{L}_x$ and $\hat{L}_y$).
In spherical polar coordinates we have

$$\begin{aligned}\hat{K} &= \sum_j \frac{p_{r_j}^2}{2m_j} + \frac{p_{\theta_j}^2}{2m_j r_j^2} + \frac{p_{\phi_j}^2}{2m_j r_j^2 \sin^2 \theta_j} \\ &= -\frac{\hbar^2}{2} \sum_j \frac{1}{m_j r_j^2} \left[\frac{\partial}{\partial r_j}\left(r_j^2 \frac{\partial}{\partial r_j}\right) + \frac{1}{\sin \theta_j} \frac{\partial}{\partial \theta_j}\left(\sin \theta_j \frac{\partial}{\partial \theta_j}\right) \right. \\ &\qquad\qquad\qquad\qquad\qquad\qquad \left. + \frac{1}{\sin^2 \theta_j} \frac{\partial^2}{\partial \phi_j^2} \right] \end{aligned} \quad (3\text{-}6d)$$

Angular momentum around the three-cartesian axes and the square of the total angular momentum are

$$\hat{L}_x = \frac{\hbar}{i} \sum \left(-\sin \phi_j \frac{\partial}{\partial \theta_j} - \cot \theta_j \cos \phi_j \frac{\partial}{\partial \phi_j} \right) \quad (3\text{-}6e)$$

$$\hat{L}_y = \frac{\hbar}{i} \sum \left(\cos \phi_j \frac{\partial}{\partial \theta_j} - \cot \theta_j \sin \phi_j \frac{\partial}{\partial \phi_j} \right) \quad (3\text{-}6f)$$

$$\hat{L}_z = \frac{\hbar}{i} \sum \frac{\partial}{\partial \phi_j} \quad (3\text{-}6g)$$

$$\hat{L}^2 = -\hbar^2 \left[\sum \frac{1}{\sin \theta_j} \frac{\partial}{\partial \theta_j}\left(\sin \theta_j \frac{\partial}{\partial \theta_j}\right) + \frac{1}{\sin^2 \theta_j} \frac{\partial^2}{\partial \phi_j^2} \right] \quad (3\text{-}6h)$$

The total energy, or Hamiltonian operator, is

$$\hat{H} = \hat{K} + \hat{U} \quad (3\text{-}7)$$

The rule, as stated, for constructing operators is ambiguous as to the order of multiplication by a coordinate and differentiation in such terms as $(1/r^2)[\partial/\partial r(r^2 \partial/\partial r)]$ of (3-6d). For some cases, this is a difficult problem. If the form of the operator is unambiguous in cartesian coordinates, a safe rule is to construct the operator in cartesian coordinates and then transform to the desired coordinate system.

3-4. Eigenstates and Eigenfunctions. We consider here operators which are not explicitly functions of time and wave functions which are not functions of time. If the wave function $\psi(\mathbf{q})$ satisfies the equation

$$\hat{B}\psi(\mathbf{q}) = b\psi(\mathbf{q}) \quad (3\text{-}8)$$

where b is a constant, ψ is said to be an eigenfunction of $\hat{B}$ and b is an eigenvalue. The expectation value $\langle B \rangle$ is then clearly

$$\langle \psi | \hat{B} | \psi \rangle = b \langle \psi | \psi \rangle = b$$

Furthermore, it can be shown that if the system is in the eigenstate ψ, all measurements of the physical quantity B will give the same value b; that is, there is no uncertainty as to the value of B.

For any operator there is actually a set of eigenfunctions ψ_i and eigenvalues b_i. It can be shown that the eigenfunctions of a Hermitian operator form a complete orthogonal set so that any arbitrary function can be expressed as an infinite linear combination of the eigenfunctions ψ_i. It can be shown that for a system in an arbitrary state $\Psi(\mathbf{q},t)$, which is not necessarily an eigenstate of the operator $\hat{B}$, a measurement of the value of B will give one of the eigenvalues b_i. But, of course, in the general case the values of b_i for a set of measurements on identically prepared systems will not all be the same.

We are especially interested in the eigenfunctions of the Hamiltonian operator

$$\hat{H}\psi_i(\mathbf{q}) = \epsilon_i \psi_i(\mathbf{q}) \tag{3-9}$$

If a system is in one of the states ψ_i, its energy is known with certainty as ϵ_i. For reasons which will become apparent shortly, it is said to be in a stationary state.

It is possible that there are several linearly independent eigenfunctions of an operator $\hat{B}$ with the same eigenvalue b. Two or more eigenfunctions of the Hamiltonian operator with the same energy are said to be degenerate. If the eigenfunctions ψ_{i1} and ψ_{i2} both have eigenvalue ϵ_i, an arbitrary linear combination $a\psi_{i1} + b\psi_{i2}$ is also an eigenfunction with eigenvalue ϵ_i.

If two operators commute, any nondegenerate eigenfunction of one operator is also an eigenfunction of the other operator. If there are several degenerate eigenfunctions, it is possible to choose a suitable set of linear combinations which are simultaneously eigenfunctions of both operators. Thus, it is possible for a system to be in a state in which the values of two dynamical quantities are both known with certainty if the two operators commute. If two operators do not commute, there is not a complete set of eigenstates in which the values of the dynamical variables for both operators are known with certainty.

3-5. Time-dependent Schrödinger Equation. The variation of the wave function of a system with time is governed by the time-dependent Schrödinger equation:

$$\hat{H}\left(\mathbf{q}, \frac{\partial}{\partial \mathbf{q}}\right) \Psi(\mathbf{q},t) = -\frac{\hbar}{i}\frac{\partial \Psi(\mathbf{q},t)}{\partial t} \tag{3-10}$$

By direct substitution, we see that the wave function

$$\Psi_j(\mathbf{q},t) = \psi_j(\mathbf{q})e^{-i\epsilon_j t/\hbar} \tag{3-11}$$

is a solution if, and only if,

$$\hat{H}\psi_j(\mathbf{q}) = \epsilon_j \psi_j(\mathbf{q}) \tag{3-12}$$

i.e., if ψ_j is an eigenfunction of $\hat{H}$ with eigenvalue ϵ_j. For such a wave function the probability of finding the system at **q** at time t is

$$\Psi_j^*(\mathbf{q},t)\Psi_j(\mathbf{q},t)\,d\tau = \psi_j^*(\mathbf{q})\psi_j(\mathbf{q})\,d\tau \tag{3-13}$$

Thus, the probability does not change with time, and the system is said to be in a stationary state. The expectation value for any operator $\langle\Psi_j|\hat{B}|\Psi_j\rangle$ is not a function of time, provided that $\hat{B}$ itself does not contain the time.

3-6. Uncertainty Principle. It has already been remarked that the operators $\hat{q}_j$ and $\hat{p}_j = (\hbar/i)(\partial/\partial q_j)$ do not commute. In this case it can be shown that a system cannot be in a state such that both a position coordinate and its conjugate momentum are known with certainty. If a set of measurements of a physical quantity, say, q_1, of a system gives a distribution of different values q_{1a}, q_{1b}, ..., q_{1n}, we say that there is an uncertainty $|\Delta q_1|$ in the value of q_1 for this particular system that can be measured, for example, by the root-mean-square (rms) deviation of the measurements from the mean. We shall consider the problem of expressing the dispersion of a statistical distribution in Chap. 5.

At present, we note that, for any reasonable measure of the uncertainties $|\Delta q_j|$ and $|\Delta p_j|$ of a position and a conjugate momentum, the uncertainty principle asserts that the minimum possible uncertainties obey the relation

$$|\Delta q|\,|\Delta p| \sim \hbar \tag{3-14}$$

(where $\sim$ means "of the order of magnitude of"). The uncertainty can be larger than this lower limit. Energy and time are conjugate quantities in quantum mechanics, and the uncertainty principle also says that

$$|\Delta\epsilon|\,|\Delta t| \sim h \tag{3-15}$$

The energy of a system can be known exactly only if the system remains in the same state for infinite time; if the lifetime of a system in a particular state is Δt, the energy of the system is uncertain by at least $\Delta\epsilon/h$.

3-7. Energy Levels of Some Simple Systems. The Free Particle. The equation

$$\hat{H}\left(\mathbf{q}, \frac{\partial}{\partial q}\right)\psi(\mathbf{q}) = \epsilon\psi(\mathbf{q}) \tag{3-16}$$

is known as the time-independent Schrödinger equation. Its solution gives eigenfunctions or wave functions $\psi(\mathbf{q})$ which are stationary states of known energy ϵ. In this and the next few sections we shall examine the stationary states for several simple and important problems.

Our first problem is a single free particle moving in three dimensions in a cubical box bounded by the six planes $x = 0$, $y = 0$, $z = 0$, $x = X$, $y = Y$, and $z = Z$. The potential energy $U(x,y,z)$ is zero inside the box; that is, $U(x,y,z) = 0$ for $0 \leq x \leq X$, $0 \leq y \leq Y$, $0 \leq z \leq Z$ and

$U(x,y,z) = \infty$ outside the box. The Schrödinger equation then is

$$-\frac{\hbar^2}{2m}\left(\frac{\partial^2\psi}{\partial x^2} + \frac{\partial^2\psi}{\partial y^2} + \frac{\partial^2\psi}{\partial z^2}\right) + U(x,y,z)\psi = \epsilon\psi \qquad (3\text{-}17)$$

The solution outside the box, where $U = \infty$, is $\psi = 0$. Inside the box the equation is

$$\frac{\partial^2\psi}{\partial x^2} + \frac{\partial^2\psi}{\partial y^2} + \frac{\partial^2\psi}{\partial z^2} = -\frac{2m\epsilon}{\hbar^2}\psi \qquad (3\text{-}18)$$

A solution of this equation is

$$\psi = A \sin\left[\left(\frac{2m\epsilon_x}{\hbar^2}\right)^{1/2} x\right] \sin\left[\left(\frac{2m\epsilon_y}{\hbar^2}\right)^{1/2} y\right] \sin\left[\left(\frac{2m\epsilon_z}{\hbar^2}\right)^{1/2} z\right] \qquad (3\text{-}19)$$

where A is a normalizing constant to make $\langle\psi|\psi\rangle = 1$ and

$$\epsilon = \epsilon_x + \epsilon_y + \epsilon_z \qquad (3\text{-}20)$$

Since outside of the box $\psi = 0$, we require that the inside solution [Eq. (3-19)] vanish at the borders,

$$\psi(0,y,z) = \psi(x,0,z) = \psi(x,y,0) = 0 \qquad (3\text{-}21a)$$
$$\psi(X,y,z) = \psi(x,Y,z) = \psi(x,y,Z) = 0 \qquad (3\text{-}21b)$$

Since $\sin\alpha = 0$ for $\alpha = 0$, condition (3-21a) is automatically satisfied by the functions of (3-19). There are also solutions of the Schrödinger equation (3-17) involving the functions $\cos[(2m\epsilon_x/\hbar^2)^{1/2}x]$, but these can be omitted because of the condition (3-21a).

To satisfy the conditions of Eq. (3-21b), it is necessary that

$$\left(\frac{2m\epsilon_x}{\hbar^2}\right)^{1/2} X = s_x\pi \qquad s_x = 1, 2, 3, \ldots$$

$$\left(\frac{2m\epsilon_y}{\hbar^2}\right)^{1/2} Y = s_y\pi \qquad s_y = 1, 2, 3, \ldots$$

$$\left(\frac{2m\epsilon_z}{\hbar^2}\right)^{1/2} Z = s_z\pi \qquad s_z = 1, 2, 3, \ldots$$

or $\quad \epsilon_x = \dfrac{h^2}{8mX^2} s_x^2 \quad \epsilon_y = \dfrac{h^2}{8mY^2} s_y^2 \quad \epsilon_z = \dfrac{h^2}{8mZ^2} s_z^2 \quad s_x, s_y, s_z$
$$= 1, 2, 3, \ldots \qquad (3\text{-}22)$$

Thus, the discrete allowed energy states are

$$\epsilon = \epsilon_x + \epsilon_y + \epsilon_z = \frac{h^2}{8m}\left(\frac{s_x^2}{X^2} + \frac{s_y^2}{Y^2} + \frac{s_z^2}{Z^2}\right) \qquad (3\text{-}23)$$

and the wave function inside the box is

$$\psi = A \sin\frac{\pi s_x x}{X} \sin\frac{\pi s_y y}{Y} \sin\frac{\pi s_z z}{Z} \qquad (3\text{-}24)$$

The wave functions in the x direction are shown in Fig. 3-1. The wavelength in the x direction is $\lambda_x = 2X/s_x$. It can be shown that the wave function (3-24) corresponds to a standing-wave pattern in a box; the wavelength λ is given by

$$\frac{1}{\lambda} = \left(\frac{1}{\lambda_x^2} + \frac{1}{\lambda_y^2} + \frac{1}{\lambda_z^2}\right)^{1/2} = \frac{1}{2}\left(\frac{s_x^2}{X^2} + \frac{s_y^2}{Y^2} + \frac{s_z^2}{Z^2}\right)^{1/2} \quad (3\text{-}25)$$

Problem 3-1. Calculate the expectation value of the linear momentum p_x for the wave function (3-24).

Problem 3-2. Calculate the expectation value of the square of the linear momentum p_x^2 for the wave function (3-24). Express $\langle p_x^2 \rangle$ in terms of $\lambda_x = 2X/s_x$ and in terms of ϵ_x. Is the wave function an eigenfunction of p_x^2? Explain in words the meaning of the results for Probs. 3-1 and 3-2.

3-8. The Rigid Rotor. We consider two mass points, m_1 and m_2, at a fixed distance r apart, rotating around their mutual center of gravity. Assume that there is no potential energy which varies with orientation. The moment of inertia is $I = \mu r^2$, with $\mu = m_1 m_2/(m_1 + m_2)$. The Hamiltonian operator is just the kinetic-energy operator in spherical coordinates (3-6d) with r constant:

$$\hat{H} = \hat{K} = \frac{1}{2I}\left(\hat{p}_\theta^2 + \frac{1}{\sin^2\theta}\hat{p}_\phi^2\right) = -\frac{\hbar^2}{2I}\left[\frac{1}{\sin\theta}\frac{\partial}{\partial\theta}\left(\sin\theta\frac{\partial}{\partial\theta}\right) + \frac{1}{\sin^2\theta}\frac{\partial^2}{\partial\phi^2}\right] \quad (3\text{-}26)$$

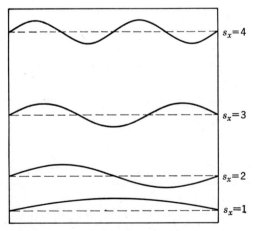

FIG. 3-1. The wave functions for a particle in a one-dimensional box. The heights of the dashed lines represent the energies of the different states. The solid lines are the wave functions. In each case, the amplitude of the wave function is zero at the walls of the box.

By comparison with (3-6h) we see that $\hat{H}$ and the total squared angular-momentum operator $\hat{L}^2$ are identical except for a constant factor, $1/2I$. An eigenfunction of one is an eigenfunction of the other, and the classical relation between energy and total angular momentum, $\epsilon = L^2/2I$, holds.*

The solutions of the Schrödinger equation,

$$\hat{H}\psi(\theta,\phi) = \epsilon\psi(\theta,\phi) \tag{3-27}$$

are a set of functions $\Theta_{jm}(\theta,\phi)$. Here j and m are integers; $j = 0, 1, 2, 3, 4, 5, \ldots$; and $m = 0, \pm 1, \pm 2, \pm 3$. However, it is required that $|m| \leq j$. [For examples of the functions $\Theta_{jm}(\theta,\phi)$, see Prob. 3-3.]

The energy levels are a function of j only.

$$\epsilon_j = \frac{j(j+1)\hbar^2}{2I} \tag{3-28}$$

The total angular momentum is, of course, $L^2 = j(j+1)\hbar^2$. The wave function can be chosen to be eigenfunctions of $\hat{L}_z$, also; then the eigenvalues are

$$L_z = m\hbar \tag{3-29}$$

Problem 3-3. Calculate the values of $\langle L^2 \rangle$ and $\langle L_z \rangle$ for the normalized wave functions:

$$\Theta_{jm} = \left(\frac{1}{4\pi}\right)^{1/2} \qquad \Theta_{jm} = \left(\frac{5}{16\pi}\right)^{1/2}(3\cos^2\theta - 1)$$

$$\Theta_{jm} = \left(\frac{3}{4\pi}\right)^{1/2}\cos\theta \qquad \Theta_{jm} = \left(\frac{15}{8\pi}\right)^{1/2}\sin\theta\cos\theta\, e^{i\phi}$$

$$\Theta_{jm} = \left(\frac{3}{8\pi}\right)^{1/2}\sin\theta\, e^{+i\phi} \qquad \Theta_{jm} = \left(\frac{15}{4\pi}\right)^{1/2}\sin\theta\cos\theta\,\frac{e^{i\phi}+e^{-i\phi}}{2}$$

$$\Theta_{jm} = \left(\frac{3}{8\pi}\right)^{1/2}\sin\theta\, e^{-i\phi} \qquad \Theta_{jm} = \left(\frac{15}{32\pi}\right)^{1/2}\sin^2\theta\, e^{-2i\phi}$$

Assign the correct values of j and m to these functions. (It is suggested that these calculations be divided up among several members of the class.)

The results reported above can be restated in the following way: For each value of the integer j there are $2j + 1$ degenerate states, with energy $\epsilon_j = j(j+1)\hbar^2/2I$, having a "total angular momentum" of $[j(j+1)]^{1/2}\hbar$. The projection of this total angular momentum along the z axis is quantized with values $m\hbar$, for $m = -j, -j+1, \ldots, j-1, j$. These results can be represented by the vector model depicted in Fig. 3-2.

It is true for mechanical systems with spherical symmetry, in general, that there are wave functions which are simultaneously eigenfunctions of $\hat{L}^2$ and $\hat{L}_z$, with eigenvalues $j(j+1)\hbar^2$ and $m\hbar(|m| \leq j)$, just as for the rigid rotor. Thus, $\hbar$ is the fundamental unit of angular momentum in quantum mechanics.

3-9. The Harmonic Oscillator. A classical one-dimensional harmonic oscillator is a mass m oscillating in the potential $U(x) = \frac{1}{2}fx^2$. The

---

* We shall not be very consistent about the matter, but we shall ordinarily use L as in this chapter or M as in Chap. 2 as the symbol for angular momentum, depending on what other symbols are in use in the particular context.

restoring force is $-\partial U/\partial x = -fx$, which is Hooke's law, and the force constant is f. The classical equation of motion is

$$m\ddot{x} = -fx$$

The solution is

$$x = x_0 \sin\left(\frac{f}{m}\right)^{\frac{1}{2}} t \qquad (3\text{-}30)$$

(there are similar cosine solutions also). The frequency is

$$\nu = \frac{1}{2\pi}\left(\frac{f}{m}\right)^{\frac{1}{2}} \qquad (3\text{-}31)$$

and the total energy is related to the amplitude x_0 of oscillation

$$\epsilon = \tfrac{1}{2} k x_0^2$$

The corresponding Schrödinger equation is

$$\left(-\frac{\hbar^2}{2m}\frac{\partial^2}{\partial x^2} + \tfrac{1}{2}fx^2\right)\psi = \epsilon\psi \qquad (3\text{-}32)$$

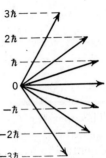

FIG. 3-2. Vector model for angular momenta for $J = 3$. The length of the arrow is the value of $\langle L^2 \rangle^{\frac{1}{2}}$ or $[J(J+1)]^{\frac{1}{2}}\hbar$ or $3.46\hbar$. The allowed values of L_z are $\pm 3\hbar$, $\pm 2\hbar$, $\pm \hbar$, 0.

The wave functions which are the solution of Eq. (3-32) are the so-called Hermite orthogonal functions. The exact formulae do not particularly concern us; solutions are possible only for discrete values of the energy

$$\epsilon = (v + \tfrac{1}{2})h\nu \qquad v = 0, 1, 2, \ldots \qquad (3\text{-}33)$$

where ν is the classical vibration frequency $(1/2\pi)(f/m)^{\frac{1}{2}}$ [(3-31)] and v is an integer, the vibrational quantum number.

The wave functions and energy levels for the first few states are shown in Fig. 3-3.

In the lowest state, $v = 0$, there is a zero-point energy of $\tfrac{1}{2}h\nu$ with respect to the bottom of the potential-energy curve. This is, in a sense, a consequence of the uncertainty principle. With energy equal to zero, both the position of the particle ($x = 0$) and the momentum of the particle ($p_x = 0$) would be certainly known. The state $\epsilon = \tfrac{1}{2}h\nu$ is a compromise in which both the momentum and the position are somewhat uncertain. The qualitative similarity of the harmonic-oscillator wave functions (Fig. 3-3) to the particle in box wave functions (Fig. 3-1) should be noted.

3-10. Separable Coordinates. It is frequently the case that, either exactly or approximately, the Hamiltonian operator can be written as a sum of operators, each of which depends on different coordinates; i.e.,

$$\hat{H}\left(q_1, \ldots, q_{3N}; \frac{\partial}{\partial q_1}, \ldots, \frac{\partial}{\partial q_{3N}}\right)$$

$$= \hat{H}_A\left(q_1, \ldots, q_a; \frac{\partial}{\partial q_1}, \ldots, \frac{\partial}{\partial q_a}\right)$$

$$+ \hat{H}_B\left(q_{a+1}, \ldots, q_b; \frac{\partial}{\partial q_{a+1}}, \ldots, \frac{\partial}{\partial q_b}\right)$$

$$+ \hat{H}_C\left(q_{b+1}, \ldots, q_{3N}; \frac{\partial}{\partial q_{b+1}}, \ldots, \frac{\partial}{\partial q_{3N}}\right) \quad (3\text{-}34)$$

The operator $\hat{H}_A$ depends only on the coordinate $q_1, \ldots, q_a$, and the operator $\hat{H}_C$ depends only on $q_{b+1}, \ldots, q_{3N}$. If $\psi_A(q_1, \ldots, q_a)$ is an eigenfunction of H_A with eigenvalue ϵ_A and if ψ_B and ψ_C are eigenfunc-

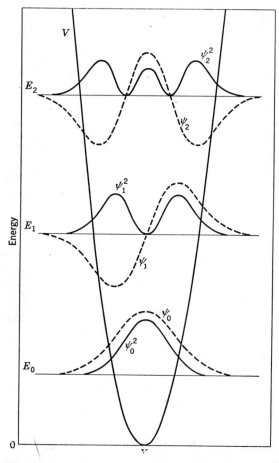

FIG. 3-3. The harmonic oscillator, showing the potential curve, the energy levels, the wave functions, and the probability distributions. (*From Kenneth S. Pitzer, "Quantum Chemistry," p. 38, Prentice-Hall, Englewood Cliffs, N.J., 1953.*)

tions of $\hat{H}_B$ and $\hat{H}_C$, it follows that

$$\psi_A(q_1, \ldots, q_a)\psi_B(q_{a+1}, \ldots, q_b)\psi_C(q_{b+1}, \ldots, q_{3N})$$

is an eigenfunction of the total Hamiltonian, with the total energy for this state being $\epsilon_A + \epsilon_B + \epsilon_C$. This remark is readily verified by direct substitution.

For the system of two particles of masses m_1 and m_2 and coordinates $x_1, y_1, z_1, x_2, y_2, z_2$ with a potential energy which is a function only of the distance r between the particles, we have already observed that the classical Hamiltonian function can be separated as a sum of terms. One, H_{cm}, refers to the motion of the center of mass, with coordinates $x_{\text{cm}}, y_{\text{cm}}, z_{\text{cm}}$ and mass $m_1 + m_2$,

$$x_{\text{cm}} = \frac{m_1 x_1 + m_2 x_2}{m_1 + m_2} \quad \text{etc.}$$

and

$$\hat{H}_{\text{cm}} = -\frac{\hbar^2}{2(m_1 + m_2)}\left(\frac{\partial^2}{\partial x_{\text{cm}}^2} + \frac{\partial^2}{\partial y_{\text{cm}}^2} + \frac{\partial^2}{\partial z_{\text{cm}}^2}\right) \quad (3\text{-}35)$$

The Hamiltonian for the internal motion depends on the reduced mass $\mu = m_1 m_2/(m_1 + m_2)$ and on the coordinates $x = x_2 - x_1$, etc.; therefore,

$$\hat{H}_{\text{int}} = -\frac{\hbar^2}{2\mu}\left(\frac{\partial^2}{\partial x^2} + \frac{\partial^2}{\partial y^2} + \frac{\partial^2}{\partial z^2}\right) + U(r)$$

or

$$\hat{H}_{\text{int}} = -\frac{\hbar^2}{2\mu r^2}\left[\frac{\partial}{\partial r}\left(r^2 \frac{\partial}{\partial r}\right) + \frac{1}{\sin\theta}\frac{\partial}{\partial \theta}\left(\sin\theta \frac{\partial}{\partial \theta}\right) + \frac{1}{\sin^2\theta}\frac{\partial^2}{\partial \phi^2}\right] + U(r) \quad (3\text{-}36)$$

The solution of the Schrödinger equation for the operator H_{cm} gives a wave function and an energy level for a free particle of mass $m_1 + m_2$ and coordinates $x_{\text{cm}}, y_{\text{cm}}, z_{\text{cm}}$. The solution of the equation for H_{int} depends on the nature of $U(r)$, but the result gives the wave functions and energy levels for the internal motion of the system. The total energy is the sum of the internal energy and the energy of the center-of-mass point, and the over-all wave function is the product of the two wave functions.

3-11. Perturbation Theory. It is frequently the case that the Hamiltonian operator for a system can be written as

$$\hat{H} = \hat{H}_0 + \hat{H}' \quad (3\text{-}37)$$

where $\hat{H}_0$ describes a system for which the wave functions and energy levels are known and for which $\hat{H}'$ is a small perturbation. The perturbation theory provides a method for making an approximate calculation for the energy levels and wave functions of the real system (the perturbed system) in terms of the energy levels ϵ_k° and wave functions ψ_k° of the unperturbed Hamiltonian. This technique of calculation does not particularly concern us here, but three general features of the results are of interest.

We first consider perturbations which are not functions of time. The first and most obvious point is that the wave functions and energy levels for the perturbed system will be only slightly different from those of the unperturbed system. The second point is that, for many cases, one effect of a perturbation is to remove all or some of the degeneracy that existed in the solutions of the unperturbed problem.

For example, for a free rigid rotor of energy $\epsilon_j = j(j+1)\hbar^2/2I$, there are $2j+1$ wave functions $\Theta_{jm}(\theta,\phi)$ corresponding to the $2j+1$ different values of the quantum number m. If now there is a potential energy of orientation $U(\theta)$, so that the energy of the rotor depends on its orientation, and if $U(\theta)$ is small compared with the unperturbed energy ϵ_j, we can add $U(\theta)$ to the Hamiltonian equation (3-26) as a perturbation. The energy levels will then be shifted slightly, and it will be found that states of different $|m|$ now have slightly different energies, but the states $+m$ and $-m$ are still degenerate.

The third point relates to perturbations $\hat{H}'$ which are explicit functions of time. The effect of the perturbation in shifting the energies of the stationary states of the operator $\hat{H}_0$ may be very small, but it is found that the time-dependent perturbation will serve to cause transitions between one state and another. Thus, if we examine the system at any one time, we will find that it is a pretty good approximation to say that it is in one of the eigenstates of $\hat{H}_0$; however, the system is not in a truly stationary state, and if we look at it at some other time, it may be in a state which is (approximately) a different eigenstate of $\hat{H}_0$.

3-12. Spin. The fundamental particles of modern physics—the electron, the proton, the neutron, the various mesons, the neutrino, and indeed even the light quantum—all have a quantized, intrinsic angular momentum in addition to the various properties associated with a wave function which depends on the space coordinates of the particle. A composite particle such as a deuteron also has an intrinsic angular momentum. We need not develop the formalism for quantum-mechanical calculations of spin systems in detail. The spin of the particle is characterized by a spin quantum number S. For electrons, protons, neutrons, neutrinos, and positrons, $S = \frac{1}{2}$. For π-mesons $S = 0$. For μ-mesons, $S = \frac{1}{2}$. For deuterons, $S = 1$. In any case the total squared angular momentum is $S(S+1)\hbar^2$. The particle can be in an eigenstate of the operator $\hat{S}_z$ for spin angular momentum along the z axis so that the component of spin angular momentum along the z axis is any one of the values $-S\hbar$, $-(S-1)\hbar$, ..., $S\hbar$. The spin state of a particle is specified by a wave function. For example, for a particle with $S = \frac{1}{2}$ the wave functions α and β assert that the components of angular momentum along the z axis are $+\hbar/2$ and $-\hbar/2$, respectively. If a system consists of a proton (particle 1) and an electron (particle 2), the spin function $\alpha(1)\beta(2)$ asserts that the system is in a state with the proton spin-angular-momentum component of $+\hbar/2$ and the electron spin-angular-momentum component of $-\hbar/2$. If we have two protons, the wave function $\alpha(1)\alpha(2)$ asserts that they are both in a state with $+\hbar/2$ of angular momentum along the z axis. The wave function $\alpha(1)\beta(2)$ asserts that the first proton has a component $+\hbar/2$ and the second one a component $-\hbar/2$ along the axis. The wave function $\beta(1)\alpha(2)$ makes a converse assertion. As we shall see later, the protons are indistinguishable and we must write either

$$\frac{1}{2^{1/2}}[\alpha(1)\beta(2) + \beta(1)\alpha(2)]$$

or

$$\frac{1}{2^{1/2}}[\alpha(1)\beta(2) - \beta(1)\alpha(2)]$$

for our wave functions, instead of $\alpha(1)\beta(2)$ or $\beta(1)\alpha(2)$.

3-13. Wave Functions for Systems of Identical Particles. Suppose for simplicity and definiteness that we have a system of N identical particles in a container such that, to a first approximation, the particles do not interact with each other. There may be an external field so that the potential energy is $U(x_1,y_1,z_1) + U(x_2,y_2,z_2) + \cdots + U(x_n,y_n,z_n)$. The total Hamiltonian is then

$$\hat{H}(x_1, \ldots ,z_n;p_1, \ldots ,p_n) = \sum_{i=1}^{n}\left[\frac{p_{x_i}^2 + p_{y_i}^2 + p_{z_i}^2}{2m} + U(x_iy_iz_i)\right]$$
$$+ \hat{H}'(x_1, \ldots ,z_n;p_1, \ldots ,p_n) \quad (3\text{-}38)$$

where the first two terms are the unperturbed Hamiltonian operator in the approximation in which the particles do not interact with each other and H' includes the weak interactions which must exist between the particles. For convenience, write $U(1)$ instead of $U(x_1,y_1,z_1)$ and $\hat{H}(1)$ instead of $\hat{H}(x_1, y_1, z_1, \partial/\partial x_1, \partial/\partial y_1, \partial/\partial z_1)$. The Hamiltonian is then

$$\hat{H} = \hat{H}(1) + \hat{H}(2) + \hat{H}(n) + \hat{H}'(1, \ldots ,n) \quad (3\text{-}39)$$

Let $\phi_j(1)$ be a wave function for the jth eigenstate of the single particle operator $\hat{H}(1)$, and bear in mind that $\hat{H}(2)$ is identical with $\hat{H}(1)$, except that the coordinates are x_2, y_2, z_2 instead of x_1, y_1, z_1.

Consider for definiteness a system of three particles. Assume that the interaction Hamiltonian $\hat{H}'$ is negligibly small. Then the solutions of the Schrödinger equation for the system are products of the single-particle wave functions. The wave function $\phi_j(1)\phi_k(2)\phi_l(3)$ is called a system wave function, whereas any one ϕ is a single-particle wave function. The system wave function $\phi_j(1)\phi_k(2)\phi_l(3)$ asserts that particle 1 is in state j, particle 2 is in state k, particle 3 is in state l, and the energy is

$\epsilon_j + \epsilon_k + \epsilon_l$. There is a degenerate system wave function $\phi_k(1)\phi_j(2)\phi_l(3)$ which has the same energy but in which particle 1 is in state k and particle 2 in state j. Clearly, if ϕ_j, ϕ_k, and ϕ_l are different single-particle states, there are $3! = 6$ different ways of assigning particles 1, 2, 3 to states j, k, l.

It is a fundamental postulate of quantum mechanics that all these wave functions are not acceptable and that there is only one wave function for such a system of identical particles. The acceptable system wave function must be either symmetric or antisymmetric for interchange of the coordinates of any two identical particles.

For the example at hand, the system wave function can be

Symmetric:

$$\psi_s(1,2,3) = \phi_j(1)\phi_k(2)\phi_l(3) + \phi_j(1)\phi_k(3)\phi_l(2) + \phi_j(2)\phi_k(3)\phi_l(1)$$
$$+ \phi_j(2)\phi_k(1)\phi_l(3) + \phi_j(3)\phi_k(1)\phi_l(2) + \phi_j(3)\phi_k(2)\phi_l(1) \quad (3\text{-}40)$$

Antisymmetric:

$$\psi_a(1,2,3) = \phi_j(1)\phi_k(2)\phi_l(3) - \phi_j(1)\phi_k(3)\phi_l(2) - \phi_j(2)\phi_k(1)\phi_l(3)$$
$$+ \phi_j(2)\phi_k(3)\phi_l(1) - \phi_j(3)\phi_k(2)\phi_l(1) + \phi_j(3)\phi_k(1)\phi_l(2) \quad (3\text{-}41)$$

That the function ψ_s is unchanged and the function ψ_a changes sign on interchange of the coordinates of any two particles is apparent by inspection. [It should also be remarked that the system wave functions in (3-40) and (3-41) may have to be multiplied by a constant in order to be properly normalized.]

In the wave functions $\psi_s(1,2,3)$ and $\psi_a(1,2,3)$ of (3-40) and (3-41) we cannot say which particle is in state ϕ_j, which is in ϕ_k, and which is in ϕ_l, although the wave function does assert that the system is in a state in which the three single-particle states ϕ_j, ϕ_k, and ϕ_l are occupied. Contrary to one's naïve classical intuition, identical particles are truly indistinguishable.

If the identical particles each have half-integral spins, the system wave function must be antisymmetric. This is true for electrons, protons, and Cl^{35} nuclei (spin $= 3/2$), for example. Such particles are called fermions. For particles of integral spin, such as light quanta ($S = 1$), H^2 nuclei ($S = 1$), He^4 nuclei, or O^{16} nuclei ($S = 0$), the wave function must be symmetric. Particles with symmetric system wave functions are called bosons.

It is convenient at times to talk about hypothetical particles which are distinguishable. We coin the term *boltzons* for such particles.

The system wave function can be expressed as a product of single-particle wave functions only if the interactions between the particles are negligibly weak. When the interactions between the particles are strong, the wave functions cannot be expressed as a product of single-particle wave functions; but the system wave function must still be either symmetric or antisymmetric for the interchange of identical particles.

When the interactions are weak, the single-particle approximation is satisfactory. It is convenient to refer to a single-particle wave function as an orbital. A set of N orbitals $\phi_{j_1}, \ldots, \phi_{j_N}$ for a system of N particles is called a configuration.

For N bosons in a configuration where the N orbitals $\phi_{j_1}, \ldots, \phi_{j_N}$ are all different, the normalized symmetric wave function is

$$\psi_s = \frac{1}{(N!)^{1/2}} \sum_P \phi_{j_1}(1), \ldots, \phi_{j_N}(N) \tag{3-42}$$

where $\sum_P$ denotes the sum over all $N!$ permutations of the N particles among the N orbitals.*

A general method for constructing an antisymmetric wave function from N different single-particle wave functions is to construct the determinant

$$\psi_a = \frac{1}{N^{1/2}} \begin{vmatrix} \phi_{j_1}(1) & \cdots & \phi_{j_1}(N) \\ \vdots & & \vdots \\ \phi_{j_N}(1) & \cdots & \phi_{j_N}(N) \end{vmatrix} \tag{3-43}$$

It is clear that for fermions it is not possible to have a configuration in which two orbitals are the same; the wave function would be unaltered upon interchanging the coordinates of the two particles in these two orbitals but must change sign. Thus, for fermions the only possible configurations are ones in which all the particles are in different orbitals. There is no analogous restriction for bosons.

The single-particle wave functions, or orbitals, include spin. Thus, if $\phi_i(\mathbf{r})$ is a space wave function [where $\mathbf{r}$ is the vector (x,y,z)], there are two orbitals $\phi_i\alpha$ and $\phi_i\beta$. The Pauli principle then implies that two fermions cannot occupy the same spatial orbitals unless they have opposite spin. For a system of two fermions with a single space orbital, the system wave

---

* The orbitals, or single-particle wave functions, are assumed to be mutually orthogonal.

function constructed according to (3-43) is

$$\psi(1,2) = \phi_i(\mathbf{r}_1)\phi_i(\mathbf{r}_2)[\alpha(1)\beta(2) - \beta(1)\alpha(2)] \qquad (3\text{-}44)$$

We may also note at this point that, in general, additional degeneracies exist because of spin. Thus, for two fermions with two different space orbitals ϕ_i and ϕ_j, the system wave function based on the configuration $\phi_i\alpha$, $\phi_j\beta$ is

$$\psi_\mathrm{I}(1,2) = \frac{1}{2^{1/2}}[\phi_i(1)\alpha(1)\phi_j(2)\beta(2) - \phi_j(1)\beta(1)\phi_i(2)\alpha(2)] \qquad (3\text{-}45)$$

and the system wave function based on the configuration $\phi_i\beta$, $\phi_j\alpha$ is

$$\psi_\mathrm{II}(1,2) = \frac{1}{2^{1/2}}[\phi_i(1)\beta(1)\phi_j(2)\alpha(2) - \phi_j(1)\alpha(1)\phi_i(2)\beta(2)] \qquad (3\text{-}46)$$

If the space coordinates and the spin coordinates are separable for the Hamiltonian operator, the wave functions ψ_I and ψ_II are degenerate. The correct linear combinations prove to be as follows:

$$\psi_T(1,2) = \frac{1}{2^{1/2}}(\psi_\mathrm{I} + \psi_\mathrm{II})$$
$$= \tfrac{1}{2}[\phi_i(1)\phi_j(2) - \phi_j(1)\phi_i(2)][\alpha(1)\beta(2) + \beta(1)\alpha(2)] \qquad (3\text{-}47)$$
$$\psi_S(1,2) = \frac{1}{2^{1/2}}(\psi_\mathrm{I} - \psi_\mathrm{II})$$
$$= \tfrac{1}{2}[\phi_i(1)\phi_j(2) + \phi_j(1)\phi_i(2)][\alpha(1)\beta(2) - \beta(1)\alpha(2)] \qquad (3\text{-}48)$$

The wave functions ψ_T and ψ_S are in fact triplet and singlet wave functions as regards the spin angular momentum. It can be seen that the space factors in the wave functions are different; when weak interactions between the particles are considered, these two functions do not have the same energy.

For our purposes, the main point is that there are still two system wave functions possible, whether they be ψ_I and ψ_II or ψ_T and ψ_S.

A similar but more complicated situation will prevail for systems containing more particles. The number of allowed states, however, is not affected by the question of which spin functions are associated with which space functions for a set of degenerate configurations. The only reason for mentioning the complications due to spin degeneracy here is to remark that they do not affect the number of allowed states and the fundamental statistical mechanical situation.

Let us return to our example of three particles and three orbitals, ϕ_i, ϕ_j, ϕ_k. It is frequently convenient to consider the number of wave functions for a system of hypothetical distinguishable particles or boltzons. If the three orbitals are different, $\phi_i \neq \phi_j \neq \phi_k$, there would be $3! = 6$ system wave functions for boltzons, 1 for bosons, and 1 for fermions. If two of the three orbitals are identical, $\phi_i = \phi_j \neq \phi_k$, there are 3 system wave functions for boltzons, $\phi_i(1)\phi_i(2)\phi_k(3)$, $\phi_i(1)\phi_i(3)\phi_k(2)$, and $\phi_i(3)\phi_i(2)\phi_k(1)$, 1 for bosons, and none for fermions. If all three orbitals are the same, there is one system state for boltzons, one for bosons, and none for fermions. The simple illustration is clearly general, as expressed in paragraphs (b) and (c) below.

From the point of view of statistical mechanics, the essential points arising out of the basic indistinguishability of particles are the following:

(a) For a system of N weakly interacting identical particles in a configuration with N distinct orbitals, there is only one system wave function. We cannot say which particle is in a particular orbital ϕ_i and which is in another ϕ_k, but we can say that the N particular orbitals $\phi_1, \ldots, \phi_i, \ldots, \phi_j, \ldots, \phi_N$ are occupied.

(b) If the particles were distinguishable, there would be $N!$ different wave functions corresponding to the $N!$ possible permutations of the N particles among the N different orbitals.

(c) If some of the orbitals are identical, the number of possible system states for boltzons is less than $N!$ and greater than one; while, there is one state for bosons and none for fermions.

3-14. The Quantum-mechanical Perfect Gas. A single-particle state for a free particle in a rectangular box is specified by the three integral quantum numbers s_x, s_y, s_z. The energy for this state is

$$\epsilon_{s_x s_y s_z} = \frac{h^2}{8m}\left(\frac{s_x^2}{X^2} + \frac{s_y^2}{Y^2} + \frac{s_z^2}{Z^2}\right) \tag{3-49}$$

In general, for any system with energy levels $\epsilon_i(XYZ)$ that depend on the dimensions of the container, the force exerted in the x direction (that is, on the YZ wall) by a particle in the ith quantum state is

$$F_{xi} = -\frac{\partial \epsilon_i(X,Y,Z)}{\partial X} \tag{3-50}$$

From (3-49) for a particle in a box

$$F_{x,s_x,s_y,s_z} = -\frac{\partial \epsilon_{s_x s_y s_z}}{\partial X} = \frac{h^2 s_x^2}{4mX^3} \tag{3-51}$$

From a classical point of view (cf. Probs. 3-1 and 3-2), the state s_x, s_y, s_z is to be described as one in which the particle is moving back and forth along the x axis (back and forth so that the average momentum $\langle p_x \rangle = 0$) with energy $s_x^2 h^2/8mX^2$, etc. The velocity in the x direction would be $(2\epsilon_x/m)^{1/2}$ or $|v_x| = s_x h/2mX$. The momentum change on striking the wall is $2mv_x$, and the number of collisions with the YZ wall in unit time is $v_x/2X$. The force excited on the wall in this classical picture is momentum change per unit time, or mv_x^2/X, or $s_x^2 h^2/4mX^3$ in agreement with Eq. (3-51).

As we shall see (Prob. 3-4), the important values of s are very large.

At room temperature in an ordinary-sized container for an ordinary gas, they are of the order of 10^8. Take $s_x = 10^8$. A small fractional interval of kinetic energy in the x direction, $d\epsilon_x/\epsilon_x = 10^{-5}$, corresponds to $ds_x/s_x = d\epsilon_x/2\epsilon_x$ and $ds_x = 500$. If there is an interval $ds_x = 500$, $ds_y = 500$, and $ds_z = 500$, the number of states in this interval is $500^3 = 1.25 \times 10^8$. All these states have almost exactly the same energy and exert almost exactly the same value of F_x, F_y, F_z on the respective walls. But, at least in principle, these states are distinguishable, and it is possible to determine whether a particle is in the state or orbital (1.00×10^8, 1.02×10^8, and 1.01×10^8) or in the state (1.01×10^8, 1.00×10^8, and 1.02×10^8).

Let the vector $\mathbf{s}$ stand for the triple of numbers (s_x, s_y, s_z) so that $\phi_\mathbf{s}$ is the corresponding single-particle wave function. We group together all the states which are in the interval s_x to $s_x + ds_x$, s_y to $s_y + ds_y$, and s_z to $s_z + ds_z$ as a set of degenerate states. There are $g_s = ds_x ds_y ds_z$ distinguishable single-particle states or orbitals included in this group of degenerate states.

The exact nature of the grouping is unimportant. But it is important to realize that there are a large number of distinguishable single-particle states with almost exactly the same values of the energy and almost exactly the same values of the forces F_x, F_y, F_z and that we can group these together as being approximately degenerate.

An idealized model for a quantum-mechanical perfect gas is a set of N particles in a box, with negligible interactions between the particles so that the single-particle energy levels given by Eq. (3-49) are correct. A configuration of the system is specified by the N vectors $\mathbf{s}_1, \ldots, \mathbf{s}_N$ with the corresponding orbitals $\phi_{\mathbf{s}_1}, \phi_{\mathbf{s}_2}, \ldots, \phi_{\mathbf{s}_N}$. We can construct one and only one system wave function for this configuration by the rules already stated. The system wave function will depend on whether the particles are fermions or bosons. But in either case there is one function; we cannot say which particle is in state $\mathbf{s}_1$ and which is in $\mathbf{s}_j$, but we can say that the physically distinct and recognizable orbitals $\mathbf{s}_1, \mathbf{s}_2, \ldots, \mathbf{s}_N$ are occupied. Of course, for fermions, no two of the orbitals can be the same. (The orbitals should include spin also, but this will not change the general features of the situation.)

For statistical mechanical purposes, we are interested in a less detailed description which we call a distribution. As indicated above, we group together all the g_s states in a narrow interval. A distribution is characterized by the set of numbers $N_{\mathbf{s}_1}, N_{\mathbf{s}_2}, \ldots, N_{\mathbf{s}_j}, \ldots$, with $\sum_{\mathbf{s}_i} N_{\mathbf{s}_i} = N$, in which it is said that there are $N_{\mathbf{s}_1}$ orbitals from the group of $g_{\mathbf{s}_1}$ degenerate states $\mathbf{s}_1$ which are occupied, $N_{\mathbf{s}_j}$ orbitals of the group of $g_{\mathbf{s}_j}$ degenerate states $\mathbf{s}_j$ which are occupied, etc. It is natural and usual to say that there are $N_{\mathbf{s}_1}$ particles in the states of energy $\epsilon_{\mathbf{s}_1}$ and $N_{\mathbf{s}_j}$ particles in the states of energy s_j. Since the particles are indistinguishable, it is better to say that there are N_{s_j} orbitals of energy ϵ_{s_j} which are occupied, but either way of describing the situation is acceptable.

There are g_{s_j} possible orbitals in the group of states s_j. If the particles are fermions, an allowed distribution must have $g_{s_j} \geq N_{s_j}$. For fermions with $g_{s_j} > N_{s_j}$, or for bosons in any case, there are a large number of ways of selecting the N_{s_j} orbitals from among the g_{s_j} available of energy ϵ_{s_j}. That is, there are a large number of system states or wave functions that correspond to the distribution $N_{s_1}, \ldots, N_{s_j}, \ldots$. The number of system wave functions for a distribution is of fundamental importance for statistical mechanics, and we shall calculate this number in Chap. 6.

Our model for a perfect gas is highly oversimplified. It is the analogue in quantum mechanics of the model of a perfect gas in classical statistical mechanics in which the molecules move in straight lines, without any intermolecular collisions, and are specularly reflected on striking the walls, without exchanging energy with the walls. There are actually short-range attractive and repulsive forces between molecules. In quantum mechanics, these interactions could be described as a perturbation. There are several effects due to such perturbations.

One effect is to change the average energy of the gas. The gas is then no longer a perfect gas. The theory that we shall study later, as well as our practical knowledge, tells us that this effect is not important for a sufficiently dilute gas; for example, it is not very important for the ordinary permanent gases at room temperature and atmospheric pressure.

But for an ordinary gas at room temperature and atmospheric pressure, the average time between collisions for any one gas molecule is about 10^{-10} sec, and the mean free path between collisions is about 5×10^{-6} cm. The description of the gas in terms of single-particle wave functions that are standing waves extending from wall to wall is obviously absurd; however, the statistical thermodynamic properties of the gas can be correctly calculated with this model. Fundamentally, the reason for this is that the existence of collisions changes the wave functions that should be used to describe the system but does not change the number of wave functions for a given energy interval.

A gas with intermolecular collisions cannot rigorously be said to be in a particular configuration at any one time. However, we can give an approximate description of the system by saying that at any one time the gas is in a definite, stationary system state—with all the particles having prescribed single-particle states—and that the effect of collisions is to induce transitions from one configuration to another. Thus, collisions provide a mechanism for allowing the gas to sample all the possible configurations available to it; therefore, the collisions are a mechanism for producing statistical equilibrium.

There is one other interesting effect of the collisions. In a classical description a molecule in a dilute gas after a collision moves with a constant vector velocity until the next collision. In a quantum-mechanical description, can it be said to be in a definite quantum state with a known energy? The uncertainty principle asserts that

$$|\delta\epsilon|\,|\delta t| \sim h \tag{3-52}$$

where δt is the lifetime of a state and $\delta\epsilon$ is the uncertainty of the energy of the state. Thus, if we say that a molecule undergoes a transition from one single-particle state to another on collision, the energy of the state is uncertain, as given by Eq. (3-52). Thus, it was not unreasonable for us to group together a set of states in a narrow energy interval as being degenerate even if there is a slight spread in the values of s_x, s_y, s_z.

PROBLEM

3-4. Calculate the value of s_x, s_y, s_z for the case $s_x = s_y = s_z$ for a hydrogen atom (atomic weight 1.00) in a box of dimensions 1 cc if the particle has a kinetic energy of $\tfrac{3}{2}\,kT$, for $T = 300°K$. What significant fact does this calculation illustrate?

While you are at it, calculate and record the value of the quantity $h^2/8m$ (for atomic weight 1.00) which occurs in such calculations and the ratio $h^2/8mkT$ for $T = 300°K$.

4

Thermodynamics

4-1. Introduction. A major objective of our studies is to derive the laws of thermodynamics from statistical mechanics and to derive statistical mechanical formulae for the calculation of the thermodynamic functions of a substance by using molecular-structure data. The purpose of the present chapter is to review those aspects of thermodynamics which are needed for this program.*

4-2. Work and Heat. We assume that the concepts of work and heat are familiar to the reader. For the examples that we study, the most important form of work is $P\,dV$ work, but we shall consider electrical and magnetic work in later chapters. The work done by a system on its surroundings in changing from state A to state B is given by

$$w = \int_A^B P\,dV \tag{4-1}$$

where P is the pressure exerted by the surroundings on the system and dV is the increase in volume of the system.

An empirical temperature t can be established by means of a mercury thermometer or a gas thermometer. Two systems in thermal contact have the same value of t.

Heat can be considered to be defined by the usual sort of calorimetric experiment, in which one measures the temperature change of a fixed mass of water, for example. The heat leaving the surroundings during the change of the system from state A to state B is

$$q = \int_A^B dq \tag{4-2}$$

Whether we take q as positive for heat leaving the surroundings or for heat entering the surroundings is a question of convention as regards sign. It

---

* To a large extent, the material in the early parts of this chapter is based on the mimeographed lecture notes by J. H. Irving from the lectures by Prof. J. G. Kirkwood, Selected Topics in Statistical Mechanics, delivered at Princeton University in the spring term, 1947. These notes are not generally available, but the author wishes to acknowledge his indebtedness to them.

is important, however, that one define work and heat in terms of changes occurring in the surroundings, not changes in the system. For an arbitrary change, the system can be rather complicated. It could have temperature gradients, pressure gradients, and concentration gradients; and chemical reactions could be occurring. Under these circumstances, the heat change in the system may be difficult to define. There is no objection, however, to making the surroundings simple—for example, a water bath—and the heat change in the surroundings is easily defined, operationally.

4-3. The First Law of Thermodynamics. The first law states that the energy change ΔE defined by

$$\Delta E = q - w = \int_A^B dq - \int_A^B dw \tag{4-3}$$

is independent of the path over which the change from A to B is carried out. The quantities q and w in general do depend on the path.

In differential form,

$$dE = dq - dw \tag{4-4}$$

and dE is an exact differential, although dq and dw are not. The thermodynamic energy is a function of state which can be measured by (4-4), apart from an additive constant which can be fixed by assigning a value to E in a chosen reference state.*

4-4. Reversible and Irreversible Changes. "If a system is brought from state A to state B so slowly that all intermediate states may be considered equilibrium states (i.e., there are no finite mass or heat currents) then the path is reversible, i.e., it is possible to slowly return to the initial state with no net change in the surroundings. The reversible path is an idealization which is never reached in actual processes but may be closely approached with care."†

It is difficult to give a discussion of irreversible processes which is simple, general, and complete. One very obvious kind of irreversible process occurs when a system with a fixed volume at one temperature is immersed

---

* The thermodynamic function of state which we call E, the energy, is sometimes called the total energy and sometimes the internal energy. The latter usage causes confusion because we shall speak of the energy of a molecule as being composed of its internal energy (associated with its internal coordinates—vibration, rotation, electronic excitation) plus the translational and potential energy associated with the motion and position of the center of mass of the molecule. For a collection of molecules in a perfect gas, the thermodynamic energy E is actually the sum of the "translational" energy plus the "internal" energy.

When there is a possibility of confusion, the function of state E can be clearly identified by calling it the thermodynamic-energy function.

We ordinarily use the symbol E; but in some cases, E will be the electric field, and we shall use U for the thermodynamic energy function.

† From J. G. Kirkwood, Selected Topics in Statistical Mechanics.

in surroundings at a different temperature. There is then an irreversible flow of heat with no work done.

Consider a fluid system in a cylinder separated from its surroundings by a massless, frictionless piston. Let P be the pressure of the system and P_0 the pressure of the surroundings. In any process, the work done on the surroundings is $\int P_0 \, dV$. If P is only infinitesimally greater than P_0, the system will expand reversibly and the reversible work is $w_{rev} = \int P \, dV$. If P_0 is decreased so that $P > P_0$, the piston will be accelerated. The detailed analysis of the changes in the system is a complicated problem in hydrodynamics. If the velocity of the piston is comparable with the velocity of sound in the system, there will be pressure, temperature, and velocity gradients in the system. But in any case one can see that the fluid in the system will acquire kinetic energy. To some extent, because of the viscosity of the fluid, the kinetic energy will be degraded into heat. (This may not be the only source of irreversibility in the process.) The work done on the surroundings in the irreversible expansion is

$$w_{irrev} = \int P_0 \, dV$$

Clearly $w_{rev} > w_{irrev}$. In the same way, if there is to be an irreversible compression, the external pressure must be raised so that $P_0 > P$. The piston will then accelerate inward. The rapid compression of the fluid in the system will generate a shock wave. A shock wave is intrinsically an irreversible compression, even in an inviscid fluid. There will also be irreversible processes associated with viscous dissipation of the kinetic energy. The work done on the surroundings is negative, $w_{irrev} = \int P_0 \, dV$, since the volume change is negative. Since $P_0 > P$, we have $-w_{rev} < -w_{irrev}$, or, as before, $w_{rev} > w_{irrev}$.

From the point of view of a chemist, there are other more interesting kinds of irreversible processes. Suppose that the system is a gas which is at equilibrium for the reaction

$$N_2 + 3H_2 \rightleftharpoons 2NH_3$$

Let the system expand isothermally at a slow speed such that it is hydrodynamically reversible, but at such a rate that the chemical composition does not change. After expansion, in accordance with Le Châtelier's principle, the chemical reaction slowly proceeds to the left toward its new equilibrium point. During this process, the pressure increases. If chemical equilibrium had been maintained during the expansion, the pressure of the system would have been greater than otherwise and again $w_{rev} > w_{irrev}$.

One point of general significance is then the following: In the change of state from A to B, E has a definite value.

$$E = q_{rev} - w_{rev} \quad \text{and} \quad E = q_{irrev} - w_{irrev}$$

It appears by examination of some simple cases that

$$w_{rev} > w_{irrev} \tag{4-5a}$$

If this be the case,

$$q_{rev} > q_{irrev} \tag{4-5b}$$

Among the common causes of irreversible phenomena, we recognize temperature gradients and heat conduction, velocity gradients and viscous-momentum transfer, concentration gradients and diffusion, and lack of chemical equilibrium.

4-5. The Second Law. The formal statement of the second law comprises statements (a) and (b) below:*

(a) "On any reversible path, there exists an integrating factor, $T(t)$, common to all systems and dependent only on the empirical temperature, t, such that

$$dS = \frac{dq}{T(t)} \tag{4-6}$$

defines an exact differential form."

The entropy S is a function of state which is defined by Eq. (4-6) except for a constant of integration, the entropy in a reference state. The entropy change in going from state A to B is

$$S_B - S_A = \int_A^B \frac{dq}{T} \tag{4-7}$$

over a reversible path. The quantity T is the absolute temperature.

(b) "For all natural processes (non-reversible)

$$S_B - S_A = \Delta S > \int_A^B \frac{dq}{T} \tag{4-8}$$

where T is the temperature of the surroundings."

Given Eq. (4-7), (4-8) is seen to be in agreement with the observation in the previous section that $q_{rev} > q_{irrev}$.

4-6. Some Thermodynamic Derivations. We consider a single-phase system with a fixed amount of matter (a closed system). In the absence of any applied fields—electrical, magnetic, or gravitational—the state of the system is determined by two variables, for example, T and V or T and P. We choose T and V to start with. Since E and S are functions of state, they are completely determined by T and V.

We can write

$$dE = \left(\frac{\partial E}{\partial V}\right)_T dV + \left(\frac{\partial E}{\partial T}\right)_V dT \tag{4-9}$$

$$dS = \left(\frac{\partial S}{\partial V}\right)_T dV + \left(\frac{\partial S}{\partial T}\right)_V dT \tag{4-10}$$

* J. G. Kirkwood, Selected Topics in Statistical Mechanics.

But $dE = dq - dw$, and, for a reversible change, $dq = T\,dS$, $dw = P\,dV$, so that
$$dE = T\,dS - P\,dV \qquad (4\text{-}11)$$

Substitute for dS from (4-10):
$$dE = \left[T\left(\frac{\partial S}{\partial V}\right)_T - P\right]dV + T\left(\frac{\partial S}{\partial T}\right)_V dT \qquad (4\text{-}12)$$

so that
$$\left(\frac{\partial E}{\partial V}\right)_T = T\left(\frac{\partial S}{\partial V}\right)_T - P \qquad (4\text{-}13a)$$

and
$$C_V = \left(\frac{\partial E}{\partial T}\right)_V = T\left(\frac{\partial S}{\partial T}\right)_V \qquad (4\text{-}13b)$$

The first statement of Eq. (4-13b) is the definition of the heat capacity C_V; the second follows from the comparison of (4-12) and (4-9). Since dE is a perfect differential,
$$\frac{\partial^2 E}{\partial T\,\partial V} = \frac{\partial^2 E}{\partial V\,\partial T}$$

By application of this so-called cross-differentiation identity to (4-13a) and (4-13b), we obtain
$$\frac{\partial}{\partial T}\left(T\frac{\partial S}{\partial V} - P\right) = \frac{\partial}{\partial V}\left(T\frac{\partial S}{\partial T}\right)$$

or
$$T\frac{\partial^2 S}{\partial T\,\partial V} + \left(\frac{\partial S}{\partial V}\right)_T - \left(\frac{\partial P}{\partial T}\right)_V = T\frac{\partial^2 S}{\partial V\,\partial T}$$

or
$$\left(\frac{\partial S}{\partial V}\right)_T = \left(\frac{\partial P}{\partial T}\right)_V \qquad (4\text{-}14)$$

The quantity $T(\partial S/\partial V)_T$ is the coefficient of the latent heat of isothermal expansion. From (4-13a) and (4-14), we obtain another useful relation:
$$\left(\frac{\partial E}{\partial V}\right)_T = T\left(\frac{\partial P}{\partial T}\right)_V - P \qquad (4\text{-}15)$$

The enthalpy $H = E + PV$ is a function of state, since E, P, and V are; with its use we can derive an analogous relation to (4-14) for $(\partial S/\partial P)_T$ (Prob. 4-1).

4-7. The Third Law of Thermodynamics. The entropy change for a substance between two states can be calculated by integrating the differential form (4-10). For a constant-pressure reversible process, the entropy change is
$$\Delta S = \int_{T_1}^{T_2} \frac{C_p\,dT}{T} + \sum \Delta S_{\text{trans}}$$

where $\sum \Delta S_{\text{trans}}$ is the sum of the entropies of transitions for any phase changes between temperatures T_1 and T_2.

The third law of thermodynamics asserts that the entropies of all substances at thermodynamic equilibrium at $T = 0$ have a common value. This common value is arbitrarily chosen to be zero.

At the thermodynamic level, the third law is simply an extremely useful generalization from observations. The more elegant mathematical aspects of thermodynamic manipulations are not usually required for its application. In statistical mechanics, however, the third law is of deep and fundamental significance.

A number of substances which are believed to be substances at thermodynamic equilibrium are observed all to have the same value for their entropies at $T = 0$, and we take this value as zero entropy. This is the third law. There are other preparations which, it is observed experimentally, have a positive entropy at $T = 0$. These are all cases for which there are reasons for believing that the preparations are not at thermodynamic equilibrium. A liquid which congeals to a glass upon cooling, instead of crystallizing, is found to have a positive entropy at $T = 0$. A solid solution, for example, an alloy, is not at true equilibrium at $T = 0$; it should separate into several pure phases or form an ordered superlattice. Ordinary crystalline carbon monoxide at $T = 0$ is observed to have a residual positive entropy, which is evidently due to a frozen-in disorder in the orientation of the molecules—the orientations CO and OC along the appropriate crystallographic axis both occurring; at thermodynamic equilibrium at $T = 0$, the carbon and the oxygen atoms would be at definite, different lattice sites.

4-8. The Free-energy Functions. We start with the relations

$$dE = T\, dS - P\, dV \tag{4-16}$$
$$dH = T\, dS + V\, dP \tag{4-17}$$

The free energies A and F are defined by

$$A = E - TS \quad \text{Helmholtz free energy} \tag{4-18}$$
$$F = H - TS \quad \text{Gibbs free energy} \tag{4-19}$$

In Eq. (4-16) we write $T\, dS = d(ST) - S\, dT$ and transpose the perfect differential $d(ST)$; thus we obtain

$$d(E - ST) = -S\, dT - P\, dV$$
$$dA = -S\, dT - P\, dV \tag{4-20}$$

$$\left(\frac{\partial A}{\partial T}\right)_V = -S \quad \left(\frac{\partial A}{\partial V}\right)_T = -P \tag{4-21}$$

By a similar manipulation

$$dF = -S\, dT + V\, dP \tag{4-22}$$

$$\left(\frac{\partial F}{\partial T}\right)_P = -S \quad \left(\frac{\partial F}{\partial P}\right)_T = V \tag{4-23}$$

Two similar free-energy functions, A/T and F/T, are useful. By straightforward manipulation we find

$$d\left(\frac{A}{T}\right) = -\frac{E}{T^2}dT - \frac{P}{T}dV \qquad (4\text{-}24)$$

$$\left[\frac{\partial(A/T)}{\partial T}\right]_V = -\frac{E}{T^2} \qquad \left[\frac{\partial(A/T)}{\partial V}\right]_T = -\frac{P}{T} \qquad (4\text{-}25)$$

$$d\left(\frac{F}{T}\right) = -\frac{H}{T^2}dT + \frac{V}{T}dP \qquad (4\text{-}26)$$

$$\left[\frac{\partial(F/T)}{\partial T}\right]_P = -\frac{H}{T^2} \qquad \left[\frac{\partial(F/T)}{\partial P}\right]_T = \frac{V}{T} \qquad (4\text{-}27)$$

4-9. The Chemical Potential. Thermodynamic Functions for Open Systems. The derivations of the two preceding sections apply to a closed system, i.e., a system with a fixed amount of material. We now consider an open system, that is, one which can exchange matter with the surroundings. Let the system contain r components, and denote by N_i the number of molecules of the ith component. The differential form (4-16) indicates that for a closed system we can conveniently regard E as a function of S and V; for an open system, E could depend on S, V, $N_1, \ldots, N_r$. We can therefore write

$$dE = T\,dS - P\,dV + \sum_{i=1}^{r}\left(\frac{\partial E}{\partial N_i}\right)_{S,V,N_j \ne i} dN_i \qquad (4\text{-}28)$$

We set

$$\mu_i = \left(\frac{\partial E}{\partial N_i}\right)_{S,V,N_j \ne i}$$

and write

$$dE = T\,dS - P\,dV + \sum_{i=1}^{r} \mu_i\,dN_i \qquad (4\text{-}29)$$

By writing $T\,dS = d(ST) - S\,dT$ and $P\,dV = d(PV) - V\,dP$ and transposing $d(ST)$ and/or $d(PV)$ as needed, we obtain

$$dH = T\,dS + V\,dP + \sum \mu_i\,dN_i \qquad (4\text{-}30)$$

$$dA = -S\,dT - P\,dV + \sum \mu_i\,dN_i \qquad (4\text{-}31)$$

$$dF = -S\,dT + V\,dP + \sum \mu_i\,dN_i \qquad (4\text{-}32)$$

$$d\frac{A}{T} = -\frac{E}{T^2}dT - \frac{P}{T}dV + \sum \frac{\mu_i}{T}dN \qquad (4\text{-}33a)$$

$$d\frac{F}{T} = -\frac{H}{T^2}dT + \frac{V}{T}dP + \sum \frac{\mu_i}{T}dN_i \qquad (4\text{-}33b)*$$

so that
$$\mu_i = \left(\frac{\partial E}{\partial N_i}\right)_{S,V,N_j \neq i} = \left(\frac{\partial H}{\partial N_i}\right)_{S,P,N_j \neq i} = \left(\frac{\partial A}{\partial N_i}\right)_{T,V,N_j \neq i}$$
$$= \left(\frac{\partial F}{\partial N_i}\right)_{T,P,N_j \neq i} \qquad (4\text{-}34)$$

The quantity μ_i is called the chemical potential of component i. Operationally, μ_i is the increase in F when one molecule of component i is added to a system at constant T and P or the change in A when one molecule of component i is added to a system at constant T and V.

For any function of state X the derivative $(\partial X/\partial N_i)_{T,P,N_j \neq i}$ is called the partial molecular value of X and denoted by $\tilde{X}_i$. We see that μ_i is the partial molecular free energy $\tilde{F}_i$. The partial molecular volume is $\tilde{V}_i = (\partial V/\partial N_i)_{T,P,N_j \neq i}$.

4-10. A Comment on Thermodynamic Derivations. The differential expressions (4-29) to (4-33) are all equivalent statements of the basic relations between the thermodynamic functions. One expression is readily derived from the others from the definitions of H, A, and F in terms of E, PV, and S, with the trick of writing $x\,dy = d(yx) - y\,dx$ and then transposing the perfect differential $d(yx)$.

A number of useful thermodynamic formulae are directly obtainable from these relations. By inspection, such relations as

$$\left(\frac{\partial E}{\partial S}\right)_{V,N_i} = T$$

and
$$\left[\frac{\partial(A/T)}{\partial T}\right]_{V,N_i} = -\frac{E}{T^2} \quad \text{or} \quad \left[\frac{\partial(A/T)}{\partial(1/T)}\right]_{V,N_i} = E$$

are obtained. By using the cross-differentiation identity, other relations such as

$$\left(\frac{\partial S}{\partial V}\right)_{T,N_i} = \left(\frac{\partial P}{\partial T}\right)_{V,N_i} \qquad \text{from (4-31)}$$

or
$$\left[\frac{\partial(\mu_i/T)}{\partial(1/T)}\right]_{V,N_i} = \left(\frac{\partial E}{\partial N_i}\right)_{T,V,N_j \neq i} \qquad \text{from (4-33c)}$$

are deduced. The first relation was previously obtained as (4-14) by a more cumbersome derivation.

---

* For some purposes, it is more expedient to write

$$d\frac{A}{T} = E\,d\frac{1}{T} - \frac{P}{T}dV + \sum \frac{\mu_i}{T}dN_i \qquad (4\text{-}33c)$$

$$d\frac{F}{T} = H\,d\frac{1}{T} + \frac{V}{T}dP + \sum \frac{\mu_i}{T}dN_i \qquad (4\text{-}33d)$$

with $1/T$ as the temperature variable.

In view of the differential form (4-29), E, the thermodynamic energy, is a convenient thermodynamic function (or, it is said, the "natural" thermodynamic function) to use for an open system when S, V, and the N_i's are chosen as independent variables. In the same sense, several other natural thermodynamic functions for particular choices of independent variables are as follows:

Thermodynamic function	Independent variable
H	S, P, N_i
A or A/T	T, V, N_i
F or F/T	T, P, N_i
S	E, V, N_i

As a practical suggestion, it is strongly recommended that the student first consider the possibility of using one of the expressions (4-29) to (4-33) when confronted with a problem of making a thermodynamic derivation.

We simply mention two other devices which crop up constantly in thermodynamic arguments. In transforming from one system of independent variables to another, relations like

$$\left(\frac{\partial E}{\partial T}\right)_P = \left(\frac{\partial E}{\partial T}\right)_V + \left(\frac{\partial E}{\partial V}\right)_T \left(\frac{\partial V}{\partial T}\right)_P$$

are often useful. In the other typical derivation, we start, for example, from

$$dS = \left(\frac{\partial S}{\partial T}\right)_V dT + \left(\frac{\partial S}{\partial V}\right)_T dV$$

and deduce that

$$\left(\frac{\partial T}{\partial V}\right)_S = -\frac{(\partial S/\partial V)_T}{(\partial S/\partial T)_V}$$

4-11. Euler's Theorem and the Gibbs-Duhem Relation. A homogeneous function of order n of the p variables, $x_1, \ldots, x_p$, is a function $f(x_1, \ldots, x_p)$ such that $f(\lambda x_1, \lambda x_2, \ldots, \lambda x_p) = \lambda^n f(x_1, \ldots, x_p)$ for an arbitrary number λ. For example, the function $f(x,y) = x^3 + 3x^2y + y^3$ is homogeneous of order 3, the function $f(x,y) = ax + by$ is homogeneous of order 1, $f(x,y) = (ax + by)/(cx + dy)$ is homogeneous of order 0, whereas $f(x,y) = ax + by + c$ and $f(x,y) = \sin x \cos y$ are not homogeneous.

Euler's theorem asserts that, for a homogeneous function of order n,

$$nf(x_1, \ldots, x_p) = x_1 \frac{\partial f(x_1, \ldots, x_p)}{\partial x_1} + \cdots + x_i \frac{\partial f}{\partial x_i} + \cdots + x_p \frac{\partial f}{\partial x_p} \quad (4\text{-}35)$$

The theorem can be readily verified for the examples cited.

Temperature, pressure, and density are intensive properties. Volume, internal energy, enthalpy, entropy, and the free energies A and F are extensive properties—their values are proportional to the mass of the system at constant temperature and pressure (also, we may note, at constant temperature and density). That is, if, at constant temperature and pressure, we increase the amount of each component by the same factor λ, the volume is increased by the factor λ;

$$V(T,P,\lambda N_1, \ldots, \lambda N_r) = \lambda V(T,P,N_1, \ldots, N_r)$$

Therefore, by Euler's theorem

$$V(T,P,N_1, \ldots, N_r) = \sum N_i \left(\frac{\partial V}{\partial N_i}\right)_{T,P,N_j \neq i} = \sum N_i \tilde{V}_i \quad (4\text{-}36)$$

Similarly for the Gibbs free energy, for which the "natural" variables are T, P, and the N_i's

$$F(T,P,\lambda N_1, \ldots, \lambda N_r) = \lambda F(T,P,N_1, \ldots, N_r)$$

so that
$$F(T,P,N_1, \ldots, N_r) = \sum N_i \left(\frac{\partial F}{\partial N_i}\right)_{T,P,N_j \neq i}$$

$$= \sum N_i \mu_i(T,P,N_1, \ldots, N_r) \quad (4\text{-}37)$$

(It should be remarked that volume and the thermodynamic-energy functions are extensive properties only for systems sufficiently large so that surface-tension effects are not important.)

The notation of (4-37) reminds us that the chemical potential itself is a function of T, P, and the composition variables. From (4-37), for an arbitrary variation in the independent variables,

$$dF = \Sigma N_i \, d\mu_i + \Sigma \mu_i \, dN_i$$

But, at constant T and P, by (4-32),

$$dF)_{T,P} = \Sigma \mu_i \, dN_i$$

Therefore, $\quad \Sigma N_i \, d\mu_i = 0 \quad T$ and P constant $\quad (4\text{-}38)$

This is the Gibbs-Duhem relation. In the same way,

$$\Sigma N_i \, d\tilde{V}_i = 0 \quad T \text{ and } P \text{ constant} \quad (4\text{-}39)$$

If the variation in (4-38) is only a variation dN_j in one component,

$$\sum_i N_i \left(\frac{\partial \mu_i}{\partial N_j}\right)_{T,P,N_k \neq j} = 0 \quad \text{for all } j \quad (4\text{-}40)$$

Now each μ_j itself is a function of T, P, $N_1, \ldots, N_r$. Furthermore, the μ_j's are intensive properties, since increasing the amount of each component by the same proportional factor λ does not change the value of μ_j; that is, the μ's are homogeneous functions of order 0 as regards the composition variables, or

$$\mu_j(T,P,\lambda N_1, \ldots ,\lambda N_r) = \mu_j(T,P,N_1, \ldots ,N_r)$$

Therefore, by Euler's theorem,

$$0 = \sum_i N_i \left(\frac{\partial \mu_j}{\partial N_i}\right)_{T,P,N_k \neq i} \quad \text{for all } j \tag{4-41}$$

From (4-32) we have, by the cross-differentiation identity,

$$\left(\frac{\partial \mu_i}{\partial N_j}\right)_{T,P,N_k \neq j} = \left(\frac{\partial \mu_j}{\partial N_i}\right)_{T,P,N_k \neq i} \tag{4-42}$$

In view of Eq. (4-42), (4-40) and (4-41) are seen to be equivalent.

We can conclude this section with the remark that the relations derived are not restricted to a multicomponent system but are also significant assertions for a one-component system. For an open one-component system, we can write

$$dE = T\,dS - P\,dV + \mu\,dN \tag{4-43a}$$
$$dF = -S\,dT + V\,dP + \mu\,dN \tag{4-43b}$$
$$\mu = \left(\frac{\partial F}{\partial N}\right)_{T,P} \tag{4-43c}$$
$$F = N\mu \tag{4-43d}$$

In a one-component system, the chemical potential is the free energy per molecule and the statement (4-43d) is obvious. A formulation in terms of the chemical potential may seem to be trivial for the one-component case, but it is in fact useful for elementary statistical mechanics.

4-12. Criteria of Equilibrium. If a closed system undergoes a spontaneous (irreversible) change, $dS > dq/T$. A system is therefore at equilibrium if, for every conceivable change that could occur,

$$dS \leq \frac{dq}{T} \tag{4-44}$$

It is to be emphasized that in an equation such as (4-44) [or (4-45), (4-46), and (4-48)] the quantities dS and dq represent virtual variations. We imagine a change in the system and calculate dq and dS for this change; if (4-44) is obeyed, the change could not occur spontaneously. But according to the first law, $dq = dE + P\,dV$; so the condition (4-44) for equilibrium can be restated as

$$dS \leq \frac{dE + P\,dV}{T} \tag{4-45}$$

The symbol $dS)_{E,V}$ will mean the change in the function S for a change in which the functions E and V are held fixed. From (4-45) an isolated (constant-E) constant-volume system is at equilibrium if, and only if, for all conceivable changes

$$dS)_{E,V} \leq 0 \tag{4-46}$$

A number of equivalent criteria of equilibrium can be readily derived. The one that interests us the most concerns the variation in free energy at constant T and P. Since $F = E + PV - ST$, for a change at constant T and P,

$$dF)_{T,P} = dE + P\,dV - T\,dS \tag{4-47}$$

But, according to (4-45) the function on the right-hand side of (4-47) is greater than zero for all variations in a system at equilibrium. A condition for equilibrium is therefore

$$dF)_{T,P} \geq 0 \tag{4-48}$$

4-13. Chemical Reactions and Chemical Equilibrium. A generalized chemical reaction can be written

$$\nu_A A + \nu_B B \cdots \rightarrow \nu_D D + \nu_E E \cdots \tag{4-49a}$$

where the ν's are small integers or simple fractions and are called the stoichiometric coefficients and A, B, D, etc., are the various chemical substances. The reaction can also be written as

$$0 = \nu_D D + \nu_E E \cdots - \nu_A A - \nu_B B \cdots \tag{4-49b}$$

It will often be convenient to write the reaction as

$$0 = \sum_{i=1}^{r} \nu_i \{i\} \tag{4-50}$$

where $\{i\}$ stands for the ith chemical species and it is understood that ν_i is positive for products and negative for reactants.

If λ is a progress variable for a reaction, the changes in the number of molecules of each component due to reaction in the forward ($d\lambda > 0$) or backward ($d\lambda < 0$) direction are given by

$$dN_i = \nu_i\,d\lambda \tag{4-51}$$

The Gibbs free-energy change in general is

$$dF = -S\,dT + V\,dP + \Sigma \mu_i\,dN_i$$

At constant temperature and pressure, when the changes dN_i are all due to the chemical reaction,

$$dF)_{T,P} = \left(\sum_{i=1}^{r} \mu_i \nu_i \right) d\lambda \tag{4-52}$$

If the system is to be at equilibrium, $dF)_{T,P} \geq 0$ for all conceivable changes. The conceivable changes are $d\lambda > 0$ (reaction goes forward) and $d\lambda < 0$ (reaction goes backward). If $dF)_{T,P} \geq 0$ for both possible variations, we must have $dF)_{T,P} = 0$ and

$$\Sigma \mu_i \nu_i = 0 \tag{4-53}$$

as the condition for chemical equilibrium.

4-14. Equilibrium between Phases. If, in a system containing several components, there are two phases which are in complete equilibrium with each other, they must have the same temperature and pressure. Let A denote one phase and B the other. If dN_i molecules of component i are transferred from phase A to phase B at constant T and P, the free-energy change in phase A is

$$dF^A = -\mu_i^A \, dN_i$$

whereas the free-energy change in phase B is

$$dF^B = \mu_i^B \, dN_i$$

The total free-energy change is

$$dF)_{T,P} = dF^A + dF^B = (\mu_i^B - \mu_i^A) \, dN_i$$

Since dN_i can be either positive or negative, and since the argument applies to each independent component, the requirement for equilibrium between two phases is

$$\mu_i^B = \mu_i^A \quad \text{for all } i \tag{4-54}$$

If two systems A and B are separated by a rigid, immovable, selectively permeable membrane and are at equilibrium, they must be at the same temperature. They need not be at the same pressure since the membrane is rigid. However, by the same argument as used above, for each component for which the membrane is permeable, it must be true that

$$\mu_i^B = \mu_i^A \tag{4-55}$$

There is no condition on the chemical potentials of components to which the membrane is impermeable.

4-15. The Perfect Gas. The perfect (or ideal) gas is the most important single example for the chemical applications of elementary statistical mechanics. We accept, as an empirical law, the equation of state

$$PV = rT \tag{4-56}$$

where T is the thermodynamic absolute temperature, which is defined by the second law, and r is a constant. (The proportionality factor for defining the temperature numerically is then fixed by taking the triple point of ice as 273.160°K.)*

* *Compt. rend. dixieme conf. gén. poids mesures,* 1945.

A general equation, valid for any fluid, is [(4-15)]

$$\left(\frac{\partial E}{\partial V}\right)_T = T\left(\frac{\partial P}{\partial T}\right)_V - P$$

From (4-56) it then follows that for a perfect gas

$$\left(\frac{\partial E}{\partial V}\right)_T = 0 \tag{4-57}$$

The thermodynamic-energy function of a perfect gas depends on the temperature only.

An additional nonthermodynamic fact about a perfect gas (which is, however, a consequence of the fundamental assumptions of statistical mechanics) is Avogadro's hypothesis that equal volumes at the same pressure and temperature of different perfect gases contain the same numbers of molecules. In the equation,

$$PV = NkT \tag{4-58a}$$

where N is the number of molecules, k is a universal constant which is the same for all gases. In terms of the number of moles, n,

$$PV = nRT \tag{4-58b}$$

where $R = N_0 k$, and N_0 is Avogadro's number, the number of molecules per mole.

For a one-component gas,

$$dF = -S\, dT + V\, dP + \mu\, dN \tag{4-43b}$$

and, by the cross-differentiation identity,

$$\left(\frac{\partial \mu}{\partial P}\right)_{T,N} = \left(\frac{\partial V}{\partial N}\right)_{T,P} = \tilde{V} \tag{4-59}$$

But, from the perfect-gas law,

$$\left(\frac{\partial V}{\partial N}\right)_{T,P} = \frac{kT}{P} = \tilde{V} \tag{4-60}$$

and so
$$\left(\frac{\partial \mu}{\partial P}\right)_{T,N} = \frac{kT}{P} \tag{4-61}$$

On integration at constant T and N this gives

$$\mu(T,P) = kT \ln\frac{P}{P'} + \mu(T,P') \tag{4-62}$$

where $\mu(T,P')$ is the chemical potential at temperature T and pressure P'.
If P_0 is the pressure in a defined standard state, we can write (4-62) as

$$\mu(T,P) = kT \ln\frac{P}{P_0} + \mu^\circ(T) \tag{4-63}$$

where $\mu°(T)$ is the chemical potential in the defined standard state of pressure P_0.

It is possible to become somewhat confused about pressure units. A simple and straightforward, albeit not the most general, way to think about the situation is to assume that we start with the cgs system with volume in cubic centimeters and

$$k = 1.3804_4 \times 10^{-16} \text{ erg deg}^{-1} \text{ molecule}^{-1}$$

The pressure as calculated from (4-58a) will then be in dynes per square centimeter. Regard all pressures as being measured in these units. If we take as our standard state a pressure of 1 atm, then

$$P_0 = 1.01325 \times 10^6 \text{ dynes cm}^{-2} \tag{4-64}$$

The ratio P/P_0 which appears in (4-63) is a dimensionless number, but its numerical value is that of the pressure P measured in atmospheres. As written in (4-63), we see that the quantity of which we are taking the logarithm is a dimensionless number.

It is sometimes convenient to choose as standard state a state with a given concentration. If the gas law is written $P = (N/V)kT$, we recognize that (N/V) is concentration in units of atoms per cubic centimeter. Let the concentration in the defined standard state be N_c/V_c. By direct substitution in (4-63),

$$\mu\left(T, \frac{N}{V}\right) = kT \ln \frac{N/V}{N_c/V_c} + kT \ln \frac{N_c kT}{V_c P_0} + \mu°(T) \tag{4-64a}$$

$$\mu\left(T, \frac{N}{V}\right) = kT \ln \frac{N/V}{N_c/V_c} + \mu^c(T) \tag{4-64b}$$

$\mu^c(T)$, the chemical potential in the standard state of concentration N_c/V_c, is defined by comparison of (4-64a) and (4-64b). If the standard state is $N_c/V_c = 1$ atom cc^{-1}, it is natural to write (4-64b) as

$$\mu\left(T, \frac{N}{V}\right) = kT \ln \frac{N}{V} + \mu^c(T) \tag{4-65}$$

but we should remember that the expression $\ln (N/V)$ is really the logarithm of the dimensionless ratio $(N/V)/(N_c/V_c)$.

For a perfect gas which is a mixture containing $N_1, \ldots, N_i, \ldots, N_r$ molecules of the several components

$$PV = (N_1 + \cdots + N_i + \cdots + N_r)kT = \left(\sum_{i=1}^{r} N_i\right) kT \tag{4-66}$$

The partial pressure of the ith component is

$$P_i = N_i \frac{kT}{V} \tag{4-67}$$

and the mole fractions are given by

$$X_i = \frac{N_i}{\Sigma N_i} \tag{4-68}$$

The free-energy equation is

$$dF = -S\,dT + V\,dP + \Sigma \mu_i\,dN_i \tag{4-69}$$

so that

$$\left(\frac{\partial \mu_i}{\partial P}\right)_{T,N_j} = \left(\frac{\partial V}{\partial N_i}\right)_{T,P,N_j \neq i} = \tilde{V}_i = \frac{kT}{P} \tag{4-70}$$

By integration, we obtain an expression of the form

$$\mu_i(T,P) = kT \ln P + \text{constant}$$

At a fixed composition, the total pressure P is proportional to the partial pressure P_i, $P_i = X_i P$, and a particularly useful expression for the chemical potential of the ith component is

$$\mu_i(T,P_i) = kT \ln \frac{P_i}{P_0} + \mu_i^\circ(T,N_1, \ldots, N_r) \tag{4-71}$$

where $\mu_i^\circ(T,N_1, \ldots, N_r)$ is the chemical potential of the ith component when its partial pressure is the standard pressure P_0 and the composition is $N_1, \ldots, N_r$. Let this multicomponent system A be equilibrated with a system B containing only the ith component through a membrane which is permeable to this component only. We make the plausible assumption, which is verified by experiment, that at equilibrium the partial pressure P_i in the mixture is equal to the pressure of this particular component on the other side of the membrane. The chemical potential of the pure component i in system B is, however, by (4-63)

$$\mu_i(T,P_i) = kT \ln \frac{P_i}{P_0} + \mu_i^\circ(T) \tag{4-72}$$

At equilibrium the chemical potentials μ_i are the same on both sides of the membrane. It therefore follows that in (4-71) the quantity $\mu_i^\circ(T,N_1, \ldots, N_r)$ is not in fact a function of the composition, and the chemical potential of the ith component of a perfect-gas mixture is given by

$$\mu_i(T,P_i) = kT \ln \frac{P_i}{P_0} + \mu_i^\circ(T) \tag{4-73}$$

where $\mu_i^\circ(T)$ is the chemical potential of the ith component at a pressure P_0. In a perfect-gas mixture the chemical potential of each component depends only on its partial pressure and is independent of the amounts and partial pressures of the other components.

For the chemical reaction $\Sigma \nu_i \{i\} = 0$, the condition for chemical equilibrium is $\Sigma \nu_i \mu_i = 0$. If the reactants and products are components of a perfect gas, we use the expression (4-73) for the chemical potentials.

Sec. 4-15] THERMODYNAMICS 59

$$0 = \sum \nu_i \mu_i = kT \sum \nu_i \ln \frac{P_i}{P_0} + \sum \nu_i \mu_i^\circ(T) \tag{4-74a}$$

or

$$0 = kT \ln \left[\prod_i \left(\frac{P_i}{P_0} \right)^{\nu_i} \right] + \sum \nu_i \mu_i^\circ(T) \tag{4-74b}$$

or

$$\prod_i \left(\frac{P_i}{P_0} \right)^{\nu_i} = e^{-[\Sigma \nu_i \mu_i^\circ(T)]/kT} \tag{4-74c}$$

$$\prod_i \left(\frac{P_i}{P_0} \right)^{\nu_i} = K_P \tag{4-74d}$$

The quantity K_P is the equilibrium constant for partial-pressure units, and (4-74c) and (4-74d) are the familiar mass-action equilibrium law. It is to be emphasized that the stoichiometric coefficients ν_i are positive for products and negative for reactants.

By reference to Eq. (4-65) we see that the corresponding equilibrium expression in concentration units is

$$\prod_i \left(\frac{N_i/V_i}{N_c/V_c} \right)^{\nu_i} = K_c = e^{-\Sigma \nu_i \mu_i^c(T)} \tag{4-75}$$

where $\mu_i^c(T)$ is the chemical potential of component i at the standard concentration N_c/V_c.

We have gone to a good deal of trouble to point out that equilibrium expressions such as (4-74d) and (4-75) really involve products of dimensionless ratios of pressures to standard pressures or concentrations to standard concentrations. If P_0 is taken as the value in cgs units of a pressure of 1 atm, then, as already pointed out, the ratio P_i/P_0 is in effect the partial pressure in atmospheres. In practice, for transforming equilibrium constants from one set of units to another, it is equally expedient to regard the equilibrium constants as having units. Thus for the reaction

$$N_2 + 3H_2 = 2NH_3$$

the equilibrium expression with pressures expressed in atmospheres is

$$\frac{p_{NH_3}^2}{p_{N_2} p_{H_2}^3} = K_P$$

and we can say that K_P has the dimensions of reciprocal square atmospheres. If we wish to convert this to a concentration equilibrium constant, with [NH_3] representing concentration in moles per liter, we recall that

$$p_{NH_3} = [NH_3]RT$$

where $R = 0.08205$ liter atm mole^{-1} deg^{-1}. We then have
$$\frac{[\text{NH}_3]^2}{[\text{N}_2][\text{H}_2]^3} = K_c$$
and the units of K_c are mole^{-2} liter^2. Clearly, on dimensional grounds
$$K_c = K_p\,(\text{atm}^{-2})\,(RT)^2 \qquad \text{liter}^2\,\text{atm}^2\,\text{mole}^{-2}$$

4-16. The Entropy of Mixing. For a one-component gas we have, starting with
$$dF = -S\,dT + V\,dP + \mu\,dN$$
that
$$\tilde{S} = \left(\frac{\partial S}{\partial N}\right)_{T,P} = -\left(\frac{\partial \mu}{\partial T}\right)_{N,P} \tag{4-76}$$
and for a perfect gas, by differentiation of the expression (4-63) for μ,
$$\tilde{S} = -k \ln \frac{P}{P_0} - \frac{\partial \mu^\circ(T)}{\partial T}$$
$$= -k \ln \frac{P}{P_0} + \tilde{S}^\circ(T) \tag{4-77}$$
where $\tilde{S}^\circ$ is the partial molecular entropy in the standard state of pressure P_0. The total entropy is $S = N\tilde{S}$. The higher the pressure, the less the entropy per molecule.

For a multicomponent system, we have
$$S = \sum N_i \tilde{S}_i = -\sum N_i \left(\frac{\partial \mu_i}{\partial T}\right)_{P,N_j} \tag{4-78}$$
For a perfect-gas mixture
$$\tilde{S}_i = -\frac{\partial \mu_i}{\partial T} = -k \ln \frac{P_i}{P_0} - \frac{\partial \mu_i^\circ(T)}{\partial T}$$
$$= -k \ln \frac{P_i}{P_0} + \tilde{S}_i^\circ(T)$$

To calculate the so-called entropy of mixing at constant volume or pressure for a perfect gas, we start with separated pure components, each in its own compartment, but with all components at the same pressure P. There are N_i molecules of each gas, and the volume for component i is the fraction $N_i/\Sigma N_i$ of the total volume. The total entropy is then
$$S = -k \sum N_i \ln \frac{P}{P_0} + \sum N_i \tilde{S}_i^\circ(T) \tag{4-79a}$$
If the partitions between the components are removed and the gases allowed to mix, the final state is one in which each component has a partial pressure of
$$P_i = \frac{N_i}{\Sigma N_i} P = X_i P$$

where X_i is the mole fraction of the ith component. The partial molecular entropy is

$$\tilde{S}_i = -k \ln \frac{P_i}{P_0} + \tilde{S}_i^\circ(T) = -k \ln \frac{PX_i}{P_0} + \tilde{S}_i^\circ(T) \qquad (4\text{-}79b)$$

and the total entropy is

$$S = \sum N_i \tilde{S}_i = -k \sum N_i \ln \frac{PX_i}{P_0} + \sum N_i \tilde{S}_i^\circ(T) \qquad (4\text{-}79c)$$

The entropy of mixing is obtained as the difference between (4-79c) and (4-79a).

$$S = -k\Sigma N_i \ln X_i \qquad (4\text{-}80)$$

This is the desired result for the positive entropy of isothermal mixing of pure perfect gases at constant volume (and therefore at constant total pressure). In a sense, it is not the fact that the gases are mixed which

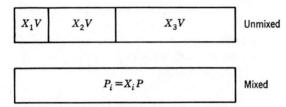

FIG. 4-1. In the unmixed state, each gas is in its own compartment. The pressures are all the same, and the volume of each compartment is X_iV, where V is the total volume and X_i is the mole fraction of the ith component. After mixing, all the components are distributed uniformly throughout V, each at a partial pressure of $P_i = X_iP$.

leads to the entropy increase, but simply the fact that each gas is allowed to expand from pressure P to pressure X_iP. The entropy of each component in the mixture is the same as it would be in the pure state at the same partial pressure.

4-17. Notation, Thermodynamic Functions, and Thermodynamic Tables. Thermodynamic functions or other extensive functions of state written in large capital letters, S, F, C_P, will refer to the value of the extensive function for the given system with whatever amount of material it contains. Symbols for functions of state written in small capital letters, s, f, v, c_P, represent gram-molal quantities.

In this book we shall almost always use partial molecular quantities such as the chemical potential per molecule, which is the partial molecular free energy, or the partial molecular volume $\tilde{V}_i = (\partial V/\partial N_i)_{T,P}$. In thermodynamics texts, the same symbols are used for the corresponding partial molal quantities. Where necessary, we can specify in words which meaning is intended.

The equilibrium constant in pressure units was given by

$$K_P = e^{-\Sigma \nu_i \mu_i^\circ / T}$$

The quantity $\Sigma \nu_i \mu_i^\circ(T)$ is the standard free-energy change in the reaction "per molecule" (that is, for the reaction with ν_i molecules of component i where ν_i is a small rational number). The molar standard free-energy change is $\Delta \mathrm{F}^\circ = N_0(\Sigma \nu_i \mu_i^\circ)$, and the equilibrium constant can also be written as $K_P = e^{-\Delta \mathrm{F}^\circ / RT}$, with $R = N_0 k$.

It is customary in thermodynamic tabulations to use a subscript T to indicate the temperature at which the value of the thermodynamic function is given; thus, one finds F_T, F_{298}, and s_T instead of $\mathrm{F}(T)$, $\mathrm{F}(298)$, and $\mathrm{s}(T)$. The accepted standard state in the tabulations is 1 normal atmosphere. This is indicated by a degree sign; H_T° is the molal enthalpy at temperature T and a pressure of 1 atm. We shall use the superscripts a and c to denote the standard states of 1 atom or molecule cc^{-1} and 1 mole liter^{-1}; for example, F_T^a, H_T^c, $\mu^a(T)$, etc. The degree sign will be used to indicate both the usual standard state of 1 atm and, in other contexts, for a defined standard state in general.

Of course, the perfect-gas law is often not a sufficiently accurate approximation to the real-gas behavior. For a real gas, the defined standard state is the state of unit fugacity, for which the thermodynamic functions for the substance are those for a hypothetical perfect-gas at 1 atm pressure. The values for the thermodynamic properties are measured at a low pressure, where the gas is perfect. By using the formulae for a perfect gas [$(\partial H / \partial P)_T = 0$, $(\partial F / \partial P)_T = N(kT/P)$, etc.] the values are computed for the hypothetical perfect gas at unit pressure. The real-gas behavior is discussed in Chap. 15, where a more rigorous definition of fugacity is given.

The molal-enthalpy increment for a substance between 0°K and T is often tabulated. It can be calculated from heat-capacity data,

$$\mathrm{H}_T^\circ - \mathrm{H}_0^\circ = \int_0^T c_P \, dT + \sum \Delta \mathrm{H}_{\text{transitions}}^\circ$$

and, in some cases, calculated by statistical mechanics.

The standard Gibbs free energy per mole of a substance at 0°K is $\mathrm{F}_0^\circ = \mathrm{H}_0^\circ - T \mathrm{s}_0^\circ = \mathrm{H}_0^\circ$ (since both T and s_0° are zero). The free-energy increment to temperature T is

$$\mathrm{F}_T^\circ - \mathrm{F}_0^\circ = \mathrm{H}_T^\circ - \mathrm{H}_0^\circ - T \mathrm{s}_T^\circ$$

The free-energy increment can also be calculated directly from one of the equations

$$\frac{\partial \mathrm{F}}{\partial T} = -\mathrm{s} \qquad \frac{\partial (\mathrm{F}/T)}{\partial T} = -\frac{\mathrm{H}}{T^2}$$

As we shall see the quantity $(\mathrm{F}_T^\circ - \mathrm{F}_0^\circ)/T$, which is called the free-energy function, has an especially simple significance in statistical mechanics. Furthermore, it is very useful in calculations of chemical equilibria. For a chemical reaction, we have, in an obvious notation,

$$\ln K_P = -\frac{\Delta F_T^\circ}{RT} = -\frac{\sum_i \nu_i F_{Ti}^\circ}{RT}$$

$$= -\sum \nu_i \frac{F_{Ti}^\circ - F_{0i}^\circ}{RT} - \frac{\Sigma \nu_i F_{0i}^\circ}{RT}$$

$$= -\sum \nu_i \frac{F_{Ti}^\circ - F_{0i}^\circ}{RT} - \frac{\Delta F_0^\circ}{RT} \qquad (4\text{-}81)$$

Thus, if we know the standard free-energy change at 0°K, ΔF_0° (which equals ΔH_0°), and the free-energy functions $(F_T^\circ - F_0^\circ)/T$ for the components, we can directly calculate the equilibrium constant at any temperature.

General references 24 to 29 are particularly useful as compilations of thermodynamic data.* Many of the tabulations list the heat of formation ($\Delta H f^\circ$), the free energy of formation, and the equilibrium constant of formation of a given substance from the elements in their standard state, but they list the absolute entropy, not the entropy of formation from the elements.

Thus for iodine atoms and iodine molecules we find in Series I of the National Bureau of Standards (NBS) tables [25]

	$\Delta H f_{298}^\circ$	$\Delta F f_{298}^\circ$	$\log K_f(298)$	S_{298}°
$I(g)$	25.482	16.766	-12.893	43.1841
$I_2(g)$	14.876	4.63	-3.3937	62.280
$I_2(c)$	0.000	0.000	0.000	27.9

The standard state for iodine has been selected as the crystal $I_2(c)$. The equilibrium constant $\log K_f$ for I atoms is the equilibrium constant of the reaction

$$\tfrac{1}{2} I_2(c) = I(g) \qquad K = p_I = 10^{-12.893} \text{ atm}$$

but the entropy change for this reaction is $\Delta s = 43.2 - (27.9/2) = 29.2$ entropy units. If we are interested in the reaction

$$I_2(g) = 2I(g) \qquad K_D = \frac{p_I^2}{p_{I_2}}$$

we have $\log K_D = 2 \log K_f(\text{I atoms}) - \log K_f(I_2 \text{ gas}) = -22.39$.

---

* There is available a useful compilation of measured and estimated heats of formation, T. L. Cottrell, "The Strength of Chemical Bonds," 2d ed., Butterworth, London, 1958, and a simple discussion of the use of thermodynamic tables, J. L. Margrave, J. Chem. Educ., **32**:520 (1955).

It is worthwhile to learn how to use the various thermodynamic compilations efficiently and accurately.

PROBLEMS

4-1. Derive relations for $(\partial H/\partial P)_T$ and $(\partial S/\partial P)_T$ which are the analogues of Eqs. (4-15) and (4-14).

4-2. Write out a differential form similar to (4-29), (4-30), (4-31), or (4-32) in which T, V, and the μ_i's are chosen as the independent variables (*Note:* In thinking about this, it must be understood that these relations apply to an open system. If, for example, in a one-component system, μ is changed at constant T and V, the number of molecules in the system must be changed.) What is the natural thermodynamic function as defined in Sec. 4-10 for this choice of independent variables? Obtain derivative expressions for S, P, and N_i for this choice of independent variables.

Now write out a similar differential form, using T, V, and all the μ_i/T as independent variables, and obtain corresponding thermodynamic relations therefrom.

4-3. Use the NBS Series I tables to calculate $\Delta H°$, $\Delta S°$, $\Delta F°$, and log K_P for the reaction at 298°:

$$Br_2(g) \rightleftharpoons 2Br(g)$$

Estimate the equilibrium constant at 1000°K, assuming that $\Delta H°$ is constant.

Use the NBS Series III tables (if available), and calculate $\Delta H°_{1000°}$, $\Delta S°_{1000°}$, and log K_P at 1000°K.

4-4. Use the American Petroleum Institute (API) tables [25] to determine the thermodynamic functions and equilibrium constant for the reaction

$$CO \rightleftharpoons C(g) + O(g)$$

at 298 and 3000°K.

4-5. Use the API tables [25] to calculate the thermodynamic functions and equilibrium constant at 298 and 1500°K for a reaction of organic compounds that interests you. For example, the following reactions are of technical and theoretical interest:

$$C(graphite) + 2H_2 \rightleftharpoons CH_4$$
$$2C(graphite) + 2H_2 \rightleftharpoons C_2H_2$$
$$2CH_4 \rightleftharpoons C_2H_2 + 3H_2$$
$$C_2H_6 \rightleftharpoons C_2H_4 + H_2$$
$$C_7H_{16}(n) \rightleftharpoons C_6H_5(CH_3) + 4H_2$$

4-6. From thermodynamic tables, compile the standard entropy changes per mole at 298° for some or all of the following dissociation reactions. Formulate an empirical generalization as to the magnitude of the entropy changes of dissociation reactions:

$$H_2 \rightleftharpoons 2H \quad O_2 \rightleftharpoons 2O \quad Cl_2 \rightleftharpoons 2Cl \quad H_2O(g) \rightleftharpoons H + OH$$
$$HCl \rightleftharpoons H + Cl \quad N_2O_4 \rightleftharpoons 2NO_2 \quad C_2H_6 \rightleftharpoons C_2H_4 + H_2$$

5
Mathematics

5-1. Introduction. This chapter is devoted to a miscellaneous assortment of mathematical methods and results which it is expedient to consider before embarking on the study of statistical mechanics. The results obtained will be useful in later chapters.

5-2. Combinational Problems

Lemma I. The number of permutations of N distinguishable objects is $N!$.

The first object can be chosen in any one of N ways. For any one of these selections, there are $N - 1$ choices for the second member of the array. There are thus $N(N - 1)$ ways of selecting the first two objects and $N(N - 1)(N - 2) \cdots 1 = N!$ possible arrays or permutations of all N objects. Each such permutation is distinguishable from all the others.

Lemma II. The number of ways of assigning N distinguishable objects into r distinguishable containers so that there are $N_1, N_2, \ldots, N_i, \ldots, N_r$ objects in the respective containers is

$$t = \frac{N!}{N_1! N_2! \cdots N_i! \cdots N_r!} = \frac{N!}{\prod_{i=1}^{r} N_i!} \qquad (5\text{-}1)$$

Note that if, for example, $N_1 = 3$, we are concerned with the fact that objects A, B, D are in the first container, rather than C, E, F; but we are not concerned whether the order of selection was ABD or BAD, etc. There are $N_1!$ possible permutations of the N_1 objects in the first container. Then the product of the number of ways of selecting the groups, t, times the number of permutations within each group gives the total number of possible permutations of the N objects, or

$$tN_1! N_2! \cdots N_r! = N!$$

which leads immediately to (5-1).

Lemma III. The number of ways of selecting N distinguishable objects from a set of g distinguishable objects is

$$t = \frac{g!}{N!(g-N)!} \qquad (5\text{-}2)$$

This follows immediately from (5-1), since we are dividing the g objects into two groups—the selected group of N objects and the omitted group of $g - N$ objects.

Lemma IV. Given g distinguishable containers and N indistinguishable objects, the number of ways of putting the N objects into the g containers when there is no restriction as to the number of objects in any one container is

$$t = \frac{(g+N-1)!}{(g-1)!N!} \qquad (5\text{-}3)$$

Thus, in one arrangement there are 3 objects in the first container and 5 objects in the second container, etc., but the objects are indistinguishable, and we cannot say which 3 of the N objects are in the first container. In a different arrangement, there are 5 objects in the first container and 3 objects in the second container, etc. We construct an array of N crosses and $g-1$ vertical partitions, as illustrated:

$$\text{xxx}|\text{xxxxx}|\text{x}|\ |\text{x}|\text{xx}\ \cdot\ \cdot\ \cdot\ |\text{xx}$$

The number of crosses before the first partition is the number of objects in the first container; the number of crosses between the $(i-1)$th and the ith partitions is the number of objects in the ith container, and the number of crosses after the $(g-1)$th partition is the number of objects in the gth container. In the illustration above, there are 3 objects in the first container, 5 in the second, 1 in the third, 0 in the fourth, etc., and 2 in the last partition. We begin by thinking of the N crosses or objects as labeled and distinguishable and of the $g-1$ partitions as labeled and distinguishable. There are then $(g + N - 1)!$ permutations of the assembly of crosses and partitions. Each of these permutations is a possible configuration of the system. But since the objects are indistinguishable, there are $N!$ permutations of the crosses which correspond to the same configuration. There are also $(g-1)!$ permutations of the partitions for a given configuration, since the g distinguishable containers are specified by their order in the array, not by the particular partition at the boundaries of the container. The total number of arrangements is therefore

$$t = \frac{(g+N-1)!}{N!(g-1)!}$$

Lemma V. The number of ways of putting N distinguishable objects into g distinguishable boxes is g^N.

The first object can be assigned in any one of g ways. The second object can be assigned in any one of g ways, independently of the way in which the first object was assigned. There are then g^2 ways of assigning the first two objects and g^N ways of assigning the N objects.

5-3. Lagrange Multipliers. For a function of n variables, $f(x_1, \ldots, x_n)$, the solution of the n equations

$$\frac{\partial f(x_1, \ldots, x_n)}{\partial x_i} = 0 \qquad i = 1, \ldots, n \qquad (5\text{-}4)$$

defines a point which is an extremum for the function $f(x_1, \ldots, x_n)$. At this point, the function has either a maximum, a minimum, a saddle point, or some sort of inflection point for all possible variations in x_1, $x_2, \ldots$, and x_n.

Our problem is somewhat different. We are given a function $f(x_1, \ldots, x_n)$. However, the variables $x_1, \ldots, x_n$ are not all independent but are constrained by some conditions. For illustrative purposes, we take two such conditions:

$$g(x_1, \ldots, x_n) = 0 \qquad (5\text{-}5a)$$
$$h(x_1, \ldots, x_n) = 0 \qquad (5\text{-}5b)$$

The problem is to find the extremum points of $f(x_1, \ldots, x_n)$, when the allowed variations in the x's are restricted by the two conditions (5-5a) and (5-5b). A straightforward but pedestrian way to do this is to solve Eqs. (5-5) for two variables x_{n-1} and x_n in terms of $x_1, \ldots, x_{n-2}$. These solutions when inserted in $f(x_1, \ldots, x_n)$ give a function of $n - 2$ independent variables, and the extremum can then be found from the $n - 2$ equations analogous to (5-4).

The method of Lagrange multipliers or undetermined multipliers is more elegant and more useful for our purposes. It asserts that there are two constants α and β such that the solution of the $n + 2$ equations

$$\frac{\partial f(x_1, \ldots, x_n)}{\partial x_i} + \alpha \frac{\partial g(x_1, \ldots, x_n)}{\partial x_i} + \beta \frac{\partial h(x_1, \ldots, x_n)}{\partial x_i} = 0$$
$$i = 1, \ldots, n \qquad (5\text{-}6a)$$
$$g(x_1, \ldots, x_n) = 0 \qquad (5\text{-}6b)$$
$$h(x_1, \ldots, x_n) = 0 \qquad (5\text{-}6c)$$

determines an extremum of f, subject to the constraints (5-5). Note that (5-6) consists of $n + 2$ equations, for the $n + 2$ "unknowns" α, β, $x_1, \ldots, x_n$.

An extremum having been found, the question of whether it is a maximum, a minimum, a saddle point, etc., is often decided by assumption or common sense, rather than by formal mathematical criteria.

A partial proof of the conditions (5-6) is of interest because it closely resembles what we shall actually do with this technique. The variation in f for any variation in the x's is given by

$$df = \frac{\partial f}{\partial x_1} dx_1 + \cdots + \frac{\partial f}{\partial x_i} dx_i + \cdots + \frac{\partial f}{\partial x_n} dx_n$$

$$= \sum_{i=1}^{n} \frac{\partial f}{\partial x_i} dx_i \qquad (5\text{-}7a)$$

The allowed variations $dx_1, \ldots, dx_n$ are restricted to those which satisfy the conditions (5-6b) and (5-6c), i.e., to those for which

$$\sum \frac{\partial g}{\partial x_i} dx_i = 0 \qquad \sum \frac{\partial h}{\partial x_i} dx_i = 0 \qquad (5\text{-}7b)$$

If, now, Eqs. (5-6a) are satisfied, we can write

$$\frac{\partial f}{\partial x_i} = -\alpha \frac{\partial g}{\partial x_i} - \beta \frac{\partial h}{\partial x_i} \qquad \text{for all } i$$

Thus, upon substituting in (5-7a),

$$df = -\alpha \sum \frac{\partial g}{\partial x_i} dx_i - \beta \sum \frac{\partial h}{\partial x_i} dx_i$$

and if the restrictions (5-6b) and (5-6c) are satisfied, $df = 0$.

We have actually shown that, for differentiable, well-behaved functions, a solution of Eqs. (5-6) gives an extremum of f under the prescribed constraints. With a little more argument, it can also be shown that for any extremum under the prescribed constraints there are constants α and β so that Eqs. (5-6) are satisfied.

5-4. Distribution Functions and Averages. We shall deal repeatedly with the concept of a distribution and a distribution function. We are given a population or sample consisting of N elements or molecules. There are a set of states $1, \ldots, i, \ldots, r, \ldots$, and each element of the population must be in one of these states. The set of numbers $N_1, N_2, \ldots, N_i, \ldots, N_r, \ldots$ which specifies the number of elements in each state is a distribution. The number of states available to the elements of the population may be either finite or infinite; in the latter case, of course, not all the states can be occupied.

The distribution must satisfy the equation

$$\sum_i N_i = N \qquad (5\text{-}8)$$

We define a quantity p_i as the fractional number of elements in the state i,

$$p_i = \frac{N_i}{\Sigma N_i} = \frac{N_i}{N} \qquad (5\text{-}9a)$$

and we may speak of p_i as the probability of finding an element in the state i in the given distribution.

Clearly
$$\Sigma p_i = 1 \tag{5-9b}$$

Let b_i be the value of some physical property that the molecules or elements of the population have if they are in the state i. We denote the average value of this property in the given distribution either by $\bar{b}$ or by $\langle b \rangle$. We shall most frequently use the latter notation, at least in the early chapters of this book.* The average value is given by

$$\langle b \rangle = \frac{\Sigma N_i b_i}{N} = \Sigma p_i b_i \tag{5-10}$$

If the property is one which is additive, such as energy, force on a wall, or mass, then the total value for the entire population or system is

$$B = \Sigma N_i b_i = N \langle b \rangle \tag{5-11}$$

We seek a significant measure of whether the distribution of the quantities b_i is sharply peaked around its mean value or is a rather broad one. The quantity $b_i - \langle b \rangle$ is the deviation of the value in state i from the mean. The average deviation from the mean, $\langle b_i - \langle b \rangle \rangle$, is given by

$$\langle b_i - \langle b \rangle \rangle = \Sigma p_i (b_i - \langle b \rangle) = \Sigma p_i b_i - \langle b \rangle = 0 \tag{5-12}$$

and is zero. The mean-square deviation $\langle (b_i - \langle b \rangle)^2 \rangle$, which is commonly called σ^2, is of necessity positive or zero and is a significant measure of the sharpness of the distribution:

$$\begin{aligned} \sigma^2 &= \langle (b_i - \langle b \rangle)^2 \rangle = \Sigma p_i (b_i - \langle b \rangle)^2 \\ &= \Sigma p_i b_i^2 - 2 \langle b \rangle \Sigma p_i b_i + \langle b \rangle^2 \Sigma p_i \\ &= \langle b^2 \rangle - 2 \langle b \rangle^2 + \langle b \rangle^2 = \langle b^2 \rangle - \langle b \rangle^2 \end{aligned} \tag{5-13}$$

For any distribution, the mean-square deviation is the mean of the square minus the square of the mean.

The mean-square deviation σ^2 is also called the variance; the rms deviation σ is called the standard deviation.

Instead of the running index i, which ranges over a set of discrete states, the state of an element of the population can be specified by a continuous variable x. We denote by $dN(x)$ the number of elements which have values of x in the interval x, $x + dx$, and we define the function $p(x)$ by

$$dN(x) = N p(x) \, dx \tag{5-14}$$

---

* The symbol $\langle b \rangle$ is also used for the quantum-mechanical expectation value of the quantity in a prescribed quantum-mechanical state. With a little care, confusion between these two meanings can be avoided.

Since the total number of elements is N,

$$\int dN(x) = \int_R Np(x)\,dx = N \tag{5-15}$$

so that

$$\int_R p(x)\,dx = 1 \tag{5-16}$$

where R is the range of values over which x can vary.

For this formulation in terms of continuous functions to be applicable, it is necessary that the number of elements, $dN(x)$, in the interval x, $x + dx$ be reasonably large for a small interval dx, so that $dN(x)$ can be regarded as changing almost continuously with the size of the interval dx, or as we move from one interval x, $x + dx$ to a neighboring one.

Either of the functions $Np(x)$ or $p(x)$ may be called a distribution function; $p(x)$ is also called the probability density function. The probability of finding a system in the interval x, $x + dx$ is $p(x)\,dx$.

Alternatively we may say that the probability, on looking at an element of the system, that it will be found to be characterized by a value of x in the interval x, $x + dx$ is $p(x)\,dx$. The average value of x for the population is then

$$\langle x \rangle = \frac{\int x\,dN(x)}{N} = \int_R xp(x)\,dx \tag{5-17}$$

The average value of x^2 is

$$\langle x^2 \rangle = \int_R x^2 p(x)\,dx \tag{5-18}$$

Just as before, the average deviation from the mean is $\langle x - \langle x \rangle \rangle = 0$, and the mean-square deviation is

$$\sigma^2 = \langle (x - \langle x \rangle)^2 \rangle = \langle x^2 \rangle - \langle x \rangle^2 \tag{5-19}$$

If $b(x)$ is the value of some property for those elements which are in the interval x, $x + dx$, then the average value of b for the population is

$$\langle b \rangle = \frac{\int b(x)\,dN(x)}{N} = \int_R b(x)p(x)\,dx \tag{5-20}$$

We transform to a new coordinate system with its origin at $\langle x \rangle$:

$$y = x - \langle x \rangle \tag{5-21}$$

The variable y is the deviation from the mean. We call the distribution function with respect to the new variable $f(y)$, and $f(y) = p(y + \langle x \rangle)$;

$$dN(y) = Nf(y)\,dy \tag{5-22}$$

Clearly $\quad \langle y \rangle = 0 \quad \langle y^2 \rangle = \sigma^2 = \langle (x - \langle x \rangle)^2 \rangle = \langle x^2 \rangle - \langle x \rangle^2$

A common distribution function is given by

$$f(y) = Ae^{-\alpha y^2} \tag{5-23}$$

The quantity a is regarded as given and characterizing the distribution; the constant A is determined by the normalizing condition

$$\int_R f(y)\, dy = 1 \qquad (5\text{-}24)$$

If, for example, the range of variation of x was $0 \leq x \leq \infty$, the range for y is $-\langle x \rangle \leq y \leq \infty$; however, if $e^{(-a\langle x\rangle^2)}$ is very small, the integral in (5-24) can be approximated:

$$\int_{-\langle x \rangle}^{\infty} f(y)\, dy \approx \int_{-\infty}^{+\infty} f(y)\, dy = 1 = \int_{-\infty}^{+\infty} A e^{-ay^2}\, dy = A \left(\frac{\pi}{a}\right)^{1/2} \qquad (5\text{-}25)$$

Therefore, we can write

$$f(y) = \left(\frac{a}{\pi}\right)^{1/2} e^{-ay^2} \qquad (5\text{-}26)$$

The mean-square deviation is

$$\sigma^2 = \int y^2 f(y)\, dy = \left(\frac{a}{\pi}\right)^{1/2} \int_{-\infty}^{+\infty} y^2 e^{-ay^2}\, dy = \frac{1}{2a} \qquad (5\text{-}27)$$

Therefore, the distribution function can be rewritten as

$$f(y) = \left(\frac{1}{2\pi\sigma^2}\right)^{1/2} e^{-y^2/2\sigma^2} \qquad (5\text{-}28)$$

This is known as the Gaussian distribution function.

5-5. Stirling's Approximation. An approximate formula for $N!$ which is valid for large values of N is

$$N! \approx (2\pi N)^{1/2} N^N e^{-N} \qquad (5\text{-}29)$$

We are really more interested in $\ln N!$, and

$$\ln N! \approx (N + \tfrac{1}{2}) \ln N - N + \tfrac{1}{2} \ln 2\pi \qquad (5\text{-}30)$$

For large values of N, this can be further approximated as

$$\ln N! = N \ln N - N \qquad (5\text{-}31)$$

This is the form of Stirling's approximation which we shall use most frequently. The quantity N is supposed to take on integral values only. For large values of N, however, we can treat it as a continuous variable. This enables us to differentiate (5-30):

$$\frac{d(\ln N!)}{dN} \approx \ln N + \frac{1}{2N} \qquad (5\text{-}32)$$

and, for large N,

$$\frac{d(\ln N!)}{dN} \approx \ln N \qquad (5\text{-}33)$$

It is pointed out in Prob. 5-4 that this formula for the derivative is not really dependent on Stirling's approximation.

5-6. Replacing a Sum by an Integral. We shall encounter a number of important problems in which it is required that we evaluate the sum of a series such as

$$S_0 = \sum_{n=0}^{\infty} n^p e^{-\alpha n^2} = \sum_{n=0}^{\infty} f(n) \qquad (5\text{-}34a)$$

where $\quad f(n) = n^p e^{-\alpha n^2} \quad$ or $\quad S_1 = \sum_{n=1}^{\infty} n^p e^{-\alpha n^2} = \sum_{n=1}^{\infty} f(n) \qquad (5\text{-}34b)$

where p is a positive number (usually a small integer) and $\alpha > 0$.

In all cases, when $\alpha \ll 1$, both sums (5-34a) and (5-34b) may be approximated by the same integral:

$$S_0 \approx S_1 \approx I = \int_0^{\infty} n^p e^{-\alpha n^2} \, dn = \int_0^{\infty} f(n) \, dn \qquad (5\text{-}35)$$

In the integrals, we are regarding n as a continuous variable.

An intuitive justification of the procedure is as follows: The sum S_0 above may be written (n being regarded as a continuous variable) as

$$S_0 = f(0) \int_0^1 dn + f(1) \int_1^2 dn + \cdots f(j) \int_j^{j+1} dn + \cdots \qquad (5\text{-}36a)$$

whereas I can be written as

$$I = \int_0^1 f(n) \, dn + \int_1^2 f(n) \, dn + \cdots \int_j^{j+1} f(n) \, dn + \cdots \qquad (5\text{-}36b)$$

If α is small, the function $n^p e^{-\alpha n^2}$ changes but slightly as n changes by one integer.* Then the individual terms in the sums for S_0 [(5-36a)] and I [(5-36b)] are almost equal, and $S_0 \approx I$. Furthermore, from the definitions (5-34b) and (5-34a), $S_0 = S_1 + f(0)$. But if α is small, there are many terms comparable with $f(0)$ that contribute to the sums, and $S_0 \approx S_1$. Therefore

$$S_0 \approx S_1 \approx I \qquad (5\text{-}35)$$

As an illustration, consider the simple case

$$S_0 = \sum_{n=0}^{\infty} e^{-\alpha n^2} \qquad S_1 = \sum_{n=1}^{\infty} e^{-\alpha n^2} \qquad (5\text{-}37)$$

For this case, $S_0 = 1 + S_1$. The integral is

$$I = \int_0^{\infty} e^{-\alpha n^2} \, dn = \frac{1}{2} \left(\frac{\pi}{\alpha} \right)^{1/2} \qquad (5\text{-}38)$$

---

* Even for small α, there is a large difference between the first few terms $f(0)$, $f(1)$, $f(2)$ when $p \neq 0$. However, the maximum of the function $n^p e^{-\alpha n^2}$ occurs at $n^2 = p/2\alpha$. For small α, this maximum occurs at a large value of n, where $f(n)$ is a slowly varying function. The first few terms $f(0) + f(1)$, etc., then make a negligible contribution to S_0, S_1, and I.

For this example, since the integrand is a monotonically decreasing function, it is easy to see that $S_0 > I > S_1$. But $S_0 = 1 + S_1$; so

$$1 + S_1 > \frac{1}{2}\left(\frac{\pi}{\alpha}\right)^{1/2} > S_1$$

If $\alpha \ll 1$, the integral $\frac{1}{2}(\pi/\alpha)^{1/2}$ is a large number, and

$$S_0 \approx I \approx S_1$$

The essential point is again that, for $\alpha \ll 1$, the function $e^{-\alpha n^2}$ changes but slightly as the argument n changes by unity.

The approximation of replacing the sum by an integral can be justified by the Euler-Maclaurin summation formula,* which also provides a series expansion for obtaining a more accurate result when necessary.

The accuracy of the approximation increases as α decreases. Thus, in evaluating the translational partition function (Chap. 6), a typical value of α is 10^{-16}, and the replacement of a sum by an integral is highly accurate. In the evaluation of the rotational partition function (Chap. 8), larger values of α occur, and there are problems for which the approximation, while useful, is not sufficiently accurate for the practical computation of thermodynamic functions.

5-7. Some Useful Integrals

$$\int_{-\infty}^{+\infty} e^{-\alpha x^2}\, dx = \left(\frac{\pi}{\alpha}\right)^{1/2} \tag{5-39a}$$

$$\int_{-\infty}^{+\infty} x^2 e^{-\alpha x^2}\, dx = \frac{1}{2\alpha}\left(\frac{\pi}{\alpha}\right)^{1/2} \tag{5-39b}$$

$$\int_{-\infty}^{+\infty} x^4 e^{-\alpha x^2}\, dx = \frac{3}{4\alpha^2}\left(\frac{\pi}{\alpha}\right)^{1/2} \tag{5-39c}$$

$$\int_{0}^{\infty} x e^{-\alpha x^2}\, dx = \frac{1}{2\alpha} \tag{5-40a}$$

$$\int_{0}^{\infty} x^3 e^{-\alpha x^2}\, dx = \frac{1}{2\alpha^2} \tag{5-40b}$$

$$\int_0^\infty x^{2n+1} e^{-ax^2} = \frac{n!}{2a^{n+1}}$$

PROBLEMS

5-1. Evaluate $N!$ for $N = 10$ and $N = 100$ from (a) a handbook, (b) Eq. (5-29) and (c) by taking the antilogarithm of Eq. (5-31), i.e.,

$$N! \approx N^N e^{-N}$$

What is the per cent error in the several approximations?

5-2. Evaluate $\ln N!$ for $N = 10$ and $N = 100$ from (a) a handbook, (b) Eq. (5-30), and (c) Eq. (5-31). What is the per cent error in the several approximations?

5-3. Compare formulae (5-30) and (5-31) for $\ln N!$ for $N = 10^{10}$.

* Margenau and Murphy [23], p. 457.

5-4. Consider the evaluation of $d(\ln N!)/dN$, using the approximate formula

$$\frac{d(\ln N!)}{dN} \approx \ln (N+1)! - \ln N!$$

or

$$\frac{d(\ln N!)}{dN} \approx \ln N! - \ln (N-1)!$$

but without using Stirling's approximation.

5-5. The sum

$$S = \sum_{n=0}^{\infty} e^{-\alpha n}$$

is the sum of a geometrical series, and its value is readily seen to be

$$S = \frac{1}{1 - e^{-\alpha}}$$

Compare this with the integral

$$I = \int_0^{\infty} e^{-\alpha n}\, dn$$

and discuss the conditions under which $I \approx S$.

6

The Statistical Mechanics of a System of Independent Particles

6-1. Introduction. This is one of the most important chapters in the text. The fundamental statistical mechanical formulae for a system of independent particles are derived. The relation between statistical mechanical functions and thermodynamic functions for such systems is also established.

In a later chapter (Chap. 13), two more general treatments of statistical mechanics, which are applicable to systems of interacting particles, will be developed.

6-2. The Concept of a Distribution. The system contains N independent identical particles. There are a set of single-particle wave functions or states, or orbitals. In principle, these orbitals, or states, are all distinguishable, and it is possible to determine whether, in a given configuration of the system, a particular orbital is being used or not.

The states, or orbitals, are collected together in groups which we call levels. All the states in the ith group, or level, are degenerate; that is, they have the same energy. Depending on our purposes, the states in one level may be required to also have some other property in common; for example, they may all have the same values for $\partial \epsilon_i/\partial X$, $\partial \epsilon_i/\partial Y$, $\partial \epsilon_i/\partial Z$. This particular situation was illustrated in Sec. 3-14 for the translational levels of a particle in a box. For a diatomic molecule, we shall choose to group together only those states which have the same vibrational energy and the same rotational energy as well as the same total energy.

There are g_i single particle states in the ith level, and we speak of this as the degeneracy of the level. The g's are supposed to be large numbers; as indicated in Chap. 3, a large number of quantum states that are very close together in their properties can be lumped together as being approximately degenerate in order to achieve large g's.

The occupation number of a level is the number of particles that are in that level. A distribution is a specification of an entire set of occupation numbers. The situation is displayed as follows:

Energy levels: $\epsilon_1, \epsilon_2, \ldots, \epsilon_i, \ldots, \epsilon_k, \ldots$
Degeneracies: $g_1, g_2, \ldots, g_i, \ldots, g_k, \ldots$
Occupation no.: $N_1, N_2, \ldots, N_i, \ldots, N_k, \ldots$

The set of numbers $N_1, N_2, \ldots, N_i, \ldots$ is a distribution. If we move, say, five particles from the second energy level to the first, thus increasing N_1 and decreasing N_2 by 5, we obtain a new distribution.

6-3. The Number of Wave Functions for a Distribution. Fermi-Dirac Case. Consider a set of N fermions and a distribution $N_1, N_2, \ldots, N_i, \ldots$. How many different system wave functions (or configurations) are there for this distribution? There are g_1 orbitals of energy ϵ_1. There are to be N_1 of these g_1 orbitals occupied, and, for fermions, it is necessary that $N_1 \leq g_1$. In how many ways can we select N_1 orbitals from the set of g_1 distinguishable orbitals? We call this number t_1. This is the problem treated as Lemma III of Sec. 5-2. The answer is

$$t_1 = \frac{g_1!}{N_1!(g_1 - N_1)!} \tag{6-1}$$

If there are t_1 ways of selecting the N_1 orbitals for occupation in the first set of g_1 levels and t_i ways of selecting the N_i orbitals for occupation in the ith level, the total number of configurations, or system wave functions, that are possible for the given distribution is

$$t_D = \prod_i t_i = \prod_i \frac{g_i!}{N_i!(g_i - N_i)!} \tag{6-2}$$

It is to be reemphasized that the particles are indistinguishable, and we cannot say which particles are in which particular energy level. However, in principle one can discover which particular orbitals are occupied in a given configuration, and t_D is the number of configurations that are possible for the given distribution.

We convert the infinite product to a sum by taking logarithms and use the Stirling approximation

$$\ln t = \sum \left[g_i \ln g_i - N_i \ln N_i - (g_i - N_i) \ln (g_i - N_i) \right.$$
$$\left. + \frac{1}{2} \ln \frac{g_i}{2\pi N_i (g_i - N_i)} \right] \tag{6-3}$$

For large g_i and N_i, the last term will be completely negligible compared with the others, which may be rearranged to give

$$\ln t = \sum_i \left(-g_i \ln \frac{g_i - N_i}{g_i} + N_i \ln \frac{g_i - N_i}{N_i} \right) \tag{6-4}$$

This is the desired result.

6-4. The Number of Systems Wave Functions for a Distribution. Bose-Einstein Case. For bosons, we have the problem of assigning N_1 indistinguishable particles to g_1 distinguishable orbitals when there is no restriction as to the number of particles that can occupy any one orbital. A preferable but equivalent way of putting the matter is to say that we have the problem of selecting N_1 orbitals for occupancy from the g_1 orbitals available, and any one of the orbitals may be used as often as desired. This is the problem treated in Lemma IV of Sec. 5-2, and the number of configurations is

$$t_1 = \frac{(g_1 + N_1 - 1)!}{(g_1 - 1)!N_1!} \tag{6-5}$$

Then, for the number of system wave functions available to the system in the distribution $N_1, N_2, \ldots, N_i, \ldots$, we have

$$t_D = \prod_i t_i = \prod_i \frac{(g_i + N_i - 1)!}{(g_i - 1)!N_i!} \tag{6-6}$$

Using the same approximations as before and also $g - 1 \approx g$,

$$\ln t = \sum \left(g_i \ln \frac{g_i + N_i}{g_i} + N_i \ln \frac{g_i + N_i}{N_i} \right) \tag{6-7}$$

6-5. Distinguishable Particles. It turns out to be useful to treat the problem of a hypothetical collection of distinguishable particles where the wave functions are subject to no symmetry restrictions. According to Lemma II of Sec. 5-2, there are

$$\frac{N!}{\prod_i N_i!}$$

ways of assigning the N distinguishable particles so that there are $N_1, N_2, \ldots, N_i, \ldots$ in the various energy levels. According to Lemma V, there are $g_i^{N_i}$ ways of selecting orbitals for the N_i particles in the ith level. Therefore, if there were no restrictions due to the identity of the particles,

$$t_D = N! \prod_i \frac{g_i^{N_i}}{N_i!} \tag{6-8a}$$

$$\ln t_D = N \ln N - \sum N_i \ln N_i + \sum N_i \ln g_i = N \ln N + \sum N_i \ln \frac{g_i}{N_i} \tag{6-8b}$$

This equation, which leads to the so-called Boltzmann statistics, was a fundamental equation in the development of statistical mechanics prior to the discovery of the basic indistinguishability for identical particles. We call these hypothetical particles boltzons. In the next section, we refer to the expression for t in (6-8) as t_{boltz}.

6-6. Corrected Boltzons. Suppose that for a system of fermions or bosons, $g_i \gg N_i$ for all i. The system is then dilute in the sense that most of the orbitals are not occupied. But, for fermions, each term t_i in the infinite product (6-1) was

$$t_i = \frac{g_i(g_i - 1) \cdots (g_i - N_i + 1)}{N_i!}$$

But if $g_i \gg N_i$, the smallest factor in the numerator is $g_i - N_i + 1$, which is almost equal to but slightly smaller than g_i. There are N_i factors in the numerator; so $t_{i(\text{FD})} \leq g_i^{N_i}/N_i!$, with the equality sign approximately applicable for a dilute system. In the same way, from (6-5) for bosons, $t_{i(\text{BE})} \geq g_i^{N_i}/N_i!$. Then, since $t = \prod_i t_i$, we have in general

$$t_{\text{FD}} \leq t_{\text{boltz}}/N! \leq t_{\text{BE}} \tag{6-9}$$

(This result was illustrated for the case of three particles in Sec. 3-13.) For dilute systems, where it is unlikely on the grounds of probability that there will be two particles in a given single-particle wave function, the difference between FD statistics, where such configurations are excluded, and BE statistics, where they are permitted, is negligible. In this case, we have

$$t_{\text{FD}} \approx t_{\text{BE}} \approx \frac{t_{\text{boltz}}}{N!} = \prod_i \frac{g_i^{N_i}}{N_i!} \tag{6-10}$$

We can correct for the fundamental indistinguishability of the particles without worrying about symmetric or antisymmetric wave functions by dividing the Boltzmann enumeration by $N!$. We shall say, when Eq. (6-10) is applicable, that we are dealing with corrected Boltzmann statistics and call the particles corrected boltzons.

We can summarize our results with Eqs. (6-11) and (6-12). In the general case,

$$\ln t = \sum \left(\pm g_i \ln \frac{g_i \pm N_i}{g_i} + N_i \ln \frac{g_i \pm N_i}{N_i} \right) \quad \begin{cases} \text{BE, upper sign} \\ \text{FD, lower sign} \end{cases} \tag{6-11}$$

For dilute systems, i.e., corrected boltzons,

$$\ln t = \sum \left(N_i \ln \frac{g_i}{N_i} + N_i \right) \tag{6-12}$$

It is convenient for the derivations that follow to treat the N_i as continuous variables. This approximation is permissible because the important values of N_i are large numbers. Then, by differentiation, $d(\ln t) = \Sigma[\partial(\ln t)/\partial N_i]\, dN_i$,

$$d(\ln t) = \sum \ln \frac{g_i \pm N_i}{N_i} dN_i \qquad \begin{cases} \text{BE, upper sign} \\ \text{FD, lower sign} \end{cases} \quad (6\text{-}13)$$

$$d(\ln t) = \sum \ln \frac{g_i}{N_i} dN_i \qquad \text{corrected boltzons} \quad (6\text{-}14)$$

6-7. A Fundamental Assumption for Statistical Mechanics. We assume that we are dealing with an isolated system which has a fixed total energy E (or, more precisely, that the energy of the system is in a rather narrow energy band between E and $E + \delta E$) and that there are a fixed number of particles. Thus, the distributions that we shall consider are restricted by the conditions or constraints.

$$\Sigma N_i = N \quad (6\text{-}15)$$
$$\Sigma N_i \epsilon_i = E \quad (6\text{-}16)$$

Now, in accordance with the discussion in Sec. 3-14, when the effects of collisions are considered, the state of the system at any one time can be approximately represented as a configuration of free-particle orbitals; but, because of collisions, the system is constantly undergoing transitions from one such configuration, or system state, to another. What will its properties then be?

It should be realized that, for most practical cases, the numbers of states t [in expressions (6-2) and (6-6) or (6-10) and (6-11)] are very large for some distributions and less for some other distributions. What we are going to do is to find that distribution which maximizes t (or $\ln t$), subject to the constraints (6-15) and (6-16), and assert that the properties of the actual system at equilibrium are, to a high degree of approximation, the properties of this most probable distribution.

The above procedure, which we shall actually use, may be regarded as an approximation. We can suppose that a more fundamental assumption about the system is that, as it undergoes transitions from one system quantum state to another, it will spend over a long period of time an equal amount of time in each of the possible quantum states of the system. This is one form of the assumption of equal a priori probability for individual quantum states. Thus, we should calculate the properties of the system by averaging over the properties of each system quantum state with unit weight. This is what is done in the so-called "method of steepest descents."* This is mathematically more abstruse and in some ways less illuminating than the procedure used here (and in most other texts). The results are the same. This is because, as we shall see in

---

* This method, due to Darwin and Fowler, is described by Fowler [3] and by Fowler and Guggenheim [1].

Sec. 14-20, the most probable distribution (set of quantum states) contains most of the states.*

6-8. The Fundamental Distribution Law. Without further apology, we now calculate that distribution which gives the largest value of ln t, subject to the constraints (6-15) and (6-16). We shall assume that the properties of an actual equilibrium system are the same as the properties of this most probable distribution. As already remarked, the important values of N_i are large numbers; so we can make the approximation of treating them as continuous variables. Introduce Lagrange multipliers α and $-\beta$ for Eqs. (6-15) and (6-16); the maximum is then found from the equations

$$\frac{\partial(\ln t)}{\partial N_i} + \alpha \frac{\partial N}{\partial N_i} - \beta \frac{\partial E}{\partial N_i} = 0 \quad \text{for all } i \qquad (6\text{-}17)$$

plus (6-15) and (6-16). Note that there is one equation in (6-17) for each energy level plus the two equations (6-15) and (6-16), i.e., just enough to determine all the N_i and α and β.

Note also that $\partial N/\partial N_i = 1$ and $\partial E/\partial N_i = \epsilon_i$. The values of $\partial(\ln t)/\partial N_i$ were calculated in (6-13). Equation (6-17) then becomes

$$\ln \frac{g_i \pm N_i}{N_i} + \alpha - \beta\epsilon_i = 0 \quad \begin{cases} \text{BE, upper sign} \\ \text{FD, lower sign} \end{cases} \qquad (6\text{-}18)$$

or

$$\frac{N_i}{g_i \pm N_i} = e^\alpha e^{-\beta\epsilon_i} \quad \begin{cases} \text{BE, upper sign} \\ \text{FD, lower sign} \end{cases} \qquad (6\text{-}19)$$

or

$$\frac{N_i}{g_i} = \frac{e^\alpha e^{-\beta\epsilon_i}}{1 \mp e^\alpha e^{-\beta\epsilon_i}} \quad \begin{cases} \text{BE, upper sign} \\ \text{FD, lower sign} \end{cases} \qquad (6\text{-}20a)$$

or

$$\frac{N_i}{g_i} = \frac{1}{e^{-\alpha}e^{\beta\epsilon_i} \mp 1} \qquad (6\text{-}20b)$$

The similar procedure using (6-14) for corrected boltzons gives

$$\frac{N_i}{g_i} = e^\alpha e^{-\beta\epsilon_i} \qquad (6\text{-}21)$$

---

* There is another form of the assumption of equal a priori probability of individual system quantum states. We suppose that we prepare a large number of separate systems (an ensemble) all with the same number of particles, N, and the same total energy. At a particular time, we determine the quantum state that each system is in. The fundamental assumption, then, is that by examining this ensemble we shall find that any one system quantum state is as probable as any other. The average over the properties of all systems (ensemble average) is then the average over the properties of all quantum states with unit weight.

But in an experiment we measure the time average properties of one particular system in a particular container in the laboratory. We assume that the time average properties of a particular system are the same as the ensemble averages described above. That is, we assume that the particular system in its motion moves through all the possible quantum states of the system consistent with (6-14) and (6-15) and spends on the average equal time in each. The hypothesis that such is the case is called the "ergodic hypothesis." The hypothesis is plausible but difficult to prove, and it has been the subject of a great deal of abstruse and inconclusive discussion.

It is easy to see that (6-21) holds for the distinguishable Boltzmann case also.

Equation (6-21) is commonly known as the Boltzmann distribution law. Equations (6-21) and (6-19) or (6-20) are the fundamental distribution laws for systems of independent particles.

It must be clearly remembered that Eqs. (6-18) through (6-21) give values of N_i for the most probable distribution, which we identify as the equilibrium distribution, and not for an arbitrary distribution. We shall almost always be dealing with this equilibrium distribution, and it is not usually necessary to denote it with a special symbol. Occasionally, we shall have to write $N_{i,\mathrm{mp}}$ and $\ln t_{\mathrm{mp}}$ to make it clear when we are discussing the equilibrium distribution rather than an arbitrary distribution.

6-9. The Parameters α and β. In general, α and β are to be evaluated by substituting the appropriate expression for N_i [(6-20) or (6-21)] into the equations

$$\Sigma N_i = N \qquad \Sigma N_i \epsilon_i = E$$

The results are simple only for the case of Boltzmann statistics.

In this case,

$$N = \Sigma N_i = \Sigma g_i e^\alpha e^{-\beta \epsilon_i} = e^\alpha \Sigma g_i e^{-\beta \epsilon_i} \qquad (6\text{-}22a)$$

or

$$e^\alpha = \frac{N}{\Sigma g_i e^{-\beta \epsilon_i}} \qquad (6\text{-}22b)$$

The molecular-partition function q is defined by

$$q = \Sigma g_i e^{-\beta \epsilon_i} \qquad (6\text{-}23)$$

We shall see that it is the fundamental function needed for calculating the statistical mechanical properties for a system of independent corrected boltzons. The quantity q is a function of β and of the parameters which determine the single-particle energy levels of the system—the volume, the electric field, the magnetic field, etc. For many problems, q is a function of β and V only.

We have

$$e^\alpha = \frac{N}{q} \qquad N_i = \frac{N}{q} g_i e^{-\beta \epsilon_i} \qquad (6\text{-}24)$$

The energy equation is

$$E = \sum N_i \epsilon_i = \frac{N}{q} \sum \epsilon_i g_i e^{-\beta \epsilon_i} = -\frac{N}{q} \left(\frac{\partial q}{\partial \beta}\right)_V = -N \left[\frac{\partial (\ln q)}{\partial \beta}\right]_V \qquad (6\text{-}25)*$$

We assume for simplicity that the only parameter affecting the energy levels is V, but the extension to other cases is obvious.

* We can also write that the average value of the energy per particle is

$$\langle \epsilon \rangle = \frac{\Sigma N_i \epsilon_i}{N} = \frac{E}{N} = -\left(\frac{\partial \ln q}{\partial \beta}\right)_V \qquad (6\text{-}25a)$$

Equation (6-25) can, in principle, be used to determine β from E; we shall see later that we usually regard β as known in an experiment and use (6-25) to calculate E.

As the reader probably already knows, the quantity β will turn out to be $1/kT$. Now we shall show that β is a function of temperature only. Consider two systems, each with its own set of particles and single-particle energy levels. The two systems are capable of exchanging energy. A mixture of argon atoms and helium atoms in the same box is an example. Another example would be two separate gas samples, in two different compartments, separated by a thin, strong partition. The partition is impermeable to molecules but is a heat conductor, so that energy can flow from one system to the other. It is immaterial which statistics (FD, BE, CB) the separate systems obey.

Let the various quantities for the first system be denoted by N'_i, g'_i, ϵ'_i, t'; those of the second are N''_j, g''_j, ϵ''_j, t''. The conservation constraints and Lagrange multipliers are $\Sigma N'_i = N'$, (α'); $\Sigma N''_j = N''$, (α''); $\Sigma N'_i \epsilon'_i + \Sigma N''_j \epsilon''_j = E$, $(-\beta)$. The essential point is that, whi'e there are separate equations with separate Lagrange multipliers α' and α'' for the conservation of the two different kinds of particles, there is only one equation with one Lagrange multiplier $-\beta$ for the conservation of the total energy. The number of system wave functions for the whole system in a distribution $N'_1, \ldots, N'_i, \ldots, N'_k, \ldots; N''_1, \ldots, N''_j, \ldots, N''_m, \ldots$ is the product of the number of system wave functions for the distribution N'_i of the first system and the number of system wave functions for the N''_j distribution for the second system. That is,

$$t = t't''$$
$$\ln t = \ln t' + \ln t'' \qquad (6\text{-}26)$$

The general conditions for the most probable distribution are

$$\frac{\partial(\ln t)}{\partial N'_i} + \alpha' \frac{\partial N'}{\partial N'_i} + \alpha'' \frac{\partial N''}{\partial N'_i} - \beta \frac{\partial E}{\partial N'_i} = 0 \qquad \text{for all } i$$

$$\frac{\partial(\ln t)}{\partial N''_j} + \alpha' \frac{\partial N'}{\partial N''_j} + \alpha'' \frac{\partial N''}{\partial N''_j} - \beta \frac{\partial E}{\partial N''_j} = 0 \qquad \text{for all } j$$

But $\partial N'/\partial N''_j = 0 = \partial N''/\partial N'_i$. The conditions therefore are

$$\frac{\partial(\ln t')}{\partial N'_i} + \alpha' - \beta \epsilon'_i = 0 \qquad \text{for all } i$$
$$\frac{\partial(\ln t'')}{\partial N''_j} + \alpha'' - \beta \epsilon''_j = 0 \qquad \text{for all } j \qquad (6\text{-}27)$$

These equations are solved exactly as before.

$$\frac{N'_i}{g'_i} = \frac{e^{\alpha'} e^{-\beta \epsilon'_i}}{1 \mp e^{\alpha'} e^{-\beta \epsilon'_i}} \qquad \frac{N''_j}{g''_j} = \frac{e^{\alpha''} e^{-\beta \epsilon''_j}}{1 \mp e^{\alpha''} e^{-\beta \epsilon''_j}} \qquad (6\text{-}28)$$

Sec. 6-9] STATISTICAL MECHANICS OF INDEPENDENT PARTICLES

The two distribution laws for the separate kinds of particles are just as before, and independent of each other, except that the two systems have a common β. But if two systems are in contact to the extent that they can exchange energy but do not otherwise interact with each other, the only macroscopic property that they have in common is the temperature. Therefore, β is a function of the temperature only. It cannot depend on the nature of the particles, the volume of the system, etc.

A similar argument, in which we allow the two systems to exchange particles as well as energy, is illuminating as to the nature of the parameter α. Suppose that there are two systems I and II, as, for example, two boxes separated by a semipermeable membrane which is permeable to molecules of type A only. There could be molecules of type A and molecules of type B in box I and molecules of type A and molecules of type C in box II. The systems therefore are capable of exchanging energy and A particles. There are sets of energy levels, degeneracies, and occupation numbers for the several particles in the several boxes:

$$\epsilon_{iA}^{I}, g_{iA}^{I}, N_{iA}^{I}; \epsilon_{jB}^{I}, g_{jB}^{I}, N_{jB}^{I}$$
$$\epsilon_{kA}^{II}, g_{kA}^{II}, N_{kA}^{II}; \epsilon_{pC}^{II}, g_{pC}^{II}, N_{pC}^{II}$$

There is one energy-conservation equation with one Lagrange multiplier $-\beta$; there is one conservation equation for type A particles,

$$\sum_i N_{iA}^{I} + \sum_k N_{kA}^{II} = N_A$$

with a Lagrange multiplier α_A. Note that

$$\frac{\partial N_A}{\partial N_{iA}^{I}} = 1 \qquad \frac{\partial E}{\partial N_{kA}^{II}} = \epsilon_{kA}^{II} \qquad \text{etc.}$$

There are separate conservation equations for type B particles and type C particles. In a given distribution, the total number of system wave functions is the product of the number of system wave functions for the given subsystems:

$$t = t_A^{I} t_A^{II} t_B^{I} t_C^{II} \tag{6-29}$$

We seek that distribution which maximizes t subject to the different constraints. We reemphasize that these constraints permit energy and particles of type A to flow between systems I and II.

Proceeding exactly as before, it will be found that the distribution laws are exactly as before:

$$N_{iA}^{I} = \frac{g_{iA}^{I} e^{\alpha_A - \beta \epsilon_{iA}^{I}}}{1 \mp e^{\alpha_A - \beta \epsilon_{iA}^{I}}}$$

$$N_{kA}^{II} = \frac{g_{kA}^{II} e^{\alpha_A - \beta \epsilon_{kA}^{II}}}{1 \mp e^{\alpha_A - \beta \epsilon_{kA}^{II}}} \tag{6-30}$$

$$N_{jB}^{I} = \frac{g_{jB}^{I} e^{\alpha_B - \beta \epsilon_{jB}^{I}}}{1 \mp e^{\alpha_B - \beta \epsilon_{jB}^{I}}}$$

(and similarly for particles of type C). Each kind of particle obeys Fermi-Dirac or Bose-Einstein statistics depending on its intrinsic nature.

The essential point is that there is a common β and a common Lagrange multiplier α_A for A-type particles in the two boxes. In thermodynamics, if one component is distributed between several phases or systems at equilibrium, the chemical potential of this component is the same in all the phases. Thus, we suspect that the parameter α is related to the chemical potential. The exact relation will be developed shortly.

6-10. The Perfect Gas and β. It is convenient now to make the identification $\beta = 1/kT$ by calculating the properties of a perfect gas.

We take a system of particles in a rectangular box as in Sec. 3-14 and assume that corrected Boltzmann statistics apply. The force in the x direction exerted by an atom in the ith quantum state is $F_{xi} = -\partial \epsilon_i / \partial X$. The force exerted by a system in a particular distribution is $F_x = \Sigma N_i F_{xi}$. For this purpose the quantum states of the system must be grouped together in such a way that the g_i states of the ith level all have approximately the same value of ϵ_i and the same value of $\partial \epsilon_i / \partial X$.

As discussed in Sec. 6-7, we take as the force exerted by a system at equilibrium the force exerted by the most probable distribution. $F_x = \Sigma N_i F_{xi}$, where, for the equilibrium distribution, $N_i = g_i N e^{-\beta \epsilon_i}/q$; so

$$F_x = \frac{N}{q} \sum g_i \left(-\frac{\partial \epsilon_i}{\partial X}\right) e^{-\beta \epsilon_i} = \frac{N}{\beta q}\left(\frac{\partial q}{\partial X}\right)_{\beta,Y,Z}$$

$$= \frac{N}{\beta}\left[\frac{\partial (\ln q)}{\partial X}\right]_{\beta,Y,Z} \tag{6-31}$$

The partition function is $\Sigma g_i e^{-\beta \epsilon_i}$. There are, for example, g_j single-particle states of energy ϵ_j. We could just as well write

$$q = \sum_i e^{-\beta \epsilon_i} \tag{6-32}$$

where the index i now ranges over all the individual quantum states. There will still be g_j terms of magnitude $e^{-\beta \epsilon_j}$ in the sum, but they are now exhibited separately.

The individual levels for a perfect gas are given by

$$\epsilon_{s_x s_y s_z} = \frac{h^2}{8m}\left(\frac{s_x^2}{X^2} + \frac{s_y^2}{Y^2} + \frac{s_z^2}{Z^2}\right)$$

We then have

$$q = \sum_{s_x=1}^{\infty} \sum_{s_y=1}^{\infty} \sum_{s_z=1}^{\infty} e^{-(\beta h^2/8m)(s_x^2/X^2 + s_y^2/Y^2 + s_z^2/Z^2)} \tag{6-33a}$$

$$q = \sum_{s_x,s_y,s_z} e^{-(\beta h^2 s_x^2/8mX^2)} e^{-(\beta h^2 s_y^2/8mY^2)} e^{-(\beta h^2 s_z^2/8mZ^2)} \tag{6-33b}$$

$$q = \left(\sum_{s_x} e^{-(\beta h^2 s_x^2/8mX^2)}\right)\left(\sum_{s_y} e^{-(\beta h^2 s_y^2/8mY^2)}\right)\left(\sum_{s_z} e^{-(\beta h^2 s_z^2/8mZ^2)}\right) \tag{6-33c}$$

$$q = q_x q_y q_z \tag{6-33d}$$

Sec. 6-10]   STATISTICAL MECHANICS OF INDEPENDENT PARTICLES   85

where
$$q_x = \sum_{s_x=1}^{\infty} e^{-\beta h^2 s_x^2/8mX^2}, \text{ etc.} \tag{6-33e}$$

Equation (6-33c) is the crucial one in this derivation; the unrestricted sum over all values of s_x, s_y, s_z is the product of the three separate sums.

We now evaluate q_x. If $\beta h^2/8mX^2 \ll 1$, as it is for a typical gas (cf. Prob. 3-4), the sum can be replaced by an integral,

$$q_x = \int_{s_x=0}^{\infty} e^{-(\beta h^2 s_x^2/8mX^2)} \, ds_x = \left(\frac{2\pi m}{\beta h^2}\right)^{1/2} X$$

$$q_y = \left(\frac{2\pi m}{\beta h^2}\right)^{1/2} Y \tag{6-34}$$

$$q_z = \left(\frac{2\pi m}{\beta h^2}\right)^{1/2} Z$$

$$q = q_x q_y q_z = \left(\frac{2\pi m}{\beta h^2}\right)^{3/2} XYZ = \left(\frac{2\pi m}{\beta h^2}\right)^{3/2} V \tag{6-35}$$

where V is the volume of the system.

The force in the x direction on the rectangular box is

$$F_x = \frac{N}{\beta} \frac{\partial (\ln q)}{\partial X} = \frac{N}{\beta X} \tag{6-36}$$

The pressure is the force per unit area, F_x/YZ,

$$P = \frac{F_x}{YZ} = \frac{N}{\beta XYZ} = \frac{N}{\beta V} \tag{6-37}$$

A perfect gas, however, obeys the equation $PV = NkT$. If our quantum-mechanical model of the energy levels for a perfect gas is satisfactory and if our statistical procedure of taking the properties of the most probable distribution as the properties of the system is acceptable, the statistical quantity β must be identified with $1/kT$, where T is the thermodynamic temperature.

We further note that the total energy E is given by

$$E = -N \left[\frac{\partial (\ln q)}{\partial \beta}\right]_V = \frac{3}{2}\frac{N}{\beta} = \frac{3}{2}NkT = \frac{3}{2}PV \tag{6-38}$$

We shall return to other properties of the perfect gas later. Our main purpose now is to continue the development of the fundamental relations of statistical mechanics.

Before moving on, we should note that in a cubical box the translational energy levels can be written as $\epsilon_{s_x s_y s_z} = (h^2/8mV^{2/3})(s_x^2 + s_y^2 + s_z^2)$. We can then talk about the pressure of a particular quantum state $P_i = -\partial \epsilon_i/\partial V$. Corresponding to (6-31), there is a relation

$$P = \frac{N}{\beta}\left[\frac{\partial (\ln q)}{\partial V}\right] \tag{6-39}$$

The relation $P = N/\beta V$ then follows directly from the expression for q in terms of V [(6-35)]. This is the way the derivation is usually presented. The relation (6-39) is useful and valid in general, even though it was derived here only for a cubical box.

We now know that the parameter β of the fundamental distribution laws is $1/kT$ for all systems of independent particles. For convenience in calculations and for brevity of notation, we shall often use β as a variable instead of T in our derivations.

6-11. The Statistical Mechanical Expression for the Heat Change. For any system of independent particles, $E = \Sigma N_i \epsilon_i$. The energy levels are a function of some external parameters of the system, and the generalized force in the ith quantum state due to the variation of a particular parameter l is $F_{il} = -\partial \epsilon_i/\partial l$. The parameter l can represent the volume V or one of the three dimensions X, Y, Z of a container, the value of an applied electric or magnetic field, or the position in a gravitational field or in a centrifugal field. The work done by a particle in the ith quantum state if the parameter l is changed is $dw_i = F_{il}\, dl = (-\partial \epsilon_i/\partial l)\, dl$. The total force for changing the parameter l exerted by the system in a given distribution is $F_l = \Sigma N_i F_{il}$.

For an arbitrary small change imposed upon the system,

$$dE = \Sigma N_i\, d\epsilon_i + \Sigma \epsilon_i\, dN_i \qquad (6\text{-}40)$$

In this arbitrary change, the energy levels may have been changed by changing the parameters l; this gives the $\Sigma N_i\, d\epsilon_i$ term. There may also have been a change in the distribution of particles among the energy levels; this is the $\Sigma \epsilon_i\, dN_i$ term.

If the change is carried out reversibly (cf. the discussion in Sec. 4-4), the thermodynamic work done on the surroundings dw_rev is equal to the force exerted by the system times the displacement dl:

$$dw_\text{rev} = F_l\, dl = \sum_i N_i F_{il}\, dl = \sum N_i \left(-\frac{\partial \epsilon_i}{\partial l}\right) dl = -\sum N_i\, d\epsilon_i \qquad (6\text{-}41)$$

It is important to appreciate that the only cause of a change $d\epsilon_i$ in the energy levels is the change in the parameter l so that $d\epsilon_i = (\partial \epsilon_i/\partial l)\, dl$.

Therefore, we can rewrite (6-40):

$$dE = \Sigma \epsilon_i\, dN_i - dw_\text{rev} \qquad (6\text{-}42)$$

But from thermodynamics, $dE = dq_\text{rev} - dw_\text{rev}$ (q is now the heat, not the molecular-partition function). Therefore,

$$dq_\text{rev} = \Sigma \epsilon_i\, dN_i \qquad (6\text{-}43)$$

Heat is a concept which is intrinsically associated with the nonmechanical concept of temperature and is not defined in pure mechanics. In statistical mechanics, the heat absorbed by a system from its surroundings in a reversible change is identified with that part of the energy change of a system $\Sigma \epsilon_i \, dN_i$ which results from the redistribution of the particles among the several energy levels.

Several minor points may be noted about this significant result. We are particularly interested in the case that the system remains in its equilibrium or most probable distribution during the change. However, the equation for the heat absorbed is also applicable to a nonequilibrium distribution, provided the forces of the system are almost balanced by the forces of the surroundings so that work is done reversibly on the surroundings. Furthermore, if the volume and other mechanical parameters of the system do not change ($d\epsilon_i = 0$), then no work is done and the heat absorbed is $\Sigma \epsilon_i \, dN_i$, even for an irreversible flow of heat.

6-12. The Statistical Mechanical Entropy. We define

$$S_{\text{sm}} = k \ln t \tag{6-44}$$

This is the Boltzmann definition for the statistical mechanical entropy S_{sm}. In (6-44), t is the total number of system wave functions available to the system, which is subject to the constraints imposed [(6-15) and (6-16)]. We shall continue to assume (and shall justify later) that $\ln t \approx \ln t_{\text{mp}}$, where t_{mp} refers to the most probable distribution. Then our working equation is

$$S_{\text{sm}} = k \ln t_{\text{mp}} \tag{6-45}$$

Consider an arbitrary reversible variation in the system. The energy levels may be changed, and the total energy may be changed. By the hypothesis of reversibility, the system remains in an equilibrium state during the change; that is, it is always in the most probable distribution. But t is a function only of the distribution parameters N_i so that

$$dS_{\text{sm}} = k \, d(\ln t) = k \sum_i \frac{\partial (\ln t)}{\partial N_i} \, dN_i \tag{6-46}$$

But if the system is at equilibrium, that is, in the most probable distribution, all the $\partial(\ln t)/\partial N_i$ have been chosen to satisfy the equations $[\partial(\ln t)/\partial N_i] + \alpha - \beta \epsilon_i = 0$. Remember also that $\Sigma dN_i = 0$. Then

$$dS_{\text{sm}} = k \sum_i (-\alpha + \beta \epsilon_i) \, dN_i = -\alpha k \sum dN_i + \beta k \sum \epsilon_i \, dN_i$$

$$= \beta k \sum \epsilon_i \, dN_i = \frac{dq_{\text{rev}}}{T} \quad \text{by (6-43)} \tag{6-47}$$

Thus, the differential of the statistical mechanical entropy is equal to the differential of the thermodynamic entropy for an arbitrary reversible change. The two quantities can then differ only by an additive constant for any one substance.

As we shall see, if the statistics of independent particles are applicable at all at very low temperatures, the corrected boltzon approximation fails and the particles must be treated as fermions or bosons. According to the distribution law (6-20b) for very large β, the molecules will all be in the lowest single-particle levels available to them (this point is developed in detail in Sec. 6-16); and, for practical cases, there will be one or at most a few system states corresponding to this distribution. Thus, the statistical mechanical entropy will either be rigorously zero or, at most, of the order of a few k. But, at ordinary temperatures, the entropy of a system containing 1 mole of particles is of the order of $10^{24}k$. Compared with this value, the entropy at $T = 0$ is either rigorously zero or so small as to be negligible and effectively zero.

Actually, at low temperatures, the assumption that the particles are independent and noninteracting is usually not justified. But when one inquires into all the interactions for a quantum-mechanical system of N particles, one finds that there is a single nondegenerate lowest quantum state for the entire system so that at absolute zero only one system wave function is occupied.

Thus, from every point of view, we find that $S_{sm} = 0$ at $T = 0$. But, by the third law, the thermodynamic entropy is zero at $T = 0$. Since we have proved that the thermodynamic entropy and the statistical entropy can differ at most by a constant, this constant must be zero and the two quantities are identical.

We also want to prove that the statistical mechanical entropy obeys that part of the second law of thermodynamics which states that $dS > dq_{irrev}/T$. An equivalent statement in thermodynamics is that for an isolated constant-volume system, $dS)_{E,V} > 0$ for any spontaneous change. Consider such an isolated constant-volume system which consists of two subsystems, the prime system and the double-prime system. Each part is in its equilibrium (most probable) distribution, but the two parts may not be in equilibrium with each other. (For example, they might have different pressures or different temperatures.) The total number of system wave functions for the entire system is $t't''$ and

$$S = k \ln t't'' = k \ln t' + k \ln t'' = S' + S''$$

If now the two subsystems are allowed to interact, then, by our fundamental hypothesis as to the identity of the equilibrium state with the most probable state, the distributions will either not change at all or change in such a way as to increase the value of $\ln t = \ln t' + \ln t''$, that is, to go into a more probable distribution. Thus, the entropy will increase for any spontaneous change in an isolated constant-volume system. For an analysis of the simpler problem of the entropy change when only heat flows from one system to another, see Prob. 6-4. Thus, our fundamental hypothesis as to the nature of the equilibrium state guarantees that the statistical mechanical entropy will satisfy the condition $dS)_{E,V} > 0$ for any spontaneous process.

We therefore adopt the following point of view: Statistical mechanics, with its basic assumption that the properties of the most probable distribution are the properties of the system at equilibrium, permits us to calculate the mechanical and thermal properties of a system from the laws of mechanics. We can define a function of state, the statistical mechanical entropy, and show that the second and third laws of thermodynamics are derivable from the laws of statistical mechanics. (The first law, the over-all conservation of energy, is essentially assumed in the pure mechanics.) The laws of thermodynamics are therefore a consequence of statistical mechanics (the generalization to systems of dependent particles will be quite easy!). We can now go ahead and find out how to calculate other thermodynamic functions in statistical mechanics.

6-13. The Thermodynamic Functions for a System of Corrected Boltzons. Consider, for simplicity, that the only parameter determining the energy levels is the volume. For corrected boltzons, from (6-12) and (6-45)

$$S = k \ln t = k \sum \left(N_i \ln \frac{g_i}{N_i} + N_i \right) \qquad (6\text{-}48)$$

Substitute $N_i/g_i = (N/q)e^{-\beta \epsilon_i}$ into the logarithm term, but keep the other N_i's.

$$\frac{S}{k} = \left(\sum N_i \right) \ln \frac{q}{N} + \beta \sum N_i \epsilon_i + \sum N_i \qquad (6\text{-}49)$$

$$S = kN \ln \frac{q}{N} + \frac{E}{T} + kN$$

For the Helmholtz free energy A,

$$A = E - TS = -kNT \ln \frac{q}{N} - kNT \qquad (6\text{-}50)$$

The chemical potential is given by $\mu = (\partial A/\partial N)_{T,V}$. Recall that q is a function of T and V, but not of N.

$$\mu = \left(\frac{\partial A}{\partial N} \right)_{T,V} = -kT \ln \frac{q}{N} - kNT \frac{\partial [\ln (q/N)]}{\partial N} - kT = -kT \ln \frac{q}{N} \qquad (6\text{-}51a)$$

For the Gibbs free energy $F = N\mu$; so

$$F = -kNT \ln \frac{q}{N} \qquad (6\text{-}51b)$$

Since $F = A + PV$, from (6-50) and (6-51b) we derive $PV = kNT$. The perfect-gas law holds for any system of independent particles which obey corrected Boltzmann statistics and for which the volume is a meaningful variable.*

Thus, we see that all the thermodynamic functions of a system of corrected boltzons can be calculated if the energy-level system is known and the partition function is calculated.

Problem 6-1. Derive explicit formulae for the free energies, the entropy, the thermodynamic energy, and the enthalpy of a perfect monatomic gas of corrected boltzons from the partition function (6-35) and the general formulae just described. Choose T and V as independent variables. Transform to T and P as independent variables.

6-14. Mixtures of Particles. As shown in the derivation leading to (6-28), in a mixture of two kinds of particles in one box, each kind of particle obeys its own distribution law, but with a common β. The number of wave functions for the system is the product of the number of wave functions for each kind of particle [Eq. (6-26)].† Therefore $S = S' + S''$. Since $E = E' + E''$ and $P = P' + P''$, it follows that $A = A' + A''$, $F = F' + F''$ and all the thermodynamic functions are additive. For the chemical potentials,

$$\mu' = \left(\frac{\partial A}{\partial N'}\right)_{T,V,N''} = -kT \ln \frac{q'}{N'}$$
$$\mu'' = \left(\frac{\partial A}{\partial N''}\right)_{T,V,N'} = -kT \ln \frac{q''}{N''} \qquad (6\text{-}52)$$

Thus, all the relations just derived for one-component systems hold, with obvious extensions, for multicomponent systems.

This section, although short, is significant. For notational simplicity most of our derivations are given for one-component systems. Many of the interesting problems in chemistry, especially the problem of chemical equilibrium, pertain to mixtures of several components, and our results are directly applicable to these problems.

---

* This is a curious result. If the identification $\beta = 1/kT$ is somehow made, the perfect-gas law can be derived without any specific consideration of the particle in box energy levels.

† This statement is not strictly true. It is not true for the totality of distributions available to the combined system, but it is true in the approximation that the number of system states for each subsystem is equal to the number in the most probable distribution. If the total energy is fixed at E, let one system have energy E_1 and the other $E - E_1$; the number of system wave functions for this division of the energy is $t = t'(E - E_1)t''(E_1)$. The total number of states available is

$$t = \sum_{E_1} t'(E - E_1)t''(E_1)$$

The only important term in this double sum is when E_1 is the most probable amount of energy in the double-prime system; then $t = t't''$.

6-15. Comments on the Boltzmann Distribution and on Corrected Boltzmann Statistics.

The distribution law is

$$\frac{N_i}{N} = \frac{g_i}{q} e^{-\beta \epsilon_i} \tag{6-53a}$$

so that, for two different levels i and j,

$$\frac{N_i}{N_j} = \frac{g_i}{g_j} e^{-\beta(\epsilon_i - \epsilon_j)} \tag{6-53b}$$

At a temperature sufficiently high so that $\beta(\epsilon_i - \epsilon_j) \ll 1$, $N_i/N_j \approx g_i/g_j$; the population of a level is proportional to the number of states, g_i, therein. At lower temperatures, the Boltzmann factor $e^{-\beta(\epsilon_i - \epsilon_j)}$ favors the population of states of lower energy. In general, statistical equilibrium represents the balance between the randomizing forces of thermal agitation, which tend to populate all states equally, and the tendency of a mechanical system to go into the state of lowest possible energy. The greater the temperature, the greater the relative effect of the randomizing forces of thermal agitation. The student should not allow the various complicated calculations he becomes involved in to obscure this simple, basic situation.

The single-particle partition function $q = \Sigma g_i e^{-\beta \epsilon_i}$ is also called by the appropriate name of *sum over states*. It is a weighted sum of the number of states available to the particles, the weighting factors being the Boltzmann factors $e^{-\beta \epsilon_i}$. The higher the temperature, the larger this factor and the greater the number of states which are effectively available to the system.

The chemical potential is given by $\mu = -kT \ln(q/N)$. The larger the partition function, the more negative the chemical potential. In mechanics a particle always seeks a position of lowest potential energy. A real particle, which is subject to the randomizing forces of thermal agitation, seeks to be in the situation which gives it the lowest chemical potential.

It is of the utmost importance to observe that the free energy and chemical potential are given by

$$F = -NkT \ln\left(\frac{q}{N}\right) \quad \text{and} \quad \mu = -kT \ln\left(\frac{q}{N}\right)$$

They depend on q/N, not on q. We shall see that it is q/N, not q, which is an intensive quantity.*

---

* The occurrence of q/N rather than of q arises essentially out of the fundamental indistinguishability of the particles, which requires that in dilute systems we must divide the number of wave functions for distinguishable particles by $N!$ to obtain the actual number of wave functions available to the system.

6-16. Fermi-Dirac and Bose-Einstein Statistics.

We again write the fundamental distribution law (6-20) and several rearrangements thereof,

$$\frac{N_i}{g_i} = \frac{e^\alpha e^{-\beta \epsilon_i}}{1 \mp e^\alpha e^{-\beta \epsilon_i}} = \frac{1}{e^{-\alpha} e^{\beta \epsilon_i} \mp 1}$$

$$\frac{N_i}{g_i \pm N_i} = e^\alpha e^{-\beta \epsilon_i} \qquad \frac{g_i \pm N_i}{g_i} = \frac{1}{1 \mp e^\alpha e^{-\beta \epsilon_i}} \tag{6-54}$$

(upper sign, BE; lower sign, FD, as usual).

We know that $\beta = 1/kT$ and that the parameter α is to be determined from the condition $\Sigma N_i = N$. From (6-11),

$$\frac{S}{k} = \ln t = \sum \left(\pm g_i \ln \frac{g_i \pm N_i}{g_i} + N_i \ln \frac{g_i \pm N_i}{N_i} \right) \tag{6-11}$$

$$\frac{S}{k} = \sum [\mp g_i \ln (1 \mp e^\alpha e^{-\beta \epsilon_i})] - N\alpha + \beta E \tag{6-55}$$

Then
$$A = E - TS = NkT\alpha \pm kT\Sigma g_i \ln (1 \mp e^\alpha e^{-\beta \epsilon_i}) \tag{6-56}$$

In the derivations that follow, it is expedient to introduce the function

$$y_i = e^{\alpha - \beta \epsilon_i} \tag{6-57a}$$

Observe that $N_i = g_i y_i / (1 \mp y_i)$ and that

$$A = NkT\alpha \pm kT\Sigma g_i \ln (1 \mp y_i) \tag{6-58}$$

Furthermore, recognizing that α is a function of N,

$$\left(\frac{\partial y_i}{\partial N} \right)_{\beta,V} = e^\alpha e^{-\beta \epsilon_i} \left(\frac{\partial \alpha}{\partial N} \right)_{\beta,V} = y_i \left(\frac{\partial \alpha}{\partial N} \right)_{\beta,V} \tag{6-57b}$$

and
$$d \left(\frac{y_i}{1 \mp y_i} \right) = \frac{1}{(1 \mp y_i)^2} dy_i \tag{6-57c}$$

The parameter α is to be regarded as a function of N, T, V. It is determined by

$$N = \sum N_i = \sum \frac{g_i e^\alpha e^{-\beta \epsilon_i}}{1 \mp e^\alpha e^{-\beta \epsilon_i}} = \sum \frac{g_i y_i}{1 \mp y_i} \tag{6-59}$$

The energy equation is

$$E = \sum N_i \epsilon_i = N \langle \epsilon \rangle = \sum \frac{g_i \epsilon_i y_i}{1 \mp y_i} \tag{6-60}*$$

* If f_i is any property that the system has in the ith state, then

$$\langle f \rangle = \frac{\Sigma f_i N_i}{N} = \frac{1}{N} \left(\sum \frac{g_i f_i y_i}{1 \mp y_i} \right) \tag{6-61}$$

This notation is convenient for problems such as Prob. 6-9. For example, by differentiating (6-59) with respect to N,

$$\left(\frac{\partial N}{\partial N} \right)_{\beta,V} = 1 = \sum \frac{g_i y_i}{(1 \mp y_i)^2} \left(\frac{\partial \alpha}{\partial N} \right)_{\beta,V} = N \left(\frac{\partial \alpha}{\partial N} \right)_{\beta,V} \left\langle \frac{1}{1 \mp y_i} \right\rangle \tag{6-62}$$

which is a useful relation for $(\partial \alpha / \partial N)_{\beta,V}$.

Sec. 6-16] STATISTICAL MECHANICS OF INDEPENDENT PARTICLES

We now return to expressions (6-58) and (6-59) with the object of evaluating α. By differentiation of (6-58),

$$\mu = \left(\frac{\partial A}{\partial N}\right)_{\beta,V} = kT\alpha + NkT\left(\frac{\partial \alpha}{\partial N}\right)_{\beta,V} - kT\sum \frac{g_i y_i}{1 \mp y_i}\left(\frac{\partial \alpha}{\partial N}\right)_{\beta,V}$$

But from (6-59) the last two terms in the above expression add to zero; so

$$\mu = kT\alpha \qquad \alpha = \frac{\mu}{kT} \qquad (6\text{-}63)$$

The important fact that $\alpha = \mu/kT$ has now been established for fermions, bosons, and corrected boltzons.

Since $F = N\mu$ and $F - A = PV$, we have from (6-63) and (6-56)

$$PV = \mp kT\Sigma g_i \ln(1 \mp e^{\beta(\mu-\epsilon_i)}) = \mp kT\Sigma g_i \ln(1 \mp y_i) \qquad (6\text{-}64)$$

A number of minor but interesting derivations may be made from these relations. For example, from (6-64),

$$\left(\frac{\partial(PV/kT)}{\partial \alpha}\right)_{\beta,V} = N \qquad (6\text{-}65)$$

which is actually the thermodynamic relation (Prob. 4-2)

$$\left[\frac{\partial(PV/T)}{\partial(\mu/T)}\right]_{T,V} = N \qquad (6\text{-}66)$$

We shall not make a serious study of the important applications of the Fermi-Dirac and Bose-Einstein statistics [other than for the statistics of light quanta which are bosons (Chap. 12)]. At this point, however, we shall consider the nature of the distributions at $T = 0$.

For the Fermi-Dirac case,

$$\frac{N_i}{g_i} = \frac{1}{e^{\beta(\epsilon_i - \mu)} + 1}$$

Our object is to evaluate μ as $T \to 0$ or $\beta \to \infty$. For those levels for which $\epsilon_i > \mu$, $e^{\beta(\epsilon_i - \mu)} \gg 1$, and $N_i/g_i \approx 0$ for very large β. For the levels for which $\epsilon_i < \mu$, $e^{\beta(\epsilon_i - \mu)} \approx 0$, and $N_i/g_i \approx 1$. For $\epsilon_i = \mu$, $N_i/g_i = \frac{1}{2}$. We thus have

$$\epsilon_i < \mu \quad \frac{N_i}{g_i} \approx 1; \quad \epsilon_i = \mu \quad \frac{N_i}{g_i} = \frac{1}{2}; \quad \epsilon_i > \mu \quad \frac{N_i}{g_i} \approx 0$$

This distribution is shown as the solid line in Fig. 6-1. All the levels up to $\epsilon_i = \mu$ are completely filled, and the higher states are completely empty.

The value of μ is determined by the condition that

$$N = \sum N_i = \sum_{\epsilon_i = 0}^{\epsilon_i = \mu} g_i \qquad (6\text{-}67)$$

That is, the value of μ is determined so that $\Sigma g_i = N$ for the sum over all the states $\epsilon_i < \mu$. When the temperature is raised slightly, the distribution is changed as indicated by the dashed line in Fig. 6-1.

For Bose-Einstein statistics we choose the energy of the lowest state ϵ_0 as zero; the ratio of the occupation numbers N_1/N_0 is

$$\frac{N_1}{N_0} = \frac{g_1}{g_0} \frac{e^{-\alpha} - 1}{e^{-\alpha}e^{\beta\epsilon_1} - 1}$$

We shall see that, at low temperatures, $\alpha \approx 0$. We choose T sufficiently small (β large) such that $e^{\beta\epsilon_1} \gg 1$. Then $N_1/N_0 \approx 0$. We can choose β

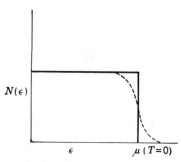

FIG. 6-1. The distribution functions for fermions at $T = 0$ (solid line) and at a slightly higher temperature (dashed line). For T slightly greater than zero, the width of the energy interval over which $N(\epsilon)$ deviates significantly from the $T = 0$ step function is of the order of kT.

so large that the same is true for all the higher states whatever the ratio g_i/g_0. Then all the atoms will be in the lowest state, and $N_0 = N$. The parameter α is then determined from

$$\frac{N}{g_0} = \frac{1}{e^{-\alpha} - 1} \quad \text{or} \quad \alpha = -\ln\frac{N + g_0}{N} \qquad (6\text{-}68)$$

If, as is typical, $N \gg g_0$, $\alpha \approx 0$. The main point is that, at low temperatures in a system of independent bosons, all the particles will be in the lowest energy state.

When the temperature is raised, the upper states begin to be populated. For free-particle translational wave functions in a cubical box, the degeneracies vary with energy so that $g_i = c\epsilon_i^{1/2}$. Under these circumstances it is found that there is a transition temperature below which almost all the atoms are in the lowest state and above which very few are. This is the Bose-Einstein condensation phenomenon.*

6-17. The Conditions for the Applicability of Boltzmann Statistics. If, in the distribution laws,

$$\frac{N_i}{g_i} = \frac{e^{\beta(\mu-\epsilon_i)}}{1 \mp e^{\beta(\mu-\epsilon_i)}} \qquad (6\text{-}69)$$

* For further discussion of Fermi-Dirac and Bose-Einstein statistics, see Kittel, pp. 86–106 [12].

for bosons and fermions, the condition

$$e^{\beta(\mu-\epsilon_i)} \ll 1 \tag{6-70}$$

is satisfied for all i, the second term in the denominator may be neglected, and the Boltzmann distribution

$$\frac{N_i}{g_i} = e^{\beta(\mu-\epsilon_i)} = \frac{N}{q} e^{-\beta\epsilon_i} \tag{6-71}$$

results. Note that this condition (6-70) also guarantees that $N_i/g_i \ll 1$, which we have already recognized as the condition for the applicability of corrected Boltzmann statistics. For convenience, we take the energy of the lowest energy level as zero, $\epsilon_0 = 0$; then all the other ϵ_i's are positive. The largest value of the left-hand side of Eq. (6-71) is then $N_0/g_0 = e^{\beta\mu} = N/q$. Therefore, if the condition $N/q \ll 1$ is satisfied, corrected Boltzmann statistics are applicable.

For a perfect gas, $N/q = (h^2/2\pi mkT)^{3/2}(N/V)$. (Note that, as previously remarked, q/N is an intensive quantity.) For a system of hydrogen atoms at 1 atm at 300°K, this relation gives $N/q = 2.5 \times 10^{-5}$. For helium gas (at. wt 4) at its boiling point of 4°K, $N/q = 0.154$. If we regard sodium metal at room temperature as an electron gas with 1 electron atom^{-1}, $N/V = 2.6 \times 10^{22}$ electrons cc^{-1}. The fraction N/q is proportional to $m^{-3/2}$; and by comparison with the calculation for hydrogen atoms, we easily see that $N/q = 2.2 \times 10^3$. Thus, corrected Boltzmann statistics apply to any ordinary gas except the light gases at low temperature (Ne, He, and H_2 especially[*]) and apply approximately even to these gases. The electrons in an ordinary metal (and often even in a semiconductor) at room temperature must be treated by Fermi-Dirac statistics. The properties of such an electron gas are radically different from the properties of a perfect boltzon gas even aside from the effects of the electrical interactions.

PROBLEMS

6-2. It will be shown later that the average kinetic energy for a perfect monatomic gas is $kT/2$ in the x direction, $kT/2$ in the y direction, and $kT/2$ in the z direction. The corresponding mean-square momenta are $\langle p_x^2 \rangle = 2m \langle \epsilon_x \rangle$. Define an average wavelength in each direction by

$$\langle \lambda \rangle = \frac{h}{\langle p_x^2 \rangle^{1/2}}$$

$\langle \lambda \rangle$ is a function of T.

Express the partition function for a perfect gas, using $\langle \lambda \rangle$ and V as independent variables.

[*] Hirschfelder, Curtiss, and Bird, pp. 419–424 [15].

6-3. Obtain a statistical mechanical formula for $(\partial \langle \epsilon \rangle / \partial \beta)_V$ for a system of corrected boltzons [where $\langle \epsilon \rangle = E/N = (\Sigma N_i \epsilon_i)/N$]. What is the sign of $\partial \langle \epsilon \rangle / \partial \beta$?

6-4. Consider two systems which are at different temperatures β_1 and β_2 (where we know $\beta = 1/kT$). Place them in thermal contact so that energy can flow from one to another, but let the volume of each be fixed. Obtain by statistical mechanics alone an expression for the entropy change when dE of energy flows from one system to another. (The result is identical with what thermodynamic relation?) In which direction will energy flow?

6-5. In the thermodynamics of a two-dimensional phase, area a plays the role that volume V does for a three-dimensional phase, and film pressure π replaces pressure. The work done in increasing the area is $\pi\, da$. A fundamental thermodynamic equation for a one-component system is

$$dE = T\, dS - \pi\, da + \mu\, dN$$

The differential of the Helmholtz free energy $(A = E - TS)$ is

$$dA = -S\, dt - \pi\, da + \mu\, dN$$

The energy levels of a two-dimensional perfect gas are given by

$$\epsilon = \frac{h^2}{8m^2}\left(\frac{s_x^2}{X^2} + \frac{s_y^2}{Y^2}\right)$$

(where area $a = XY$). Use this result to calculate the partition function, the energy, the film pressure, the entropy, and the chemical potential of a two-dimensional perfect gas which satisfies corrected Boltzmann statistics.

6-6. Show that the expression for $\ln t$ for corrected boltzons [(6-12)] can be obtained from the expressions for $\ln t$ for bosons or fermions [(6-11)] by series expansion for the case $N_i/g_i \ll 1$. (*Hint:* Expand $\ln[1 + (N_i/g_i)]$, and consistently retain all terms of the order of N_i or greater, but reject terms of the order of $N_i(N_i/g_i)$.)

6-7. Show that the thermodynamic relation $P = -(\partial A/\partial V)_{T,N}$ is satisfied with the expressions for A [(6-56)] and P [(6-64)] for Fermi-Dirac and Bose-Einstein statistics.

6-8. Is the energy per particle a function of the density of particles for a system of corrected boltzons? Is the free energy per particle a function of the density of particles for such a system? Discuss the same two questions qualitatively for a system of fermions or boltzons.

6-9. Obtain a statistical mechanical expression for $(\partial \alpha/\partial \beta)_{N,V}$ for Fermi-Dirac and/or Bose-Einstein statistics in the spirit of Eq. (6-62) for $(\partial \alpha/\partial N)_{\beta,V}$. This derivative $(\partial \alpha/\partial \beta)_{N,V}$ is equal to what other derivative that you can obtain by a similar procedure? [*Hint:* What does $(\partial \alpha/\partial \beta)_{N,V}$ equal in general thermodynamics?]

Using the above result for $\partial \alpha/\partial \beta$, show that $(\partial E/\partial \beta)_{N,V} < 0$ for fermions and/or bosons. [*Hint:* Use Schwarz's inequality: $\langle f^2 \rangle \langle g^2 \rangle \geq \langle fg \rangle^2$ (Margenau and Murphy, p. 135 [23]).]

6-10. Show that, with the pressure in a single quantum state i, defined by $P_i = -\partial \epsilon_i/\partial V$, the relation $PV = \tfrac{2}{3}E$ holds for any system of independent particles (fermions, bosons, or boltzons) for which $\epsilon_i = a_i/V^{2/3}$, where a_i depends on the quantum number i. (Note that the relation $\epsilon_i = a_i/V^{2/3}$ does hold for the particle in a box energy levels.)

7

Statistical Mechanics and Chemical Equilibrium

7-1. The Law of Mass Action. We shall investigate the distribution of atoms between reactants and products in a chemical equilibrium by the general statistical method of the preceding chapter. It is sufficient to consider a simple dissociation reaction

$$AB_e \rightleftharpoons A + eB \tag{7-1}$$

where e is a small integer. We shall talk of A and B as atoms, but they could also be polyatomic radicals. For simplicity denote the compound AB_e by C. Each molecular species C, A, and B has its own set of energy levels, degeneracies, and occupation numbers for a particular distribution, as displayed below:

Energy levels................	$\epsilon_i'^C$	ϵ_j^A	ϵ_k^B
Degeneracies................	g_i^C	g_j^A	g_k^B
Occupation numbers.........	N_i^C	N_j^A	N_k^B

For convenience in what follows, the energy levels of C are primed, $\epsilon_i'^C$. The different sets of energy levels ϵ^A, ϵ^B, and ϵ'^C must be measured from a common base. Thus, for A, we take as the zero of energy its lowest energy level in which it has zero translational energy* and is in its lowest energy state as regards whatever internal degrees of freedom it possesses—electronic, vibrational, and rotational. We make the same choice for the energy levels of B.

If the compound C is stable, there is a positive energy of dissociation into $A + eB$ atoms. Let $\Delta\epsilon_0$ be the energy of dissociation of C from its lowest level, in which it has no translational, rotational, vibrational, or electronic excitation, into the atoms A and B, with the atoms in their lowest levels. Then, with respect to the dissociated atoms, the lowest

---

* For an atom, the lowest translational energy level is actually one with $s_x = s_y = s_z = 1$ and $\epsilon = 3h^2/8mV^{2/3}$. This is so small that it may be called zero.

level of molecule C, $\epsilon_0'^C$, is given by

$$\epsilon_0'^C = -\Delta\epsilon_0 \tag{7-2}$$

The quantity $\Delta\epsilon_0$ is positive and $\epsilon_0'^C$ is negative.

It is necessary to choose $\epsilon_0'^C$ in this way for the following reason: When an A atom in its lowest state reacts with e B atoms, each one in its lowest state, to form a C molecule, there is a release of $\Delta\epsilon_0$ of chemical energy. Thus the equation which expresses the conservation of total energy when the number of molecules of A, B, and C can vary owing to the chemical reaction must include the chemical energy. This objective is achieved by measuring the energy levels of C with respect to the separated atoms.

The total-energy equation is therefore

$$E = \Sigma N_j^A \epsilon_j^A + \Sigma N_k^B \epsilon_k^B + \Sigma N_i^C \epsilon_i'^C$$

From the point of view of chemical equilibrium, we are principally interested in the total number of C-type molecules, $\Sigma N_i^C = N_C$, and the number of free atoms of each kind, $\Sigma N_j^A = N_A$ and $\Sigma N_k^B = N_B$. The total number of A atoms, free and in the compound, we call n_A, and similarly the total number of B atoms is called n_B. That is,

$$n_A = N_A + N_C \tag{7-3a}$$
$$n_B = N_B + eN_C \tag{7-3b}$$

Equation (7-3b), for example, says that the total number of B atoms, n_B, is the sum of the free B atoms, N_B, plus e times the number of AB_e molecules.

The A atoms are indistinguishable, and one cannot tell which A atoms are free and which are in the compound C. For simple results, it is necessary to assume that all the particles are corrected boltzons.

We propose to maximize the total number of system wave functions, t, available to a system with fixed energy and fixed n_A and n_B. Clearly $t = t_A t_B t_C$, where t_A is the number of possible ways of having the distribution $N_1^A, N_2^A, \ldots, N_j^A, \ldots$ of N_A A atoms among the energy levels of A; the same is true for t_B and t_C. Then, as shown by Eq. (6-10), for corrected boltzons,

$$t = \prod_j \frac{(g_j^A)^{N_j^A}}{N_j^A!} \prod_k \frac{(g_k^B)^{N_k^B}}{N_k^B!} \prod_i \frac{(g_i^C)^{N_i^C}}{N_i^C!} \tag{7-4a}$$

$$\ln t = \sum_j N_j^A \left(\ln \frac{g_j^A}{N_j^A} + 1\right) + \sum_k N_k^B \left(\ln \frac{g_k^B}{N_k^B} + 1\right)$$
$$+ \sum_i N_i^C \left(\ln \frac{g_i^C}{N_i^C} + 1\right) \tag{7-4b}$$

Sec. 7-1]   STATISTICAL MECHANICS AND CHEMICAL EQUILIBRIUM

The crux of the problem is the constraints, as stated below:

Equation	Comment	Lagrange multiplier	
$\sum_i N_i^C + \sum_j N_j^A = n_A$	Conservation of A atoms	α^A	(7-5a)
$e \sum_i N_i^C + \sum_k N_k^B = n_B$	Conservation of B atoms	α^B	(7-5b)
$\sum_i N_i^C \epsilon_i'^C + \sum_j N_j^A \epsilon_j^A + \sum_k N_k^B \epsilon_k^B = E$	Conservation of energy	$-\beta$	(7-5c)

We have already discussed and justified these three equations for E, n_A and n_B.

The three sets of equations for the most probable distribution are

$$\frac{\partial(\ln t)}{\partial N_j^A} + \alpha^A - \beta \epsilon_j^A = 0 \qquad \text{for all } j$$

$$\frac{\partial(\ln t)}{\partial N_k^B} + \alpha^B - \beta \epsilon_k^B = 0 \qquad \text{for all } k \qquad (7\text{-}6)$$

$$\frac{\partial(\ln t)}{\partial N_i^C} + (\alpha^A + e\alpha^B) - \beta \epsilon_i'^C = 0 \qquad \text{for all } i$$

or

$$\ln \frac{g_j^A}{N_j^A} + \alpha^A - \beta \epsilon_j^A = 0$$

$$\ln \frac{g_k^B}{N_k^B} + \alpha^B - \beta \epsilon_k^B = 0 \qquad (7\text{-}7)$$

$$\ln \frac{g_i^C}{N_i^C} + (\alpha^A + e\alpha^B) - \beta \epsilon_i'^C = 0$$

or

$$\begin{aligned} N_j^A &= g_j^A e^{\alpha^A} e^{-\beta \epsilon_j^A} \\ N_k^B &= g_k^B e^{\alpha^B} e^{-\beta \epsilon_k^B} \\ N_i^C &= g_i^C e^{\alpha^A} e^{e\alpha^B} e^{-\beta \epsilon_i'^C} \end{aligned} \qquad (7\text{-}8)$$

We know from general arguments that $\beta = 1/kT$. Note that the factor $e^{\alpha^A + e\alpha^B}$ occurs in the expression for N_i^C. By summing each of the three expressions in (7-8), we obtain

$$\begin{aligned} N_A &= \sum_j N_j^A = e^{\alpha^A} \sum_j g_j^A e^{-\beta \epsilon_j^A} = e^{\alpha^A} q_A \\ N_B &= \sum_k N_k^B = e^{\alpha^B} \sum_k g_k^B e^{-\beta \epsilon_k^B} = e^{\alpha^B} q_B \\ N_C &= \sum_i N_i^C = e^{\alpha^A} (e^{\alpha^B})^e \sum_i g_i^C e^{-\beta \epsilon_i'^C} = e^{\alpha^A} (e^{\alpha^B})^e q_C' \end{aligned} \qquad (7\text{-}9)$$

The partition functions, such as $q_A = \sum_j g_j^A e^{-\beta \epsilon_j^A}$, have been substituted in Eqs. (7-9). We can eliminate the α's by combining Eqs. (7-9) in such a way as to form the function $N_A(N_B)^e/N_C$:

$$\frac{N_A(N_B)^e}{N_C} = \frac{q_A(q_B)^e}{q'_C} = \frac{(\Sigma g_j^A e^{-\beta \epsilon_j^A})(\Sigma g_k^B e^{-\beta \epsilon_k^B})^e}{\Sigma g_i^C e^{-\beta \epsilon_i'^C}} \quad (7\text{-}10)$$

Equation (7-10) is essentially the law of mass action, which we have now derived from statistical mechanics. As written, it deals with total amounts, N_A, etc., rather than concentrations or partial pressures, but we shall see that the equation is readily transformed to these more familiar forms.

It should also be noted that, according to the distribution laws (7-8), the relative distribution of, for example, A atoms among the various levels ϵ_j^A is a Boltzmann distribution, which is not influenced by the presence of B- and C-type particles.

7-2. A Change in the Zero of Energy. The partition function q'_C is defined in (7-9),

$$q'_C = \sum_{i=0}^{\infty} g_i^C e^{-\beta \epsilon_i'^C} \quad (7\text{-}11a)$$

Let $\epsilon_i^C = \epsilon_i'^C - \epsilon_0'^C$; ϵ_i^C is the height of the ith level above the ground state $\epsilon_0'^C$ of the molecule C. By substitution in (7-11a)

$$q'_C = e^{-\beta \epsilon_0'^C} \Sigma g_i^C e^{-\beta \epsilon_i^C} \quad (7\text{-}11b)$$

We set $q_C = \Sigma g_i^C e^{-\beta \epsilon_i^C}$; q_C is the partition function for C molecules, with the lowest level of C taken as zero energy. Now set

$$\Delta \epsilon_0 = -\epsilon_0'^C$$
then
$$q'_C = e^{\beta \Delta \epsilon_0} q_C \quad (7\text{-}12)$$

This is the expression for the partition function of the molecule, the separated atoms being taken as the zero of energy, in terms of the binding energy $\Delta \epsilon_0$ and the partition function of the molecule, q_C, evaluated with respect to its own zero of energy.

With this relation, Eq. (7-10) becomes

$$\frac{N_A(N_B)^e}{N_C} = e^{-\beta \Delta \epsilon_0} \frac{q_A(q_B)^e}{q_C} \quad (7\text{-}13)$$

7-3. Partition Functions for Translational and Internal Degrees of Freedom. As noted in Sec. (3-10), the Hamiltonian and the energy for a molecule or an atom are strictly and rigorously separable into the sum of a translational term and an internal term, $H = H_{tr} + H_{int}$; $\epsilon_{ij} = \epsilon_{i,tr} + \epsilon_{j,int}$. There are a set of translational energy levels $\epsilon_{i,tr}$ and a set of internal energy levels $\epsilon_{j,int}$. The total energy level is denoted by a double subscript, ϵ_{ij}. A very important point is that any value of i can occur with any value of j; therefore, $g_{ij} = g_i g_j$. For the total partition function

$$q = \sum_{i,j} g_{ij} e^{-\beta \epsilon_{ij}} = \sum_{i,j} g_i g_j e^{-\beta \epsilon_i} e^{-\beta \epsilon_j} = \left(\sum_i g_i e^{-\beta \epsilon_i}\right)\left(\sum_j g_j e^{-\beta \epsilon_j}\right) = q_{tr} q_{int} \quad (7\text{-}14)$$

7-4. Transformation of the Equilibrium Expression.

According to (7-14), the partition function is a product of the translational and internal partition functions. Recall that $q_{\text{trans}} = (2\pi m kT/h^2)^{3/2} V$. For any particular species, say, C, we can then write

$$q_C = q_{C,\text{trans}} q_{C,\text{int}} = \left(\frac{2\pi m_C kT}{h^2}\right)^{3/2} V q_{C,\text{int}} \tag{7-15}$$

By substitution in (7-13), and bringing all the V factors to the left-hand side,

$$\frac{(N_A/V)(N_B/V)^e}{N_C/V} = e^{-\beta \Delta \epsilon_0} \frac{[(2\pi m_A kT/h^2)^{3/2} q_{A,\text{int}}][(2\pi m_B kT/h^2)^{3/2} q_{B,\text{int}}]^e}{(2\pi m_C kT/h^2)^{3/2} q_{C,\text{int}}} \tag{7-16a}$$

$$\frac{(N_A/V)(N_B/V)^e}{N_C/V} = K_a \tag{7-16b}$$

We recognize that we now have a conventional equilibrium expression with concentrations expressed in units of atoms per cubic centimeter, and the equilibrium constant K_a is given by the right-hand side of (7-16a).

To obtain an equilibrium expression in pressure units, we write

$$\frac{N_A}{V} = \frac{P_A}{kT} = \frac{P_A}{P_0} \frac{P_0}{kT} \quad \text{etc.} \tag{7-17}$$

where P_A is the pressure in cgs units (dynes per square centimeter) and, as previously discussed, the ratio P_A/P_0 is numerically equal to the pressure in atmospheres if P_0 is the value of the standard atmosphere in cgs units. Then, by substitution in (7-16a),

$$\frac{(P_A/P_0)(P_B/P_0)^e}{P_C/P_0}$$
$$= e^{-\beta \Delta \epsilon_0} \frac{[(2\pi m_A kT/h^2)^{3/2}(kT/P_0) q_{A,\text{int}}][(2\pi m_B kT/h^2)^{3/2}(kT/P_0) q_{B,\text{int}}]^e}{(2\pi m_C kT/h^2)^{3/2}(kT/P_0) q_{C,\text{int}}} \tag{7-18a}$$

$$\frac{(P_A/P_0)(P_B/P_0)^e}{P_C/P_0} = K_P \tag{7-18b}$$

We thus have an expression for the equilibrium constant in pressure units.

Let $[A]$ represent concentration in units of moles per liter; then

$$[A] = \frac{1{,}000}{N_0} \frac{N_A}{V} \tag{7-19}$$

where N_0 is Avogadro's number. The equilibrium expression is now transformed to

$$\frac{[A][B]^e}{[C]} = K_C$$
$$= e^{-\beta \Delta \epsilon_0} \frac{[(2\pi m_A kT/h^2)^{3/2}(1{,}000/N_0) q_{A,\text{int}}][(2\pi m_B kT/h^2)^{3/2}(1{,}000/N_0) q_{B,\text{int}}]^e}{(2\pi m_C kT/h^2)^{3/2}(1{,}000/N_0) q_{C,\text{int}}}$$
$$\tag{7-20}$$

We have explicitly exhibited the expression for the equilibrium constant for several systems of units which are commonly used; it is clear that similar expressions for any other pressure or concentration units can be readily obtained.

7-5. The Free Energy and q/N for Different Standard States. The partition function is a product of the translational and internal partition functions. The thermodynamic functions depend on the ratio q/N, and we write

$$\frac{q}{N} = \frac{q_{tr}q_{int}}{N} = \frac{q_{tr}}{N} q_{int} = \left[\left(\frac{2\pi mkT}{h^2}\right)^{3/2} \frac{V}{N}\right] q_{int} \quad (7\text{-}21)$$

We choose to associate the factor $1/N$ with q_{tr}; it is clear by inspection of (7-21) that the quantity q_{tr}/N is an intensive quantity which depends on the concentration N/V, whereas q_{tr} itself is proportional to V and independent of the number of particles therein.

The free energy is given by

$$F = -NkT \ln \frac{q}{N} = -NkT \ln \frac{q_{tr}}{N} - NkT \ln q_{int} = F_{tr} + F_{int} \quad (7\text{-}22)$$

The translational and internal free energies are implicitly defined by (7-22). In a similar way, the other thermodynamic functions can be expressed as a sum of contributions by the translational and internal energy modes.

From thermodynamics, we could write the equilibrium expression for the reaction of the ideal gases at temperature T:

$$AB_e = A + eB$$

$$K = e^{-\Delta E_0°/RT} \frac{e^{-(F_{TA}° - F_{0A}°)/RT} e^{-e(F_{TB}° - F_{0B}°)/RT}}{e^{-(F_{TC}° - F_{0C}°)/RT}} \quad (7\text{-}23)$$

In a conventional, purely thermodynamic discussion we would take $\Delta E_0°$ as the energy of reaction per mole of the actual substances AB_e, A, and B, presumably solids, at $0°K$. The free-energy function $F_{TA}° - F_{0A}°$ would then represent the change in the free energy of component A in going from the solid at $0°K$ to the chosen standard state of the ideal gas at temperature T.

We can, however, let $\Delta E_0°$ represent the hypothetical energy of reaction of the gases at $0°K$, that is, the energy change when 1 mole of gaseous AB_e with no translational, rotational, or electronic energy dissociates to A and eB atoms, each with no energy of any sort. Clearly, for this choice of $\Delta E_0°$, $\Delta E_0° = N_0 \Delta \epsilon_0$, where $\Delta \epsilon_0$ is the corresponding energy of dissociation per molecule, and the factors $e^{-\Delta E_0°/RT}$ and $e^{-\Delta \epsilon_0/kT}$ in the equilibrium expressions are identical.

With this latter choice of $\Delta E_0°$, the free-energy functions $F_{TA}° - F_{0A}°$ represent the free-energy change in heating the gas from its condition of zero translational and internal energy at 0°K to its condition in the standard state at temperature T. If the partition function q is calculated by taking the lowest energy state of the molecule as zero, then the free-energy function defined with reference to the gas at 0°K is given by

$$F_{TA}^\circ - F_{0A}^\circ = N_0 kT \ln \frac{q^\circ}{N} \tag{7-24}$$

In general

$$\frac{q}{N} = \left(\frac{2\pi mkT}{h^2}\right)^{3/2} \frac{V}{N} q_{int} \tag{7-25}$$

We denote q/N for the standard state of 1 atom cc^{-1} by q^a/N. Since $V/N = 1$ for this standard state,

$$\frac{q^a}{N} = \left(\frac{2\pi mkT}{h^2}\right)^{3/2} q_{int} \tag{7-26}$$

For the standard state of 1 atm, $V/N = kT/P_0$, and

$$\frac{q^\circ}{N} = \left(\frac{2\pi mkT}{h^2}\right)^{3/2} \frac{kT}{P_0} q_{int} \qquad \Rightarrow K_p \tag{7-27}*$$

For the standard state of 1 mole liter^{-1}, $V/N = 1{,}000/N_0$, and

$$\frac{q^c}{N} = \left(\frac{2\pi mkT}{h^2}\right)^{3/2} \frac{1{,}000}{N_0} q_{int} \tag{7-28}$$

We then have

$$F_T^a - F_0^a = -N_0 kT \ln \frac{q^a}{N} \tag{7-29a}$$

$$F_T^\circ - F_0^\circ = -N_0 kT \ln \frac{q^\circ}{N} \tag{7-29b}$$

$$F_T^c - F_0^c = -N_0 kT \ln \frac{q^c}{N} \tag{7-29c}$$

When such equations for the free-energy functions are substituted into the thermodynamic-equilibrium-constant expression (7-23), we obtain

$$K = e^{-\Delta E_0^\circ/RT} \frac{(q_A^\circ/N)(q_B^\circ/N)^e}{q_C^\circ/N} \tag{7-30}$$

(where the degree sign now stands for any chosen standard state). We see that the expressions for the equilibrium constant for several different concentration units previously derived [(7-16), (7-18), and (7-20)] are examples of this general relation.

7-6. Other Chemical Reactions. The results obtained in this chapter are not limited to dissociation reactions. By appropriate combination

---

* In our preceding remarks, we have used the degree sign to indicate an arbitrary standard state: q°/N, F°, etc. We now also use it in the conventional way for the standard state of 1 atm.

of such dissociation reactions, one can obtain any other chemical reaction. For the general chemical reaction $\Sigma \nu_i \{i\} = 0$ (returning now to the notation of Sec. 4-13), where $\{i\}$ denotes the ith chemical species and ν_i is a stoichiometric coefficient, we have

$$\prod_i [i]^{\nu_i} = K = e^{-\Delta E_0°/RT} \prod_i \left(\frac{q_i°}{N}\right)^{\nu_i} \qquad (7\text{-}31)$$

where $[i]$ is concentration or pressure of species $\{i\}$ in some suitable units and $q_i°/N$ is evaluated at the standard state at which $[i] = 1$.

PROBLEM

7-1. Let the energy states of a molecule, $\epsilon_0, \epsilon_1, \ldots, \epsilon_i$ and $\epsilon_{i+1}, \ldots$, be divided into the A subgroup, $\epsilon_0, \ldots, \epsilon_i$, and the B subgroup, $\epsilon_{i+1}, \ldots, \epsilon_\infty$. Derive an expression for the probability of finding a particle in subgroup A or in subgroup B in terms of the partition functions q_A and q_B for the subgroups of states. (This simple problem illustrates an important point of view about the meaning of the partition function.)

8

Distribution Laws, Partition Functions, and Thermodynamic Functions for Atoms and Diatomic Molecules

8-1. The Distribution Law and Thermodynamic Functions for Separable Degrees of Freedom. The energy of a molecule or an atom can usually be expressed as a sum of energies in different degrees of freedom—translational, rotational, vibrational, and electronic energy for the case of a diatomic molecule. Denote a set of such separable kinds of motion by a, b, c and the energy levels by ϵ_{ia}, ϵ_{jb}, ϵ_{kc}. The total energy is given by $\epsilon_{ijk} = \epsilon_{ia} + \epsilon_{jb} + \epsilon_{kc}$. The quantum numbers i, j, and k are independent, so that any value of i can occur with any values of j and k. The degeneracy or number of wave functions for the composite state ijk is the product of the degeneracies of the states i, j, k for the separate degrees of freedom.

$$g_{ijk} = g_i g_j g_k$$

It has been already shown in Sec. 7-3 that the partition function is the product of the partition functions for the separate degrees of freedom:

$$q = q_a q_b q_c \tag{8-1}$$

Having emphasized the generality of the situation, we can now, for simplicity, write formulae for only two kinds of energy.

The distribution law is

$$\frac{N_{ij}}{N} = \frac{g_{ij} e^{-\beta \epsilon_{ij}}}{q} = \frac{g_i e^{-\beta \epsilon_i} g_j e^{-\beta \epsilon_j}}{q_a q_b} \tag{8-2}$$

By summing over all values of j

$$\frac{\sum_j N_{ij}}{N} = \frac{g_i e^{-\beta \epsilon_i} \sum_j g_j e^{-\beta \epsilon_j}}{q_a q_b} = \frac{g_i e^{-\beta \epsilon_i}}{q_a} \tag{8-3}$$

The quantity $\sum_j N_{ij}$ is the total number of molecules in the ith state for the

a degree of freedom irrespective of their energy in the b degree of freedom; the fractional number of such molecules is the conventional Boltzmann expression, $g_i e^{-\beta \epsilon_i}/q_a$, for the a degree of freedom.

By summing (8-2) over all values of i, we can then form the function $\sum_i N_{ij}/N$, analogous to (8-3), then divide N_{ij}/N by $\sum_i N_{ij}/N$, and obtain

$$\frac{N_{ij}}{\sum_i N_{ij}} = \frac{g_i e^{-\beta \epsilon_i}}{q_a} \quad \text{(for all } j\text{)} \tag{8-4}$$

We are now dealing with all the molecules that are in a particular state j of the b degree of freedom; the fractional number of such molecules which are in the state i for the a degree of freedom is also given by $g_i e^{-\beta \epsilon_i}/q_a$.

Thus, to the extent that the energy of a molecule can be expressed as a sum of energies of translation, rotation, vibration, and electronic energy, the probability that a molecule will be in a certain energy level with respect to one degree of freedom is entirely independent of the energy it has in other degrees of freedom.

To obtain expressions for the thermodynamic functions, we first note that, for the problem at hand, the molecule has translational energy and internal energy. As noted in Sec. 7-3, the partition function can be factored and written as

$$q = q_{tr} q_a q_b \tag{8-5}$$

where $q_{tr} = (2\pi mkT/h^2)^{3/2} V$ and q_a and q_b are partition functions for internal degrees of freedom.

The thermodynamic functions depend on q/N; as already remarked in Sec. 7-5, we associate the factor $1/N$ with q_{tr} and write

$$\frac{q}{N} = \frac{q_{tr}}{N} q_a q_b \tag{8-6}$$

The free energy is

$$F = -NkT \ln \frac{q}{N} = -NkT \ln \frac{q_{tr}}{N} - NkT \ln q_a - NkT \ln q_b$$
$$= F_{tr} + F_a + F_b \tag{8-7}$$

where
$$F_{tr} = -NkT \ln \frac{q_{tr}}{N} \tag{8-8a}$$

$$F_a = -NkT \ln q_a \qquad F_b = -NkT \ln q_b \tag{8-8b}$$

The Helmholtz free energy is

$$A = -NkT \left(\ln \frac{q}{N} + 1 \right) = -NkT \left(\ln \frac{q_{tr}}{N} + 1 \right) - NkT \ln q_a$$
$$- NkT \ln q_b$$
$$= A_{tr} + A_a + A_b \tag{8-9}$$

where $A_{tr} = -NkT\left(\ln\dfrac{q_{tr}}{N} + 1\right)$ $\quad A_a = -NkT \ln q_a = F_a$

$$A_b = -NkT \ln q_b = F_b \tag{8-10}$$

The total energy (or thermodynamic energy function) is

$$E = \tfrac{3}{2}NkT + NkT^2 \frac{d(\ln q_a)}{dT} + NkT^2 \frac{d(\ln q_b)}{dT}$$
$$= E_{tr} + E_a + E_b \tag{8-11}$$

where $\quad E_{tr} = \tfrac{3}{2}NkT \quad E_a = NkT^2 \dfrac{d(\ln q_a)}{dT} \quad E_b = NkT^2 \dfrac{d(\ln q_b)}{dT}$
$$\tag{8-12}$$

The entropy is

$$S = \frac{E - A}{T} = S_{tr} + S_a + S_b \tag{8-13}$$

where $\quad S_{tr} = \tfrac{5}{2}Nk + Nk \ln \dfrac{q_{tr}}{N} \quad S_a = NkT \dfrac{d(\ln q_a)}{dT} + Nk \ln q_a$
$$\tag{8-14}$$

$$S_b = NkT \frac{d(\ln q_b)}{dT} + Nk \ln q_b$$

It should be noted that the formulae for F_{tr}, A_{tr}, and S_{tr} are in terms of q_{tr}/N and differ from the formulae for F_a, A_a, and S_a, which involve $\ln q_a$.

8-2. Energy Levels in Wave Numbers. The frequency ν_{ij} of the light quantum emitted in a radiative transition between states i and j is given by $h\nu_{ij} = \epsilon_i - \epsilon_j$. The wave number

$$\omega \text{ (reciprocal centimeters)} = \frac{1}{\lambda} = \frac{\nu}{c}$$

(where c = velocity of light) is given by $hc\omega_{ij} = \epsilon_i - \epsilon_j$. In most tabulations, the values of ϵ_j and ϵ_i are themselves given in wave numbers, and $\epsilon_j = hc\omega_j$. The Boltzmann factor involves ϵ/kT or $hc\omega/kT$. The quantity hc/k is known as the second radiation constant; its value is 1.4388 deg cm.

In spectroscopic books, one finds both ω and ν (reciprocal centimeters) used as the symbol for a quantity in units of wave numbers.

8-3. The Electronic Partition Function of Atoms. The only internal energy that an atom has is electronic, and the internal partition function is obtained by summing over these levels. (We shall see later that there may also be a set of nuclear spin energy levels but that usually these can be consistently ignored.) In general, the energy levels have been determined by spectroscopy and are tabulated in suitable references.*

* Moore [30].

We briefly review spectroscopic notation, to aid in the use of these tables. A more complete discussion of atomic energy levels is given in several of the general references.*

The electronic energy levels are called terms and have a term symbol $^{2S+1}L_J$. S is the electron spin quantum number, and the square of the spin angular momentum is $S(S+1)\hbar^2$. L is the orbital-angular-momentum quantum number; the square of the orbital angular momentum is $L(L+1)\hbar^2$. Terms with $L = 0, 1, 2, 3, 4, 5$ are called S, P, D, F, G, H, respectively. If both S and L are nonzero, they add together as vectors in a quantized way to give a resulting total angular-momentum quantum number J, the square of the total angular momentum being $J(J+1)\hbar^2$. There are $2J + 1$ allowed components of the angular momentum along

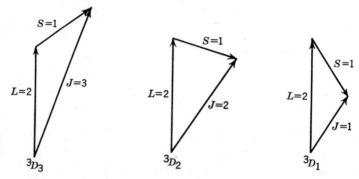

FIG. 8-1. Vector addition of $L = 2$ and $S = 1$ to give $^3D_3$ ($J = 3$), $^3D_2$ ($J = 2$), and $^3D_1$ ($J = 1$) terms.

an axis fixed in space, $J\hbar$, $(J-1)\hbar$, ..., $-J\hbar$. If the external electric or magnetic field is vanishingly weak, these various components have almost the same energy; so the degeneracy of a state with a given J is $g_J = 2J + 1$.

L and S can add together in several different quantized ways. If $L > S$, the possible values of J are $L + S, L + S - 1, \ldots, L - S$. If $S > L$, the values of J are $L + S, L + S - 1, S - L$. This so-called vector model of the atom is depicted schematically in Fig. 8-1.

Because of weak electromagnetic interactions called *spin-orbit* coupling, the different relative orientations of L and S which give rise to the different values of J have slightly different energies. Thus, in the figure, $^3D_3$, $^3D_2$, and $^3D_1$ are different terms of the spin-orbit multiplet with different energies. These differences are small for light atoms but larger as the atomic number increases. The allowed values of L are integers; the values of S are integers or half integers, for even and odd numbers of electrons, respectively. The J's are integers or half integers depending on the S's. In all cases for a free atom it is found that the wave functions

* Especially Herzberg [16].

are eigenfunctions of $(\hat{\mathbf{L}} + \hat{\mathbf{S}})^2$ (where $\hat{\mathbf{L}}$ and $\hat{\mathbf{S}}$ are the vector angular-momentum operators) and can be chosen to be eigenfunctions of $\hat{L}_z + \hat{S}_z$; that is, J and J_z are "good" quantum numbers, and the total squared angular momentum and the component along an axis fixed in space can be known with certainty. However, when the spin-orbit coupling is large, the wave functions are not eigenfunctions of $\hat{L}^2$ and $\hat{L}_z$, $\hat{S}^2$ and $\hat{S}_z$; that is, S and L are not separately "good" quantum numbers, but the term symbols are still given because they have approximate meaning.

For many atoms, there is often a low-lying set of terms which are the members of a spin-orbit multiplet; the next energy level is quite high and need not be included in the calculation of the partition function except at very high temperatures (cf. Prob. 8-3). In spectroscopic notation, Li(I) means a neutral lithium atom, Li(II) means singly ionized lithium, Li(III) means the double-charged ion, etc. The ionization potentials are given in the tabulations and are needed for calculating ionization equilibria.

To illustrate these features, the low-lying terms of some of the light atoms are displayed in Table 8-1.

The electronic partition function is $q = \Sigma g_i e^{-\epsilon_i/kT}$. At low temperatures only several terms in the sum need be considered. For the simple case of two levels each with $g = 1$, $q = 1 + e^{-\epsilon_1/kT}$; Prob. 8-2 is concerned with the variation with temperature of the partition function and associated thermodynamic functions. This is an important example, and the student should work it out.

A given L and S give rise to a spin-orbit multiplet with a set of J values from $|L + S|$ to $|L - S|$. For each value of J there are $2J + 1$ wave functions. The total number of wave functions for all the states of the spin-orbit multiplet is

$$\text{No. of wave functions} = \sum_{|L-S|}^{|L+S|} (2J + 1) = (2L + 1)(2S + 1)$$

The equality above can be proved mathematically (since it is an arithmetical series) or understood physically as follows. If the spin-orbit coupling is weak, the L and the S vectors can be uncoupled by a strong magnetic field. They are then independently oriented in the field. Since there are $2S + 1$ ways of orienting the spin vector and $2L + 1$ ways of orienting the orbital vector, the total number of such wave functions is $(2S + 1)(2L + 1)$. For the high-temperature case, where $\epsilon/kT \ll 1$ for all the levels in a multiplet, $e^{-\epsilon/kT} \approx 1$ for all these terms in the partition function. If, furthermore, the next energy level is very high ($\epsilon/kT \gg 1$), its contribution to the electronic partition function is negligible, and the simple relation $q = (2S + 1)(2L + 1)$ holds. This limiting behavior can be observed in several of the problems.

Table 8-1. Atomic Energy States*

Atom	Electron configuration	Term symbol	$g = 2J + 1$	Energy, cm^{-1}
H(I)	$1s^1$	$^2S_{1/2}$	2	0
	$2p^1$	$^2P_{1/2}$	2	82,258.907
	$2s^1$	$^2S_{1/2}$	2	82,258.942
	$2p^1$	$^2P_{3/2}$	4	82,259.272
	Ionization potential = 109,678.758 cm^{-1} = 13.595 volts			
He(I)	$1s^2$	$^1S_0$	1	0
	$1s2s$	$^3S_1$	3	159,850.318
		$^1S_0$	1	166,271.70
	Ionization potential = 198,305 cm^{-1} = 24.580 volts			
Li(I)	$1s^2 2s^1$	$^2S_{1/2}$	2	0
	$1s^2 2p^1$	$^2P_{1/2}$	2	14,903.66
	$1s^2 2p^1$	$^2P_{3/2}$	4	14,904.00
	Ionization potential = 43,487.19 cm^{-1} = 5.390 volts			
C(I)	$1s^2 2s^2 2p^2$	$^3P_0$	1	0
		$^3P_1$	3	16.4
		$^3P_2$	5	43.5
	$1s^2 2s^2 2p^2$	$^1D_2$	5	10,193.70
	Ionization potential = 90,878.3 cm^{-1} = 11.264 volts			
O(I)	$1s^2 2s^2 2p^4$	$^3P_2$	5	0
		$^3P_1$	3	158.5
		$^3P_0$	1	226.5
	$1s^2 2s^2 2p^4$	$^1D_2$	5	15,867.7
	Ionization potential = 109,836.7 cm^{-1} = 13.614 volts			

* Data from C. E. Moore, Atomic Energy States, *Natl. Bur. Standards Circ.* 467, vols. I–III, 1949, 1952, 1958.

8-4. Diatomic Molecules—Introduction. The diatomic molecule AB contains two nuclei and n electrons. The Schrödinger equation for $n + 2$ particles cannot in general be solved. A simplifying approximation is possible because of the great differences in masses of the nuclei and the electrons. Let r_{AB} be the internuclear distance. As the atoms A and B are brought together from infinity, the electrons may interact in such a way that the electronic energy is the function $U(r_{AB})$ plotted in Fig. 8-2. Thus, for larger r_{AB}, the two atoms attract each other; there is a minimum in the potential-energy curve at the equilibrium internuclear distance $r_e(AB)$. At shorter distances, A and B repel each other. The curve $U(r)$ is the electronic potential-energy curve for one particular electronic state. The motion of the nuclei in the potential $U(r)$ then gives a set of quantized energy levels depicted by the horizontal lines in the figure. These nuclear motions can be approximately represented as the rotation of a rigid dumbbell rotor plus the relative vibration of the

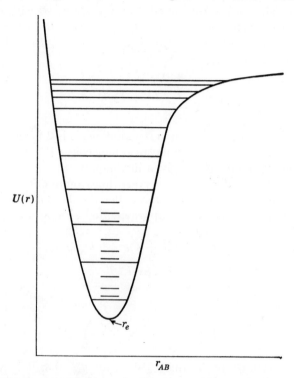

FIG. 8-2. Electronic energy $U(r)$ as a function of the internuclear distance r_{AB}. The quantized energy levels due to nuclear motions are depicted by the horizontal lines. The long lines represent the vibrational energy levels; the short lines stand for the rotational energy levels. Only a few of the rotational energy levels are shown, and these only for the three lowest vibrational levels.

masses m_A and m_B in the potential curve $U(r_{AB})$. The quantum-mechanical problem is treated in almost all books on quantum mechanics.

8-5. The Rigid Rotor. Two masses m_A and m_B at a fixed internuclear distance r_e have a moment of inertia $I_e = \mu r_e^2$, where μ is the reduced mass $m_A m_B/(m_A + m_B)$. The allowed quantized values for the square of the angular momentum of such a free rotor in three dimensions are $J(J+1)\hbar^2$, where $J = 0, 1, 2, 3, \ldots$. The corresponding energy levels are given by

$$\epsilon_J = \frac{J(J+1)\hbar^2}{2I_e} \qquad (8\text{-}15)$$

The angular momentum can have components along a particular axis in space of $M\hbar$, where $M = J, J - 1, \ldots, -J$. The rotational state with energy ϵ_J is therefore $(2J + 1)$-degenerate. There are corresponding wave functions $\Theta_{JM}(\theta)e^{iM\phi}$, of which the first few were illustrated in Prob. 3-3.

Spectroscopists use the rotational-energy constant B with dimensions of reciprocal centimeters, $\epsilon_J = hcB_eJ(J + 1)$, where $B_e = h/8\pi^2 I_e c$. We write $\epsilon_J/kT = yJ(J + 1)$, where $y = \hbar^2/2I_e kT = hcB_e/kT = 1.4388B_e/T$. The partition function is then

$$q_{\text{rot}} = \frac{1}{\sigma} \sum_{J=0}^{J=\infty} (2J + 1)e^{-yJ(J+1)} \tag{8-16}$$

The factor σ in (8-16) is the symmetry number; it is 2 for a symmetrical molecule AA and 1 for an unsymmetrical molecule AB. It is introduced arbitrarily now; justification is given in Chap. 9.

In Fig. 8-3 the heights of the vertical lines represent the individual terms of the sum in (8-16); the area under the dashed lines is the sum itself. If y is small, this may be approximated by the integral

$$q_{\text{rot}} = \frac{1}{\sigma} \int_{J=0}^{J=\infty} (2J + 1)e^{-J(J+1)y} \, dJ = \frac{1}{\sigma y} \tag{8-17}$$

$$q_{\text{rot}} = \frac{1}{\sigma} \frac{T}{1.4388 B_e} = \frac{8\pi^2 I_e kT}{\sigma h^2} \tag{8-18}$$

The expansion

$$q_{\text{rot}} = \frac{1}{\sigma y}\left(1 + \frac{y}{3} + \frac{y^2}{15} + \cdots\right) \tag{8-19}$$

gives a more accurate result than (8-17); for $y < 0.3$, it is accurate to 0.1 per cent or better. Expression (8-19) is derived in Prob. 8-8. For most diatomic molecules not containing a hydrogen atom, $B \leq 2$, and (8-17) is adequate at room temperature ($y \leq 0.01$). For HX molecules (X heavy) $B \sim 10$, and (8-19) may be needed at room temperature. For H_2, $B = 60$, and the best procedure is to sum the terms directly in the rotational partition function (but first read Chap. 9, and learn about ortho and para hydrogen).

When (8-18) applies,

$$\left(\frac{F - F_0^\circ}{T}\right)_{\text{rot}} = R \ln \sigma y \qquad \left(\frac{H - H_0}{T}\right)_{\text{rot}} = R$$
$$S_{\text{rot}} = R(1 - \ln \sigma y) \qquad C_{\text{rot}} = R \tag{8-20}$$

The distribution law for unsymmetrical molecules is

$$\frac{N_J}{N} = y(2J + 1)e^{-J(J+1)y} \tag{8-21}$$

(See Chap. 9 for the modifications for symmetrical molecules.) The student should become familiar with the way in which the population varies with J, with T, and with the moment of inertia by doing problems such as Probs. 8-4 to 8-6. The heights of the vertical lines in Fig. 8-3 are proportional to the populations of the various levels.

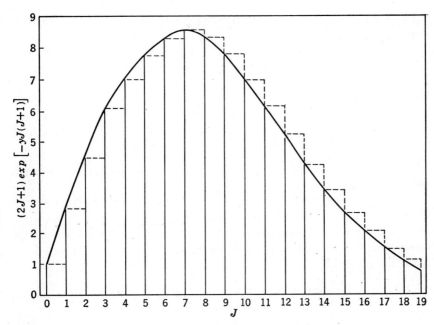

FIG. 8-3. Comparison of the sum $\sum_{j=0}^{\infty} (2J+1)e^{-yJ(J-1)}$ and the integral

$$\int_0^\infty (2J+1)e^{-J(J+1)y}\,dJ$$

The heights of the vertical lines represent the individual terms $(2J+1)e^{-yJ(J+1)}$. The area under the smooth curve is the integral; the area under the dashed lines is the sum. The curve was calculated (roughly) for $y = 0.01$.

Note that the enthalpy and heat capacity are the same (RT and R) for all rigid rotors in this approximation; however, the greater the moment of inertia, the greater the number of states which are significantly occupied, and the greater the partition function, the entropy, and the negative of the free energy.

8-6. Vibrational Energy Levels. We now consider the vibrational motion of the nuclei in the potential $U(r)$ of Fig. 8-2. In the region close to the minimum at $r = r_e$, the variation of the potential can be represented by the first nonzero term in a Taylor series,

$$U(r) = U(r_e) + \frac{1}{2}\left(\frac{\partial^2 U}{\partial r^2}\right)_{r=r_e}(r-r_e)^2$$
$$= U(r_e) + \frac{f}{2}(r-r_e)^2 \tag{8-22}$$

For a diatomic molecule with rotational-angular-momentum quantum number J and with the harmonic potential function

$$U(r) - U(r_e) = \tfrac{1}{2}f(r-r_e)^2$$

the radial wave equation is*

$$-\frac{\hbar^2}{2\mu_{AB}}\frac{1}{r^2}\frac{d}{dr}\left(r^2\frac{d\psi}{dr}\right) + \left[\tfrac{1}{2}f(r-r_e)^2 + \frac{\hbar^2}{2\mu_{AB}}\frac{J(J+1)}{r^2} - \epsilon\right]\psi = 0$$

The term involving $J(J+1)/r^2$ is the centrifugal-force term; it is small for small values of J.

If the potential function is reasonably steep (that is, f is large) so that the wave function is mainly confined to a small region around $r = r_e$, the above equation can be approximately transformed to the wave equation for a one-dimensional harmonic oscillator [Eq. (3-32)]. The energy levels are given by

$$\epsilon_v = h\nu(v + \tfrac{1}{2}) \qquad v = 0, 1, 2, \ldots \tag{8-23a}$$

with
$$\nu = \frac{1}{2\pi}\left(\frac{f}{\mu_{AB}}\right)^{1/2} \tag{8-23b}$$

Note that ν is the vibration frequency of a classical harmonic oscillator, with force constant f and mass μ. In the harmonic-oscillator approximation, there is a set of equally spaced energy levels, with the spacing $h\nu$.

For $v = 0$, the oscillator has the zero-point energy $\epsilon_0 = \tfrac{1}{2}h\nu$. Except for problems involving isotopic molecules, it is convenient to take, as energy zero, the state $v = 0$.

$$\epsilon_v - \epsilon_0 = h\nu v = hc\omega_e v \qquad v = 0, 1, 2, \ldots \tag{8-24}$$

ω_e is the vibration frequency in reciprocal centimeters; the subscript emphasizes that this is the harmonic-oscillator approximation and hence related to the curvature of $U(r)$ at $r = r_e$.

---

* For a further discussion of this equation, see any of the general references on quantum mechanics and/or molecular spectra.

The partition function is a geometrical series and hence is readily evaluated. Set $(\epsilon_v - \epsilon_0)/kT = uv$, where $u = h\nu/kT = 1.4388\omega_e/T$.

$$q = \sum_{v=0}^{\infty} e^{-vu} = \frac{1}{1 - e^{-u}} \quad (8\text{-}25)$$

$$\left(\frac{F - F_0^\circ}{T}\right)_{vib} = R \ln(1 - e^{-u}) \quad \left(\frac{H - H_0^\circ}{T}\right)_{vib}$$

$$= R \frac{u}{e^u - 1} = \frac{N_0 h\nu/T}{e^u - 1} \quad (8\text{-}26)$$

$$C_{vib} = R \frac{u^2 e^u}{(e^u - 1)^2} \quad S_{vib} = R\left[\frac{u}{e^u - 1} - \ln(1 - e^{-u})\right]$$

For computations, it is convenient to look up the values of the harmonic-oscillator thermodynamic functions in tabulations.* There is a very

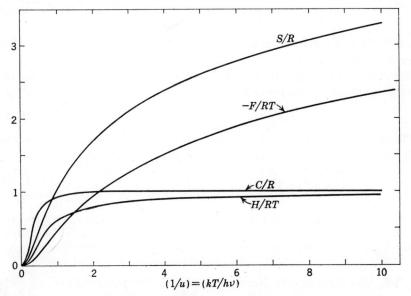

FIG. 8-4. The thermodynamic functions of a harmonic oscillator as a function of temperature.

short table in Appendix 2. The variation with temperature of the several functions is plotted in Fig. 8-4. At low T, the energy and heat capacity are both zero and the entropy zero. The first two then rise to their high-temperature limits, RT and R; but C rises faster than H/T (why?). At the high-temperature (classical) limits, C and H are independent of ω_e, but S and F are not.

* References in Sec. 8-10.

Useful approximate formulae for $u \ll 1$ are

$$q(u) = \frac{1}{u}\left(1 + \frac{u}{2} + \cdots\right)$$

$$-\frac{(F - F_0^\circ)}{T} = R \ln q = -R \ln u + \frac{Ru}{2} + \cdots \qquad (8\text{-}27)$$

$$H - H_0^\circ = \frac{\partial(F/T)}{\partial(1/T)} = RT\left(1 - \frac{u}{2} + \cdots\right)$$

For arbitrary u, the distribution among energy levels is given by

$$\frac{N_v}{N} = e^{-vu}(1 - e^{-u}) \qquad (8\text{-}28)$$

Very roughly speaking, the states which are significantly populated are all those states for which

$$vu \leq 1 \qquad \frac{\epsilon_v}{kT} = \frac{vh\nu}{kT} \leq 1 \qquad (8\text{-}29)$$

8-7. The Rigid-rotor Harmonic-oscillator Approximation for a Diatomic Molecule. To a first approximation, the energy levels of a diatomic molecule are the sum of the rotational and vibrational energy levels discussed in the previous sections.

$$\epsilon_{vJ} - \epsilon_0 = hc[\omega_e v + B_e J(J+1)] \qquad (8\text{-}30)$$

The partition function and the thermodynamic functions for the internal degrees of freedom of rotation plus vibration are

$$q = q_{\text{vib}} q_{\text{rot}} \qquad (8\text{-}31)$$
$$F = F_{\text{vib}} + F_{\text{rot}} \qquad E = E_{\text{vib}} + E_{\text{rot}} \qquad S = S_{\text{vib}} + S_{\text{rot}} \qquad \text{etc.}$$

Problem 8-1. The vibration frequency and rotational constant of HCl are $\omega_e = 2{,}990$ cm^{-1} and $B_e = 10.6$ cm^{-1}. Give a very rough qualitative sketch of the heat-capacity curve for the hypothetical perfect gas of this substance between 5 and 4000°K. Similarly, depict the qualitative behavior of the free-energy function.

8-8. The Anharmonic Nonrigid Rotor.* The potential-energy curve of a molecule in a particular electronic state as qualitatively sketched in Fig. 8-2 is not a parabola, accurately represented by $U(r - r_e) = \frac{1}{2}f(r - r_e)^2$. It is unsymmetrical around $r - r_e$, and it levels off at a finite height equal to the dissociation energy as $r - r_e$ increases. Correspondingly, the vibrational energy levels are not equally spaced; in actual cases, the spacing between the levels decreases as the vibrational quantum number increases. To illustrate these trends, the experimentally determined spacing of the vibrational energy levels for H_2 is displayed in Fig. 8-5.

The harmonic-oscillator approximation is still a pretty good representation of the lower energy levels. A semiempirical method of representing

---
* We essentially follow the treatment given by Pitzer, Appendix 14 [10].

the actual spacing of the energy levels is to write a power series in the quantum number $v + \tfrac{1}{2}$.

$$\frac{\epsilon_v}{hc} = \omega_e(v + \tfrac{1}{2}) - \omega_e x_e(v + \tfrac{1}{2})^2 + \cdots \tag{8-32}$$

The leading term ω_e is still given by $\omega_e = (1/2\pi c)(f/\mu)^{\frac{1}{2}}$, where $\tfrac{1}{2}f(r - r_e)^2$ is the leading term in the expansion of the potential-energy

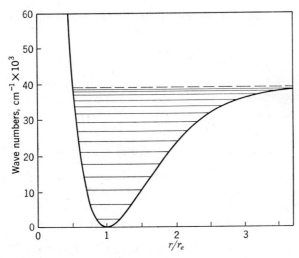

FIG. 8-5. The potential curve and the vibrational energy levels for the H_2 molecule. The equilibrium internuclear distance is $r_e = 0.742$ Å.

function, $U(r) = \tfrac{1}{2}f(r - r_e)^2 + g(r - r_e)^3 + \cdots$. Therefore ω_e is related to the curvature of the potential curve at the minimum.

The zero-point energy is now

$$\frac{\epsilon_0}{hc} = \omega_e\left(\frac{1}{2}\right) - \omega_e x_e\left(\frac{1}{4}\right) + \cdots \tag{8-33}$$

The height of the levels above the level ϵ_0 is

$$\frac{\epsilon_v - \epsilon_0}{hc} = \omega_e v - \omega_e x_e(v^2 + v) + \cdots$$
$$= (\omega_e - 2\omega_e x_e)v - \omega_e x_e v(v - 1) + \cdots \tag{8-34}$$

We now define

$$\omega_0 = \omega_e - 2\omega_e x_e \tag{8-35}$$

In (8-33), since $x_e \ll 1$, we may write $\omega_e x_e \approx \omega_0 x$ (with $x = x_e$); then

$$\frac{\epsilon_v - \epsilon_0}{hc} = \omega_0 v - \omega_0 x v(v - 1) + \cdots \tag{8-36}$$

The equation has been arranged so that the first vibrational spacing, which is often directly observed in infrared spectroscopy, is $\epsilon_1 - \epsilon_0/hc = \omega_0$. Other definitions of ω_0 are used by some authors; in particular, Herzberg ("Diatomic Molecules" [17]) defines ω_0 as $\omega_e - \omega_e x_e$. The energy-level equation he uses is then different from (8-36).

The other approximations in Eq. (8-30) arise because the separation between rotation and vibration is only approximate. The molecule actually vibrates as it rotates, and so its moment of inertia is not constant; furthermore, there are centrifugal stretching effects. When account is taken of these effects, the energy-level equation becomes

$$\frac{\epsilon_{v,J}}{hc} = \omega_e\left(v + \frac{1}{2}\right) - \omega_e x_e \left(v + \frac{1}{2}\right)^2 + B_e J(J + 1)$$
$$- DJ^2(J + 1)^2 - \alpha\left(v + \frac{1}{2}\right)J(J + 1) \quad (8\text{-}37)$$

$$B_e = \frac{h}{8\pi^2 I_e c} \qquad D = \frac{4B_e^3}{\omega_e^2}$$

The $\omega_e x_e$ term is due to the anharmonicity of the potential curve; the D term is due to centrifugal stretching, and the α term is related to the fact that the molecule vibrates while it rotates and does not have a constant moment of inertia. The correction terms are all usually fairly small.

Upon introducing ω_0 as before, this equation becomes*

$$\frac{\epsilon_{v,J} - \epsilon_0}{hc} = \omega_0 v - \omega_0 xv(v - 1) + J(J + 1)[B_0 - DJ(J + 1) - \alpha v] \quad (8\text{-}38)$$

$$\omega_0 = \omega_e - 2\omega_e x_e \qquad B_0 = B_e - \frac{\alpha}{2} \qquad \omega_0 x = \omega_e x_e$$

For thermodynamic calculations (8-38) is rewritten in dimensionless form.

$$\frac{\epsilon - \epsilon_0}{kT} = vu[1 - (v - 1)x] + J(J + 1)y[1 - J(J + 1)\gamma - v\delta] \quad (8\text{-}39)$$

$$u = \frac{hc\omega_0}{kT} \qquad y = \frac{hcB_0}{kT} \qquad \gamma = \frac{D}{B} \qquad \delta = \frac{\alpha}{B_0}$$

By expanding the partition function it can then be written $q = q_{\text{ideal}} q_{\text{corr}}$, where†

$$q_{\text{ideal}} = \frac{1}{\sigma y} \frac{1}{1 - e^{-u}}$$

$$q_{\text{corr}} = 1 + \frac{2\gamma}{y} + \frac{\delta}{e^u - 1} + \frac{2xu}{(e^u - 1)^2} + \frac{y}{3} + \frac{y^2}{15} \quad (8\text{-}40)$$

* We are now following Pitzer, Appendix 14 [10], in detail.

† In q_{corr}, we have included the terms $(y/3) + (y^2/15)$, which correct for the approximation of replacing the sum (8-16) by the integral (8-17), in evaluating the rotational partition function. These terms are particularly important at low temperature; the other correction terms are most important at high temperature, where large amplitude vibrations and high energy rotations are probable.

There are corresponding additive correction terms in the thermodynamic functions. Expand $\ln q_{\text{corr}}$, and keep only the first-order terms. By straightforward laborious calculation,

$$\frac{F_{\text{corr}}}{RT} = -\ln q_{\text{corr}} = -\frac{2\gamma}{y} - \frac{2xu}{(e^u - 1)^2} - \frac{y}{3} - \frac{\delta}{e^u - 1} \quad (8\text{-}41)$$

$$\frac{H}{R} = \frac{\partial(F/RT)}{\partial(1/T)}$$

Note that $\dfrac{\partial y}{\partial(1/T)} = \dfrac{hcB_0}{k} \qquad \dfrac{\partial y}{\partial T} = -\dfrac{k}{hcB_0} y^2$

$$\frac{\partial u}{\partial(1/T)} = \frac{hc\omega_0}{k} \qquad \frac{\partial u}{\partial T} = -\frac{k}{hc\omega_0} u^2$$

Therefore
$$\frac{E_{\text{corr}}}{R} = \frac{H_{\text{corr}}}{R} = \frac{hcB_0}{k}\left(\frac{2\gamma}{y^2} - \frac{1}{3}\right) + \frac{hc\omega_0}{k}\left[\frac{\delta e^u}{(e^u - 1)^2}\right]$$
$$+ 2x\,\frac{hc\omega_0(2ue^u - e^u + 1)}{k(e^u - 1)^3} \quad (8\text{-}42)$$

$$\frac{C_{\text{corr}}}{R} = \frac{\partial}{\partial T}\left(\frac{H_{\text{corr}}}{R}\right) = \frac{4\gamma}{y} + \frac{\delta u^2 e^u(e^u + 1)}{(e^u - 1)^3}$$
$$+ 2xu^2 e^u \frac{4ue^u - 4e^u + 2u + 4}{(e^u - 1)^4} \quad (8\text{-}43)$$

The correction terms all make the free energy more negative.

The molecular constants for the ground electronic states of several molecules are listed in Table 8-2.

Table 8-2. Molecular Constants for Several Molecules*

Molecule	Electronic state	ω_e, cm^{-1}	$\omega_e x_e$, cm^{-1}	B_e	α_e	r_e, Å
H_2	$^1\Sigma_g^+$	4,395.2	117.9	60.809	2.993	0.7417
HF	$^1\Sigma^+$	4,138.5	90.1	20.94	0.771	0.9171
O_2	$^3\Sigma_g^-$	1,580.36	12.07	1.4457	0.0158	1.2074
I_2	$^1\Sigma_g^+$	214.57	0.6127	0.0374	0.000117	2.667

* Data from G. Herzberg, "Molecular Spectra and Molecular Structure, I. Spectra of Diatomic Molecules," table 39, 2d ed., Van Nostrand, Princeton, N.J., 1950.

The anharmonic correction terms are more important for molecules containing light atoms, essentially because, in a given potential curve, the amplitude of the vibration of a light atom is greater than that for a heavier atom.

8-9. Electronic States. A diatomic molecule has a number of electronic states. Some of the curves for O_2 are exhibited in Fig. 8-6. As can be seen for this example, the vibration frequencies and the moments of inertia are different in the several states. At high temperatures, the upper states may be significantly populated and may contribute significantly to the thermodynamic functions.

Let A, B, C represent the ground state (A) and two excited states (B and C). Let ϵ_B and ϵ_C represent the electronic excitation energies from the lowest vibrational states of A to the lowest vibrational states of B and C, respectively. Let the electronic degeneracies of the several states be g_A, g_B, g_C. The rotational partition functions are denoted by

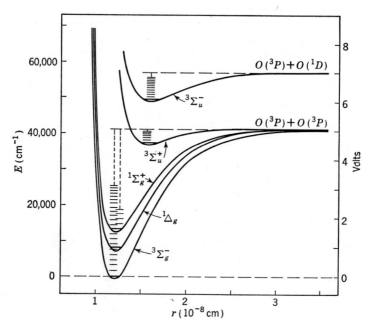

FIG. 8-6. Some of the lower potential energy curves of the O₂ molecule. There are many still higher electronic states. Some of the vibrational states for each of the several electronic states are shown. (*From G. Herzberg, "Molecular Spectra and Molecular Structure, I. Spectra of Diatomic Molecules," 2d ed., p. 446, Van Nostrand, Princeton, N.J., 1950; for further information, see this reference.*)

q_{rot}^A, q_{rot}^B, q_{rot}^C and the vibrational partition functions by q_{vib}^A, q_{vib}^B, q_{vib}^C. The correction factors for the failure of the rigid-rotor harmonic-oscillator approximation are q_{corr}^A, q_{corr}^B, q_{corr}^C. By direct summation, the total partition function is seen to be

$$q = \left(\frac{2\pi mkT}{h^2}\right)^{3/2} \frac{V}{\sigma} [g_A q_{\text{rot}}^A q_{\text{vib}}^A q_{\text{corr}}^A + g_B e^{-\epsilon_B/kT} q_{\text{rot}}^B q_{\text{vib}}^B q_{\text{corr}}^B \\ + g_C e^{-\epsilon_C/kT} q_{\text{rot}}^C q_{\text{vib}}^C q_{\text{corr}}^C] \quad (8\text{-}44)$$

The molecular constants for the various electronic states can often be found in reference works.*

* The molecular constants of all known electronic states of diatomic molecules, as of September, 1949, are tabulated in Herzberg, "Diatomic Molecules," table 39, pp. 501–581 [17]. There is also a compilation in vol. I, pt. 3, pp. 1–52, 691–702, of

An electronic state is given a term symbol $^{2S+1}\Lambda^{(+,-)}_{(g,u)\Omega}$. S is the electron spin quantum number, which is, as usual, one-half the number of unpaired electrons. Λ is the orbital angular-momentum quantum number; the symbols for $\Lambda = 0, 1, 2, 3$ are $\Sigma, \Pi, \Delta, \Phi$, respectively. For an atom, an angular-momentum vector L can have $2L + 1$ orientations in space. However, for a diatomic molecule, the electronic angular-momentum vector cannot have $2\Lambda + 1$ independent orientations; it is quantized along the internuclear axis and can have components only of $\pm \Lambda \hbar$ with a corresponding degeneracy of 2. The total electronic degeneracy is therefore $2S + 1$ for a Σ state, and $2(2S + 1)$ for all other states. The $2(2S + 1)$-degenerate states ($\Lambda \neq 0, S \neq 0$) are split into $2S + 1$ levels, each twofold-degenerate, by spin-orbit coupling. The electron spin is thus quantized along the internuclear axis, and the component of total electronic angular momentum along this axis has values $\pm \Omega \hbar$, where $\Omega = \Lambda + S, \Lambda + S - 1, \ldots, |\Lambda - S|$. The $\pm$ sign arises because the angular momentum can be oriented in either direction along the internuclear axis. Thus, the ground electronic state of NO is $^2\Pi$, its two components are $^2\Pi_{1/2}$ and $^2\Pi_{3/2}$. The $^2\Pi_{3/2}$ state is 121.1 cm^{-1} above the $^2\Pi_{1/2}$ state (see the example in Prob. 8-12). Each of the states $^2\Pi_{1/2}$ and $^2\Pi_{3/2}$ is twofold-degenerate because of the two possible orientations of the total angular momentum, $\Omega \hbar$, with respect to the NO axis. Thus, all told, the $^2\Pi$ configuration gives rise to four states.

The subscript g or u and the superscript $+$ or $-$ are symmetry symbols. The subscript g or u refers only to symmetrical diatomic molecules; g means that the electronic wave function is symmetrical with respect to an inversion through the origin; u means that the wave function changes sign under this operation. The superscript $+$ or $-$ applies to any linear

---

Landolt-Börnstein's "Zahlenwerte und Funktionen," Sechste Auflage, Springer-Verlag, Berlin, 1951. In the Herzberg compilation, the height of the minimum of the potential-energy curve for each excited electronic state above the minimum of the potential curve of the ground state is listed as T_e. To obtain the actual separation of the lowest vibrational levels of the two states which we have called ϵ_B above, a zero-point correction is necessary. Thus, in wave numbers

$$\frac{\epsilon_B}{hc} = T_e + \frac{\omega_B}{2} - \frac{\omega_A}{2} \tag{8-45}$$

since the zero-point energies of the two states in wave numbers are $\omega_A/2$ and $\omega_B/2$. Thus for the $^1\Delta_g$ state of O_2, $T_e = 7{,}918.1$ cm^{-1}; however, the vibration frequencies of the ground state $^3\Sigma_g^-$ and the $^1\Delta_g$ state are $\omega_0 = 1{,}556$ cm^{-1} and 1,484 cm^{-1}, respectively. Therefore, the actual separation of the lowest levels is 36 cm^{-1} less than T_e, or 7,882 cm^{-1}. In this particular example, if we ignore the small differences in the vibrational and rotational partition functions of the two states, the relative populations are given by $^1\Delta_g/^3\Sigma_g^- = \frac{2}{3} \epsilon^{(-1.4388 \times 7{,}882/T)}$. (The degeneracy factor of $\frac{2}{3}$ is explained below.) When $T = 3200°K$, this ratio is 0.021. Incidentally, this is just about the temperature where O_2 at ordinary pressures begins to dissociate significantly. (Compare Prob. 8-11.)

molecule, but in a Σ state only. For such states, the wave function either is invariant or changes sign for a reflection in a plane containing the internuclear axis. These symmetry symbols are not important for statistical calculations, except as regards the nuclear spin statistics discussed in Chap. 9.

Electronic states are more fully discussed in Herzberg, "Diatomic Molecules," chap. V [17]. Actually, by very elementary considerations involving molecular orbital wave functions, it is possible to visualize the wave functions with their symmetries for diatomic molecules.*

For electronic states which have orbital angular momentum, the interaction with rotation can be rather complicated. This is a specialized subject into which we cannot enter deeply, but a brief discussion of several simple cases is profitable. The problem was treated by Hund and is now usually discussed in terms of "Hund's cases, a, b, c, d, e."† Usually, but not always, it is possible to calculate a sufficiently accurate partition function without going into all the details of the couplings and the splittings.

The total squared angular momentum of a molecule is always a constant of the motion with value $J(J+1)\hbar^2$. In Hund's case a, the electronic angular momentum $\pm\Omega\hbar$ is tightly coupled to the internuclear axis. The angular momentum of nuclear rotation N is not a constant of the motion, but it adds vectorially to the electronic angular momentum to give the resultant squared angular momentum $J(J+1)\hbar^2$. The energy levels for this case can be written as

$$\frac{\epsilon(J)}{hc} = B[J(J+1) - \Omega^2] \qquad (8\text{-}46a)$$

where $B = h/8\pi^2 Ic$ as usual and the allowed values of J are Ω, $\Omega + 1$, $\Omega + 2$, ... In dimensionless form,

$$\frac{\epsilon(J)}{kT} = y[J(J+1) - \Omega^2] \qquad y = \frac{Bhc}{kT} \qquad (8\text{-}46b)$$

For example, for the $^2\Pi_{3/2}$ state of NO, the allowed values of J are $3/2$, $5/2$, $7/2$, . . . and

$$q_{\text{rot}} = e^{9y/4} \sum_{J=3/2}^{\infty} e^{-yJ(J+1)} \qquad (8\text{-}47)$$

In this case, $B = 1.70$, and at 300°K, $y = 0.0079$. The factor $e^{9y/4}$ is then approximately unity, and the sum from $J = 3/2$ to $J = \infty$ can be replaced by the integral for $J = 0$ to $J = \infty$; so $q_{\text{rot}} = 1/y$, as usual.

---
* C. A. Coulson, "Valence," p. 97, Oxford, New York, 1952.
† Herzberg, "Diatomic Molecules," chap. V.2 [17].

In the same way for the $^2\Pi_{1/2}$ case, the partition function is

$$q_{\text{rot}} = e^{y/4} \sum_{J=1/2}^{J=\infty} e^{-yJ(J+1)} \approx \frac{1}{y} \tag{8-48}$$

The $^2\Pi_{3/2}$ state is 121.1 cm^{-1} above $^2\Pi_{1/2}$, and the combined electronic and rotational partition function is

$$q = \frac{2}{y}(1 + e^{-1.4388 \times 121.1/T}) \tag{8-49}$$

(See the example in Prob. 8-12.)

On the other hand, for OH, which also has a $^2\Pi$ ground state, the problem is more complicated. The value of B is 10.01 cm^{-1}; at 300°K, $y = 0.048$. The separation of the electronic states $^2\Pi_{1/2} - {}^2\Pi_{3/2}$ is listed as 139.7 cm^{-1}, but when account is also taken of the rotational energy [(8-46a)], the separation of the lowest states ($J = 1/2$ for $^2\Pi_{1/2}$ and $J = 3/2$ for $^2\Pi_{3/2}$) is 119.7 cm^{-1}.

Furthermore for states of OH with moderately large J ($J > 7/2$), Hund's case a no longer applies; the electron spin angular momentum begins to interact strongly with the nuclear rotational angular momentum rather than with the orbital angular momentum (Hund's case b). Quite different formulae for the energy levels result. Since the electron spin is no longer coupled to the electronic orbital angular momentum, the states cannot be classified as $^2\Pi_{1/2}$ and $^2\Pi_{3/2}$.

The discussion of OH above is presumably too brief to be very helpful to the student who is not acquainted with the subject; however, it should serve to emphasize that, for complicated cases such as this, especially with rotational spacings that are comparable with kT, a reasonably detailed understanding of the energy-level system is necessary for an accurate calculation of the partition function.*

8-10. Practical Computations. This section is a compendium of working formulae for the rigid-rotor harmonic-oscillator case, for atoms, diatomic molecules, and polyatomic molecules. The anharmonic corrections for a diatomic molecule are given in Sec. 8-8.

Fundamental Thermodynamic Equations

$$\frac{F - F_0}{T} = -R \ln \frac{q}{N} \qquad E - E_0 = -R \left\{ \frac{\partial [\ln(q/N)]}{\partial(1/T)} \right\}_V$$

$$H - H_0 = -R \left\{ \frac{\partial [\ln(q/N)]}{\partial(1/T)} \right\}_P$$

$$S = R\left(1 + \ln \frac{q}{N}\right) + \frac{E - E_0}{T}$$

$$\frac{q}{N} = \left(\frac{q_{\text{tr}}}{N}\right) q_{\text{rot}} q_{\text{vib}} q_{\text{elect}}$$

* See D. H. Dawson and H. L. Johnston, *J. Am. Chem. Soc.*, **55**: 2744 (1933).

For an internal degree of freedom, such as rotation,

$$\left(\frac{F - F_0}{T}\right)_{rot} = -R \ln q_{rot}$$

$$(E - E_0)_{rot} = (H - H_0)_{rot} = -R\left[\frac{\partial(\ln q_{rot})}{\partial(1/T)}\right]$$

$$S_{rot} = R \ln q_{rot} + \left(\frac{E - E_0}{T}\right)_{rot}$$

Chemical Equilibrium

$$aA + bB \rightleftharpoons cC + dD$$

$$\frac{[D]^d[C]^c}{[A]^a[B]^b} = K = e^{-\Delta E_0^\circ/RT} \frac{(q_D^\circ/N)^d(q_C^\circ/N)^c}{(q_A^\circ/N)^a(q_B^\circ/N)^b}$$

q_B°/N is the molecular partition function divided by N evaluated at the chosen standard state, in which $[B] = 1$, and using the ground state of molecule B at $0°K$ as energy zero.

Physical Constants. The values of the physical constants used are given in Appendix 1. In general, in the formulae below, the numbers have been rounded off to one part in 10,000 or the next place thereafter.

For practical calculations, the quantities $\tfrac{3}{2}R$, $\tfrac{5}{2}R$, $R \ln 10$, $\tfrac{3}{2}R \ln 10$, are useful.

$$R = 1.9873 \text{ cal mole}^{-1} \text{ deg}^{-1} \qquad \tfrac{3}{2}R = 2.9809 \qquad \tfrac{5}{2}R = 4.9681_5$$
$$2.30259R = 4.5758_5 \qquad \times \tfrac{3}{2} = 6.8638$$
$$\times \tfrac{5}{2} = 11.4396 \text{ cal mole}^{-1} \text{ deg}^{-1}$$

Translation

$$\frac{q}{N} = \left(\frac{2\pi mkT}{h^2}\right)^{3/2} \frac{V}{N}$$

The expressions for q°/N for several common choices of the standard state are:

Standard state	Atoms cc^{-1}	Moles liter^{-1}	Atm
$\dfrac{q^\circ}{N}$	$\left(\dfrac{2\pi mkT}{h^2}\right)^{3/2}$ $1.8791_5 \times 10^{20} M^{3/2}T^{3/2}$	$\left(\dfrac{2\pi mkT}{h^2}\right)^{3/2} \dfrac{1{,}000}{N_0}$ $0.31198 M^{3/2}T^{3/2}$	$\left(\dfrac{2\pi mkT}{h^2}\right)^{3/2} \dfrac{kT}{P_0}$ $0.02560 M^{3/2}T^{5/2}$

M = molecular weight (chemical scale); $P_0 = 1.01325 \times 10^6$ dynes cm^{-2} atm^{-1}.

$$\frac{F^\circ - F_0^\circ}{T} = -\tfrac{3}{2}R \ln M - \tfrac{5}{2}R \ln T + 7.2836 \text{ cal deg}^{-1}$$

for standard state of 1 atm

$$E^\circ - E_0^\circ = \tfrac{3}{2}RT \qquad H^\circ - H_0^\circ = \tfrac{5}{2}RT \qquad C_v = \tfrac{3}{2}R \qquad C_p = \tfrac{5}{2}R$$

$$S^\circ = \tfrac{5}{2}R - \frac{F^\circ - F_0^\circ}{T}$$

Rotation (diatomic molecules)

$$q_{\text{rot}} = \frac{1}{\sigma}\frac{2IkT}{\hbar^2} = \frac{1}{\sigma}\frac{8\pi^2 IkT}{h^2} = \frac{1}{\sigma}\frac{T}{1.4388B}$$

$$B(\text{cm}^{-1}) = \frac{h}{8\pi^2 cI} = \frac{27.989 \times 10^{-40}}{I \text{ (g cm}^2)} = \frac{16.858}{I'(\text{awu} \times \text{Å}^2)}$$

(awu means chemical atomic-weight units)

$$I' (\text{awu} \times \text{Å}^2) = I \text{ (g cm}^2) \times 0.60232 \times 10^{40}$$
$$q_{\text{rot}} = 0.041227 I'T = 2.4832 \times 10^{38} IT$$

$$-\frac{F - F_0}{T} = R \ln \frac{IT \times 10^{39}}{\sigma} - 2.7683 \text{ cal deg}^{-1}$$

$$= R \ln \frac{T}{\sigma B} - 0.72298$$

$$= R \ln \frac{I'T}{\sigma} - 6.3367$$

$$(\text{E} - \text{E}_0)_{\text{rot}} = (\text{H} - \text{H}_0)_{\text{rot}} = RT$$

Rotation (nonlinear polyatomic molecules) (as shown in Chap. 11)

$$q_{\text{rot}} = \frac{\pi^{1/2}}{\sigma}\left(\frac{8\pi^2 I_x kT}{h^2}\right)^{1/2}\left(\frac{8\pi^2 I_y kT}{h^2}\right)^{1/2}\left(\frac{8\pi^2 I_z kT}{h^2}\right)^{1/2}$$

where I_x, I_y, I_z are the principal moments of inertia.

$$\sigma q_{\text{rot}} = 0.014837(I'_x I'_y I'_z)^{1/2} T^{3/2} = 6.9358 \times 10^{57}(I_x I_y I_z)^{1/2} T^{3/2}$$

(As before, I' is in awu times angstroms squared, I in grams times centimeters squared.)

$$-\frac{F - F_0}{T} = R \ln \frac{(I'_x I'_y I'_z)^{1/2} T^{3/2}}{\sigma} - 8.3676 \text{ cal deg}^{-1}$$

$$= R \ln \frac{(I_x I_y I_z)^{1/2} \times 10^{57} T^{3/2}}{\sigma} + 3.8487$$

$$\text{H} - \text{H}_0 = \text{E} - \text{E}_0 = \tfrac{3}{2} RT$$

Vibration

$$u = \frac{h\nu}{kT} = 1.4388\omega \text{ (cm}^{-1})/T$$

$$q = \frac{1}{1 - e^{-u}}$$

Tables of harmonic-oscillator thermodynamic functions are given by Pitzer, "Quantum Chemistry," pp. 457–467 [10], and by Torkington, *J. Chem. Phys.*, **18**: 1373 (1956). There is a very short table in Appendix 2.

Isotopic Molecules. A detailed discussion of the partition function for isotopic molecules is given in Chap. 9. The following remark is helpful for practical calculations of thermodynamic functions from spectroscopic data if only moderate accuracy is desired, such as in Prob. 8-12. For example, for a molecule such as Cl_2, Herzberg ("Diatomic Molecules" [17]) gives spectroscopic data for Cl^{35}-Cl^{35} only. But ordinary chlorine has an atomic weight of 35.457 and contains about 24 per cent Cl^{37}. To calculate the partition function of the ordinary chlorine, use 2×35.457 for the molecular weight for the translational partition function, multiply the B_e given by Herzberg by $35.00/35.46$, and multiply the vibration frequency ω_e by the factor $(35.00/35.46)^{1/2}$. A more accurate and more laborious procedure is to calculate separate partition functions for the several isotopic molecules; but for heavy atoms, the above approximation would be an improvement over the procedure of simply using the molecular data for Cl^{35}-Cl^{35}, and sufficiently accurate for most purposes.

PROBLEMS

8-2. Consider the atoms in a perfect gas with two internal energy levels ϵ_1 ($=0$) and ϵ_2 so that the internal partition function is

$$q_{\text{int}} = g_1 + g_2 e^{-\beta \epsilon_2}$$

Consider the case $g_1 = g_2 = 1$.

Denote the internal energy, heat capacity, free energy, and entropy per particle as ϵ, C, F, and S. Calculate them as a function of T^* ($T^* = kT/\epsilon_2$, a dimensionless temperature) for $T^* = 0.1, 0.3, 0.6, 0.9, 1.1, 1.5, 2.0, 5.0, 10.0$. Plot the dimensionless quantities C/k, S/k, ϵ/ϵ_2 and F/kT as a function of T^*. Explain the significance of the various trends. (The calculations may be divided up among several members of the class.)

8-3. Given a 1-cc sample of oxygen atoms containing 2.69×10^{19} atoms (standard density). At $1000°K$, what is q_{int}? How many O atoms are in each of the states $^3P_0$, $^3P_1$, $^3P_2$, $^1D_2$? (Data in Table 8-1.)

8-4. B is the rotational constant, and J is the rotational quantum number. Assume that the temperature is such that $Bhc/kT \ll 1$. Regard J as a continuous variable, and determine what J is the most probable rotational state.

8-5. Use data from Herzberg ("Diatomic Molecules" [17]). Plot a distribution function among the different rotational energy levels for CO at $T = 200°K$; at $T = 2000°K$. Calculate q_{rot}.

8-6. The same as Prob. 8-5, but for HCl.

8-7. The discrete energy levels of a hydrogen atom relative to a separate proton and electron are given by

$$E_n = -\frac{2\pi^2 m e^4}{n^2 h^2} \qquad n = \text{principal quantum no.} = 1, 2, 3, \ldots$$

The degeneracy is $g_n = 2n^2$. The radius of the Bohr orbit is $r_n = n^2 a_0$, where $a_0 =$ Bohr radius $= 0.530 \times 10^{-8}$ cm $= h^2/4\pi^2 me^2$.

The ionization potential is clearly $2\pi^2 me^4/h^2 = 13.63$ ev molecule^{-1} (kT at $300°K = 0.0258$ ev molecule^{-1}).

Take the ground state as energy zero, so that the energy levels are $E_n = 13.63(1 - 1/n^2)$ ev. Evaluate the internal partition function for a hydrogen atom, considering the discrete states only.

ATOMS AND DIATOMIC MOLECULES

If you think your answer is somewhat disconcerting, calculate the appropriate value of n that makes the "occupation number" $2n^2 e^{-E_n/kT} = 1$ at 300°K, and the value of the corresponding Bohr radius. Then calculate the approximate volume $4\pi r^3/3$ for such an atom. Remember that the formula for the energy levels is correct only for noninteracting atoms; it does not apply to those values of n for which the volume of the atom $4\pi r^3/3$ is greater than V/N, the volume available per atom.

For a further discussion of this interesting paradox, see H. C. Urey, *Astrophys. J.*, **59**: 1 (1924).

8-8. Derive expression (8-18) by using the Euler-Maclaurin summation formula.

8-9(*a*). Assuming that the simple harmonic-oscillator approximation is rigorously applicable, obtain an expression for the fraction of all the molecules which are in the quantum state v and higher, that is, $(N_v + N_{v+1} + N_{v+2} + \cdots + N_\infty)/N$.

(*b*) In the same way, assuming that the rigid-rotor approximation is strictly accurate, calculate the fractional number of molecules which are in the rotational state J and higher. (Assume that, for the rotation parameter, $y \ll 1$.)

8-10. To get a feeling for typical values, take a molecule XY with molecular weights $M_X = M_Y = 20$, $r_e = 2$ Å, and $\omega = 2{,}000$ cm^{-1}. Calculate the translational partition function $q°/N$ (at 1 atm pressure), and q_{rot} and q_{vib} at 298 and 2000°K.

For the dissociation reaction

$$XY \rightleftharpoons X + Y$$

express the equilibrium constant as a product of partition functions times the Boltzmann factor for dissociation, $e^{-\beta \Delta \epsilon_0}$.

If we write the equilibrium constant in pressure units as

$$K = Ze^{-\beta \Delta \epsilon_0}$$

what is the magnitude of the so-called "preexponential factor" Z at 2000°K?

8-11. For the reaction

$$O_2 \rightleftharpoons 2O$$

the dissociation constant, in atmospheres, $K_P = P_O^2/P_{O_2}$ is close to unity (actually $K_P = 0.85$ atm) at 3750°K. Thus for a total pressure of 1.7 atm, $P_{O_2} = P_O = 0.85$ atm, and the degree of dissociation is ⅓. The energy of dissociation $\Delta E_0°$ is 117.18 kcal.

Using the results of the Prob. 8-9, find what fraction of the undissociated molecules at this temperature has a vibrational energy in excess of about one-half the dissociation energy, i.e., in excess of 58.6 kcal.

Can you give a qualitative explanation of why, at moderately high temperatures and ordinary densities, there is a much higher probability of dissociation than of vibrational excitation, even to states which have only about half as much energy as the dissociation energy? In reflecting on this question, consider the magnitude of typical entropies of dissociation (Prob. 4-6) and/or the typical magnitude for the partition-function product which is the "preexponential factor" Z of Prob. 8-10.

8-12. Pick a diatomic molecule, look up and record ω_e, $x_e\omega_0$, B_e, α, B_0, and any other pertinent information. At $T = 298.16°$K, compute $(q/N)_{\text{trans}}$, q_{rot}, q_{vib}, q_{corr}, and $(q/N)_{\text{total}}$. Compute $-(F° - H_0°)/T$, $-(H° - H_0°)/T$, C_p, $S°$ at 298.16. Use a standard state of 1 atm. Compare with the NBS Series III tables or some other appropriate source of data.

Try to guess T such that $K \approx 1$ for $XY \rightleftharpoons X + Y$, where XY is the diatomic molecule. Calculate the above thermodynamic functions for XY at T. Calculate $q°/N$ total for X and for Y, and calculate the equilibrium constant for the dissociation. Compare with NBS Series III tables or with other literature data, if possible.

Table 8-3. Sample Data Sheet for Problem 8-12

Function	Trans.	Rot.	Vib.	Corr.	Elect.	Total	NBS
\multicolumn{8}{c}{NO at 298.16°K}							
$q°/N$	6.435×10^6	1.222×10^2	1.0001	1.003	3.134	2.420×10^9	
$-\dfrac{F° - H_0°}{T}$	31.16	9.541	1.97×10^{-3}	5.95×10^{-3}	2.257	42.97	42.98
$\dfrac{H° - H_0°}{T}$	4.968	1.987	2.09×10^{-3}	~ 0	0.4232	7.380	7.359
$S°$	36.13	11.53	4.08×10^{-3}	5.95×10^{-3}	2.680	50.35	50.34
$C_p°$	4.968	1.987	2×10^{-3}	~ 0	0.1566	7.114	7.137
\multicolumn{8}{c}{NO at 4750°K}							
$q°/N$	6.518×10^9	1.947×10^3	2.310	1.036	3.928	1.193×10^{14}	
$-\dfrac{F° - H_0°}{T}$	44.92	15.04	1.66	5.48×10^{-2}	2.719	64.42	64.43
$\dfrac{H° - H_0°}{T}$	4.968	1.987	1.48	~ 0	0.036	8.48	8.602
$S°$	49.89	17.03	3.14	5.48×10^{-2}	2.755	72.90	73.03
$C_p°$	4.968	1.987	1.937	~ 0	6.7×10^{-4}	8.893	9.183
\multicolumn{8}{c}{^{14}N at 4750°K}							
$q°/N$	2.079×10^9				4.03	8.378×10^9	
$-\dfrac{F° - H_0°}{T}$	42.64				2.77	45.41	45.415
$\dfrac{H° - H_0°}{T}$	4.968				0.08614	5.054	5.0577
$S°$	47.61				2.86	50.46	50.472
$C_p°$	4.968				0.4979	5.466	5.4977
\multicolumn{8}{c}{^{16}O at 4750°K}							
$q°/N$	2.538×10^9				8.832	2.242×10^{10}	
$-\dfrac{F° - H_0°}{T}$	43.03				4.33	47.36	47.370
$\dfrac{H° - H_0°}{T}$	4.968				0.0885	5.057	5.0576
$S°$	48.00				4.42	52.42	52.428
$C_p°$	4.968				0.204	5.172	5.1799

At 4750°K: $K_D = 0.210$
$K_D = 3.840$ taking $D_0° = 5.29$ ev
NBS: $K_D = 3.743$
NO, spectral data: $\omega_e = 1{,}903.85$ cm^{-1}; $\omega_0 = 1{,}875.92$; $\omega_e x_e = 13.97$; $\omega_e y_e = -0.00120$; $B_e = 1.7046$; $B_0 = 1.6957$; $\alpha_e = 0.0178$; $r_e = 1.1508$ Å; $\mu = 7.46881$; $D_0° = 6.48$ ev, or 149.5 kcal; $^2\Pi_{3/2} \rightarrow {}^2\Pi_{1/2}$, 121.1 cm^{-1}
source: W. F. Dove, January, 1959.

The results should be turned in on a neat tabular form suitable for duplication and distribution to the class.

An example of such a computation for NO by W. F. Dove is given in Table 8-3. In this particular case, the data in the NBS Series III tables used by the student were based on the heat of dissociation of nitrogen of 7.37 ev. It is now known that the correct value is 9.756 ev (225.1 kcal mole^{-1}). The corresponding heat of dissociation of NO used in the NBS tables is 5.29 ev; the correct value, used by Dove, is 6.48 ev (149.5 kcal mole^{-1}). It may be seen that he was able to duplicate the NBS calculations by using the incorrect D and by using the correct D to (presumably) calculate the correct K.

9

Nuclear Spin Statistics. Isotope Effects

9-1. Introduction. A symmetry number of 2 was introduced arbitrarily into the calculation of the partition function of a symmetrical diatomic molecule in the preceding chapter. The fundamental reason for this factor is the symmetry requirement on the wave function for a system containing a pair of identical nuclei, whether they be bosons or fermions. We shall study this subject in the next few sections.

Apart from the factors relating to nuclear spin statistics and symmetry numbers, there are small differences in the thermodynamic properties of isotopic molecules, due to the small mass differences. This subject is treated in Sec. 9-5.

9-2. Ortho and Para Hydrogen. The hydrogen molecule contains several electrons (two, but this fact is not essential here) and two protons. The electronic wave function is constructed so as to be antisymmetric with respect to the interchange of the coordinates of the two electrons. Since protons are fermions, the total wave function must also be antisymmetric with respect to the interchange of the coordinates of the two protons.

If we transform to coordinates relative to the center of mass of the system, the coordinates of the two electrons can be specified by the vectors $\mathbf{r}_1$ and $\mathbf{r}_2$. The nuclear coordinates are specified by the vector $\mathbf{r}_{AB}$ pointing from proton A to proton B (cf. Probs. 2-1 to 2-4). The vector $\mathbf{r}_{AB}$ can be expressed in spherical polar coordinates in terms of the distance r_{AB} and the polar and azimuthal angles θ and ϕ. The distance r_{AB} is a positive number. The operator for interchanging the protons may be called the permutation operator P_{AB}. The effect of this interchange is the same as a rotation of the axis of the molecule by an angle of 180°, which is equivalent to the transformation $(\theta,\phi) \to (\pi - \theta, \phi + \pi)$ (see Fig. 9-1).

Call the spin quantum number of a nucleus I. For protons, $I = \frac{1}{2}$. The total molecular wave function includes, as a factor, a nuclear spin wave function. The spin wave function for proton A is $\alpha(A)$ or $\beta(A)$, which asserts that this proton has spin angular momentum of $+\frac{1}{2}\hbar$ or $-\frac{1}{2}\hbar$, respectively, along an axis of quantization. The nuclear spin

Sec. 9-2] NUCLEAR SPIN STATISTICS. ISOTOPE EFFECTS

wave functions for a system of two protons are

$$\alpha(A)\alpha(B)$$
$$(\tfrac{1}{2})^{1/2}[\alpha(A)\beta(B) + \beta(A)\alpha(B)] \quad (\tfrac{1}{2})^{1/2}[\alpha(A)\beta(B) - \beta(A)\alpha(B)] \quad (9\text{-}1)$$
$$\beta(A)\beta(B)$$

Call the total nuclear spin quantum number for the molecule T and its component along the axis of quantization T_z. The three spin functions on the left are symmetrical with respect to the permutation operator

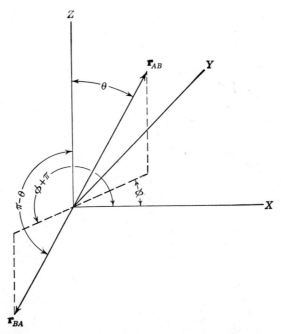

FIG. 9-1. The vector $\mathbf{r}_{BA}$ pointing from B to A is obtained from the vector $\mathbf{r}_{AB}$ by the transformation $(\theta,\phi) \rightarrow (\pi - \theta, \phi + \pi)$.

P_{AB}; it can readily be shown that they correspond to $T = 1$ [i.e., to a total squared angular momentum of $T(T+1)\hbar^2 = 2\hbar^2$] and $T_z = 1, 0, -1$, respectively. These are the three components of the triplet nuclear spin state. The antisymmetrical spin function on the right corresponds to the singlet state, $T = 0$.

The total wave function for H$_2$ can be written

$$\psi(\mathbf{r}_1,\mathbf{r}_2,\mathbf{r}_{AB}) = \psi_{\text{elect}}(\mathbf{r}_1,\mathbf{r}_2,\mathbf{r}_{AB})\psi_{\text{vib}}(r_{AB})\psi_{\text{rot}}(\theta,\phi)\psi_{\text{spin}}(A,B) \quad (9\text{-}2)$$

We require that $P_{AB}\psi = -\psi$. The electronic wave function is expressed in terms of the coordinates of the electrons with respect to the nuclei, and therefore it depends on $\mathbf{r}_{AB}$. For the $^1\Sigma_g^+$ electronic ground state of

H_2, $P_{AB}\psi_{\text{elect}} = \psi_{\text{elect}}$. The vibrational wave function ψ_{vib} is a function of the positive quantity r_{AB}, which is unaffected by P_{AB}. [The vibrational wave function is usually written as a function of the displacement from the equilibrium internuclear distance, $r_{AB} - r_{eAB}$, where r_{eAB} is the equilibrium internuclear distance. The harmonic-oscillator wave functions are symmetric or antisymmetric with respect to a change in sign of $r_{AB} - r_{eAB}$ (cf. Fig. 3-3), but the anharmonic-oscillator wave functions do

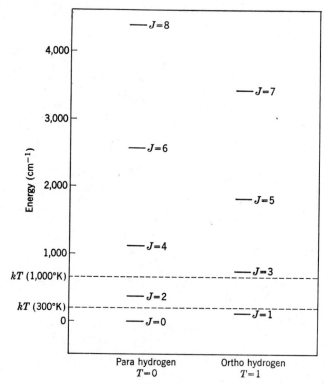

Fig. 9-2. The rotational energy levels of para and ortho hydrogen.

not have any symmetry around r_{eAB}. In any case, this has nothing to do with the symmetry with respect to the operator P_{AB}.]

The quantum numbers for the rotational wave functions are J and M (see Sec. 3-8). By inspection, for the first few wave functions (Prob. 3-3) it can be seen that $P_{AB}\psi_{\text{rot}}(\theta,\phi) = \psi_{\text{rot}}(\pi - \theta, \phi + \pi) = (-1)^J \psi_{\text{rot}}(\theta,\phi)$. It is true in general that the rotational wave function is symmetrical for even J and antisymmetrical for odd J. Therefore to satisfy the over-all antisymmetry, the even-J rotational states can occur only with the antisymmetrical singlet spin states $T = 0$; the states with J odd can occur only with the triplet spin states $T = 1$. The situation is displayed in

Fig. 9-2. The odd-J triplet nuclear spin states are called ortho hydrogen. The even-J singlet nuclear spin states are para hydrogen.*

Collisions between molecules in the gas phase are in general effective in causing transitions between different rotational states and therefore in maintaining thermal equilibrium as regards this degree of freedom.† However, there is only a very low probability of a transition between a triplet and a singlet nuclear spin state for a collision with a diamagnetic molecule in the liquid or gas phase. The nuclear spin transition (spin flip) is induced only by a magnetic perturbation; it is therefore caused by paramagnetic catalysts, either as molecules in the gas phase or in solution or by surface catalysis. The interconversion can also be caused by dissociation and recombination and, in the presence of hydrogen atoms, by the exchange reaction $H + H_2 \rightarrow H_2 + H$.‡

The main point is that, in the absence of paramagnetic catalysts, the rate of interconversion between ortho and para hydrogen is usually small, and they can be regarded as separate substances.

We now write partition functions for the various kinds of hydrogen, including nuclear spin degeneracy. It is sufficient to consider only the nuclear spin factor and the rotational partition function since the vibrational and electronic partition functions are the same for ortho and para hydrogen.

Consider first any diatomic molecule AB, where A and B are not identical. For the nucleus of A, with spin I_A, there are $2I_A + 1$ possible orientations of the spin along an axis of quantization. All these orientations have essentially the same energy,§ and the nuclear spin partition function for A is $2I_A + 1$. For the unsymmetrical diatomic molecule

---

* In general, for symmetrical diatomic molecules, the nuclear spin states are divided into two groups: symmetrical and antisymmetrical with respect to the interchange of the two nuclei. The group of states of higher nuclear spin degeneracy is called ortho; the group of lower spin degeneracy is called para. For the scientist who has not had the benefit of a classical education, the words *orthodox* and *paradox* are a useful mnemonic device for this distinction.

† For a discussion of the probability of a rotational transition in a collision and the rate of establishment of rotational equilibrium, see K. F. Herzfeld, Relaxation Phenomena in Gases, in F. D. Rossini (ed.), "High Speed Aeronautics and Jet Propulsion," vol. 1, Thermodynamics and Physics of Matter, Princeton University Press, Princeton, N.J., 1955. It is estimated that about 1 collision in every 200 in H_2 gas at room temperature results in a change in the rotational quantum number J. For other molecules, such as N_2 or O_2, the probability of a rotational transition on collision is still greater.

‡ See A. Farkas, "Orthohydrogen, Parahydrogen, and Heavy Hydrogen," chap. IV, Cambridge, New York, 1935; S. Glasstone, K. J. Laidler, and H. Eyring, "The Theory of Rate Processes," pp. 107–112, 211–213, McGraw-Hill, New York, 1941.

§ In the presence of a magnetic field there are differences in the energies of the different spin orientations, but these differences are negligibly small compared with kT and can be ignored in our present discussion (cf. Chap. 19).

AB, the product of the rotational partition function and the nuclear spin partition functions is

$$\frac{1}{y}(2I_A + 1)(2I_B + 1) \tag{9-3}$$

where $y = h^2/8\pi^2 IkT$. Note that, in the previous chapters, the spin degeneracy factor $(2I_A + 1)(2I_B + 1)$ was omitted. The reason for this omission will be considered shortly.

For the H_2 molecule, only certain combinations of nuclear spin wave functions and rotational wave functions are allowed, and the partition function for an equilibrium mixture of ortho and para hydrogen, including nuclear spin degeneracy, is

$$q = \sum_{J=0}^{J=\infty}(2T + 1)(2J + 1)e^{-J(J+1)y}$$

with $T = 0$ for J even and $T = 1$ for J odd; or

$$q = \sum_{\text{even } J}(2J + 1)e^{-J(J+1)y} + 3\sum_{\text{odd } J}(2J + 1)e^{-J(J+1)y} \tag{9-4}$$

The distribution law for such an equilibrium mixture is

$$n_J = \begin{cases} \dfrac{(2J + 1)e^{-J(J+1)y}}{q} & J \text{ even} \\ \dfrac{3(2J + 1)e^{-J(J+1)y}}{q} & J \text{ odd} \end{cases} \tag{9-5}$$

At high temperatures, where $y \ll 1$, the integral approximation applies, so that

$$\sum_{\text{even } J}(2J + 1)e^{-J(J+1)y} \approx \sum_{\text{odd } J}(2J + 1)e^{-J(J+1)y}$$

$$\approx \frac{1}{2}\int_0^\infty (2J + 1)e^{-J(J+1)y}\,dJ = \tfrac{1}{2}y$$

Therefore, according to (9-4), at high temperatures $q = 2/y$. In the absence of symmetry requirements, the partition function, according to (9-3), would be $4/y$, which is twice as great as the true value for an equilibrium mixture of ortho and para hydrogen. This is an example of the reason for using a symmetry number of 2 in calculating the partition function of a symmetrical diatomic molecule.

In a mixture of hydrogen gas at equilibrium, the ratio of ortho to para hydrogen is

$$\frac{\text{Ortho}}{\text{Para}} = \frac{3\sum_{\text{odd } J}(2J + 1)e^{-J(J+1)y}}{\sum_{\text{even } J}(2J + 1)e^{-J(J+1)y}} \tag{9-6a}$$

and, in the high-temperature approximation,

$$\frac{\text{Ortho}}{\text{Para}} = 3 \tag{9-6b}$$

The equilibrium composition of hydrogen as a function of temperature is shown in Fig. 9-3.*

By definition, "normal" hydrogen is the high-temperature equilibrium mixture which contains 25 per cent of the para species. In many experiments, when hydrogen, prepared by nuclear spin equilibration at high temperatures, is cooled to low temperatures, the composition remains fixed at this ratio. The heat capacity and the entropy (including nuclear spin) of the separate species and of normal hydrogen are shown in Fig. 9-4.

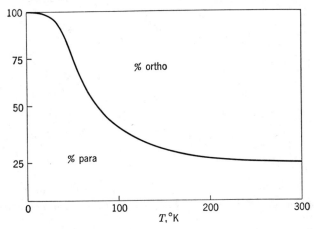

FIG. 9-3. The equilibrium percentages of ortho and para hydrogen as a function of temperature.

Because of the large rotational constant ($B_0 = 60.8$ cm^{-1}) of H$_2$, only a few states are significantly populated at room temperature (cf. Prob. 9-4), and the classical approximation to the partition function is by no means adequate. Nevertheless, the equilibrium ratio of ortho to para hydrogen is 2.99 at 300°K, which is very close to the high-temperature limit of 3.00.

When liquid hydrogen at 20°K is placed in contact with charcoal, the ortho-para equilibrium is established; the product is practically pure para hydrogen in its lowest state $J = 0$ (cf. Fig. 9-3). By removing the charcoal and then warming, a metastable preparation of pure para hydrogen is obtained. It has recently been discovered that ortho hydrogen is more strongly adsorbed by alumina than is para hydrogen; by adsorption and

---

* The thermodynamic properties of hydrogen in its various isotopic and nuclear spin modifications are compiled by H. W. Wooley, R. B. Scott, and F. G. Brickwedde, *J. Research Natl. Bur. Standards*, **41**: 379 (1948) (*Research Paper* RP 1932).

selective desorption at 20°K it is possible to prepare pure ortho (odd-J nuclear triplet) hydrogen.*

9-3. Other Examples of Nuclear Spin Statistics. The deuteron is a boson since it contains two nucleons, i.e., a proton and a neutron. The spin is 1. It can be shown that the spin states of D_2 with total nuclear spin T of 2 and 0 are symmetrical, and the state $T = 1$ is antisymmetrical with respect to P_{AB}. (Without writing out the spin eigenfunctions, it is

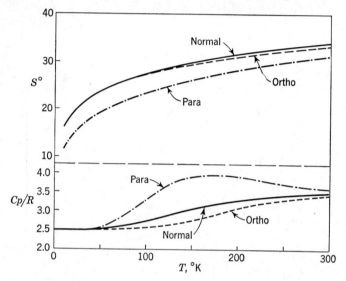

Fig. 9-4. Entropy (upper curve) and heat capacity (lower curve) of ortho, para, and normal hydrogen in the ideal-gas state.

obvious that in general the state of maximum T is symmetric, since the two spins are then parallel.) In order that the total wave function by symmetrical, the rotational states of even J must have $T = 2$ or 0 and a spin degeneracy of 6; the odd-J states have $T = 1$ and a spin degeneracy of 3. The even-J states are therefore ortho deuterium.

The combined rotational plus spin partition function is

$$q = 6 \sum_{J \text{ even}} (2J + 1)e^{-J(J+1)y} + 3 \sum_{J \text{ odd}} (2J + 1)e^{-J(J+1)y} \quad (9\text{-}7)$$

At high temperatures, this is $9/2y$. Without the symmetry restrictions it would be twice as great.

In general, an isotope with an even mass number is a boson and has an integral spin I. The nuclear spin states of even T ($T = 2I, 2I - 2, \ldots, 0$) are symmetrical. The total number of spin states for T even is

$$[2(2I) + 1] + [2(2I - 2) + 1] + \cdots + 1 = (2I + 1)(I + 1) \quad (9\text{-}8)$$

* C. M. Cunningham, D. S. Chapin, and H. L. Johnston, *J. Am. Chem. Soc.*, **80**: 2377, 2382 (1958).

Sec. 9-3] NUCLEAR SPIN STATISTICS. ISOTOPE EFFECTS 137

The states of odd T ($2I - 1, 2I - 3, \ldots, 1$) are antisymmetrical; the total number of such spin states is

$$[2(2I - 1) + 1] + [2(2I - 3) + 1] + \cdots + 3 = (2I + 1)I \quad (9\text{-}9)$$

The ratio of the statistical weights of the ortho (T-even) and para (T-odd) states is $(I + 1)/I$.

The high-temperature partition function is

$$q = (2I + 1)(I + 1) \sum_{J \text{ even}} (2J + 1)e^{-J(J+1)y}$$

$$+ (2I + 1)I \sum_{J \text{ odd}} (2J + 1)e^{-J(J+1)y} \approx (2I + 1)^2 \frac{1}{2y} \quad (9\text{-}10)$$

An isotope with an odd mass number is a fermion and has a half integral spin. The states of odd J go with the symmetrical nuclear spin functions, that is, with $T = 2I, 2I - 2, \ldots, 3, 1$. The statistical weight of this set of nuclear spin functions is

$$[2(2I) + 1] + [2(2I - 2) + 1] + \cdots + 7 + 3 = (2I + 1)(I + 1) \quad (9\text{-}11)$$

The states of even J go with the antisymmetrical spin states $T = 2I - 1, 2I - 3, \ldots, 0$. The statistical weight of these spin states is $(2I + 1)I$. The high-temperature rotation plus spin partition function is $(2I + 1)^2(1/2y)$. This result is the same for fermions (I half integral) and for bosons (I integral); in both cases, the ortho-para ratio at high temperatures is $(I + 1)/I$.

Consider now a chemical equilibrium such as $X_2 \rightleftharpoons 2X$. The translational, vibrational, and electronic factors in the equilibrium constant are unaffected by the considerations in this chapter. The nuclear spin degeneracy of the right-hand side is $(2I + 1)^2$. If the high-temperature approximation for the rotational partition function applies, the spin plus rotation partition function for X_2 is $(2I + 1)^2(1/2y)$. The factors $(2I + 1)^2$ cancel, and the ratio which appears in the equilibrium constant is just $2y$. Thus, if we omit nuclear spin degeneracy from the partition function for the atoms and the molecules, we can calculate the correct equilibrium constant and the correct change in entropy and free energy for the reaction $X_2 \rightleftharpoons 2X$ by using the symmetry number of 2 for X_2.

The same cancellation of nuclear spin factors will occur for any chemical reaction involving X_2, provided that the high-temperature approximation for the rotational partition function applies. Thus, for the reaction

$$X_2 + AB \rightleftharpoons AX + BX$$

the nuclear spin degeneracy factors will contribute the same factor $(2I_A + 1)(2I_B + 1)(2I_X + 1)^2$ to the partition-function product on both sides of the equation, but there is a factor of $\frac{1}{2}$ for X_2 due to the fact that only half the rotational states are allowed. *We can calculate the correct equilibrium constant by omitting all nuclear spin degeneracy factors, but including the symmetry number of 2 for homopolar diatomic molecules. For this reason, the thermodynamic functions listed in the standard compilations are given without the nuclear spin partition function.*

Nuclear spin partition functions must be included when processes involving the separation of different nuclear spin states are considered, as, for example, in the separation of ortho and para hydrogen or in the analysis of nuclear magnetic-resonance experiments.

It is interesting to note that for N_2^{14} ($I = 0$ for N^{14}), the rotational states of odd J do not occur since there are no antisymmetric spin states. However, for O_2^{16}, in its lowest electronic state ($I = 0$ for O^{16}) the rotational states with even rotational quantum number are forbidden. In this case, the electronic wave function changes sign when the nuclei are permuted. The nuclear spin function is of necessity symmetrical; so the rotational wave function must be antisymmetrical.*

9-4. Isotopes and Symmetry Numbers. Suppose that element X consists of two isotopes X^A and X^B. The diatomic molecule will contain the three species X_2^A, X_2^B, and $X^A X^B$. There are certain differences in the numerical values of the partition functions for the different isotopic molecules, due to the actual differences in mass. These differences are considered in the next section. Our attention here is focused only on the effects due to symmetry factors, and so we assume that the other differences are negligible.

Our purpose is to show that, for over-all chemical equilibria involving the molecule X_2, it is correct to use a symmetry number of 2, even though X_2 actually consists of the three species X_2^A, X_2^B, and $X^A X^B$.

---

* That the electronic wave function is antisymmetrical for interchange of the nuclei can be seen as follows: The term symbol for the electronic wave function is $^3\Sigma_g^-$. Let the x axis be the internuclear axis, with nuclei at $\pm x/2$. Rotation of 180° around the y axis permutes the nuclei; however, this operation also changes the coordinates of each electron from x, y, z to $\bar{x}, y, \bar{z}$. The electrons can be returned to their original position by an inversion ($\bar{x}, y, \bar{z} \to x, \bar{y}, z$) followed by a reflection in the xz plane ($x, \bar{y}, z \to x, y, z$). In the inversion, a g-state wave function does not change sign; in xz reflection a Σ^- wave function does change sign. Thus, the electronic wave function changes sign when the nuclear coordinates are permuted and the electrons are held fixed. Therefore, since the total wave function must be symmetrical in the nuclear coordinates, and since the nuclear spin function is symmetrical ($I = 0$), and since the electronic wave function is antisymmetrical, only antisymmetrical rotational functions are permitted. Incidentally, the quantum number for the rotational motion of the heavy nuclei for O_2 is called K, J being reserved for the total angular momentum which is obtained by vector addition of the rotational angular momentum and the electron spin (Herzberg, p. 167 [17]).

Consider the equilibrium

$$X_2^A + X_2^B \rightleftharpoons 2X^AX^B \qquad (9\text{-}12)$$
$$K_{12} = \frac{(X^AX^B)^2}{(X_2^A)(X_2^B)}$$

Upon omitting nuclear spin factors and neglecting the small differences in mass, but regarding the isotopes X^A and X^B as distinguishable chemical species, the rotational partition functions are

$$q(X_2^A) = \frac{1}{2y} \qquad q(X_2^B) = \frac{1}{2y} \qquad q(X^AX^B) = \frac{1}{y} \qquad (9\text{-}13a)$$

Since the vibrational and translational partition functions cancel,

$$K_{12} = 4 \qquad (9\text{-}13b)$$

Let n_A and n_B be the total number of X^A atoms and X^B atoms in unit volume of a sample when the concentrations (in molecules per cubic centimeters) of the several molecular species are (X_2^A), (X_2^B), and (X^AX^B). Let f_A and f_B be the mole fractions, or isotopic abundances of X^A and X^B atoms. Then

$$f_A = \frac{n_A}{n_A + n_B} \quad \text{and} \quad f_B = \frac{n_B}{n_A + n_B} \qquad (9\text{-}14a)$$

Furthermore,
$$n_A = 2(X_2^A) + (X^AX^B) \qquad (9\text{-}14b)$$
$$n_B = (X^AX^B) + 2(X_2^B)$$

By substitution of the relations (9-14) into the equilibrium equation (9-12) with $K_{12} = 4$, we obtain

$$(X^AX^B) = \frac{n_A n_B}{n_A + n_B} \qquad (X_2^A) = \frac{1}{2}\frac{n_A^2}{n_A + n_B} \qquad (X_2^B) = \frac{1}{2}\frac{n_B^2}{n_A + n_B} \qquad (9\text{-}15)$$

so that the mole fractions of the several kinds of molecules are

$$\frac{(X^AX^B)}{(X^AX^B) + (X_2^A) + (X_2^B)} = 2f_A f_B$$
$$\frac{(X_2^A)}{(X^AX^B) + (X_2^A) + (X_2^B)} = f_A^2 \qquad (9\text{-}16)$$
$$\frac{(X_2^B)}{(X^AX^B) + (X_2^A) + (X_2^B)} = f_B^2$$

Problem 9-1. Prove relations (9-15) and (9-16).

Now consider an over-all equilibrium, such as

$$X_2 + D_2 \rightleftharpoons 2DX \qquad (9\text{-}17)$$

where D is a monoisotopic element (not deuterium) so that we know how to set up its partition function. Considering the isotopic composition of element X, reaction (9-17) can be written as

$$X_2^A + D_2 \rightleftharpoons 2DX^A \qquad K_{18} = \frac{(DX^A)^2}{(D_2)(X_2^A)} \qquad (9\text{-}18)$$

$$X_2^B + D_2 \rightleftharpoons 2DX^B \qquad K_{19} = \frac{(DX^B)^2}{(D_2)(X_2^B)} \qquad (9\text{-}19)$$

$$X^A X^B + D_2 \rightleftharpoons DX^A + DX^B \qquad K_{20} = \frac{(DX^A)(DX^B)}{(D_2)(X^A X^B)} \qquad (9\text{-}20)$$

It is clear from the partition functions (9-13a) that

$$K_{18} = K_{19} = 2K_{20} \qquad (9\text{-}21)$$

Making this substitution into the equilibrium equations in (9-18) and (9-19) and adding (9-18), (9-19), and (9-20) in the appropriate way, we obtain

$$(DX^A)^2 + (DX^B)^2 + 2(DX^A)(DX^B) = 2K_{20}(D_2)[(X_2^A) + (X_2^B) + (X^A X^B)]$$

or $\qquad [(DX^A) + (DX^B)]^2 = 2K_{20}(D_2)[(X_2^A) + (X_2^B) + (X^A X^B)] \qquad (9\text{-}22)$

which is the same as

$$(DX)^2 = 2K_{20}(D_2)(X_2) \qquad (9\text{-}23)$$

since the total amount of DX and the total amount of X_2, irrespective of isotopic composition, are given by

$$(DX) = [(DX^A) + (DX^B)] \qquad (9\text{-}24a)$$
and $\qquad (X_2) = [(X_2^A) + (X_2^B) + (X^A X^B)] \qquad (9\text{-}24b)$

Note that the equilibrium constant K_{20} is calculated for the species $X^A X^B$, that is, without using a symmetry number of 2 for X_2 (although of course a symmetry number of 2 is used for D_2). Thus Eq. (9-23) says that the correct over-all equilibrium constant for reactions such as (9-17) involving the species X_2 is $2K_{20}$; that is, it is to be calculated with a symmetry number of 2 for X_2, even though X_2 is a mixture of several isotopic species. It is important to note that there has been no isotopic fractionation in the reaction (9-17). The ratio of X^A atoms to X^B atoms is the same in the compound DX and in the molecule X_2. This may be verified by explicit calculations using relations like (9-14) and (9-22).

There is another way of putting the matter which we shall state without explicit proof. When two distinguishable species Y and Z that form ideal solutions with each other are mixed at constant volume, there is an entropy change

$$\frac{S}{k} = -n_Y \ln \frac{n_Y}{n_Y + n_Z} - n_Z \ln \frac{n_Z}{n_Y + n_Z} \qquad (9\text{-}25)$$

(where n_Y and n_Z are the numbers of atoms) but no enthalpy change.

If we started with the pure separated species X_2^A, X_2^B, and $X^A X^B$ and mixed them to obtain the equilibrium proportions, there is an entropy increase according to (9-25). There is an entropy increase on mixing DX^A and DX^B. By straightforward calculations similar to those already performed, it can be shown that, for calculating over-all chemical equilibria, the entropy of isotope mixing can be consistently omitted, provided that the symmetry number of 2 is used in calculating the thermodynamic properties of X_2. Accordingly, the entropy of isotope mixing is omitted in thermodynamic compilations.

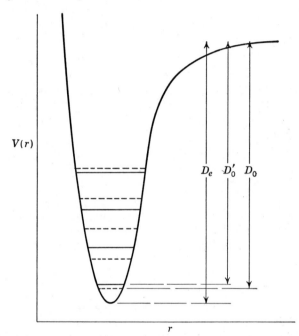

FIG. 9-5. Potential curve showing a few of the vibrational energy levels for two isotopic diatomic molecules. The solid horizontal lines are the vibrational levels for the lighter isotope, the dashed lines for the heavier isotope. D_e is the (hypothetical) energy of dissociation from the minimum of the potential curve; D_0' and D_0 are the actual energies of dissociation of the lighter and heavier isotopic molecules, respectively. Clearly, $D_e > D_0 > D_0'$; however, the spacing of the levels is closer for the heavier isotope.

9-5. Isotope Effects for Diatomic Molecules. Because of the mass difference, isotopic molecules differ slightly in their thermodynamic properties; there are corresponding isotopic fractionation effects in chemical equilibria. The classical general treatment of the subject is due to Bigeleisen and Mayer,[*] and we follow this treatment closely.

The basic cause of the differences in chemical properties of isotopic molecules is the difference in zero-point vibrational energy and in the vibrational energy spacing. This is illustrated in Fig. 9-5.

[*] J. Bigeleisen and M. G. Mayer, *J. Chem. Phys.*, **15**: 291 (1947).

It is convenient here to use a different notation than that of the preceding section and call the two isotopes B and B', B' being the lighter. The isotopic molecules are AB and AB' (where A is an atom of a different element). The electronic potential-energy curve is the same for AB and AB' since the nuclear charges and the number of electrons are the same. (The magnetic moments and quadrupole moments of the nuclei may be different, but this has a negligible effect on the molecular energy levels.) However, the vibrational energy levels are different. In the harmonic-oscillator approximation, the levels are given by $\epsilon_v = h\nu(v + \tfrac{1}{2})$, where $\nu = (\tfrac{1}{2}\pi)(f/\mu)^{1/2}$. The lighter isotopic molecule has more zero-point energy $\tfrac{1}{2}h\nu$; therefore, as indicated in the figure, it has a smaller dissociation energy D_0'. Thus at low temperatures, when all the molecules are in the lowest vibrational state, the energy of dissociation of the light molecule AB' will be less than the energy of dissociation of the heavy molecule AB.

However, the spacing of the vibrational levels is smaller for AB than for AB', so that, as the temperature is raised, AB acquires vibrational energy sooner than AB'. As regards dissociation, this tends to compensate for the difference in the dissociation energies from the states $v = 0$.

Our quantitative analysis will be based on the harmonic-oscillator rigid-rotor approximation. We take the minimum of the potential-energy curve as energy zero. The vibrational partition function is then

$$q_{\text{vib}} = \sum_{v=0} e^{-u(v+1/2)} = \frac{e^{-u/2}}{1 - e^{-u}} \qquad (9\text{-}26)$$

As usual, $u = h\nu/kT$.

This result is to be contrasted with Eq. (8-25). When the vibrational energy levels are measured with respect to the lowest vibrational state rather than with respect to the minimum in the potential-energy curve, the vibrational partition function is

$$q_{\text{vib}} = \sum_{v=0} e^{-vu} = \frac{1}{1 - e^{-u}} \qquad (8\text{-}25)$$

At high temperatures, $u \ll 1$, q_{vib} approaches its classical value $1/u$. We write, at all temperatures,

$$q_{\text{vib}} = \frac{1}{u} \frac{u e^{-u/2}}{1 - e^{-u}} \qquad (9\text{-}27)$$

Consider the dissociation reaction

$$AB \rightleftharpoons A + B \qquad (9\text{-}28)$$

The dissociation equilibrium constant for (9-28) is given by

$$K_{AB} = \left(\frac{2\pi\mu_{AB}kT}{h^2}\right)^{3/2} \frac{h^2}{8\pi^2\mu_{AB}r_e^2 kT} \frac{1}{q_{\text{vib},AB}} e^{-D_e/kT} \quad \text{atoms cc}^{-1} \quad (9\text{-}29)$$

where μ_{AB} is the reduced mass $m_A m_B/(m_A + m_B)$. The first factor in (9-29) results from the ratio of translational partition functions. Note that the last factor, the Boltzmann factor for the dissociation energy, involves D_e, the energy of dissociation from the minimum of the potential-energy curve, since the zero-point energy has been included in q_{vib}.

In the high-temperature limit, $q_{\text{vib}} \to 1/u = kT/h\nu$, and the equilibrium constant for dissociation approaches

$$K = \frac{1}{2^{5/2}\pi^{3/2}} \left(\frac{f}{kT}\right)^{1/2} \frac{1}{r_e^2} e^{-D_e/kT} \quad (9\text{-}30)$$

In the high-temperature limit, there is no mass dependence, and therefore there can be no isotope effect. Incidentally, note that there is no dependence on h, so that there are no quantal effects on the high-temperature equilibrium.

Problem 9-2. Derive Eqs. (9-29) and (9-30) in detail.

Using (9-29) and (9-27), the ratio of the dissociation constants for the two isotopic molecules AB and AB' is

$$\frac{K_{AB'}}{K_{AB}} = \left(\frac{\mu'}{\mu}\right)^{1/2} \frac{u'}{u} \frac{1-e^{-u'}}{1-e^{-u}} \frac{u}{u'} \frac{e^{-u/2}}{e^{-u'/2}} \quad (9\text{-}31)$$

But $(\mu')^{1/2}u' = \mu^{1/2}u$; so (9-31) becomes

$$\frac{K_{AB'}}{K_{AB}} = \frac{(1-e^{-u'})ue^{-u/2}}{(1-e^{-u})u'e^{-u'/2}} \quad (9\text{-}32)$$

This is the desired final result. We introduce further approximations, based on the assumption that the difference in isotopic masses is small, so that $u' = u + \Delta u$, where Δu is small. Note that, if B' is the lighter isotope, Δu is positive. By substitution and expansion one then obtains

$$\frac{K_{AB'}}{K_{AB}} = 1 + \Delta u \left(\frac{1}{2} - \frac{1}{u} + \frac{1}{e^u - 1}\right) \quad (9\text{-}33)$$

At high temperatures, when $u \to 0$, (9-33) approaches

$$\frac{K_{AB'}}{K_{AB}} = 1 + \frac{\Delta u}{u}\left(\frac{u^2}{12}\right) \quad (9\text{-}34)$$

Problem 9-3. Derive Eqs. (9-33) and (9-34).

Equation (9-33) is suitable for calculation of isotopic equilibria involving heavy atoms, where $\Delta u/u \ll 1$. For molecules containing a hydrogen or deuterium atom, $\Delta u/u$ is not small, and (9-32) must be used.

Equation (9-32) is based on the harmonic-oscillator rigid-rotor approximation, and on the integration approximation $q = 1/y$, for the rotational partition function. This last approximation is not valid for H_2, HD, D_2 (and tritium molecules), and quantal effects on the rotational partition function must be considered.

The isotope effect is largest at low temperatures where u becomes large. The limiting form of Eq. (9-33) at low temperatures is

$$\frac{K_{AB'}}{K_{AB}} = 1 + \frac{\Delta u}{2}$$

The difference in dissociation constants is then entirely due to the zero-point energy difference. At low temperatures, where the isotope effect is largest and most interesting, the harmonic-oscillator approximation is accurate, and (9-32) and (9-33) are satisfactory, except possibly for the point noted above concerning the rotational partition function.

For an isotopic exchange reaction, which is not a dissociation reaction, such as $AB + CB' \rightleftharpoons AB' + CB$, there is a Δu for AB' and AB, and a different Δu for CB' and CB; the difference between these determines the isotope effect. Clearly, the largest isotope effect occurs for dissociation reactions.

The theory for polyatomic molecules is very similar to that for diatomic molecules (cf. Sec. 11-15).

9-6. More About Ortho-Para Equilibria. To calculate the relative population of various rotational levels in spectroscopic experiments, it is frequently necessary to use the detailed considerations about nuclear spin statistics discussed previously. For thermodynamic calculations about gaseous molecules, it is sufficient simply to use the symmetry number of 2, except for the various isotopic H_2 molecules. All other gaseous substances condense to the liquid or solid state before the rotational spacings become comparable with kT. It appears, however, that there are some interesting cases of molecules trapped in solid materials at very low temperatures, where detailed considerations about nuclear spin statistics are involved. Thus the NH_2 radical can be trapped in an argon matrix at 4°K.* Spectroscopic evidence indicates that the NH_2 molecule is capable of almost free rotation in this medium. NH_2 is not a diatomic molecule, but there are similar considerations about the symmetry of its rotational states. The moments of inertia are sufficiently small so that at 4°K all the molecules are in the symmetrical lowest rotational state, with zero rotational energy and angular momentum. The electronic wave function of the NH_2 molecule is antisym-

---

* G. W. Robinson and M. McCarty, "Faraday Society Discussion on Free Radical Stabilization," p. 60, Sheffield University, September, 1958; *J. Chem. Phys.*, **28**: 349 (1958), **30**: 999 (1959). S. N. Foner, E. L. Cochran, V. A. Bowers, and C. K. Jen, *Phys. Rev. Lett.*, **1**: 91 (1958).

metric with respect to rotation around the symmetry axis (because the odd electron is mainly in a p_π orbital on the nitrogen atom, and this orbital is perpendicular to the plane of the molecule). Therefore, the allowed spin states of the protons are the symmetrical triplet states $\alpha\alpha$, $2^{-\frac{1}{2}}(\alpha\beta + \beta\alpha)$, $\beta\beta$; but the spin state $2^{-\frac{1}{2}}(\alpha\beta - \beta\alpha)$ does not occur. These considerations are necessary in order to explain the relative intensities of the several lines in the paramagnetic resonance spectrum.* It is probable that the CH_3 radical in a CH_4 matrix at 20°K is undergoing free rotation, and similar, but more complicated, considerations apply.†

PROBLEMS

9-4. By direct summation, evaluate the partition functions at $T = 298.16°$ for (a) ortho H_2; (b) para H_2. Calculate the ratio ortho/para. Calculate the populations of the various rotational states (those which are significantly populated).

9-5. (a) Show that the ratio of isotopic atoms X^A and X^B is the same in the DX species [that is, $(DX^A)/(DX^B)$] and in the X_2 species, if the equilibrium relations (9-18) to (9-20) obtain.

(b) Consider that the mole fractions of X^A and X^B are f_A and f_B, as defined in (9-14a). Start with N_0 moles of pure DX. What is the entropy change for separating this into $f_A N_0$ moles of DX^A and $f_B N_0$ moles of DX^B?

Now start with $N_0/2$ molecules of the equilibrium mixture of X_2^A, X_2^B, $X^A X^B$. Calculate the entropy change for separating this mixture into the three isotopic components.

9-6. An isotope X has nuclear spin 1, and the diatomic molecule X_2 has a $^1\Sigma_g^+$ electronic state. What nuclear spin states for the molecule are associated with the various rotational states of X_2?

The rotational constant $B = h/8\pi^2 Ic$ for X_2 is 5 cm^{-1}.

What, approximately, are the relative abundances of the several nuclear spin states of X_2 at 300°K? At 1°K (assuming gas molecules or freely rotating liquid molecules at 1°K)?

9-7. X and X' are isotopic atoms. BX is a diatomic molecule. Develop a general expression for the equilibrium constant of the reaction

$$BX + X' \rightleftharpoons BX' + X$$

under conditions that the harmonic-oscillator rigid-rotor approximation applies.

9-8. Use data in Herzberg [17] to estimate the equilibrium constant at 300°K for the reaction

$$H-Cl^{35} + Br-Cl^{37} \rightleftharpoons H-Cl^{37} + Br-Cl^{35}$$

* H. M. McConnell, *J. Chem. Phys.*, **29**: 1422 (1958).
† B. Smaller and M. Matheson, *J. Chem. Phys.*, **28**: 1169 (1958); T. Cole, H. O. Pritchard, N. R. Davidson, and H. M. McConnell, *Mol. Phys.*, **1**: 406 (1958).

10

Classical Statistical Mechanics. The Kinetic Theory of Gases

10-1. Classical Phase Space. The idea of describing the motion of a particle or of a system of particles as the motion of a point in phase space and the concept of a volume element in phase space were introduced in Sec. 2-7. If we ignore the possibility of electronic excitation, a free atom has only translational energy and potential energy determined by its position in a gravitational field. From the point of view of classical mechanics, it is a single, structureless particle with three position coordinates and three corresponding conjugate momenta. Its motion can be described as the motion of a point in a six-dimensional phase space. The configuration and motion of a gas containing N atoms can be specified by $3N$ position coordinates and $3N$ conjugate momenta; in a classical description, the state of the gas at any instant is a point in a $6N$-dimensional phase space. For a diatomic molecule (the electronic degrees of freedom being ignored) there is a 12-dimensional phase space. The phase space for a single atom or molecule is frequently called μ space (μ for molecule), while that for an entire system is called γ space (γ for gas).* At the present state of our developments, we are principally concerned with μ space.

There is a close relation between volume in phase space in classical mechanics and the number of quantum states in quantum mechanics. Consider the quantum mechanical problem of a free particle in a one-dimensional box of length L. The allowed wave functions are $(2/L)^{1/2} \sin(\pi s x/L)$, with $s = 1, 2, 3, \ldots$. The wavelength of the particle is $\lambda = 2L/s$, the square of the momentum is $\langle p^2 \rangle = h^2 s^2 / 4L^2$, and the energy is $\epsilon = h^2 s^2 / 8mL^2 = \langle p^2 \rangle / 2m$. The expectation value of the linear momentum $\langle p \rangle$ is zero, because, in a classical description, the wave function corresponds to a particle moving both backward and forward in the box.

---
* Tolman, p. 45 [2].

Sec. 10-1] KINETIC THEORY OF GASES 147

If we regard the energy ϵ as a continuous variable, then an arbitrary ϵ lies between two possible quantized values, with quantum numbers s and $s + 1$, such that

$$\frac{h^2 s^2}{8mL^2} \leq \epsilon \leq \frac{h^2(s+1)^2}{8mL^2}$$

For a large value of ϵ and a corresponding large value of s, the difference between s and $s + 1$ is unimportant. For a large value of ϵ, regarded as a continuous variable, we can say that there are about s quantum states that have an energy between 0 and ϵ; the value of s is the nearest integer to $(8m\epsilon L^2/h^2)^{1/2}$. We can write

$$\text{No. of quantum states between 0 and } \epsilon \approx s \approx \left(\frac{8m\epsilon L^2}{h^2}\right)^{1/2} \quad (10\text{-}1a)$$

In a classical treatment, the motion of the particle occurs in a two-dimensional phase space with orthogonal coordinates x and p. Corresponding to a kinetic energy ϵ, the particle has a momentum p of plus *or* minus $(2m\epsilon)^{1/2}$. The volume in phase space, ϕ, for all motions with energy less than ϵ is

$$\phi = \int_0^L dx \int_{-(2m\epsilon)^{1/2}}^{+(2m\epsilon)^{1/2}} dp = 2L(2m\epsilon)^{1/2} \quad (10\text{-}1b)$$

By comparison of (10-1b) and (10-1a) we see that, for this example, the classical volume in phase space divided by Planck's constant h is the number of quantum states therein.

This is a simple illustration of a general theorem, which we shall state and illustrate but shall not prove. Speaking generally and not very precisely, the theorem asserts that for a mechanical system, with Hamiltonian coordinates $p_1, \ldots, p_n; q_1, \ldots, q_n$, for any region in the $2n$-dimensional classical phase space of volume ϕ there correspond approximately ϕ/h^n allowed quantum states.

The theorem, as stated above, is very useful. If we wish to give a logically more precise statement, we must define more clearly the particular quantum states which are to be associated with any region in phase space.

One simple way of doing this is the following: Consider a system for which the Schrödinger equation is $\hat{H}(p_1, \ldots, p_n; q_1, \ldots, q_n)\psi = \epsilon\psi$. Consider an energy interval ϵ_k to ϵ_j sufficiently large so that there are a large number of quantum states therein.

The motion of a classical particle with the same Hamiltonian can be described as the motion of a point in $2n$-dimensional phase space. If the particle has energy ϵ_k, it is constrained to move on the $(2n-1)$-dimensional hypersurface defined by the equation

$$H(p_1, \ldots, p_n; q_1, \ldots, q_n) = \epsilon_k \quad (10\text{-}2)$$

Particles with energy between ϵ_k and ϵ_j have phase points in that volume of phase space between the two hypersurfaces defined by $H(\mathbf{p},\mathbf{q}) = \epsilon_k$ and $H(\mathbf{p},\mathbf{q}) = \epsilon_j$. (We use vector notation $\mathbf{p}$ and $\mathbf{q}$ for the n-component vectors $p_1, \ldots, p_n$ and $q_1, \ldots, q_n$.)

The general theorem is that the volume of phase space enclosed between these two surfaces is approximately h^n times the number of quantum states in this energy interval.

No. of quantum states between ϵ_k and ϵ_j

$$\approx \frac{1}{h^n} \int_{H(\mathbf{p},\mathbf{q})=\epsilon_k}^{H(\mathbf{p},\mathbf{q})=\epsilon_j} \cdots \int dp_1 \cdots dp_n \, dq_1 \cdots dq_n \quad (10\text{-}3)$$

Another simple illustration of the theorem may be helpful. For a one-dimensional harmonic oscillator, the Hamiltonian is

$$H(p,q) = \frac{p^2}{2\mu} + \frac{fq^2}{2}$$

with f the force constant. In this case, and in general for a conservative system in classical mechanics, we write $\epsilon(p,q)$ and $H(p,q)$ interchangeably. The volume in phase space for all classical motions of energy less than $\epsilon_{\max}$ is the area of the ellipse in pq space, $\epsilon_{\max} = p^2/2\mu + fq^2/2$. The area of an ellipse, $(x/a)^2 + (y/b)^2 = 1$, in xy space is πab, so that, in the present instance, the area in phase space is $2\pi\epsilon_{\max}(\mu/f)^{1/2}$. In terms of the classical vibration frequency $\nu = (1/2\pi)(f/\mu)^{1/2}$, we have

$$\text{Vol. in phase space for } \epsilon < \epsilon_{\max} = \frac{\epsilon_{\max}}{\nu} \quad (10\text{-}4)$$

For the quantum-mechanical harmonic oscillator, the energy levels are $\epsilon_v = h\nu(v + \tfrac{1}{2})$. (For comparison with the classical problem, it is appropriate to measure energy from the bottom of the potential curve and thus include the zero-point energy.) For large ϵ, the number of quantum states with energy between 0 and ϵ is of the order of $v = \epsilon/h\nu$. This is indeed $1/h$ times the volume in phase space [(10-4)].

Problem 10-1. Work out a similar illustration of the number of states and the volume in phase space for a rigid rotor in three dimensions.

The theorem is actually more general than the statement given in Eq. (10-3). As already indicated, for any region in phase space of volume ϕ, not necessarily bounded by two hypersurfaces of constant energy, one can and does say, speaking loosely, that the number of quantum states is ϕ/h^n. Speaking precisely, if the region in phase space is bounded by surfaces on which some physical quantity B is a constant, there are corresponding quantum states which are eigenfunctions of the operator $\hat{B}$ with eigenvalues b. The number of such eigenstates between

b_j and b_k is again approximately h^{-n} times the corresponding volume of phase space.

The relation between volume in phase space and number of quantum states is closely related to the uncertainty principle. Because of the uncertainty principle, a real particle or a mechanical system cannot be exactly located in classical phase space; there is an uncertainty $\delta q\, \delta p \approx h$ for each coordinate and conjugate momentum. The position of the particle in phase space is uncertain within a volume element $\delta p_1 \cdots \delta p_n\, \delta q_1 \cdots \delta q_n \sim h^n$. Since the uncertainty arises from the laws of quantum mechanics, it is natural to suspect that this volume is associated with a single quantum state.

10-2. Quantum Statistical Mechanics and Classical Statistical Mechanics. The single-particle quantum-mechanical partition function is

$$q = \sum_j g_j e^{-\beta \epsilon_j} \tag{10-5a}$$

and the fractional number of molecules with energy ϵ_j is

$$\frac{N_j}{N} = \frac{g_j e^{-\beta \epsilon_j}}{q} \tag{10-5b}$$

For certain systems, the energy levels are closely spaced, so that the functions $e^{-\beta \epsilon_j}$ and $e^{-\beta \epsilon_{j+1}}$ differ but slightly. Under these circumstances, it was possible to evaluate the partition-function sum (10-5a) as an integral. In effect, we regard ϵ_j, not as an exact number, but as a small energy interval, over which the quantity $e^{-\beta \epsilon_j}$ is almost constant and in which the number of states g_j is reasonably large.

According to the theorem stated in the preceding section, we can substitute a volume element in phase space divided by h^n for the number of states g_j:

$$g_j = \frac{dp_1 \cdots dp_n\, dq_1 \cdots dq_n}{h^n} \tag{10-6}$$

The partition function then becomes

$$q = \frac{1}{h^n} \int_{p_1} \cdots \int_{q_n} e^{-\beta \epsilon}\, dp_1 \cdots dp_n\, dq_1 \cdots dq_n \tag{10-7a}$$

For such a classical system, the energy is identical with Hamilton's function $H(p_1, \ldots, q_n)$, and we can write

$$q = \frac{1}{h^n} \int_{p_1} \cdots \int_{q_n} e^{-\beta H(p_1, \ldots, q_n)}\, dp_1 \cdots dq_n \tag{10-7b}$$

The distribution law (10-5b) becomes

$$\frac{dN}{N} = \frac{e^{-\beta H(p_1, \ldots, q_n)}\, dp_1 \cdots dq_n}{q h^n} \tag{10-7c}$$

where dN is the number of systems with coordinates in the volume element $dp_1 \cdots dq_n$ of phase space. The classical phase integral is defined as

$$\zeta = \int_{p_1} \cdots \int_{q_n} e^{-\beta H(p_1, \ldots, p_n; q_1, \ldots, q_n)} \, dp_1 \cdots dp_n \, dq_1 \cdots dq_n \quad (10\text{-}8)$$

Equations (10-7) can then be written

$$q = \frac{\zeta}{h^n} \quad (10\text{-}9a)$$

$$\frac{dN}{N} = \frac{e^{-\beta H(p_1, \ldots, p_n; q_1, \ldots, q_n)} \, dp_1 \cdots dp_n \, dq_1 \cdots dq_n}{\zeta} \quad (10\text{-}9b)$$

The integrals in (10-7) and (10-8) extend over all values of the p's and q's.

Equation (10-9b) is the Boltzmann distribution law in classical statistical mechanics.

There are two distinct ways in which one might try to use classical mechanics as an adequate approximation to quantum mechanics. One of these meanings refers purely to mechanics, not statistical mechanics. For a particular mechanical problem, say, the collision of two particles, it may or it may not be possible to calculate the behavior of the system with sufficient accuracy for the purposes at hand by classical mechanics. There are criteria for deciding this question, but they do not concern us here.

In another sense, in statistical mechanics, we speak of a classical approximation when we calculate the quantum-mechanical partition function as the classical phase integral divided by h^n, as in (10-7a). The necessary condition for the validity of this approximation is that the spacing of the adjacent discrete energy levels ϵ_j and ϵ_{j+1} should be small compared with kT. The ratio of the Boltzmann factors is then very close to unity.

$$\frac{e^{-\epsilon_{j+1}/kT}}{e^{-\epsilon_j/kT}} = e^{-(\epsilon_{j+1}-\epsilon_j)/kT} \approx 1$$

so that the sum in (10-5a) can be replaced by an integral. Such a substitution of the integral for a sum was made in evaluating the rotational partition function, although, in order to discuss the infrared rotational spectrum of a diatomic molecule, it is necessary to use the quantum-mechanical calculation of the energy levels. Thus, in this example, the classical approximation is satisfactory for calculating the partition function, but not the detailed mechanical behavior of the system.

Problem 10-2. Calculate the classical phase integral for a one-dimensional harmonic oscillator, and discuss the conditions under which it is equivalent to the quantum-mechanical partition function.

The reader will note, in the topics discussed in this and the following chapters, many instances where we are essentially using a classical approximation, in one or both of the senses described above.

We use the term *classical statistical mechanics* loosely to describe the approximation in which we use the phase integral divided by h^n for the partition function. There was, of course, a theory of statistical mechanics based on classical mechanics prior to the discovery of quantum theory. In a precise sense, we should use the term classical statistical mechanics only for this theory. Equation (10-9b) is the Maxwell-Boltzmann distribution law of that theory. The equation does not contain the quantum of action, h. The thoughtful reader can observe that under conditions where (10-9b) is a satisfactory approximation to the distribution law, thermodynamic-energy quantities, such as the energy and the enthalpy, can be calculated without reference to quantum mechanics. In quantum statistical mechanics, the entropy change in going from state A to state B is the logarithm of the ratio of the number of system wave functions that correspond to state A and the number of wave functions for state B. There is a corresponding calculation in pure classical statistical mechanics. It involves a ratio of volumes in the phase space (γ space) of the system. This classical calculation gives the correct answer provided that in both states A and B the spacing of the energy levels is small compared with kT.*

However, if state A is at 0°K, there is only one system quantum state occupied. The entropy change in going to state B is, by definition, the absolute entropy of state B. The calculation of this entropy change requires a knowledge of the actual number of wave functions available to the system in state B and of necessity involves the quantum of action, h. This entropy change could not be calculated in pure classical statistical mechanics. (From one point of view, this is because the spacing of the energy levels is not small compared with kT in state A where $T = 0°K$.) Furthermore, for any successful entropy-change calculations in classical statistical mechanics, it was necessary arbitrarily to introduce a factor of $1/N!$ (N being the number of particles) in order to get consistent results. This factor, which is due to the indistinguishability of identical particles and the symmetry requirements of wave functions, was understandable only after the discovery of quantum theory.

10-3. Equipartition of Energy. Consider, for example, a one-dimensional harmonic oscillator in classical statistical mechanics. The energy

---

* The principles of classical statistical mechanics are discussed in a number of places, most notably by J. W. Gibbs, Elementary Principles in Statistical Mechanics, in "Collected Works of J. Willard Gibbs," vol. 2, part. 1, Longmans, New York, 1931; see also Tolman, sec. 51 [2].

is given by
$$H(p,q) = \frac{p^2}{2\mu} + \tfrac{1}{2}fq^2 \qquad (10\text{-}10)$$

The phase integral is
$$\zeta = \int_{-\infty}^{+\infty}\int_{-\infty}^{+\infty} e^{-\beta p^2/2\mu} e^{-\beta f q^2/2}\, dp\, dq = \left(\frac{2\pi\mu}{\beta}\right)^{1/2} \left(\frac{2\pi}{f\beta}\right)^{1/2}$$

The fractional number of systems with coordinates in the interval p, $p + dp$, q, $q + dq$ is
$$\frac{dN}{N} = \frac{e^{-\beta H}\, dp\, dq}{\zeta}$$

The average kinetic energy of a particle is therefore given by
$$\bar{K} = \int \frac{p^2}{2\mu} \frac{dN}{N} = \frac{[\int (p^2/2\mu) e^{-\beta p^2/2\mu}\, dp][\int e^{-\beta f q^2/2}\, dq]}{(\int e^{-\beta p^2/2\mu}\, dp)(\int e^{-\beta f q^2/2}\, dq)} \qquad (10\text{-}11)$$

The integrals over q cancel. If we call
$$\zeta_K = \int e^{-\beta p^2/2\mu}\, dp = \left(\frac{2\pi\mu}{\beta}\right)^{1/2} \qquad (10\text{-}12)$$

then the integral over p in the numerator of (10-11) is $-\partial \zeta_K/\partial \beta$ and
$$\bar{K} = -\frac{\partial(\ln \zeta_K)}{\partial \beta} = \frac{1}{2\beta} = \frac{kT}{2} \qquad (10\text{-}13)$$

In the same way, the average potential energy of a particle can be written as
$$\bar{U} = \int \frac{fq^2}{2} \frac{dN}{N}$$

and proceeding just as before, it will be found that
$$\bar{U} = -\frac{\partial[\ln (\int e^{-\beta f q^2/2}\, dq)]}{\partial \beta} = \frac{1}{2\beta} = \frac{kT}{2} \qquad (10\text{-}14)$$

The results (10-13) and (10-14) are illustrations of the general theorem of equipartition of energy. Speaking loosely, the theorem says that the average energy is $kT/2$ for any Hamiltonian coordinate (a p or a q) for which the Hamiltonian can be expressed as a quadratic function of the coordinate and for which classical statistical mechanics is applicable.

A precise statement of the theorem, with a general derivation, will now be given.* Suppose that, for a molecular system, the energy is separable and there are some degrees of freedom for which the spacing of energy

---

* The derivation is taken from Fowler and Guggenheim, pp. 121–124 [1].

levels is small compared with kT so that in the statistical sense they are "classical," while other degrees of freedom can be treated quantum-mechanically. The energy can then be written as

$$\epsilon_{\text{total}} = \epsilon_{\text{cl}} + \epsilon_{\text{qu}}$$

The distribution of energy in the classical degrees of freedom is independent of the distribution of energy in the quantum-mechanical degrees of freedom. We focus our attention on ϵ_{cl} and omit the subscript. If there are n classical degrees of freedom, the kinetic energy is $K = \sum_{i=1}^{n} p_i^2/2m_i$, where the m_i are generalized masses. The m_i may be functions of the coordinates; for example, for a rigid rotor with moment of inertia I, $K = p_\theta^2/2I + p_\phi^2/2I \sin^2 \theta$, so that $m_\phi = I \sin^2 \theta$. However, the m_i are not functions of any of the p's. There are some potential-energy terms which are quadratic in a coordinate, and there may be some part of the potential energy which is not. The energy in the classical degrees of freedom can then be written as

$$\epsilon(\mathbf{p},\mathbf{q}) = H(\mathbf{p},\mathbf{q}) = \sum_{i=1}^{n} \frac{p_i^2}{2m_i} + \sum_{j=1}^{t} \frac{f_j q_j^2}{2} + V(q_{t+1}, \ldots, q_n) \quad (10\text{-}15)$$

The derivation will be applicable only for those terms for which the m_i and the f_j are not functions of $q_1, \ldots, q_t$, but they can be functions of $q_{t+1}, \ldots, q_n$. The average energy in the coordinates $p_1, \ldots, p_n$; $q_1, \ldots, q_t$ which satisfy these conditions is

$$\overline{\epsilon(p_1, \ldots, p_n; q_1, \ldots, q_t)} = \frac{\int_{p_1} \cdots \int_{q_n} \left[\sum_{1}^{n} (p_i^2/2m_i) + \sum_{1}^{t} (f_j q_j^2/2)\right] e^{-\beta \epsilon(\mathbf{p},\mathbf{q})} \, dp_1 \cdots dq_n}{\zeta} \quad (10\text{-}16)$$

where ζ is the phase integral for the n classical degrees of freedom.

$$\zeta = \int_{p_1} \cdots \int_{q_n} e^{-\beta \epsilon(\mathbf{p},\mathbf{q})} \, dp_1 \cdots dq_n \quad (10\text{-}17)$$

The average energy in (10-16) is the sum of $n + t$ terms of which the first one is typical. For compactness of notation, let $\epsilon_{2n}(p_2, \ldots, p_n;$ $q_1, \ldots, q_n)$ be the energy in all the terms other than p_1, so that, referring to (10-15),

$$\epsilon(\mathbf{p},\mathbf{q}) = \frac{p_1^2}{2m_1} + \epsilon_{2n}$$

Then the first term in (10-16) is

$$\overline{\epsilon(p_1)} = \frac{\int_{p_2} \cdots \int_{q_n} e^{-\beta \epsilon_{2n}(p_2,\ldots,q_n)} dp_2 \cdots dq_n \int_{-\infty}^{+\infty} (p_1^2/2m_1) e^{-\beta p_1^2/2m_1} dp_1}{\int_{p_2} \cdots \int_{q_n} e^{-\beta \epsilon_{2n}} dp_2 \cdots dq_n \int_{-\infty}^{+\infty} e^{-\beta p_1^2/2m_1} dp_1} \quad (10\text{-}18)$$

As indicated, we can integrate over p_1 first, since none of the other terms involves p_1. We have

$$\int_{-\infty}^{+\infty} e^{-\beta p_1^2/2m_1} dp_1 = \left(\frac{2\pi m_1}{\beta}\right)^{1/2} \qquad \int_{-\infty}^{+\infty} \frac{p_1^2}{2m_1} e^{-\beta p_1^2/2m_1} dp_1 = \frac{1}{2\beta}\left(\frac{2\pi m_1}{\beta}\right)^{1/2}$$

It may not be possible to cancel the factor $(2\pi m_1/\beta)^{1/2}$ which occurs in both the numerator and denominator integrals of (10-18) since m_1 may be a function of the variables $q_{t+1}, \ldots, q_n$. However, this factor contributes equally in the numerator and denominator to the rest of the integrals, which are therefore equal. We are left with

$$\overline{\epsilon(p_1)} = \frac{1}{2\beta} = \frac{kT}{2} \quad (10\text{-}19a)$$

and
$$\overline{\epsilon(p_1,\ldots,p_n;q_1,\ldots,q_t)} = \frac{n+t}{2} kT \quad (10\text{-}19b)$$

This is the theorem of the equipartition of energy. In classical statistical mechanics, each translational degree of freedom $p_i^2/2m_i$ and each potential-energy term of the form $f_j q_j^2/2$ has an average energy of $kT/2$, subject to the restrictions already noted on the m_i and f_j.

Problem 10-3. Illustrate the theorem of the equipartition of energy for the various forms of energy of a diatomic molecule at low temperatures ($h\nu \gg kT$) and at high temperatures ($h\nu \ll kT$).

The derivation requires that the limits of integration of the coordinates of interest be $-\infty$ to $+\infty$. Suppose that, for a polyatomic molecule, there is an angle variable ϕ with a potential energy of bending of $\frac{1}{2} f \phi^2$, but with the range of ϕ restricted, say, $-\pi/2 \leq \phi \leq \pi/2$ (cf. the discussion of restricted rotation, Chap. 11). If the force constant f is large enough so that $f(\pi/2)^2/2kT \gg 1$, then the Boltzmann factor in the phase integral is negligible at the limits of integration and

$$\int_{-\pi/2}^{+\pi/2} e^{-f\phi^2/2kT} d\phi \approx \int_{-\infty}^{+\infty} e^{-f\phi^2/2kT} d\phi = \left(\frac{2\pi kT}{f}\right)^{1/2}$$

If this approximation is valid, the equipartition theorem applies to the kinetic energy and the potential energy for this bending mode. In general, of course, the harmonic potential is an approximation to the true potential which holds for small amplitude vibrations, whether they be stretches or bends. At very high temperatures, where the average amplitude of vibration exceeds the harmonic limits, the equipartition expression for the average potential energy fails.

10-4. The Maxwell Velocity Distribution. When applied to the translational energy and coordinates of the particles of a perfect gas, the general distribution law (10-9b) becomes

$$\frac{dN(p_x,p_y,p_z,x,y,z)}{N} = \frac{e^{-(p_x^2+p_y^2+p_z^2)/2mkT} e^{-V(x,y,z)/kT} \, dp_x \, dp_y \, dp_z \, dx \, dy \, dz}{\int_p \cdots \int_z e^{-(p_x^2+p_y^2+p_z^2)/2mkT} e^{-V(x,y,z)/kT} \, dp_x \, dp_y \, dp_z \, dx \, dy \, dz} \quad (10\text{-}20)$$

where N is the total number of molecules in the system and

$$dN(p_x,p_y,p_z,x,y,z)$$

is the number which have momentum and position coordinates in the interval p_x to $p_x + dp_x$, ..., z to $z + dz$. (In describing such distributions, we shall use the concise expression "the number of molecules that have momentum and position coordinates p_x, p_y, p_z, x, y, z" when it is clear that we mean the number in the interval p_x to $p_x + dp_x$, etc.) In deriving (10-20) from (10-9b), we integrate over any internal degrees of freedom of the molecules, rotational, vibrational, and electronic, since these are separable from the translational degrees of freedom.

It is to be noted that Eq. (10-20) applies to the molecules in a perfect gas. That is, it is assumed that there are no intermolecular forces, and the energy of any one molecule is independent of the coordinates of the other molecules. The potential energy $V(x,y,z)$ is a function of the position of a molecule in space, irrespective of the positions of the other molecules, and it is supposed to represent the influence of an external field—centrifugal, gravitational, or electric, for example.

It is apparent, upon examination of (10-20), that the distribution of molecules in space (xyz) is independent of the distribution over the momenta. By integrating over the space coordinates, we obtain an expression for the number of molecules $dN(p_x,p_y,p_z)$ that have momentum coordinates p_x, p_y, p_z:

$$\frac{dN(p_x,p_y,p_z)}{N} = \frac{e^{-(p_x^2+p_y^2+p_z^2)/2mkT} \, dp_x \, dp_y \, dp_z}{\iiint e^{-(p_x^2+p_y^2+p_z^2)/2mkT} \, dp_x \, dp_y \, dp_z} \quad (10\text{-}21a)$$

$$\frac{dN(p_x,p_y,p_z)}{N} = \frac{e^{-(p_x^2+p_y^2+p_z^2)/2mkT} \, dp_x \, dp_y \, dp_z}{(2\pi mkT)^{3/2}} \quad (10\text{-}21b)$$

The indicated integration in the denominator of (10-21a) has been performed in (10-21b).

Of course, Eq. (10-21) for the probability that a molecule has certain values of p_x, p_y, p_z is equivalent to the quantum-mechanical equation derived in Chap. 6 for the probability that a molecule has the translational quantum numbers s_x, s_y, s_z.

In the most usual laboratory situation, there are no external fields—centrifugal, gravitational, or electric, for example—which significantly affect the potential energy of the molecules, $V(x,y,z)$, of Eq. (10-20). If the potential energy is constant, the distribution of molecules in space is uniform. However, the kinetic energy and the potential energy are separable; therefore, when such fields are present, they do not influence the momentum distribution as given by (10-21), although they do affect the density of molecules in space.*

We can also be interested in the number of molecules that have velocities between v_x and $v_x + dv_x$, v_y and $v_y + dv_y$, v_z and $v_z + dv_z$. But $p_x = mv_x$, etc., so that

$$\frac{dN(v_x,v_y,v_z)}{N} = \left(\frac{m}{2\pi kT}\right)^{3/2} e^{-m(v_x^2+v_y^2+v_z^2)/2kT} \, dv_x \, dv_y \, dv_z \qquad (10\text{-}22)$$

This is the usual statement of the Maxwell (or Maxwell-Boltzmann) velocity distribution.

Let the velocity be described in terms of its scalar magnitude v (with $v^2 = v_x^2 + v_y^2 + v_z^2$), and let the direction of the velocity vector be specified with the usual polar angles θ and ϕ, so that $v_z = v \cos \theta$, $v_x = v \sin \theta \cos \phi$, $v_y = v \sin \theta \sin \phi$. A volume element in velocity space is $v^2 \sin \theta \, d\theta \, d\phi \, dv$. Then, from (10-22)

$$\frac{dN(v,\theta,\phi)}{N} = \left(\frac{m}{2\pi kT}\right)^{3/2} e^{-mv^2/2kT} v^2 \sin \theta \, d\theta \, d\phi \, dv \qquad (10\text{-}23a)$$

Note that the allowed limits for v_x, v_y, and v_z are $(-\infty < v_x < +\infty)$, but v is only positive, and $0 \leq v < \infty$. The quantity $\sin \theta \, d\theta \, d\phi$ is a differential unit of solid angle, $d\omega$, and we rewrite (10-23a) as

$$\frac{dN(v,\theta,\phi)}{N} = \left(\frac{m}{2\pi kT}\right)^{3/2} e^{-mv^2/2kT} v^2 \, dv \, d\omega \qquad (10\text{-}23b)$$

which emphasizes that the velocity distribution is isotropic in space.

To obtain the distribution in the magnitude of the velocity, irrespective of direction, we integrate over θ and ϕ, thus getting a factor of 4π:

$$\frac{dN(v)}{N} = 4\pi \left(\frac{m}{2\pi kT}\right)^{3/2} e^{-mv^2/2kT} v^2 \, dv \qquad (10\text{-}24)$$

* The situation is somewhat more complicated, however, for charged particles in a magnetic field. The distribution of the generalized momenta, $\mathbf{p} - (e/c)\mathbf{A}$, where $\mathbf{A}$ is the magnetic vector potential, satisfies (10-21a), but the distribution of the mechanical momenta, $p_x = m\dot{x}$, etc., does not.

Several interesting average velocities can be computed from the velocity distribution laws. The average velocity in the x direction, which is given by

$$\bar{v}_x = \frac{\int v_x \, dN(v_x, v_y, v_z)}{N}$$

$$= \left(\frac{m}{2\pi kT}\right)^{3/2} \int\!\!\int\!\!\int_{-\infty}^{+\infty} v_x e^{-m(v_x^2 + v_y^2 + v_z^2)/2kT} \, dv_x \, dv_y \, dv_z$$

$$= 0$$

is zero by symmetry.

The mean-square velocity in any one direction is

$$\overline{v_x^2} = \left(\frac{m}{2\pi kT}\right)^{3/2} \int\!\!\int\!\!\int_{-\infty}^{+\infty} v_x^2 e^{-m(v_x^2 + v_y^2 + v_z^2)/2kT} \, dv_x \, dv_y \, dv_z$$

$$= \left(\frac{m}{2\pi kT}\right)^{1/2} \int_{-\infty}^{+\infty} v_x^2 e^{-mv_x^2/2kT} \, dv_x = \frac{kT}{m} \qquad (10\text{-}25a)$$

The mean kinetic energy of motion in the x direction is

$$\tfrac{1}{2} m \overline{v_x^2} = \tfrac{1}{2} kT \qquad (10\text{-}25b)$$

Clearly then

$$\overline{v^2} = \overline{v_x^2} + \overline{v_y^2} + \overline{v_z^2} = \frac{3kT}{m} \qquad (10\text{-}26a)$$

and

$$\tfrac{1}{2} m \overline{v^2} = \tfrac{3}{2} kT \qquad (10\text{-}26b)$$

The important results (10-25b) and (10-26b) are, of course, examples of the equipartition of energy theorem.

The mean scalar magnitude of the velocity is obtained from the distribution law (10-24).

$$\bar{v} = \int \frac{v \, dN(v)}{N} = \int_0^\infty 4\pi \left(\frac{m}{2\pi kT}\right)^{3/2} v^3 e^{-mv^2/2kT} \, dv$$

$$= \left(\frac{8kT}{\pi m}\right)^{1/2} \qquad (10\text{-}27)$$

A plot of the distribution function of (10-24),

$$\frac{1}{N} \frac{dN(v)}{dv} = 4\pi \left(\frac{m}{2\pi kT}\right)^{3/2} e^{-mv^2/2kT} v^2$$

is displayed in Fig. 10-1. The most probable velocity

$$v_{\text{mp}} = \left(\frac{2kT}{m}\right)^{1/2} \qquad (10\text{-}28)$$

is the velocity for the maximum of the curve.

Numerical expressions for the several average velocities in terms of the gram-molecular weight M are*

$$v_{mp} = \left(\frac{2kT}{m}\right)^{1/2} = \left(\frac{2RT}{M}\right)^{1/2} = 12{,}895 \left(\frac{T}{M}\right)^{1/2} \quad \text{cm sec}^{-1}$$

$$\bar{v} = \left(\frac{8kT}{\pi m}\right)^{1/2} = \left(\frac{8RT}{\pi M}\right)^{1/2} = 14{,}551 \left(\frac{T}{M}\right)^{1/2} \tag{10-29}$$

$$(\overline{v^2})^{1/2} = \left(\frac{3kT}{m}\right)^{1/2} = 15{,}794 \left(\frac{T}{M}\right)^{1/2}$$

10-5. The Number of Molecules Hitting a Surface. Effusion. Consider a sample of gas with N molecules per unit volume in a semi-infinite

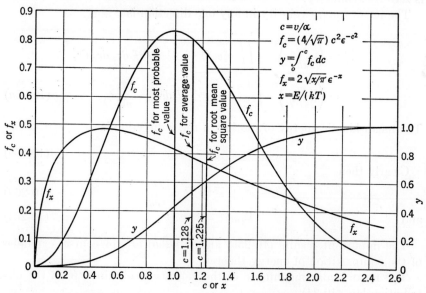

FIG. 10-1. Plots illustrating the Maxwell-Boltzmann distribution law. The function f_c shows the distribution function for the scalar magnitude of the velocity [Eq. (10-24)] in terms of the variable $c = v/v_{mp}$, where $v_{mp} = (2kT/m)^{1/2}$ [Eq. (10-28)]. In the figure, the quantity which we have called v_{mp} in the text is denoted by α. The function f_x is the distribution function for the energy in terms of the variable $x = \epsilon/kT = mv^2/2kT$. The function $y(c)$ is the fraction of molecules which have a velocity less than c. (*From S. Dushman, "Vacuum Technique," p. 11, Wiley, New York, 1949.*)

container with its top boundary the xy plane. We assume that the velocity distribution is isotropic, but, for the moment, it need not be the Maxwell distribution, that is

$$\frac{dN(v,\theta,\phi)}{N} = \frac{f(v)\, d\omega\, dv}{4\pi} \tag{10-30a}$$

$$\frac{dN(v,\theta,\phi)}{N} = \frac{f(v)\, \sin\theta\, d\theta\, d\phi}{4\pi}\, dv \tag{10-30b}$$

* S. Dushman, "Vacuum Technique," p. 10, Wiley, New York, 1949.

where $d\omega$ is a differential unit of solid angle and θ and ϕ are the polar angles of the velocity vector with respect to the z axis. By integrating over all angles, we get

$$\frac{dN(v)}{N} = f(v)\, dv \tag{10-31}$$

The distribution function $f(v)$ is normalized so that $\int_0^\infty f(v)\, dv = 1$. It is convenient to note at this time that the average velocity is

$$\bar{v} = \frac{\int v\, dN(v)}{N} = \int_0^\infty v f(v)\, dv \tag{10-32}$$

From (10-24), for the Maxwell distribution, the distribution function $f(v)$ is given by

$$f(v) = 4\pi \left(\frac{m}{2\pi kT}\right)^{3/2} v^2 e^{-mv^2/2kT} \tag{10-33}$$

We shall inquire into the rate at which molecules strike an element of surface dA located at the origin and in the xy plane. Assume for the

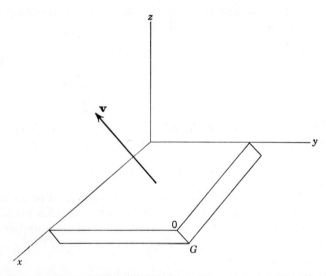

FIG. 10-2. All the molecules with velocity vector (v,θ,ϕ) in the parallelepiped with base dA in the xy plane and with the other side GO of length $v\, dt$, and direction (θ,ϕ) of the velocity vector $\mathbf{v}$, will strike dA in the time interval dt.

moment that the diameter of the molecules is vanishingly small so that there are no molecular collisions. A molecule with velocity v moves a distance $v\, dt$ during an interval of time dt. Therefore, the number of molecules $de(v,\theta,\phi,t,A)$ moving up from below with velocity v (that is, we reemphasize, with velocity in the interval v, $v + dv$) in the direction θ, ϕ that will strike dA during the interval dt are all the molecules with the specified velocity vector (v,θ,ϕ) in the little parallelepiped (Fig. 10-2) with base dA, length of the other side of $v\,dt$ (GO in the figure), and with the direction of the side GO being the direction of the motion of the molecule, that is, θ and ϕ.

The vertical height of the prism is $v\cos\theta\,dt$, and its volume is $v\cos\theta\,dt\,dA$. The number of molecules with the specified velocity vector in the prism is therefore from (10-30) $Nf(v)\,dv(d\omega/4\pi)v\cos\theta\,dt\,dA$,

$$de(v,\theta,\phi,t,A) = \frac{Nf(v)\,dv\,d\omega}{4\pi} v\cos\theta\,dt\,dA \qquad (10\text{-}34)$$

We omit dt and talk about $de(v,\theta,\phi,A)$, the number of molecules of the specified kind striking dA per unit time, or the collision rate,

$$de(v,\theta,\phi,A) = \frac{Nvf(v)\cos\theta\,d\omega\,dv\,dA}{4\pi} \qquad (10\text{-}35a)$$

or $$de(v,\theta,\phi,A) = \frac{Nvf(v)\cos\theta\sin\theta\,d\theta\,d\phi\,dv\,dA}{4\pi} \qquad (10\text{-}35b)$$

The total number of molecules $de(\theta,\phi,A)$ striking dA per unit time while moving in the direction θ, ϕ is obtained by integrating (10-35) over all values of v; we recognize from (10-32) that the resulting integral $\int vf(v)\,dv$ contributes a factor of $\bar{v}$, the average velocity.

$$de(\theta,\phi,A) = \frac{N\bar{v}\cos\theta\,d\omega\,dA}{4\pi} \qquad (10\text{-}36a)$$

$$de(\theta,\phi,A) = \frac{N\bar{v}\cos\theta\sin\theta\,d\theta\,d\phi\,dA}{4\pi} \qquad (10\text{-}36b)$$

The total rate at which molecules strike the surface, $e(A)$, is obtained by integrating over all *upward* directions ($0 \le \theta \le \pi/2; 0 \le \phi \le 2\pi$),

$$e(A) = \tfrac{1}{4}N\bar{v}\,dA \qquad (10\text{-}37)$$

Equations (10-35) to (10-37) are the desired results. For an arbitrary isotropic velocity distribution, the number of molecules striking unit area of a surface in unit time is $N\bar{v}/4$ [(10-37)], and the number striking per unit solid angle at angle θ with respect to the normal to the surface is proportional to $\cos\theta$ [(10-36a)].

For the Maxwell velocity distribution we have

$$e(A) = \tfrac{1}{4}N\left(\frac{8kT}{\pi m}\right)^{1/2} dA \qquad (10\text{-}38)$$

Molecules actually have a finite diameter and collide with each other. To a first approximation, this does not affect the number of collisions with the walls. There are several ways of arguing this point. In the first place, we can take the time interval dt small, so that the length $v\,dt$ of the prism of Fig. 10-2 is short compared with a mean-free path and there is not much probability that a molecule in the prism moving toward the wall will be deflected from its course by a collision. In the second place, even for a longer prism, for every molecule starting in the prism and directed properly which is scattered by collision into a new direction of motion, θ', ϕ', there will be, on the average, another collision in which some molecule moving in the direction θ' and ϕ' will be scattered into the direction θ and ϕ by collision and will hence strike the surface. These arguments are not valid when the molecular collision diameter becomes comparable in magnitude with the mean-free path; but then the ideal-gas law does not hold under these conditions, either.

The precise kinetic-theory formulae which we have derived for the number of molecules striking the surface at any given velocity and in any given direction can be used to obtain a kinetic-theory derivation for the pressure of a perfect gas (Prob. 10-4), which we have previously derived by more general arguments.

An important practical application of the formulae is the calculation of the number of molecules escaping from a small, thin hole in a wall of a container. If there is a hole at dA with linear dimensions small compared with the mean-free path, and a vacuum on the other side, a molecule escaping through the hole will not suffer any collisions as it goes through the hole. The flow out of the hole is then *molecular flow*, or effusion, as distinct from hydrodynamic flow. The number of molecules effusing in unit time from a small hole is given by (10-35) to (10-37). The rate of effusion of molecules through a small aperture is used to measure low vapor pressures of substances, especially at high temperatures. The effusion formulae are also important for designing a molecular beam apparatus. Our choice of the symbol e for the collision rate with an element of surface was to suggest the phrase "the number of molecules *effusing* (or escaping)."

There are several additional relationships we should like to establish for the purposes of the next section. In the first place, the calculations are not limited to collisions with a plane surface. In particular, the fundamental formulae (10-35) and (10-36) apply to each element dA of an arbitrary surface, θ and ϕ being the angles of the velocity vector with respect to the normal to that particular element of surface. A result we need in the next section is that the total number of molecules colliding with the surface of a sphere of radius D in a gas is, from (10-37),

$$e(\text{sphere}) = \tfrac{1}{4} N \bar{v} A = \tfrac{1}{4} N \bar{v} (4\pi D^2) = \pi N \bar{v} D^2 \qquad (10\text{-}39)$$

In (10-35b), the collision rate is given as a function of the angle (θ,ϕ) of the velocity vector with respect to the normal to the surface. We wish to reexpress this result in terms of the velocity components parallel to the normal to the surface, $v_\parallel$ (that is, perpendicular to the surface), and perpendicular to the normal, $v_\perp$ (or parallel to the surface). If our axes are chosen so that the z axis is the normal to the surface,

$$v_\parallel = v \cos \theta$$
$$v_\perp = v \sin \theta \quad (10\text{-}40a)$$

By straightforward use of Jacobians, or by recognizing the identity with the transformation between cartesian and two-dimensional polar coordinates, the appropriate relation for the differential unit of area in velocity space is

$$dv_\perp \, dv_\parallel = v \, dv \, d\theta \quad (10\text{-}40b)$$

Now integrate (10-35b) over all values of ϕ (since the relationships between $v_\perp$ and $v_\parallel$ with v and θ are independent of the azimuthal angle ϕ),

$$de(v,\theta,A) = \tfrac{1}{2} N v f(v) \cos \theta \sin \theta \, d\theta \, dv \, dA$$

Insert the Maxwell velocity distribution function (10-33) for $f(v)$,

$$de(v,\theta,A) = 2\pi \left(\frac{m}{2\pi kT}\right)^{3/2} N e^{-mv^2/2kT} v^2 \sin \theta \cos \theta \, v \, dv \, d\theta \, dA \quad (10\text{-}41)$$

The terms in (10-41) are grouped in a suggestive way. We now substitute for v and θ in terms of $v_\perp$ and $v_\parallel$. We recall that

$$\tfrac{1}{2} mv^2 = \tfrac{1}{2} mv_\perp^2 + \tfrac{1}{2} mv_\parallel^2$$

and write the energy either way, as convenient.

$$de(v_\perp,v_\parallel,A) = 2\pi \left(\frac{m}{2\pi kT}\right)^{3/2} N e^{-mv^2/2kT} v_\perp \, dv_\perp \, v_\parallel \, dv_\parallel \, dA \quad (10\text{-}42a)$$

A more interesting way to write (10-42a) is

$$\frac{de(v_\parallel,v_\perp,A)}{\tfrac{1}{4}(8kT/\pi m)^{1/2} N \, dA} = e^{-mv^2/2kT} \frac{v_\perp \, dv_\perp \, v_\parallel \, dv_\parallel}{(kT/m)^2} \quad (10\text{-}42b)$$

The denominator of the left-hand side is the total number of collisions on dA; so (10-42b) is an expression for the fractional number of collisions of the appropriate type. The differential unit $(v_\perp \, dv_\perp \, v_\parallel \, dv_\parallel)/(kT/m)^2$ was chosen to be dimensionless.

In terms of the *linear* momenta $p_\parallel$ and $p_\perp$, with $p_\parallel = mv_\parallel$, $p_\perp = mv_\perp$, and $p^2 = p_\perp^2 + p_\parallel^2$, (10-42b) becomes

$$\frac{de(p_\perp,p_\parallel,A)}{\tfrac{1}{4}(8kT/\pi m)^{1/2} N \, dA} = e^{-p^2/2mkT} \frac{p_\perp \, dp_\perp \, p_\parallel \, dp_\parallel}{(mkT)^2} \quad (10\text{-}43)$$

The velocities and momenta $v_\parallel$, $v_\perp$, $p_\parallel$, $p_\perp$ are intrinsically positive and vary from zero to infinity.

10-6. Molecular Collisions. We are concerned with binary collisions in a dilute gas and need to consider only interactions between one pair of molecules at a time. There are two particles with masses m_A and m_B

and coordinates x_A, y_A, z_A and x_B, y_B, z_B. As seen in Chap. 2, the equations of motion of such a system can be separated into the equations of motion of a "particle" of mass $m_A + m_B$, with the coordinates of the center of mass of the system, and the equations of motion of a "particle" representing the relative motions of the system, with the particle mass equal to the reduced mass, $\mu = m_A m_B/(m_A + m_B)$, and with coordinates, the relative coordinates, $x = x_B - x_A$, $y = y_B - y_A$, $z = z_B - z_A$. There are conjugate relative momenta p_x, p_y, p_z. Let r be the distance between the particles and $U(r)$ the potential of interaction. The simplest and crudest model for a collision is the rigid-sphere model, in which $U(r) = 0$ for $r > D$ and $U(r) = \infty$ for $r < D$ where D is the collision diameter of the two particles. As shown in Fig. 10-3, D can be taken as $(D_A + D_B)/2$, where D_A and D_B are the diameters of the atoms A and B.

We construct a sphere of *radius D* (although D is a collision diameter) around the origin and call this the collision sphere. We can say that a collision occurs whenever the mass point representing the relative coordinates strikes the surface of the collision sphere. The problem of intermolecular collisions is thus reduced to the problem of collisions with a surface considered in the previous section.

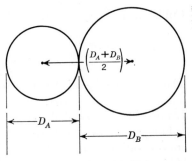

FIG. 10-3. The collision diameter or distance between centers of atoms A and B, with diameters D_A and D_B, respectively, is $D = (D_A + D_B)/2$.

The relative energy and the center-of-mass energy are separable, and at statistical equilibrium the distribution of relative energies will be a Boltzmann distribution which is independent of the distribution for the external energy. By classical statistics, the probability $df(p_x, p_y, p_z)$ of finding the "relative particle" with momenta p_x, p_y, p_z is

$$df(p_x, p_y, p_z) = \frac{e^{-(p_x^2 + p_y^2 + p_z^2)/2\mu kT} \, dp_x \, dp_y \, dp_z}{(2\pi \mu kT)^{3/2}} \qquad (10\text{-}21b)$$

The probability of finding certain velocities is given by the corresponding expressions (10-22) to (10-24).

To begin with, let there be one A atom and one B atom in unit volume. The number of collisions e_{AB} of the relative particle which has an isotropic Boltzmann distribution of velocities with the collision sphere of radius D has already been calculated in Eq. (10-39).

$$e_{AB} = \pi \bar{v} D^2 = \left(\frac{8\pi kT}{\mu}\right)^{1/2} D^2 = \left(\frac{8\pi RT}{M}\right)^{1/2} D^2 \qquad (10\text{-}44)$$

with R the gas constant and M the reduced molecular weight $M_A M_B/(M_A + M_B)$.

If there are now N_A and N_B molecules per unit volume in the gas, the number of collisions per second (the collision rate) for any one A molecule with the B molecules is

$$\text{Collision rate per } A \text{ molecule} = \left(\frac{8\pi kT}{\mu}\right)^{1/2} D^2 N_B = e_{AB} N_B \quad (10\text{-}45)$$

The total rate of collisions per unit volume is $e_{AB} N_A N_B$.

For collisions between identical molecules, this procedure counts each collision twice, and the total number of collisions is given by

$$\text{Collision rate for identical molecules} = \frac{1}{2}\left(\frac{8\pi kT}{\mu}\right)^{1/2} D^2 N_A (N_A - 1)$$

$$\approx \frac{1}{2}\left(\frac{8\pi kT}{\mu}\right)^{1/2} D^2 N_A^2 \quad (10\text{-}46)$$

The factor of $\tfrac{1}{2}$ in (10-46) corresponds to the division by a symmetry number of 2 in the partition function of a homopolar diatomic molecule. It must also be recalled that μ in (10-46) is the reduced mass, or $m_A/2$.

For practical computation, it is often convenient to use molar concentration units. Let $[B]$ be concentration in moles per liter; $[B] = 1{,}000 N_B/N_0$. For this purpose, we choose to rewrite (10-45) because the quantities equated there do not contain dimensions of concentration units;

$$\text{Collision rate per } A \text{ molecule} = \left(\frac{8\pi kT}{\mu}\right)^{1/2} \frac{D^2 N_0 [B]}{1{,}000}$$

$$= e'_{AB}[B] \quad (10\text{-}47a)$$

The collision number e'_{AB} is given by

$$e'_{AB} = \left(\frac{8\pi RT}{M}\right)^{1/2} \frac{D^2 N_0}{1{,}000}$$

$$= 2.753 \times 10^9 \left(\frac{T}{M}\right)^{1/2} D^2 \quad \text{liters mole}^{-1} \text{ sec}^{-1} \text{ with } D \text{ in Å} \quad (10\text{-}47b)$$

We are interested in subdividing the collisions into certain classes. At the moment of collision, there is an angle of the relative velocity vector with respect to the line of centers of the two atoms. In our language, this is the angle between the velocity vector and the normal to the collision sphere at the point where the "relative particle" collides with the collision sphere. We shall be concerned with the components of velocity parallel to the line of centers and perpendicular to the line of centers at the moment of impact (Fig. 10-4). Thus, for unit concentration of particles A and B, the probability of a collision on an element dA of the collision sphere with velocity components $v_\parallel$ and $v_\perp$, along and perpendicular to the line of centers, is given by (10-42b),

$$de(v_\perp,v_\|,A) = \frac{1}{4}\left(\frac{8kT}{\pi\mu}\right)^{1/2} dA \frac{e^{-\mu v^2/2kT}v_\perp \, dv_\perp \, v_\| \, dv_\|}{(kT/\mu)^2} \quad (10\text{-}42b)$$

All elements of area on the surface of the collision sphere are equivalent;

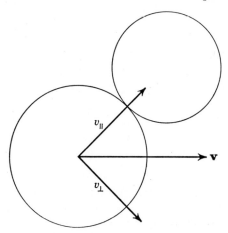

FIG. 10-4. At the moment of collision, the velocity vector **v** for the relative velocity can be resolved into components $v_\|$ along the line of centers and $v_\perp$ perpendicular to the line of centers. There are corresponding momenta $p_\|$ and $p_\perp$. The angular momentum of the system is $p_\perp D$, where D is the separation of the centers at the moment of collision.

so we multiply by its area $4\pi D^2$, and the total number of collisions characterized by $v_\perp$ and $v_\|$ is

$$de(v_\perp,v_\|) = \left(\frac{8\pi kT}{\mu}\right)^{1/2} D^2 \frac{e^{-\mu v^2/2kT}v_\| \, dv_\| \, v_\perp \, dv_\perp}{(kT/\mu)^2} \quad (10\text{-}48a)$$

which can be rewritten as

$$\frac{de(v_\perp,v_\|)}{e_{AB}} = \frac{e^{-\mu v^2/2kT}v_\| \, dv_\| \, v_\perp \, dv_\perp}{(kT/\mu)^2} \quad (10\text{-}48b)$$

The left-hand side is the fraction of all collisions with velocities $v_\perp$ and $v_\|$. The components of kinetic energy in the two directions are

$$\epsilon_\| = \tfrac{1}{2}\mu v_\|^2 \qquad \epsilon_\perp = \tfrac{1}{2}\mu v_\perp^2$$

and the distribution law for collisions becomes

$$\frac{de(\epsilon_\perp,\epsilon_\|)}{e_{AB}} = \frac{de(\epsilon_\perp,\epsilon_\|)}{(8\pi kT/\mu)^{1/2}D^2} = e^{-\epsilon_\|/kT-\epsilon_\perp/kT}\, d\left(\frac{\epsilon_\|}{kT}\right) d\left(\frac{\epsilon_\perp}{kT}\right) \quad (10\text{-}49)$$

In terms of the momenta along and perpendicular to the line of centers, we have

$$\frac{de(p_\perp,p_\|)}{(8\pi kT/\mu)^{1/2}D^2} = e^{-p^2/2\mu kT} \frac{p_\| \, dp_\| \, p_\perp \, dp_\perp}{(\mu kT)^2} \quad (10\text{-}50)$$

The component of momentum $p_\parallel$ along the line of centers can also properly be called p_r, since it is in fact the conjugate momentum to the radial coordinate r of polar coordinates. The angular momentum of the colliding pair, L_J, is related to the linear momentum perpendicular to the line of centers:

$$L_J = D p_\perp$$

(cf. the legend to Fig. 10-4). We may recall that the total kinetic energy of the colliding pair is

$$\epsilon = \frac{p_r^2}{2\mu} + \frac{L_J^2}{2\mu D^2}$$

As a matter of fact, at any separation r of the colliding pair, the energy is

$$\epsilon = \frac{p_r^2}{2\mu} + \frac{L_J^2}{2\mu r^2}$$

The total energy ϵ and the angular momentum L_J are, of course, constants of the motion. (Thus, as the two particles approach each other, ϵ and L_J are constant, but the division of ϵ into components $\epsilon_\parallel$ and $\epsilon_\perp$ varies with the distance of approach.)

There is another way of dividing up the collisions, which is needed for a problem discussed in Chap. 12. We want to calculate the collision rate for collisions in which the total kinetic energy is in the interval ϵ, $\epsilon + d\epsilon$, and the angular momentum is L_J, $L_J + dL_J$.

The transformation equations are

$$\epsilon = \epsilon_\parallel + \epsilon_\perp$$
$$L_J^2 = \epsilon_\perp (2\mu D^2)$$

The Jacobian of the transformation is

$$\begin{vmatrix} \dfrac{\partial \epsilon}{\partial \epsilon_\parallel} & \dfrac{\partial \epsilon}{\partial \epsilon_\perp} \\ \dfrac{\partial L_J}{\partial \epsilon_\parallel} & \dfrac{\partial L_J}{\partial \epsilon_\perp} \end{vmatrix} = \begin{vmatrix} 1 & 1 \\ 0 & \dfrac{\mu D^2}{L_J} \end{vmatrix} = \frac{\mu D^2}{L_J}$$

so that

$$d\epsilon_\parallel \, d\epsilon_\perp = \frac{L_J}{\mu D^2} \, d\epsilon \, dL_J \tag{10-51a}$$

From (10-49), the collision rate becomes

$$de(\epsilon, L_J) = \frac{(8\pi)^{1/2}}{(\mu k T)^{3/2}} e^{-\epsilon/kT} L_J \, dL_J \, d\epsilon \tag{10-51b}$$

Note that the collision diameter D does not appear in this formulation, but of course it is implicitly contained in the angular momentum.

For any of the expressions we multiply by $N_A N_B$ to calculate the total number of collisions per unit volume per unit time with the given characteristics when the atom concentrations are N_A and N_B.

In the collision theory of chemical kinetics, we are interested in the number of collisions in which the energy is greater than a specified amount ϵ_0. We begin with expression (10-49) for the fractional number of collisions $de(\epsilon_\parallel,\epsilon_\perp)$ with energies $\epsilon_\parallel$ and $\epsilon_\perp$.

$$\frac{de(\epsilon_\parallel,\epsilon_\perp)}{e_{AB}} = e^{-\epsilon_\parallel/kT}\, d\left(\frac{\epsilon_\parallel}{kT}\right) e^{-\epsilon_\perp/kT}\, d\left(\frac{\epsilon_\perp}{kT}\right) \tag{10-49}$$

The fraction of all the collisions in which the translational energy along the lines of centers is greater than ϵ_0, regardless of the value of $\epsilon_\perp$, is obtained by integrating (10-49) with $0 \leq \epsilon_\perp < \infty$ and $\epsilon_0 < \epsilon_\parallel < \infty$.

$$\frac{e(\epsilon_\parallel > \epsilon_0)}{e_{AB}} = e^{-\epsilon_0/kT} \tag{10-52}$$

One plausible assumption in chemical kinetics is that a reaction occurs between two molecules if and only if the energy along the line of centers is greater than the critical energy ϵ_0. Expression (10-52) gives the fraction of all collisions which satisfy this condition. Equation (10-52) with e_{AB}, the total collision rate, given by (10-44) is therefore the fundamental equation of the collision theory of chemical kinetics; it is known as the Arrhenius equation.*

10-7. The Kinetic Theory of Gases. The distribution functions developed in the preceding sections constitute only an introduction to the kinetic theory of gases. This theory treats many other properties of gases. We shall study one of these topics, the deviations from the perfect-gas law due to the existence of the intermolecular potential, in Chap. 15.

Furthermore, by using the distribution functions already derived and by considering the mechanics of molecular collisions in greater detail for more realistic intermolecular potentials than the rigid-sphere model, it is possible to make extremely successful calculations for the transport properties of simple gases—viscosity, diffusion, and heat conductivity. For these and other developments of the kinetic theory, the reader is referred to the specialized treatises.†

PROBLEMS

10-4. Assuming specular reflections, so that a molecule incident on a surface at angle θ with respect to the normal is reflected at angle θ, calculate the pressure of a perfect gas from the formula for the number of molecules striking a surface.

10-5. Calculate the average velocity of a molecule effusing from a small hole in a gas at thermal equilibrium (be careful, this is tricky). Calculate the average energy of an effusing molecule.

* A. F. Trotman-Dickinson, "Gas Kinetics," chap. 2, Butterworth, London, 1955.
† T. G. Cowling and S. Chapman, "The Mathematical Theory of Non-uniform Gases," Cambridge, New York, 1939, addendum, 1952; see also Hirschfelder, Curtiss, and Bird [15].

10-6. Obtain an expression for the fractional rate of collisions between molecules A and B in which the total relative translational energy $\epsilon = \epsilon_\perp + \epsilon_\parallel$ is greater than a specified energy ϵ_0. (*Hint:* Calculate the number of collisions in which the energy is less than ϵ_0.)

10-7. A gas at a fixed concentration of N molecules cc^{-1} is allowed to effuse through a small hole in the wall of the container under steady-state conditions. Calculate the *concentration* of molecules in the beam at a distance of L from the hole along the normal. The maximum diameter of a hole for molecular flow is of the order of the mean-free path $(1/ND^2)$, where D is the collision diameter. What size hole and what concentration N should be used to maximize the concentration in the beam?

10-8. Derive a general relation [analogous to Eq. (6-25)] for classical statistical mechanics between the phase integral and the average energy. Use this result to discuss, either for a simple example or in general, the equipartition theorem.

11
Polyatomic Molecules

11-1. Introduction. The energy of a molecule can be rigorously separated into a term for the translational motion of the center of mass plus a term for the energy due to internal motions. The latter can usually be rather accurately separated into a contribution due to the electronic motion plus a contribution due to the motions of the nuclei for the molecule in a given electronic state. The nuclear motions can be approximately represented as vibrations of small amplitude around the positions of minimum potential energy plus rotation of the molecule as a whole around its center of mass. Our principal task is the study of the distribution laws and the calculation of the thermodynamic functions for a polyatomic molecule at this level of approximation.

A molecule containing n atoms has $3n$ generalized configuration coordinates and $3n$ conjugate momenta. In classical mechanics, its motion can be described in a $6n$-dimensional phase space. The position of the center of mass is specified by three space coordinates; the translational energy of the molecule as a whole involves the three conjugate momenta. A rigid nonlinear molecule requires three generalized position coordinates to specify its orientation relative to a set of axes fixed in space, and there are three corresponding conjugate momenta to specify the kinetic energy of rotation. There then remain $3n - 6$ configuration coordinates and $3n - 6$ conjugate momenta to describe the vibrational displacements of the nuclei from their positions of minimum potential energy. We say that there are $3n - 6$ vibrational degrees of freedom. There are some molecules which are not properly described as rigid, because they have certain internal rotations for which the potential energy does not change. The effects of free or relatively free internal rotations will be taken up later.

A linear molecule requires only two coordinates and two conjugate momenta to specify its orientation and rotation; there are then $3n - 5$ vibrational degrees of freedom.*

* The vast majority of polyatomic molecules are nonlinear, and throughout the chapter we shall be talking about sums and products over the $3n - 6$ vibrational

The translational motion in three directions contributes a factor of $(2\pi MkT/h^2)^{3/2}$ to the partition function, where M is the total mass. We have already studied the distribution laws and the contributions to the thermodynamic functions for translational energy.

11-2. Moments of Inertia. We state some definitions, theorems, and remarks about the moments of inertia of a rigid molecule which are useful for the discussion of its rotational motion.

I. Let x_i'', y_i'', z_i'' be the cartesian coordinates of the ith nucleus, of mass m_i (with $i = 1, \ldots, n$) in an arbitrary frame of reference. The coordinates of the center of mass are then given by

$$Mx_{cm} = \Sigma m_i x_i''$$
$$My_{cm} = \Sigma m_i y_i'' \qquad (11\text{-}1)$$
$$Mz_{cm} = \Sigma m_i z_i''$$

where $M = \Sigma m_i$

The cartesian coordinates of the atoms, x_i', y_i', z_i', relative to the center of mass are then given by

$$x_i' = x_i'' - x_{cm} \qquad y_i' = y_i'' - y_{cm} \qquad z_i' = z_i'' - z_{cm} \qquad (11\text{-}2)$$

II. The moments and products of inertia of the molecule around the center of mass in this frame of reference are defined by

$$\left. \begin{array}{l} I_{x'x'} = \sum_{i=1}^{n} m_i(y_i'^2 + z_i'^2) \\ I_{y'y'} = \sum m_i(z_i'^2 + x_i'^2), \text{ etc.} \end{array} \right\} \quad \text{moments of inertia} \quad (11\text{-}3a)$$

$$I_{x'y'} = \Sigma m_i x_i' y_i' \qquad I_{y'z'} = \Sigma m_i y_i' z_i' \qquad \text{products of inertia} \quad (11\text{-}3b)$$

By a rotation of the axes, one can find a new set of cartesian axes with the coordinates of the atoms relative to the center of gravity denoted by x_i, y_i, z_i, in which the products of inertia all vanish; $I_{xy} = I_{yz} = I_{xz} = 0$. The moments of inertia I_{xx}, I_{yy}, I_{zz} are denoted simply by I_x, I_y, I_z and are called the principal moments of inertia.

III. For reasonably symmetrical molecules the choice of the principal axes is usually very simple and obvious. For example, for the pyramidal molecule $HCCl_3$ a line through the C—H bond is a threefold symmetry axis. This is one of the principal axes of the molecule, and we call it the z axis. The other two principal axes are not unique. Any two mutually perpendicular axes which are perpendicular to the threefold axis may be chosen as the x and the y axes; it will be found that these are principal axes and that $I_y = I_x$. The only tedious problem here is the location of the center of mass.

---

degrees of freedom and three moments of inertia. The reader can always insert "or $3n - 5$ vibrational degrees of freedom for a linear molecule plus two degrees of rotational freedom."

Sec. 11-2] POLYATOMIC MOLECULES

In general, if a molecule has a symmetry axis, this will be one of the principal axes. If the molecule has a plane of symmetry, one principal axis will be perpendicular to this plane.

IV. For a planar molecule, we choose the xy plane to contain the molecule; then

$$I_z = I_x + I_y \tag{11-4}$$

V. Choose a coordinate system which does not have the center of mass as its origin. Let the coordinates of the atoms be x_i'', y_i'', z_i'' ($i = 1, \ldots, n$) and the coordinates of the center of gravity x_{cm}, y_{cm}, z_{cm}; then the moments and products of inertia in this system, $I_{x''x''}$, etc., are related to the moments and products of inertia in a parallel coordinate system with origin at the center of gravity, by

$$\begin{aligned} I_{x'x'} &= I_{x''x''} - M(y_{cm}^2 + z_{cm}^2) \\ I_{x'y'} &= I_{x''y''} - Mx_{cm}y_{cm} \end{aligned} \tag{11-5}$$

where M is the total mass of the molecule.

The result is often useful for calculating moments of inertia around the center of mass.

VI. As we shall see, the quantity of interest for the calculation of thermodynamic functions is the product $I_x I_y I_z$. It can be shown that the determinant

$$\begin{vmatrix} I_{x'x'} & -I_{x'y'} & -I_{x'z'} \\ -I_{x'y'} & I_{y'y'} & -I_{y'z'} \\ -I_{x'z'} & -I_{y'z'} & I_{z'z'} \end{vmatrix}$$

is invariant under a rotation of the axes; therefore

$$\begin{vmatrix} I_{x'x'} & -I_{x'y'} & -I_{x'z'} \\ -I_{x'y'} & I_{y'y'} & -I_{y'z'} \\ -I_{x'z'} & -I_{y'z'} & I_{z'z'} \end{vmatrix} = \begin{vmatrix} I_x & 0 & 0 \\ 0 & I_y & 0 \\ 0 & 0 & I_z \end{vmatrix} = I_x I_y I_z \tag{11-6}$$

It is not necessary, therefore, for thermodynamic calculations to find the principal axes (although it may be convenient!).

VII. A linear molecule has two equal moments of inertia perpendicular to its axis.

Statement I is a definition; statement III is readily verified for particular cases; statements IV and V are readily proved by elementary methods; the proofs of statement VI and of that part of statement II which deals with the existence of principal axes are best done by means of matrix algebra; statement VII is obvious.

Problem 11-1. Calculate the principal moments of inertia of CCl_3Br; if the C—Br distance is 1.91 Å, the C—Cl distance is 1.77 Å, and the bond angles are all tetrahedral (109°28′). Use units of atomic weight times angstroms squared.

11-3. Energy Levels and Partition Function for the Rotation of a Symmetric-top Molecule. In a symmetric-top molecule, $I_x = I_y \neq I_z$, where the z axis is the symmetry axis. We remark in passing that a molecule with an n-fold symmetry axis, for $n \geq 3$, is a symmetric top; in general, a molecule with a twofold axis (such as H_2O) is not. For a symmetrical top, it is found that the wave function for rotation depends on three quantum numbers J, K, M. The energy levels depend only on J and K and are given by

$$\epsilon_{JKM} = \frac{\hbar^2}{2}\left[\frac{J(J+1)}{I_x} + K^2\left(\frac{1}{I_z} - \frac{1}{I_x}\right)\right] \quad (11\text{-}7)$$

$$\frac{\epsilon_{JKM}}{kT} = \frac{hc}{kT}[J(J+1)B_x - K^2(B_z - B_x)]$$

$$= \frac{1.4388}{T}[J(J+1)B_x - K^2(B_z - B_x)] \quad (11\text{-}8a)$$

where
$$B_x = \frac{h}{8\pi^2 I_x c} \qquad B_z = \frac{h}{8\pi^2 I_z c} \quad (11\text{-}8b)$$

or
$$\frac{\epsilon_{JKM}}{kT} = y_x J(J+1) + (y_z - y_x)K^2 \quad (11\text{-}9a)$$

where
$$y_x = \frac{hcB_x}{kT} = \frac{h^2}{8\pi^2 I_x kT} \qquad y_z = \frac{hcB_z}{kT} = \frac{h^2}{8\pi^2 I_z kT} \quad (11\text{-}9b)$$

The wave functions are simultaneously eigenfunctions of the total squared angular momentum $(\hat{L}^2)$, the angular momentum along the symmetry axis, and the angular momentum along an axis fixed in space. The total squared angular momentum is $J(J+1)\hbar^2$, the component along the symmetry axis is $K\hbar$, and the component along an axis fixed in space is $M\hbar$.

If the statements about angular momenta in the above paragraph are accepted, the formula (11-7) is readily derived. The kinetic energy of rotation operator is

$$\hat{K} = \frac{\hat{L}_x^2}{2I_x} + \frac{\hat{L}_y^2}{2I_y} + \frac{\hat{L}_z^2}{2I_z} \quad (11\text{-}10)$$

where x, y, z are axes in the molecule, z being the symmetry axis, and $\hat{L}_z$ is the angular-momentum operator along the z axis. If $\langle L_z \rangle = K\hbar$ and $\langle L^2 \rangle = \langle L_x^2 + L_y^2 + L_z^2 \rangle = J(J+1)\hbar^2$, it follows that

$$\langle L_x^2 + L_y^2 \rangle = \hbar^2[J(J+1) - K^2]$$

Furthermore, $I_x = I_y$. Substitution in (11-10) then gives (11-7).

The allowed values of J are $0, 1, 2, \ldots, \infty$. For each value of J there are $2J+1$ allowed values of K, namely, $-J, -J+1, \ldots, +J$; there are also $2J+1$ allowed values of M, $-J, -J+1, \ldots, +J$. Each ϵ_{KJ} level is $(2J+1)$-degenerate because of the various M values.

An equivalent point of view which is more convenient for setting up the partition-function sum is that, for a given value of K, the allowed values of J are $|K|, |K| + 1, \ldots, \infty$. Each value of the pair (J,K) is still $(2J + 1)$-fold-degenerate, for the $2J + 1$ values of M. The partition function is therefore

$$q = \frac{1}{\sigma} \sum_{K=-\infty}^{K=+\infty} e^{-(y_z-y_x)K^2} \sum_{J=|K|}^{J=\infty} (2J+1)e^{-y_x J(J+1)} \quad (11\text{-}11)$$

where σ is the symmetry number, which we discuss in the next section.

For the usual case that $y = h^2/8\pi^2 IkT \ll 1$, the sums can be replaced by integrals. Furthermore, in this approximation, the precise value of the lower limit for the integral over J does not matter, provided that we choose a value in the neighborhood of $J = |K|$. It is convenient to choose for this lower limit $J_{\min}$, the value of J which is the solution of the equation $J_{\min}(J_{\min} + 1) = K^2$. Then

$$q = \frac{1}{\sigma} \int_{K=-\infty}^{K=+\infty} e^{-(y_z-y_x)K^2} \left[\int_{J_{\min}}^{\infty} (2J+1)e^{-y_x J(J+1)} \, dJ \right] dK \quad (11\text{-}12)$$

$$q = \frac{1}{\sigma y_x} \int_{-\infty}^{+\infty} e^{-y_z K^2} \, dK = \frac{\pi^{1/2}}{\sigma} \left(\frac{1}{y_z y_x^2} \right)^{1/2} \quad (11\text{-}13)$$

$$q = \frac{\pi^{1/2}}{\sigma} \left(\frac{8\pi^2 I_z kT}{h^2} \right)^{1/2} \left(\frac{8\pi^2 I_x kT}{h^2} \right) \quad (11\text{-}14)$$

$$q = \frac{\pi^{1/2}}{\sigma} \left(\frac{T}{1.4388 B_z} \right)^{1/2} \left(\frac{T}{1.4388 B_x} \right) \quad (11\text{-}15)$$

Equation (11-14) or (11-15) is the desired result. The thermodynamic functions associated with this partition function are discussed later.

11-4. The Symmetry Number. The symmetry number, which is a factor which must be introduced in calculating the rotational partition function, is "defined as the number of different values of the rotational coordinates which all correspond to one orientation of the molecule, remembering that the identical atoms are indistinguishable."*

For example, in the pyramidal molecule $HCCl_3$, the symmetry number is 3. We label the chlorine atoms as in Fig. 11-1; there are three configurations, shown in the top row, which can be obtained from each other by rotation around the threefold axis. There are three more configurations in the second row; however, these cannot be obtained from those in the first row by rotation but require a reflection.

The symmetry number of a molecule with tetrahedral symmetry such as CH_4 is 12. Consider the tetrahedron standing with one base horizontal and the opposite vertex at the top. Any one of the four hydrogen

* From Mayer and Mayer, p. 195 [5].

atoms can be chosen to be at the top vertex; for each such choice, there are three possible positions of the other three hydrogen atoms that are obtainable by rotation. Thus, 12 different values of the rotational coordinates all correspond to an identical configuration, for indistinguishable hydrogen atoms. However, there are all told 4! or 24 possible permutations of four distinguishable objects; 12 of these are related to

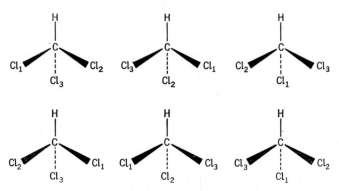

FIG. 11-1. The symmetry number of $HCCl_3$. If the chlorine atoms are labeled and regarded as distinguishable, the six configurations shown are all different. The three configurations in the bottom row are related to each other by a rotation. The three configurations in the top row are related to each other by a rotation. A reflection is necessary in order to go from a configuration in the top row to one in the bottom row. The symmetry number is 3.

each other by rotations, but the other 12 are "optical isomers" of the first group and are obtained from the first group by a reflection or an inversion.

Problem 11-2(a). What is the symmetry number of the planar trigonal molecule SO_3? (b) What is the symmetry number of a square planar molecule AB_4?

It was shown in detail in Chap. 9 for a diatomic molecule that the symmetry number is an approximate correction to the partition function which is accurate when the spacing of the rotational levels is small compared with kT $[(h^2/8\pi^2 IkT) \ll 1]$. It allows for the fact that only certain combinations of the nuclear spin wave functions and the rotational wave functions can occur together if the total wave function is going to have the correct symmetry under interchange of any two identical nuclei. The correction systematically omits the nuclear spin degeneracy of $2I + I$ per nucleus. It is applicable either to monoisotopic elements or to elements which are mixtures of several isotopes, provided that there is no isotope fractionation or fractionation on the basis of nuclear spin in the process being considered.

Similar but more complicated arguments apply when there are three or more identical nuclei in a molecule. The nuclear spin wave functions are more complicated, with more complex symmetry properties. The rotational wave functions have more complicated symmetry properties. But again, upon detailed examination of these functions, it will be found that certain products of spin functions and rotational functions are either symmetric or antisymmetric under interchange of identical nuclei; and it is only the combinations with the appropriate symmetry which are permitted.

It may also be remarked that, whereas the vibrational wave function of a diatomic molecule is always symmetric under interchange of identical nuclei, the vibrational wave functions for polyatomic molecules can be symmetric or antisymmetric or can have more complex symmetry properties for an interchange of identical nuclei. The allowed rotational and spin states are therefore different in the different vibrational states.

For spectroscopic experiments, it is necessary to inquire into the details of the symmetry properties of the various states. For the statistical mechanical calculation of thermodynamic properties, the symmetry number short-cuts this analysis and provides an automatic correction for these factors.

We shall be content with this statement of the use of the symmetry number and shall not endeavor to justify it by a detailed analysis of the symmetry properties of the rotational wave functions and of the nuclear spin wave functions.

There is a justification for the use of the symmetry number based on a classical phase-space discussion. We shall present this argument for the simple case of a homopolar diatomic molecule.*

We regard the two atoms as distinguishable, atom 1 and atom 2, with cartesian coordinates $x_1, y_1, z_1, x_2, y_2, z_2$. Now transform to a coordinate system consisting of the cartesian coordinates of the center of mass, plus spherical coordinates for the relative positions of the two atoms.

$$X_{cm} = \frac{mx_1 + mx_2}{2m} = \frac{x_1 + x_2}{2} \quad \text{etc.} \quad (11\text{-}16)$$

$$x_2 - x_1 = r \sin\theta \cos\phi \quad y_2 - y_1 = r \sin\theta \sin\phi \quad z_2 - z_1 = r \cos\theta$$

The Hamiltonian function is

$$H = H_{cm} + \frac{p_r^2}{2\mu} + \frac{p_\theta^2}{2\mu r^2} + \frac{p_\phi^2}{2\mu r^2 \sin^2\theta} + U(r) \quad (11\text{-}17)$$

where H_{cm} is the Hamiltonian for the motion of the center of mass, μ is the reduced mass, and $U(r)$ is the potential-energy function for the molecule. The classical approximation for the partition function is

$$q = q_{trans} \frac{1}{h^3} \int \cdots \int e^{-\beta H} \, dr \, d\theta \, d\phi \, dp_r \, dp_\theta \, dp_\phi \quad (11\text{-}18)$$

---

*The argument is adapted from Mayer and Mayer, pp. 195–196 [5], who give it in its general form.

We recall that the momenta p_θ and p_ϕ are given by

$$p_\theta = \mu r^2 \dot\theta \qquad p_\phi = \mu r^2 \sin^2\theta \, \dot\phi$$

The operation of permuting the coordinates of the two particles results in the transformation

$$(r,\theta,\phi,p_r,p_\theta,p_\phi) \to (r, \pi - \theta, \pi + \phi, p_r, -p_\theta, p_\phi) \qquad (11\text{-}19)$$

The Hamiltonian is invariant under this transformation.

The partition function is evaluated in (11-18) by integration over the entire volume of phase space. For each contribution from a volume element in phase space around the point $(r,\theta,\phi,p_r,p_\theta,p_\phi)$, there is an equal contribution from a volume element around the point $(r, \pi - \theta, \pi + \phi, p_r, -p_\theta, p_\phi)$. Since, however, the atoms are, in truth, indistinguishable, we should not count both these regions, which transform into each other by permuting the coordinates of the two identical particles. Thus the true partition function in the classical approximation for a symmetrical diatomic molecule involves integration over only a half of phase space and is therefore one-half of (11-18).

For an accurate evaluation of the partition function of a diatomic molecule, it is necessary to treat the vibrational motion quantum mechanically, but it is usually sufficient to treat the rotation classically. The classical partition function for rotation is the integral

$$q_{\text{rot}} = \frac{1}{h^2} \int \cdots \int e^{-\beta H} \, d\theta \, d\phi \, dp_\theta \, dp_\phi$$

for r fixed at the value such that $U(r)$ is a minimum. Permuting the two identical particles affects the variables θ, ϕ, p_θ, p_ϕ, but not r and p_r. The argument of the previous few paragraphs is still entirely valid, even though vibration is treated quantum mechanically, and the symmetry number of 2 is still an appropriate correction.

A similar but more complicated discussion of the evaluation of the classical partition function applies for a polyatomic molecule such as $HCCl_3$. In fixed relative positions of the atoms in $HCCl_3$, an integration over phase space covers all the configurations obtainable by rotation of the molecule. There are three identical regions in phase space, which should not be counted separately because of the identity of the three Cl atoms. Thus division of the partition function by a symmetry number of 3 is necessary.

To conclude this discussion, it should be remarked that the simple symmetry-number correction is not correct when the rotational-energy spacing is comparable with kT. This was illustrated in Chap. 9 for molecular hydrogen. It is then necessary to consider in detail which nuclear spin functions occur with the different rotational states and to evaluate the partition function by direct summation.

11-5. The Asymmetric Rotor.

An asymmetric rotor has three unequal principal moments of inertia. The formulae for the energy levels are quite complicated* and not suitable for the calculation of the partition function. However, for almost all practical problems, the moments of inertia are sufficiently large and the temperature sufficiently high so that the partition function can be evaluated as an integral in classical phase space.

The kinetic energy of rotation of a rigid body is given by

$$2K = I_x\omega_x^2 + I_y\omega_y^2 + I_z\omega_z^2 \qquad (11\text{-}20)$$

where ω_x, ω_y, and ω_z are the angular velocities of rotation around the three principal axes. The angular momentum around any axis is $L_x = I_x\omega_x$, etc.; the kinetic energy is then

$$K = \frac{L_x^2}{2I_x} + \frac{L_y^2}{2I_y} + \frac{L_z^2}{2I_z} \qquad (11\text{-}10)$$

This is not a directly useful expression for setting up the integral in phase space, since there are not three independent configuration coordinates which specify the orientation of the rigid body and which are conjugate to the momenta L_x, L_y, L_z.†

The description of the orientation of a rigid body is usually given in terms of the Eulerian angles.‡ When the kinetic energy is expressed in terms of these variables, the integration in phase space can be performed in a straightforward, albeit tedious, way.

We shall omit this rigorous derivation. The following somewhat intuitive argument gives the correct answer in a simple way:§

With the kinetic energy expressed in terms of the angular momenta [(11-10)], the integral over the angular momenta contributes a factor of $(2\pi I_x kT)^{1/2}(2\pi I_y kT)^{1/2}(2\pi I_z kT)^{1/2}$ to the classical phase integral.

$$\iiint_{-\infty}^{+\infty} e^{(-K/kT)}\, dL_x\, dL_y\, dL_z = (2\pi I_x kT)^{1/2}(2\pi I_y kT)^{1/2}(2\pi I_z kT)^{1/2} \qquad (11\text{-}21)$$

To integrate over the configuration coordinates, choose any one axis in the molecule. A complete rotation about this axis contributes a factor of

---

* Herzberg, pp. 44–50 [18].

† The difficulty is that the orientation of a rigid body cannot be uniquely specified by describing the amount of rotation around the three cartesian axes. For example, the reader can readily verify that the same orientation is achieved by (a) a clockwise rotation of 90° around the z axis, followed by a clockwise rotation of 90° around the x axis; and (b) a clockwise rotation of 90° around the y axis, followed by a clockwise rotation of 90° around the z axis.

‡ Margenau and Murphy, pp. 272–275 [23]; Mayer and Mayer, pp. 193–194 [5]; Wilson, pp. 142–143 [4].

§ Pitzer, pp. 443–444 [10].

2π. Integration over all possible orientations of the axis contributes a factor of 4π. This includes all possible orientations of the molecule once, and only once, and contributes a factor of $8\pi^2$. When we introduce the symmetry number and the appropriate factor of h^3, the partition function for rotation becomes

$$q = \frac{8\pi^2}{\sigma h^3}(2\pi I_x kT)^{1/2}(2\pi I_y kT)^{1/2}(2\pi I_z kT)^{1/2}$$

$$= \frac{\pi^{1/2}}{\sigma}\left(\frac{8\pi^2 I_x kT}{h^2}\right)^{1/2}\left(\frac{8\pi^2 I_y kT}{h^2}\right)^{1/2}\left(\frac{8\pi^2 I_z kT}{h^2}\right)^{1/2} \quad (11\text{-}22)$$

We note that this result agrees with that for the symmetric top [(11-14)] if we set $I_x = I_y$. The rotational partition function for a linear polyatomic molecule is the same as that for a diatomic molecule,

$$q = (1/\sigma)(8\pi^2 IkT/h^2)$$

Purely as a mnemonic device, we observe that for a three-dimensional rotor and for a two-dimensional rotor (i.e., a diatomic molecule or a linear polyatomic molecule) each degree of rotational freedom contributes a factor of $(8\pi^2 IkT/h^2)^{1/2}$; there is an additional $\pi^{1/2}$ factor for the three-dimensional rotor.

The contributions of rotation to the thermodynamic functions are readily calculated. We have

$$\left(\frac{F - F_0}{T}\right)_{\text{rot}} = -R \ln q \quad (11\text{-}23a)$$

$$\left(\frac{H - H_0}{T}\right)_{\text{rot}} = \tfrac{3}{2}R \qquad C_{\text{rot}} = \tfrac{3}{2}R \quad (11\text{-}23b)$$

The partition function is proportional to $T^{3/2}$, the energy of rotation is $\tfrac{3}{2}RT$ per mole, and the heat capacity is $\tfrac{3}{2}R$. Numerical formulae are included in the compilation at the end of Chap. 8.

A more accurate approximation for the partition function has been derived for cases where the classical approximation is not sufficiently accurate. If q_R° represents the classical approximation, the improved equation is*

$$q = q_R^\circ \left[1 + \frac{h^2}{96\pi^2 kT}\left(\frac{2}{I_x} + \frac{2}{I_y} + \frac{2}{I_z} - \frac{I_z}{I_x I_y} - \frac{I_x}{I_y I_z} - \frac{I_y}{I_z I_x}\right)\right] \quad (11\text{-}24)$$

This correction is seldom needed.

11-6. Normal Vibrations. The analysis of the normal vibrations of a polyatomic molecule is a difficult, advanced subject; it can be pursued efficiently only with the aid of the tools of matrix algebra and group theory. Our object here is to become acquainted with the general nature

* K. F. Stripp and J. G. Kirkwood, *J. Chem. Phys.*, **19**: 1311 (1951).

of the problem and the results and to become familiar with the terminology of the field. It is hoped that, with this background, the reader can use published vibrational analyses for statistical calculations.*

We consider a molecule that is not undergoing any translational or rotational motion. Let x_i°, y_i°, z_i° ($i = 1, \ldots, n$) represent the equilibrium positions of the atoms and δx_i, δy_i, δz_i be small vibrational displacements. The potential energy can be expressed as a Taylor series,

$$U(x_1^\circ + \delta x_1, y_1^\circ + \delta y_1, \ldots, z_n^\circ + \delta z_n)$$

$$= U(x_1^\circ, \ldots, z_n^\circ) + \sum_{\substack{x,y,z \\ i=1,\ldots,n}} \frac{\partial U(x_1^\circ, \ldots, z_n^\circ)}{\partial x_i} \delta x_i \quad (11\text{-}25)$$

$$+ \frac{1}{2} \sum_{\substack{x,y,z \\ i,j}} \frac{\partial^2 U}{\partial x_i\, \partial y_j} \delta x_i\, \delta y_j + \cdots$$

The displacements δx_i are to be chosen so that there is no net translation,

$$\sum_{i=1}^n m_i\, \delta x_i = 0 \qquad \sum m_i\, \delta y_i = 0 \qquad \sum m_i\, \delta z_i = 0 \quad (11\text{-}26)$$

and no net rotation†

$$\sum_i m_i(x_i^\circ\, \delta y_i - y_i^\circ\, \delta x_i) = 0$$
$$\sum m_i(y_i^\circ\, \delta z_i - z_i^\circ\, \delta y_i) = 0 \quad (11\text{-}27)$$
$$\sum m_i(z_i^\circ\, \delta x_i - x_i^\circ\, \delta z_i) = 0$$

Since we are expanding around a position of minimum potential energy, the first derivatives $\partial U/\partial x_i$ in (11-25) vanish. The important part of the potential for small displacements from the equilibrium position is the quadratic form $\Sigma(\partial^2 U/\partial x_i\, \partial y_j)\, \delta x_i\, \delta y_j$.

By use of the six equations (11-26) and (11-27), six of the variables could be eliminated, and the equations of motion for the vibrations could be written in terms of $3n - 6$ variables. Suppose that, by some process, we have found $3n - 6$ internal coordinates $q_1, \ldots, q_{3n-6}$ which represent displacements from the equilibrium position. The displacements δx_i, δy_j, etc., in cartesian coordinates are known linear functions of the q's. The potential-energy increase over that in the equilibrium position becomes

$$2U = \sum_{i,j=1}^{3n-6} f_{ij} q_i q_j \quad \text{with } f_{ij} = f_{ji} \quad (11\text{-}28)$$

---

* The standard English-language references in this field are Herzberg [18] and E. B. Wilson, Jr., J. C. Decius, and P. C. Cross, "Molecular Vibrations," McGraw-Hill, New York, 1955.

† Equation (11-27) is the condition for no rotation for small displacements δx_i, δy_i, δz_i.

and the kinetic energy is given by

$$2K = \sum_{i,j} a_{ij}\dot{q}_i\dot{q}_j \qquad (11\text{-}29)$$

For the case of generalized Lagrangian coordinates, the kinetic energy can always be written in the form (11-29), but the coefficients a_{ij} are functions of the coordinates $q_1, \ldots, q_{3n-6}$. For small vibrations around an equilibrium position, however, the a_{ij}'s can be regarded as constants with values appropriate for the equilibrium configuration.

The equations of motion

$$\frac{d}{dt}\left(\frac{\partial K}{\partial \dot{q}_i}\right) + \frac{\partial U}{\partial q_i} = 0 \qquad i = 1, \ldots, 3n-6 \qquad (11\text{-}30)$$

become

$$\sum_{j=1}^{3n-6} a_{ij}\ddot{q}_j + \sum_{j=1}^{3n-6} f_{ij}q_j = 0 \qquad i = 1, \ldots, 3n-6 \qquad (11\text{-}31)$$

We seek a solution of the form

$$q_1 = b_1 \sin 2\pi\nu t \qquad q_2 = b_2 \sin 2\pi\nu t \qquad \cdots \qquad q_{3n-6} = b_{3n-6} \sin 2\pi\nu t \qquad (11\text{-}32)$$

We then have $\ddot{q}_i = -4\pi^2\nu^2 q_i$. Set

$$\lambda = 4\pi^2\nu^2 \qquad (11\text{-}33)$$

so that

$$\ddot{q}_i = -\lambda b_i \sin 2\pi\nu t$$

For the solution (11-32), Eqs. (11-31) reduce to

$$-\lambda \sum_j a_{ij}b_j + \sum_j f_{ij}b_j = 0 \qquad i = 1, \ldots, 3n-6 \qquad (11\text{-}34)$$

or

$$\sum_{j=1}^{3n-6} (f_{ij} - \lambda a_{ij})b_j = 0 \qquad i = 1, \ldots, 3n-6 \qquad (11\text{-}35)$$

This is a set of $3n - 6$ linear homogeneous equations for the $3n - 6$ displacements b_j. If such a system of equations is to have a solution, the determinant of the coefficients must vanish:

$$\det |f_{ij} - \lambda a_{ij}| = 0 \qquad (11\text{-}36)$$

This is known as the secular equation. It is of degree $3n - 6$ in λ, and in general it has $3n - 6$ nonzero roots. Each such root is a value of λ and hence gives a value of ν by (11-33) for which the harmonic vibrations [(11-32)] are a possible solution of the equations of motion. For any one of the roots λ the displacements b_j can be determined by solution of the linear equations (11-35). It is characteristic of such linear homogeneous equations that we cannot determine the b's absolutely; but the ratios, say, b_i/b_1, can be determined.

Suppose that we fix the absolute values of the b's by some normalization condition, for example,
$$\Sigma b_i^2 = 1$$
We denote the $3n - 6$ roots of the secular equation by $\lambda_1, \ldots, \lambda_k, \ldots, \lambda_{3n-6}$. For the kth root λ_k there is a set of solutions of (11-35) for the amplitudes of the vibrations, which we now denote by b_{jk}. That is, the displacements are

$$q_1 = b_{1k} \sin 2\pi \nu_k t \quad \cdots \quad q_i = b_{ik} \sin 2\pi \nu t \quad \cdots$$
$$q_{3n-6} = b_{3n-6,k} \sin 2\pi \nu_k t \quad (11\text{-}37)$$

The motion described by Eqs. (11-37) is a normal vibration; the frequency ν_k is one of the $3n - 6$ normal vibration frequencies. For this normal mode, the relative amplitudes of the displacements of the several coordinates $q_1, \ldots, q_n$ are proportional to the quantities $b_{1k}, \ldots, b_{nk}$. All the coordinates vibrate harmonically with the frequency ν_k, and the motions are all in phase so that the displacements all go through zero at the same time.

We now define a new set of variables $Q_1, \ldots, Q_{3n-6}$ by the relation

$$q_i = \sum_{k=1}^{3n-6} b_{ik} Q_k \qquad i = 1, \ldots, 3n - 6 \qquad (11\text{-}38)$$

By substituting in the expressions for U and K [(11-28 and 11-29)] we obtain equations for these quantities in terms of the capital-Q variables. It can be shown by elementary matrix algebra that with the capital Q's defined by the transformation (11-38), K and U are now both diagonalized quadratic forms, i.e., that

$$2K = \sum_{j=1}^{3n-6} \mu_j \dot{Q}_j^2 \qquad (11\text{-}39a)$$

$$2U = \sum_{j=1}^{3n-6} \lambda_j \mu_j Q_j^2 \qquad (11\text{-}39b)$$

The equations of motion are now

$$\mu_k \ddot{Q}_k + \lambda_k \mu_k Q_k = 0 \qquad k = 1, \ldots, 3n - 6 \qquad (11\text{-}40)$$

to which the solutions obviously are

$$Q_k = Q_k^\circ \sin 2\pi \nu_k t$$
$$4\pi^2 \nu_k^2 = \lambda_k \qquad (11\text{-}41)*$$

It should be remarked that it is a general theorem in matrix algebra that the roots of a secular equation are always invariant under a linear transformation such as (11-38).

* There are $\cos 2\pi \nu_k t$ solutions also, but these simply introduce a phase factor, which is unimportant here.

The quantities Q_k are the normal coordinates of the vibrating molecule, and the frequencies ν_k are the normal vibration frequencies. The meaning of normal coordinates and normal vibrations can be clarified by the following explanation.

The displacements q_i are related to the normal coordinates Q_k by the transformation equations (11-38). The displacements expressed in cartesian coordinates, $\delta x_1, \ldots, \delta z_{3n}$, are known linear functions of the q's; therefore they are related to the Q's by a linear transformation. Suppose that this transformation is

$$\delta x_1 = \sum_{k=1}^{3n-6} a_{1k} Q_k$$
$$\cdots\cdots\cdots\cdots \quad 3n \text{ equations} \tag{11-42}$$
$$\delta z_{3n} = \sum_{k=1}^{3n-6} a_{3n,k} Q_k$$

If (in classical mechanics) a molecule is caused to vibrate by an arbitrary impact, its motions will be a complicated nonperiodic pattern. However, it will always be the case that this motion can be analyzed as the sum of $3n - 6$ harmonic vibrations with the frequencies of the normal vibrations. That is, the arbitrary motion can be represented by

$$\delta x_1 = \sum_{k=1}^{3n-6} a_{1k} Q_k^\circ \sin(2\pi\nu_k t - \phi_k)$$
$$\cdots\cdots\cdots\cdots\cdots\cdots\cdots \tag{11-43}$$
$$\delta z_n = \sum_{k=1}^{3n-6} a_{3n,k} Q_k^\circ \sin(2\pi\nu_k t - \phi_k)$$

The quantities Q_k° and ϕ_k represent the amplitude and the phase of the kth normal vibration in this particular motion that the molecule is executing.

If the molecule is excited to vibrate in a very special way (this may be difficult to do with a mechanical model by impact, but it is easy with a molecule by irradiation with infrared light of the correct frequency!), then only one normal vibration will be excited and the motion can be described by the equations

$$\delta x_1 = a_{1j} Q_j^\circ \sin(2\pi\nu_j t - \phi_j)$$
$$\cdots\cdots\cdots\cdots\cdots \tag{11-44}$$
$$\delta z_n = a_{3n,j} Q_j^\circ \sin(2\pi\nu_j t - \phi_j)$$

When the motion conforms to Eq. (11-44), only the jth normal vibration has been excited. All the atoms vibrate with the frequency ν_j; the relative amplitudes are in the ratios of the a_{ij}; the atoms all vibrate in the same phase and go through zero at the same time.

The quantum mechanics of a vibrating molecule is quite straightforward when the normal coordinates have been discovered and the expressions for the kinetic and potential energy [(11-39a) and (11-39b)] obtained.

The momentum conjugate to Q_j is

$$P_j = \frac{\partial K}{\partial \dot{Q}_j} = \mu_j \dot{Q}_j \qquad (11\text{-}45)$$

The Hamiltonian operator is

$$H = \sum_{k=1}^{3n-6} \left(\frac{\lambda_k \mu_k Q_k^2}{2} + \frac{P_k^2}{2\mu_k} \right) \qquad (11\text{-}46)$$

The corresponding Schrödinger equation is

$$\sum_k \left(-\frac{\hbar^2}{2\mu_k} \frac{\partial^2 \psi}{\partial Q_k^2} + \tfrac{1}{2}\lambda_k \mu_k Q_k^2 \psi \right) = \epsilon \psi(Q_1, \ldots, Q_{3n-6}) \qquad (11\text{-}47)$$

The equation is separable into $3n - 6$ harmonic-oscillator wave equations

$$-\frac{\hbar^2}{2\mu_j} \frac{\partial^2 \psi_j}{\partial Q_j^2} + (\tfrac{1}{2}\lambda_j \mu_j - \epsilon_j)\psi_j(Q_j) = 0 \qquad j = 1, \ldots, 3n-6 \qquad (11\text{-}48)$$

where
$$\psi = \psi_1 \cdots \psi_j \cdots \psi_{3n-6}$$

The energy ϵ_j in the jth normal vibration is given by

$$\epsilon_j = (v_j + \tfrac{1}{2})h\nu_j \qquad v_j = 0, 1, 2, \ldots \qquad (11\text{-}49)$$

where v_j is the vibrational quantum number for the jth normal mode. The vibrational energy of the molecule in the quantum state $v_1, \ldots, v_j, \ldots, v_{3n-6}$ is

$$\epsilon = (v_1 + \tfrac{1}{2})h\nu_1 + \cdots + (v_j + \tfrac{1}{2})h\nu_j + \cdots + (v_{3n-6} + \tfrac{1}{2})h\nu_{3n-6} \qquad (11\text{-}50)$$

Thus the energy of the molecule is the sum of the energies in the $3n - 6$ normal vibrations, each of which is a quantized simple harmonic oscillator.

It is instructive to remark briefly on how a normal vibration analysis proceeds if we work with cartesian coordinates. Instead of writing $\delta x_1, \delta y_1, \ldots, \delta z_n$, it is convenient to write the displacements as $\eta_1, \ldots, \eta_{3n}$, with $\eta_1 = \delta x_1, \eta_2 = \delta y_1, \ldots, \eta_{3n} = \delta z_n$.

The kinetic energy is

$$2K = \sum_{i=1}^{3n} m_i \dot{\eta}_i^2 \qquad (11\text{-}51)$$

where, of course, in this notation, $m_1 = m_2 = m_3$, $m_4 = m_5 = m_6$ (the former m_2), etc.

The potential energy is a general quadratic form,

$$2U = \Sigma f_{ij} \eta_i \eta_j \qquad (11\text{-}52)$$

We shall not impose the restrictions (11-26) and (11-27), which forbid translation and rotation, because they are analytically awkward. We seek solutions of the form

$$\eta_i = b_i \sin 2\pi \nu t \qquad i = 1, \ldots, 3n \qquad (11\text{-}53)$$

and we set $\lambda = 4\pi^2 \nu^2$. The $3n$ equations of motion then are

$$-m_i \lambda b_i \sin 2\pi \nu t + \sum_{j=1}^{3n} f_{ij} b_j \sin 2\pi \nu t = 0 \qquad i = 1, \ldots, 3n$$

or
$$-m_i \lambda b_i + \sum_{j=1}^{3n} f_{ij} b_j = 0 \qquad i = 1, \ldots, 3n \qquad (11\text{-}54)$$

The secular equation which must be satisfied if solutions for the b's are to exist is

$$\begin{vmatrix} f_{11} - \lambda m_1 & f_{12} & \cdots & \cdots \\ f_{12} & f_{22} - \lambda m_2 & \cdots & \cdots \\ \cdots & \cdots & \cdots & \cdots \\ \cdots & \cdots & \cdots & f_{3n-1,3n} \\ \cdots & \cdots & f_{3n-1,3n} & f_{3n,3n} - \lambda m_{3n} \end{vmatrix} = 0 \qquad (11\text{-}55)$$

This is an equation of degree $3n$. It will be found that there are 6 zero roots, corresponding to translation and rotation, and $3n - 6$ nonzero roots for the bona fide normal vibrations.

Problem 11-3. Calculate the normal vibrations of a linear symmetrical AB_2 molecule which is constrained to move along the axis of the molecule only. That is, let the x coordinates at the equilibrium position be $x_1^\circ = -L$ (atom B); $x_2^\circ = 0$ (atom A); $x_3^\circ = +L$ (atom B). Let the potential function depend on the two AB distances only

$$2U = k[(\delta x_1 - \delta x_2)^2 + (\delta x_3 - \delta x_2)^2]$$

If you carry out the analysis without imposing the restriction of no translation, you should get one zero root for the secular equation corresponding to translation along the x axis. It is suggested that some members of the class carry out the calculation in this way and that others start by imposing the restriction for no translation, $m_B(\delta x_1 + \delta x_3) + m_A \delta x_2 = 0$.

11-7. The Nonlinear Symmetrical AB_2 Molecule. An Example.*

We shall treat one of the simplest nontrivial normal coordinate problems, namely, the nonlinear symmetrical AB_2 molecule. This exercise is instructive because it shows that the computation is impressively formidable even for this simple case and because the results suggest the ways in which a full use of group theory simplifies the analysis.

* The study of this section is helpful but not absolutely necessary for following the main points of this chapter.

The molecular structure and coordinates are illustrated in Fig. 11-2. Atoms B and C are identical, with mass m_B. The coordinates of the equilibrium positions are

$$x_A^\circ = y_A^\circ = 0$$
$$x_B^\circ = x_C^\circ = L \cos \alpha \qquad (11\text{-}56)$$
$$y_C^\circ = -y_B^\circ = L \sin \alpha$$

The atoms are initially in the plane $z = 0$. All displacements in the z direction are either translations or rotations, and so we need not consider such motions.

Let q_1 and q_2 be the displacements from the equilibrium value of the AB and AC bond distances, respectively. Let ϕ represent the change in

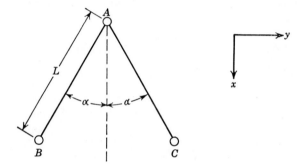

FIG. 11-2. The coordinate system for the symmetrical AB_2 molecule. Atoms B and C are identical. The bond angle is 2α, and the bond distance is L.

the bond angle from its equilibrium value of 2α. We assume a potential-energy function

$$2U = f(q_1^2 + q_2^2) + f_\phi \phi^2 \qquad (11\text{-}57)$$

where f is the stretching-force constant and f_ϕ is the bending-force constant. This is called the valence-force potential, since it assumes that the potential energy depends on the bond distances and bond angles. Instead of the angle ϕ it is convenient to introduce the corresponding linear displacement S_2; for small changes this is

$$S_2 = L\phi \qquad (11\text{-}58a)$$

so that the potential function is

$$2U = f(q_1^2 + q_2^2) + bS_2^2 \qquad (11\text{-}58b)$$

where

$$b = \frac{f_\phi}{L^2} \qquad (11\text{-}58c)$$

and the force constant b has the same dimensions (of ergs per square centimeter) as f.

We immediately introduce "symmetry coordinates" instead of q_1 and q_2,

$$S_1 = \frac{1}{2^{1/2}}(q_1 + q_2)$$
$$S_3 = \frac{1}{2^{1/2}}(q_1 - q_2) \qquad (11\text{-}59)$$
$$S_2 = S_2$$

In terms of these coordinates, the potential function is

$$2U = f(S_1^2 + S_3^2) + bS_2^2 \qquad (11\text{-}60)$$

The molecule belongs to the symmetry group C_{2v}; that is, it is symmetric under a rotation of 180° around the x axis, and the xy and xz planes are reflection planes. Note that the coordinates S_1 and S_2 are symmetric under all these symmetry operations, while S_3 changes sign for a rotation around the x axis and for a reflection in the xz plane. (Our notation deviates from the accepted group theoretical notation, in which the z axis is chosen as the symmetry axis.)

We have chosen a system of coordinates in which the potential energy has a simple expression. It is then a necessary computational chore to express the kinetic energy in these coordinates. For notational brevity we let x_B, x_A, y_B, etc., rather than δx_B, δx_A, δy_B, etc., represent the displacements from the equilibrium positions. The kinetic energy is

$$2K = m_A(\dot{x}_A^2 + \dot{y}_A^2) + m_B(\dot{x}_B^2 + \dot{y}_B^2) + m_B(\dot{x}_C^2 + \dot{y}_C^2) \quad \text{since } m_B = m_C$$
$$(11\text{-}61)$$

Refer now to Fig. 11-2. For arbitrary small displacements x_A, x_B, and x_C, the change in the BA distance, q_1, is $(x_B - x_A)\cos\alpha$, that is, the projection of the relative displacement $x_B - x_A$ along the AB direction. The change in the AC distance is $q_2 = (x_C - x_A)\cos\alpha$. The change ϕ in the angle 2α is the projection of the displacements in the direction perpendicular to the bond direction divided by the bond distances,

$$\phi = \frac{-(x_B - x_A)\sin\alpha - (x_C - x_A)\sin\alpha}{L}$$

or $\qquad S_2 = -(x_B - x_A)\sin\alpha - (x_C - x_A)\sin\alpha$

Proceeding systematically in this way, we find the following transformation equations between the cartesian displacements x_A, x_B, x_C, y_A, y_B, and y_C and q_1, q_2, and S_2:

$$q_1 = (x_B - x_A)\cos\alpha - (y_B - y_A)\sin\alpha$$
$$q_2 = (x_C - x_A)\cos\alpha + (y_C - y_A)\sin\alpha$$
$$S_2 = -(x_B - x_A)\sin\alpha - (x_C - x_A)\sin\alpha$$
$$\qquad - (y_B - y_A)\cos\alpha + (y_C - y_A)\cos\alpha \qquad (11\text{-}62)$$

The symmetry coordinates S_1 and S_3 are given in terms of q_1 and q_2 by (11-59); therefore

$$2^{1/2} S_1 = (x_B + x_C - 2x_A) \cos \alpha - (y_B - y_C) \sin \alpha$$
$$S_2 = -(x_B + x_C - 2x_A) \sin \alpha - (y_B - y_C) \cos \alpha \quad (11\text{-}63)$$
$$2^{1/2} S_3 = (x_B - x_C) \cos \alpha - (y_B + y_C - 2y_A) \sin \alpha$$

We cannot solve the three equations (11-63) for the six x's and y's in terms of the three S's; however, we add the two restrictions of no translation of the center of gravity,

$$m_A x_A + m_B(x_B + x_C) = 0 \quad (11\text{-}64)$$
$$m_A y_A + m_B(y_B + y_C) = 0 \quad (11\text{-}65)$$

and the constraint of no rotation

$$\Sigma m_i (y_i^\circ x_i - x_i^\circ y_i) = 0$$

or, since $\quad x_A^\circ = y_A^\circ = 0 \quad x_B^\circ = x_C^\circ = L \cos \alpha \quad y_C^\circ = -y_B^\circ = L \sin \alpha$ the equation for no over-all rotation is

$$m_B(-L \sin \alpha \, x_B - L \cos \alpha \, y_B) + m_B(L \sin \alpha \, x_C - L \cos \alpha \, y_C) = 0 \quad (11\text{-}66)$$

which immediately becomes

$$(x_B - x_C) \sin \alpha + (y_B + y_C) \cos \alpha = 0 \quad (11\text{-}67)$$

Equations (11-64), (11-65), and (11-67) are the three important constraints.* With these three equations and the three transformation equations (11-63), we can proceed to express the kinetic energy in terms of S_1, S_2, S_3.

Using (11-64), we find that the combination $x_B + x_C - 2x_A$ which occurs in (11-63) can be written as

$$(x_B + x_C - 2x_A) = (x_B + x_C)\left(1 + \frac{2m_B}{m_A}\right) \quad (11\text{-}68)$$

and similarly $\quad (y_B + y_C - 2y_A) = (y_B + y_C)\left(1 + \frac{2m_B}{m_A}\right) \quad (11\text{-}69)$

The quantities S_1, S_2, and S_3 can now be expressed in terms of $x_B + x_C$, $x_B - x_C$, $y_B + y_C$, and $y_B - y_C$. Furthermore (11-67) is a relation between $x_B - x_C$ and $y_B + y_C$.

Now consider the kinetic-energy term

$$2K = m_A \dot{x}_A^2 + m_B \dot{x}_B^2 + m_C \dot{x}_C^2$$

---

* We could just as well write equivalent equations for no linear or angular momentum, i.e.,

$$m_A \dot{x}_A + m_B(\dot{x}_B + \dot{x}_C) = 0 \quad (11\text{-}64a)$$
$$(\dot{x}_B - \dot{x}_C) \sin \alpha + (\dot{y}_B + \dot{y}_C) \cos \alpha = 0 \quad (11\text{-}67a)$$

Use (11-64) to express $\dot{x}_A$ in terms of $\dot{x}_B + \dot{x}_C$, and observe that $\dot{x}_B^2 + \dot{x}_C^2 = \frac{1}{2}[(\dot{x}_B + \dot{x}_C)^2 + (\dot{x}_B - \dot{x}_C)^2]$. The total kinetic energy can then be written as

$$2K = \frac{m_B}{2}\left(1 + \frac{2m_B}{m_A}\right)[(\dot{x}_B + \dot{x}_C)^2 + (\dot{y}_B + \dot{y}_C)^2]$$
$$+ \frac{m_B}{2}[(\dot{x}_B - \dot{x}_C)^2 + (\dot{y}_B - \dot{y}_C)^2] \quad (11\text{-}70)$$

The kinetic energy can now be expressed in terms of $\dot{S}_1$, $\dot{S}_2$, and $\dot{S}_3$ by the procedure indicated in the paragraph following Eqs. (11-69). By straightforward and not very tedious manipulations, we arrive at the result

$$2K = m_B \frac{m_A + 2m_B \sin^2 \alpha}{m_A + 2m_B} \dot{S}_1^2 + \frac{m_B}{2}\frac{m_A + 2m_B \cos^2 \alpha}{m_A + 2m_B} \dot{S}_2^2$$
$$+ \frac{2^{3/2} m_B^2 \sin \alpha \cos \alpha}{m_A + 2m_B} \dot{S}_1 \dot{S}_2 + \frac{m_B m_A}{m_A + 2m_B \sin^2 \alpha} \dot{S}_3^2 \quad (11\text{-}71)$$

Thus, the kinetic energy can be written as

$$2K = \mu_{11}\dot{S}_1^2 + 2\mu_{12}\dot{S}_1\dot{S}_2 + \mu_{22}\dot{S}_2^2 + \mu_{33}\dot{S}_3^2 \quad (11\text{-}72)$$

where μ_{ij} is a symbol for the appropriate coefficient in (11-71). The potential energy was

$$2U = f(S_1^2 + S_3^2) + bS_2^2 \quad (11\text{-}60)$$

The secular equation for the normal vibration frequencies ($4\pi^2\nu^2 = \lambda$) is

$$\begin{vmatrix} f - \lambda\mu_{11} & -\lambda\mu_{12} & 0 \\ -\lambda\mu_{12} & b - \lambda\mu_{22} & 0 \\ 0 & 0 & f - \lambda\mu_{33} \end{vmatrix} = 0 \quad (11\text{-}73)$$

The equation factors into a quadratic equation and a first-degree equation,

$$\lambda^2(\mu_{11}\mu_{22} - \mu_{12}^2) - \lambda(b\mu_{11} + f\mu_{22}) + fb = 0 \quad (11\text{-}74)$$
$$\lambda\mu_{33} - f = 0 \quad (11\text{-}75)$$

The third root is the solution of (11-75),

$$\lambda_3 = \frac{f}{\mu_{33}} \quad (11\text{-}76)$$

The quadratic equation can be solved explicitly for the two roots λ_1 and λ_2, but, for our purposes, it is sufficient to remark that the sum and product of the roots are

$$\lambda_1 + \lambda_2 = \frac{b\mu_{11} + f\mu_{22}}{\mu_{11}\mu_{22} - \mu_{12}^2} \quad (11\text{-}77)$$

$$\lambda_1\lambda_2 = \frac{fb}{\mu_{11}\mu_{22} - \mu_{12}^2} \quad (11\text{-}78)$$

When the expressions for the μ's are substituted into these expressions, we obtain the final results

$$4\pi^2\nu_3^2 = \lambda_3 = \frac{f}{m_B}\left(1 + \frac{2m_B}{m_A}\sin^2\alpha\right) \tag{11-79}$$

$$4\pi^2(\nu_1^2 + \nu_2^2) = \lambda_1 + \lambda_2 = \frac{f}{m_B}\left(1 + \frac{2m_B}{m_A}\cos^2\alpha\right)$$
$$+ \frac{2b}{m_B}\left(1 + \frac{2m_B}{m_A}\sin^2\alpha\right) \tag{11-80}$$

$$16\pi^4\nu_1^2\nu_2^2 = \lambda_1\lambda_2 = \frac{2fb}{m_B^2}\left(1 + \frac{2m_B}{m_A}\right) \tag{11-81}$$

If desired, the form of the normal vibrations can be calculated for each of the three roots of the secular equation.

We note that as the ratio $m_B/m_A \to 0$, $\mu_{11} \to m_B$, $\mu_{33} \to m_B$, $\mu_{22} \to m_B/2$, $\mu_{12} \to 2^{1/2}m_B \sin\alpha \cos\alpha (m_B/m_A) \approx 0$. In this limit the secular equation is already diagonalized, and

$$\lambda_1 = \frac{f}{m_B} = \lambda_3 \qquad \lambda_2 = \frac{2b}{m_B} \tag{11-82}$$

The normal coordinates are just the stretches S_1 and S_3 and the bend S_2.

In the general case S_3 is a normal coordinate, but the other two normal coordinates are linear combinations of S_1 and S_2. For small values of m_B/m_A, the forms of the normal vibrations are very similar to the displacements S_1 and S_2, however.

Note that, in the general case, the secular equation factors into separate equations for the coordinates of different symmetry types. As already remarked, S_1 and S_2 are of one symmetry type, whereas S_3 has a different behavior under the symmetry operations of the molecule. The off-diagonal elements of the secular determinant between coordinates of different types are zero.

11-8. Examples of Normal Vibrations. Further Remarks about Normal Vibrations. Figure 11-3 shows the normal vibrations for several nonlinear AB_2 molecules. The observed fundamental frequencies of

Table 11-1. Fundamental Frequencies of Representative Nonlinear AB_2 Molecules

	ν_1, cm^{-1}	ν_2, cm^{-1}	ν_3, cm^{-1}
H_2O	3,652	1,595	3,756
D_2O	2,666	1,179	2,784
SO_2	1,151	524	1,361
Cl_2O	680	330	973

several AB_2 molecules are listed in Table 11-1. The normal vibrations of CH_3Cl are illustrated in Fig. 11-4.

For molecules with symmetry elements, it is possible by the use of group theory to compute how many normal vibrations there are of any particular symmetry species (or symmetry type). One then starts the analysis by choosing symmetry coordinates which are consistent with the symmetry species of the normal vibrations. This is a great simplification in normal coordinate analysis because, as illustrated above for the

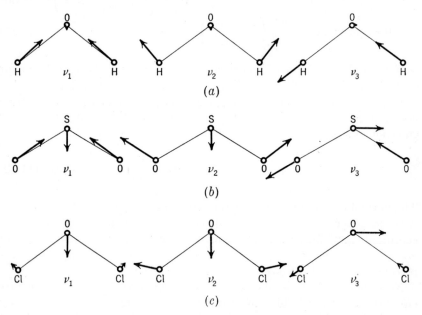

FIG. 11-3. Actual form of the normal vibrations of H_2O, SO_2, and Cl_2O. The diagrams give the correct relative amplitudes for the motion of the nuclei. The scale of amplitudes is much larger than the scale of internuclear distance if the state $v_i = 1$ is considered. No attempt is made to have the same scale in different diagrams. (*From G. Herzberg, "Infrared and Raman Spectra of Polyatomic Molecules," p. 171, Van Nostrand, Princeton, N.J., 1945.*)

AB_2 molecule, there are no cross terms in the potential-energy and kinetic-energy matrices for symmetry coordinates that belong to different symmetry species. The secular equation is automatically factored into subequations, one for each symmetry species which is represented among the normal vibrations.

The symmetry species of the normal vibrations are indicated by the letters A, B, E, F, T (in capital letters or in lower case). A and B vibrations are nondegenerate; A vibrations are symmetric with respect to rotations around the principal symmetry axis of the molecule, while B vibrations are antisymmetric with respect to rotations around this axis. The symbol E is used for vibrations which are necessarily doubly degenerate because of the symmetry of the molecule; either F or T is used for triply degenerate vibrations.

The word *degeneracy* has a somewhat different meaning as applied to normal vibrations from that which it has in quantum mechanics in general. In the latter usage, two states of equal energy are said to be degenerate. In the discussion of normal vibrations, when we say that a vibration ν_1 is nondegenerate, we mean that the molecule has only one mode of vibration with the frequency ν_1. There are a set of energy levels, with

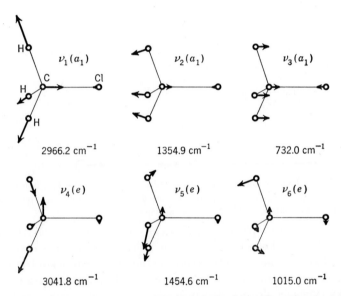

FIG. 11-4. The normal vibrations of CH_3Cl (schematic). The vibrations labeled e are doubly degenerate, but only one component is shown. (*Adapted from G. Herzberg, "Infrared and Raman Spectra of Polyatomic Molecules," p. 314, Van Nostrand, Princeton, N.J., 1945.*)

energies $\epsilon_v = h\nu(v + \frac{1}{2})$ for $v = 0, 1, 2, \ldots$. If the vibration ν_4 is doubly degenerate [in which case it is often written as in Fig. 11-4 as $\nu_4(e)$], there are two distinct normal vibrational modes, both of which have the same frequency, by virtue of the symmetry of the molecule. The simplest example of degenerate vibrations occurs for the linear symmetrical AB_2 molecule. The form of the normal vibrations is shown in Fig. 11-5. The two bending vibrations ν_{2a} and ν_{2b} are the same except for the directions of vibrations. If we denote the quantum numbers by v_a and v_b, then the three lowest energy levels are $\epsilon = 0$ ($v_a = 0, v_b = 0$), $\epsilon = h\nu_2$ ($v_a = 1, v_b = 0$; or $v_a = 0, v_b = 1$), $\epsilon = 2h\nu_2$ ($v_a = 2, v_b = 0$; or $v_a = 1, v_b = 1$; or $v_a = 0, v_b = 2$). Thus, in the ordinary sense of the word, the energy level $2h\nu_2$ is triply degenerate.

The generally accepted procedure for making a normal coordinate analysis now is formalized in the so-called "FG matrix" method.*

The normal vibration frequencies are usually determined by an analysis of the Raman and infrared spectra. The spectroscopist has the problem of deciding which of the observed absorption frequencies are the normal vibrations, rather than overtones or combination bands, and

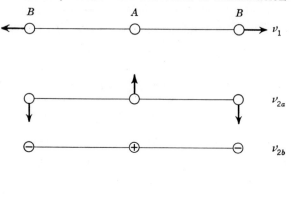

Fig. 11-5. Normal vibrations for the linear AB_2 molecule. The two bending vibrations ν_{2a} and ν_{2b} are the same except for the direction of vibration (ν_{2b} is perpendicular to the plane of the paper.) They are called degenerate.

of identifying these with the theoretical normal modes. If the vibrational analysis has been made, then for thermodynamic calculations one need only look up the values of these fundamental vibration frequencies. The degenerate modes are indicated in the letter designations of the symmetry types as already stated.

11-9. Statistical Mechanics of Molecular Vibrations. The vibrational energy of the molecule can be expressed in the form

$$\epsilon = \sum_{i=1}^{3n-6} \left(v_i + \frac{1}{2}\right) h\nu_i \tag{11-83}$$

where the sum is over the $3n - 6$ normal vibration frequencies and each vibrational quantum number v_i can have values 0, 1, 2, ...

* Described in E. B. Wilson, Jr., J. C. Decius, and P. C. Cross, "Molecular Vibrations," McGraw-Hill, New York, 1955.

The F matrix is the matrix of the coefficients for the expression of the potential energy in terms of the symmetry coordinates, as in Eq. (11-60). The G matrix is the matrix of the coefficients for the expression of the kinetic energy in terms of the conjugate momenta; the G matrix is the reciprocal of the matrix for the quadratic form which expresses the kinetic energy in terms of the velocities, $\dot{S}_i$ (where the S_i's are the symmetry coordinates). Thus, for the AB_2 molecule, the G matrix is essentially the reciprocal of the matrix of the quadratic form in (11-71).

The zero-point energy is

$$\epsilon_0 = \sum \frac{h\nu_i}{2} \tag{11-84}$$

and the excess energy over the zero-point energy is

$$\epsilon - \epsilon_0 = \sum_{i=1}^{3n-6} v_i h \nu_i \tag{11-85}$$

Since the vibrational energy is the sum of the vibrational energies of the several normal modes, and since any value of the quantum number for one mode can occur with any value of the quantum number of another mode, the vibrational partition function is the product of the vibrational partition functions for the individual normal modes.

$$q_{\text{vib}} = q(u_1)q(u_2) \cdots q(u_{3n-6}) \quad \text{where } u_i = \frac{h\nu_i}{kT} \tag{11-86}$$

If the energy levels are measured with respect to the zero vibrational state of the molecule as in (11-85),

$$q(u_i) = \frac{1}{1 - e^{-u_i}} \tag{11-87}$$

If the energy levels are measured with respect to the hypothetical minimum of the potential-energy curve as in (11-83), then

$$q(u_i) = \frac{e^{-u_i/2}}{1 - e^{-u_i}} \tag{11-88}$$

The formulae for the energy, entropy, free energy, and heat capacity in any one normal mode are the same as those for the vibration of a diatomic molecule given in Sec. 8-6. The thermodynamic functions for vibration for the molecule are the sums of the thermodynamic functions for the individual normal vibrations. A doubly degenerate vibration is treated as two independent normal vibrations for this purpose, and a triply degenerate vibration as three.

At high temperatures, where $u_i = h\nu_i/kT \ll 1$, the partition function can be expanded to give

$$q = \prod_{i=1}^{3n-6} \frac{1}{u_i} = \prod_{i}^{3n-6} \frac{kT}{h\nu_i} \tag{11-89}$$

The molar vibrational energy is then

$$\text{H}_{\text{vib}} = (3n - 6)RT \tag{11-90}$$

and the molar vibrational heat capacity is

$$\text{C}_{\text{vib}} = (3n - 6)R \tag{11-91}$$

in agreement with the classical theorem for equipartition of energy (Sec. 10-3).

There is another point to note about applying these results in practice. Rigorously speaking, normal vibrations are a property of the molecule as a whole. However, they can often be approximately represented as group vibrations. For example, for the case of CH_3Cl illustrated in Fig. 11-4, the frequencies $\nu_1(a_1)$ and $\nu_4(e_1)$ (doubly degenerate) of 2,966.2 and 3,041.8 cm^{-1} are essentially C—H bond-stretching vibrations. The frequency ν_3 of 732.1 cm^{-1} is principally a C—Cl stretch. The frequencies $\nu_6(e)$, 1,015.0 cm^{-1}, $\nu_2(a_1)$, 1,354.9 cm^{-1}, and $\nu_5(e)$, 1,454.6 cm^{-1}, are principally C—H bends. There are empirical tables of characteristic bond-stretching and -bending vibrations. The fact that these tables are of great practical significance in structural analysis by infrared spectroscopy* testifies to their approximate validity.

The usefulness of these empirical bond frequencies for statistical calculations is seriously limited, however, by the following consideration: It is usually possible to make reasonable guesses as to the high-frequency stretching vibrations for a molecule. For a complicated, "floppy" molecule, there are a large number of low-frequency skeletal, torsional, and bending vibrations. There are no well-developed empirical schemes for the estimation of these frequencies. However, it is the low frequencies which are most effectively excited at any temperature and which contribute most to the thermodynamic properties.

It should also be remarked that there are of course anharmonic effects in the vibrations of polyatomic molecules, interactions between rotation and vibration, and interactions between the several normal modes. Corrections for these effects can be made when and if justified by sufficiently accurate data about the energy levels. Because of the complexities of the spectra of polyatomic molecules, this information is less likely to be available in this case than for diatomic molecules.

Problem 11-4. The triangular molecule AB_2 has three normal vibrations—two stretches at about 2,000 cm^{-1} and one bend at 400 cm^{-1}. Make a rough, very approximate sketch of the heat capacity (C_p or C_v) of this molecule as a function of temperature. Label or explain the various features of the curve.

11-10. Restricted Internal Rotation and Torsional Vibrations.

In a molecule such as H_3C—CH_3, H_3C—OH, H_2ClC—$CHCl_2$, one of the possible internal motions that has to be considered in a vibrational analysis is the rotation of one group with respect to another around the single bond (the bond which is especially emphasized by using a line in the formulae above). For example, in H_3C—OH for which the COH bond angle is about 105° (if this angle were 180° there would be no possibility of an internal torsional motion) the potential energy will vary

---

* Herzberg, p. 195 [18].

as a function of the angle between the COH plane and one of the OCH planes. If we think of the OH group as rotating with respect to the CH_3 group, the potential energy should be periodic with a period of 120°. In symmetrical CH_3—CCl_3, the potential energy should vary as a function of the angle between the HCC and the CCCl planes; again the periodicity is 120°.

In a molecule such as C_2H_4, the resistance of the double bond to torsional motion is high, and no special problems arise because the vibration is a typical simple harmonic one. In a molecule such as H_3C—$C\equiv C$—CH_3, the barrier to relative rotation of the two CH_3 groups is very low, and the two groups can be considered to be essentially rotating freely. In a large class of compounds, including many cases of groups connected by a single bond, the potential barrier for internal rotation is of intermediate height, and some special consideration is necessary.

The problem is considerably simplified for molecules such as CH_3CCl_3 or C_2H_6, which contain symmetrical coaxial tops. We confine our discussion principally to this case. If the two tops have moments of inertia I_1 and I_2, the reduced moment of inertia for their relative motion is

$$I_r = \frac{I_1 I_2}{I_1 + I_2} \quad (11\text{-}92)$$

Let ϕ represent the angle of rotation; the potential is then $U(\phi)$. The wave equation for the relative motion of the two tops is

$$-\frac{\hbar^2}{2I_r}\frac{d^2\psi}{d\phi^2} + [U(\phi) - \epsilon]\psi(\phi) = 0 \quad (11\text{-}93)$$

The potential function must be periodic with a period of $2\pi/n$, where n is related to the symmetry of the molecules. For H_3C—OH or H_3C—CX_3, $n = 3$. This case is illustrated in Fig. 11-6.

The simplest and most commonly used potential is

$$U(\phi) = \tfrac{1}{2}U_0(1 - \cos n\phi) \quad (11\text{-}94)$$

This has the minimum value of zero at $\phi = 0$, $\pm 2\pi/n$, $\pm 4\pi/n$, etc., and maxima at $\phi = \pm \pi/n$, $\pm 3\pi/n$, etc., of U_0.*

---

* The Fourier expansion of a periodic potential of period $2\pi/n$ which is to be zero at $\phi = 0$, $\pm 2\pi/n$, $\pm 4\pi/n$, etc., and otherwise positive is

$$U(\phi) = a_0 + \sum_{j=1}^{\infty} a_j \cos jn\phi$$

with

$$a_0 = -\sum_{1}^{\infty} a_j$$

Expression (11.94) is the first term of this expansion. In practice, it seems to be a remarkably accurate representation of the actual potential curves.

The boundary condition on the solutions of the wave equation (11-93) is that they should be periodic with the symmetry of the internal rotor, i.e., that $\psi(\phi) = \psi(\phi + 2\pi/n)$. Such solutions can be found.*

It is instructive to consider the two limiting cases of very high and very low barriers.

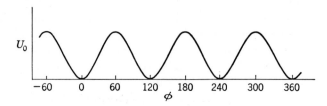

Fig. 11-6. Schematic representation of the potential curve for internal rotation for a potential barrier with threefold symmetry.

11-11. High Barriers. The Harmonic-oscillator Approximation. For a high barrier, the wave function will be confined to small values of ϕ, and we expand $\cos n\phi$,

$$\cos n\phi = 1 - \frac{n^2\phi^2}{2} + \frac{n^4\phi^4}{4!} + \cdots$$

so that, from (11-94),

$$U(\phi) = U_0 \frac{n^2\phi^2}{4} - U_0 \frac{n^4\phi^4}{48} + \cdots \tag{11-95a}$$

The first approximation is

$$U(\phi) = \frac{U_0 n^2 \phi^2}{4} \tag{11-95b}$$

With this potential, the wave equation is that for a simple harmonic oscillator. The energy levels are given by

$$\epsilon_k = (k + \tfrac{1}{2})h\nu \qquad k = 0, 1, 2, \ldots \tag{11-96}$$

with
$$\nu = \frac{n}{2\pi}\left(\frac{U_0}{2I_r}\right)^{1/2} \tag{11-97}$$

The requirements for this approximation to be valid are (a) that the positions of the first few levels should be well below the top of the barrier, i.e., that $h\nu \ll U_0$, with ν given by (11-97); and (b) that the temperature is sufficiently low so that not many molecules are excited into levels close to the top of the barrier, i.e., that $U_0/kT \gg 1$.

The calculation of thermodynamic properties for this case is straightforward.

There is one tricky point connected with symmetry numbers and "tunneling" which can be faced at this point. Consider a molecule

---

* Cf. Pitzer, pp. 239–243 [10].

Sec. 11-11] POLYATOMIC MOLECULES 197

CX_3—CY_3 as illustrated in Fig. 11-7. The case of CF_3—CCl_3 would be particularly apt as an example to which the harmonic oscillator applied (the barrier is probably reasonably high—that for C_2Cl_6 being 10 to 15 kcal—and the moments of inertia are large).

The potential curve has threefold symmetry; the potential minima in such compounds occur in the staggered configuration with the C—C—F planes bisecting two C—C—Cl planes. Fluorine F_1 can be situated in one of three equivalent positions, between Cl_1 and Cl_2, or between Cl_2 and Cl_3, or between Cl_3 and Cl_1.

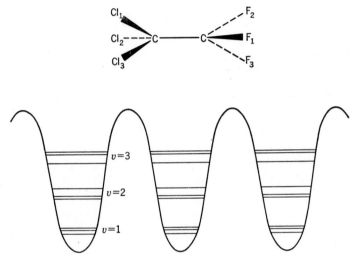

FIG. 11-7. Potential curve and energy levels (very schematic) for CCl_3—CF_3. Each torsional harmonic-oscillator level ($v = 1, 2, 3$ in the figure) is split into three sublevels as indicated. Actually, in the nonrotating molecule, a threefold degenerate level of the strict harmonic-oscillator approximation is split into a doubly degenerate level and a single level, but interactions with external rotation produce the additional splitting. For further details, see G. Herzberg, pp. 225–227 [18].

There are two possible views to take in calculating the thermodynamic properties. In viewpoint R (for rigid), the barrier is very high, and the molecule is fixed in one particular configuration—for example, the one shown. Let q_{rot} represent the external rotational partition function calculated from the moments of inertia, but not including the symmetry number.

The symmetry number is 3, and we can take the point of view that this is basically connected with the calculation of the external rotational partition function of the molecule. There is one set of vibrational energy levels for the torsional vibration, with

$$\epsilon_v = (v + \tfrac{1}{2})h\nu$$

The vibrational partition function for torsional vibration is

$$q = \frac{1}{1 - e^{-h\nu/kT}}$$

Therefore the partition-function factor for rotation plus the internal torsional motion is

$$q = \tfrac{1}{3} q_{\text{rot}} \frac{1}{1 - e^{-h\nu/kT}} \quad (11\text{-}98)$$

In viewpoint T (for tunneling), we consider the fact that the potential curve has three minima. When account is taken of this threefold minimum and of the fact that the potential function is not truly harmonic (that is, it is not really represented by $U = U_0 n^2 \phi^2/4$, and the range of ϕ is not $-\infty < \phi < \infty$), it is found by straightforward quantum mechanics that each simple harmonic energy level consists of three sublevels as indicated in Fig. 11-7. If the barrier is high, the splittings of the levels are very small, but the splittings increase as the energy level approaches the top of the barrier. In viewpoint T, not only is F_1 between Cl_2 and Cl_3, but it "tunnels" to the other equivalent positions. In the true stationary state with an infinitely long lifetime, the wave function for F_1 occurs in all three potential minima. This is fundamentally a more satisfactory point of view toward the energy levels and structure of the molecule.

If the splitting is small, the energy levels are approximately given by

$$\epsilon_v = (v + \tfrac{1}{2}) h\nu$$

as before, but each such level is threefold-degenerate (or, more precisely, split into three sublevels). In viewpoint T, the vibrational partition function (for small splitting) is given by

$$q = \frac{3}{1 - e^{-h\nu/kT}} \quad (11\text{-}99a)$$

However, in viewpoint T, the symmetry number of the molecule is 9 because, with tunneling permitted, there are now nine superimposable configurations of the molecule, whereas in viewpoint R the symmetry number was 3. In viewpoint T, the over-all partition function for external rotation plus torsional vibration is

$$q = \tfrac{1}{9} q_{\text{rot}} \frac{3}{1 - e^{-h\nu/kT}} = \tfrac{1}{3} q_{\text{rot}} \frac{1}{1 - e^{-h\nu/kT}} \quad (11\text{-}99b)$$

The final result is the same as for viewpoint R. In the harmonic-oscillator approximation for the torsional levels, with the splittings due to tunneling small, either point of view consistently used predicts the same over-all partition function for the molecule. If the splitting of the levels is large enough actually to affect the partition-function calculation, viewpoint T is obviously necessary.

11-12. Completely Free Rotation. For the wave equation

$$\frac{-\hbar^2}{2I_r} \frac{d^2\psi}{d\phi^2} + [U(\phi) - \epsilon]\psi(\phi) = 0 \tag{11-93}$$

we now study the case in which the energy of the level of interest is much greater than the barrier height,

$$\epsilon \gg U_0 \tag{11-100}$$

We can then replace $U(\phi)$ by its average value, $\tfrac{1}{2}U_0$. The wave equation becomes

$$\frac{d^2\psi}{d\phi^2} + \frac{8\pi^2 I_r}{h^2}(\epsilon - \tfrac{1}{2}U_0)\psi = 0 \tag{11-101}$$

to which the solution is

$$\psi = e^{\pm i\alpha\phi} \tag{11-102a}$$

with

$$\alpha^2 = \frac{8\pi^2 I_r}{h^2}(\epsilon - \tfrac{1}{2}U_0) \tag{11-102b}$$

(Solutions in terms of $\sin \alpha\phi$ and $\cos \alpha\phi$ are of course also permissible.)

The boundary condition is that ψ should be periodic with period of $2\pi/n$,

$$\psi\left(\phi + \frac{2\pi}{n}\right) = \psi(\phi) \tag{11-103}$$

which requires that

$$\alpha = jn \qquad j = 0, \pm 1, \pm 2, \ldots \tag{11-104}$$

The energy levels are therefore

$$\epsilon_j = \frac{h^2}{8\pi^2 I_r} n^2 j^2 + \tfrac{1}{2}U_0 \qquad j = 0, \pm 1, \pm 2 \tag{11-105}$$

The wave functions are also eigenfunctions of the angular momentum around the axis of rotation, and the eigenvalues are $jn\hbar$.

The above formulae always apply to some high energy levels. For low barriers, they may apply to all or almost all the levels (free rotation). For this case, the partition function is

$$q = \sum_{j=-\infty}^{j=+\infty} e^{-\beta(h^2 n^2 j^2/8\pi^2 I_r + \tfrac{1}{2}U_0)} \tag{11-106}$$

$$q \approx e^{-\beta U_0/2} \int_{j=-\infty}^{j=+\infty} e^{(-\beta h^2 n^2 j^2/8\pi^2 I_r)}\, dj$$

$$= e^{-\beta U_0/2}\left(\frac{8\pi^3 I_r kT}{n^2 h^2}\right)^{1/2} \tag{11-107}$$

$$q = \frac{1}{n} e^{-\beta U_0/2}\left(\frac{8\pi^3 I_r kT}{h^2}\right)^{1/2} \tag{11-108}$$

By assumption, $\beta U_0 \ll 1$, so that $e^{-\beta U_0/2} \approx 1$, and we take as the partition function for completely free internal rotation

$$q_{\text{fr}} = \frac{1}{n}\left(\frac{8\pi^3 I_r kT}{h^2}\right)^{1/2} \tag{11-109}$$

The corresponding contribution to the molar energy is $RT/2$ and to the heat capacity is $R/2$.

In the previous section, we discussed the symmetry-number problem in the calculation of the partition function for CCl_3—CF_3, assuming a high barrier so that the harmonic-oscillator approximation applied. What would be the correct calculation for a very low barrier? (The discussion might actually be applicable, for example, to CF_3—C≡C—CH_3, which would probably have a low barrier.) There is a factor of $\frac{1}{3}$ included in the partition function q_{fr} for free internal rotation [(11-109)]. The correct symmetry number to use for the external rotation is then 3, not 9. That is, the over-all partition function is

$$q = q_{\text{trans}} \frac{\pi^{1/2}}{9}\left(\frac{8\pi^2 I_x kT}{h^2}\right)^{1/2}\left(\frac{8\pi^2 I_y kT}{h^2}\right)^{1/2}\left(\frac{8\pi^2 I_z kT}{h^2}\right)^{1/2}\left(\frac{8\pi^3 I_r kT}{h^2}\right)^{1/2} q_{\text{vib}} \tag{11-110}$$

where I_x, I_y, and I_z are the external moments of inertia (in this case, $I_x = I_y$ if the symmetry axis is the z axis) and I_r is the reduced moment of inertia for internal rotation. The factor q_{vib} refers to the other normal modes of the molecule.

We might have argued that the "external" symmetry number is 9, since with free internal rotation there are nine superimposable configurations of the atoms. Then there would be a factor of 27 rather than 9 in the partition function [(11-110)]. This would be wrong. By imposing the requirement $\psi(\phi) = \psi(\phi + 2\pi/3)$ on the wave function for internal rotation [(11-103)], we have already imposed symmetry requirements on the molecule due to the identity of the configurations obtained by internal rotation of $2\pi/3$; it would be incorrect to use this requirement again in calculating the external symmetry number.

11-13. More about Restricted Rotation. For an actual problem, the wave equation (11-93) must be solved to determine the energy levels. In general, the solutions are intermediate between the harmonic-oscillator approximation and the free-rotor approximation. In some cases, the lower energy levels can be approximately represented by the harmonic-oscillator equation and the high levels by the free-rotor equation.

Figure 11-8 illustrates the relation between harmonic-oscillator energy levels, free-rotor energy levels, and the exact solution of Eq. (11-93) for the energy levels for one particular case. It is noteworthy that the transition from the harmonic-oscillator approximation to the free-rotor approximation takes place over a rather small energy interval near the top of the barrier.

With the potential function of (11-94), the wave equation (11-93) becomes

$$-\frac{\hbar^2}{2I_r}\frac{\partial^2 \psi}{\partial \phi^2} + \tfrac{1}{2}U_0(1 - \cos n\phi)\psi - \epsilon\psi = 0$$

which is readily transformed to

$$\frac{\partial^2 \psi}{\partial(n\phi)^2} + \frac{U_0 I_r}{n^2\hbar^2}(1 - \cos n\phi)\psi - \frac{2\epsilon I_r}{n^2\hbar^2}\psi = 0 \qquad (11\text{-}111a)$$

The energy levels of a free rotor depend on the parameter $n^2\hbar^2/2I_r$. It

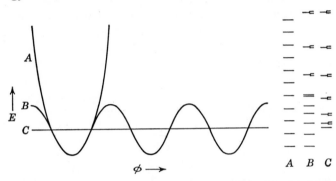

FIG. 11-8. Potential functions (on the left) and energy levels (on the right) for a restricted rotator B, an ideal harmonic oscillator A, and free rotator C. Cases A and C are approached by B at low and high energies, respectively. The splitting shown for C is schematic, to indicate the $\pm$ degeneracy of the free-rotation states. (*From Kenneth S. Pitzer, "Quantum Chemistry," p. 243, Prentice-Hall, Englewood Cliffs, N.J., 1953.*)

is evident by inspection of (11-111a) that the energy levels of a restricted rotor depend on this parameter and on the dimensionless parameter

$$\Theta = \frac{2I_r U_0}{n^2 \hbar^2} \qquad (11\text{-}111b)$$

Tables from which the energy levels can be calculated when these quantities are given are available.* From the solutions of (11-111a) for the energy levels, the contribution of the restricted rotor to the partition function and the thermodynamic functions have been calculated. These have been tabulated as functions of the dimensionless quantities U_0/kT and the partition function for free rotation $q_{fr} = (8\pi^3 I_r kT/n^2\hbar^2)^{1/2}$.†
When the barrier to rotation and the moments of inertia are known,

---

* E. B. Wilson, Jr., *Chem. Revs.*, **27**: 17 (1940).

† Pitzer, pp. 494–500 [10]; K. S. Pitzer and W. D. Guinn, *J. Chem. Phys.*, **10**: 428 (1942).

the calculation of thermodynamic properties by the use of these tables is a straightforward matter.

The analysis of problems of internal rotation with tops that are not symmetrical and coaxial is much more difficult, because of the coupling between internal and over-all rotation. The reader is referred to Pitzer's book for further references* and discussion. There is an interesting and useful approximate method for treating such problems that has been recommended for order-of-magnitude calculations.†

There has recently been a great deal of progress in the determination of barrier heights by the analysis of the splittings due to tunneling (mentioned in Sec. 11-11) by microwave spectroscopy. A large number of experimental data are now available.‡

Barrier heights have also been determined from infrared and Raman spectra, electron diffraction, dipole moment, and thermodynamic measurements. A general review of the field is available.§

There is a very wide range of values for the potential barriers which have been encountered in practice. Illustrative examples (in kilocalories per mole) are CH_3—CH_3, 2.7 to 3.0;∥ ¶ CH_3—NH_2, 1.94;∥ CH_3—SiH_3, 1.70;∥ CH_3—NO_2, 0.006;∥ C_2Cl_6, 10 to 15;** and

$$H_3CN-\overset{O}{\underset{CH_3}{C}}H, 7††$$

11-14. The Product Rule.‡‡ When an atom in a molecule is replaced by an isotopic atom, the potential-energy function and the configuration are not significantly changed. However, as already illustrated for diatomic molecules, the vibration frequencies may change by much greater amounts because of the changes in the reduced mass. These changes are useful diagnostic tools in the analysis of spectra. The thermodynamic consequences are particularly important for isotopic fractionation reactions.

For example, if an X—H bond is changed to an X—D bond, where X is a heavy atom, any vibration frequency which is principally a vibration

---

* Pitzer, pp. 239–243, 494–500 [10].

† D. Herschbach, H. Johnston, K. Pitzer, and R. Powell, *J. Chem. Phys.*, **25**: 738–739 (1956).

‡ E. B. Wilson, *Proc. Natl. Acad. Sci.*, **43**: 816 (1957); in I. Prigogine (ed.), "Advances in Chemical Physics," vol. II, pp. 367–395, Interscience, New York, 1959.

§ S. Mizushima, "Structure of Molecules and Internal Rotation," Academic Press Inc., New York, 1954.

∥ E. B. Wilson, *Proc. Natl. Acad. Sci.*, **43**: 816 (1957).

¶ D. Lide, *J. Chem. Phys.*, **29**: 1426 (1958).

** S. Mizushima, "Structure of Molecules and Internal Rotation," Academic Press Inc., New York, 1954.

†† H. S. Gutowsky and C. H. Holm, *J. Chem. Phys.*, **25**: 1228 (1956).

‡‡ E. B. Wilson, Jr., J. C. Decius, and P. C. Cross, "Molecular Vibrations," p. 183, McGraw-Hill, New York, 1955; Herzberg, pp. 231–236 [18].

of the hydrogen or deuterium atom is changed by a factor of about $2^{1/2}$. A more exact analysis of the frequency change requires a more detailed analysis of the effective reduced masses for the various normal modes. The Teller-Redlich product rule is an aid in simplifying this analysis. We state the rule without proof.

Let prime quantities refer to the isotopic molecule; the masses of the individual atoms are m_i and m_i' ($i = 1, 2, \ldots, n$). The normal vibration frequencies are ν_j and ν_j' ($j = 1, 2, \ldots, 3n - 6$), the molecular weights are M and M', and the moments of inertia are $I_x, I_x', I_y, I_y', I_z, I_z'$. The product rule is then

$$\prod_{j=1}^{3n-6} \frac{\nu_j}{\nu_j'} = \prod_{i=1}^{n} \left(\frac{m_i'}{m_i}\right)^{3/2} \left(\frac{M}{M'}\right)^{3/2} \left(\frac{I_x I_y I_z}{I_x' I_y' I_z'}\right)^{1/2} \qquad (11\text{-}112)$$

For a linear molecule, there are $(3n - 5)$ normal vibrations in the product on the left-hand side of (11-112) and only two moments of inertia (which are equal) in the product on the right side.*

Problem 11-5. Verify the product rule for the case of a diatomic molecule.

In addition to its application in isotopic equilibria problems, the product rule is useful as a check in conventional normal vibrations analysis, in that it predicts that the function

$$\frac{\prod_{j=1}^{3n-6} \nu_j \prod_{i=1}^{n} m_i^{3/2}}{M^{3/2}(I_x I_y I_z)^{1/2}} = f(f_{11}, \ldots, f_{ij}, \ldots, f_{nn}) \qquad (11\text{-}113)$$

should be independent of the masses of the atoms. As we shall see, this function can and does depend on the potential-energy surface of the molecule and on the bond angles and bond distances, but the product rule shows that it is independent of the masses of the atoms in the molecule.

In addition to the product rule, there are some "sum rules" that apply to the frequencies of a set of isotopic molecules, such as HOH, HOD, DOD.†

11-15. Isotopic Equilibria. The principles for calculating isotopic equilibrium constants were discussed in Chap. 9 for the case of diatomic

---

* The Teller-Redlich product rule is actually somewhat more detailed and powerful than (11-112), in that it can be applied to subsets of normal frequencies which all belong to the same symmetry type. This application requires some knowledge of group theory, in order to decide the extent to which the translations and rotations are included in the symmetry type under consideration, i.e., the number of times the factors M'/M and I_x'/I_x are to be included in the equations. The choice of which masses to include in the product $\Pi(m_i/m_i')$ for a subset of frequencies must also be settled by group theory.

† J. C. Decius and E. B. Wilson, Jr., *J. Chem. Phys.*, **19**: 1409 (1951).

molecules. Essentially nothing new is introduced in the consideration of polyatomic molecules. However, the opportunity is provided to review the subject and introduce a more general notation.*

In comparing the partition functions of isotopic molecules, we take as the reference point of zero energy the minimum in the potential energy rather than the actual lowest vibrational state of the molecule. The potential-energy curves of the two isotopic molecules are then the same, since to a very high degree of approximation the electronic potential-energy curves are unaffected by the difference in mass of the two isotopes (and the differences in the magnetic moments and quadrupole moments of the nuclei are equally unimportant!).

With the zero of energy taken at the minimum in the potential-energy curve, the partition function of an oscillator is

$$q(u_i) = e^{-u_i/2} \frac{1}{1 - e^{-u_i}} \qquad (11\text{-}114)$$

where $u_i = h\nu_i/kT$.

Primes being used to indicate the properties of one of the isotopic molecules, the ratio of the total partition functions of two isotopic molecules is

$$\frac{q}{q'} = \frac{\sigma'}{\sigma} \left(\frac{I_x I_y I_z}{I'_x I'_y I'_z}\right)^{1/2} \left(\frac{M}{M'}\right)^{3/2} \prod_{j=1}^{3n-6} e^{(u_j' - u_j)/2} \frac{1 - e^{-u_j'}}{1 - e^{-u_j}} \qquad (11\text{-}115)$$

Problem 11-6. Show the derivation of Eq. (11-115) in detail.

With the help of the product rule (11-112) the factors in front concerned with mass ratios can be transformed to factors containing the masses of the individual atoms and the vibration frequencies.

$$\frac{q}{q'} = \frac{\sigma'}{\sigma} \prod_{i=1}^{n} \left(\frac{m_i}{m'_i}\right)^{3/2} \prod_{j=1}^{3n-6} \frac{u_j}{u'_j} e^{(u_j' - u_j)/2} \frac{1 - e^{-u_j'}}{1 - e^{-u_j}} \qquad (11\text{-}116)$$

We bring the first two factors on the right over to the left side, and this expression becomes

$$\frac{\sigma q}{\sigma' q'} \prod_{i=1}^{n} \left(\frac{m'_i}{m_i}\right)^{3/2} = \prod_{j=1}^{3n-6} \frac{u_j}{u'_j} e^{(u_j' - u_j)/2} \frac{1 - e^{-u_j'}}{1 - e^{-u_j}} \qquad (11\text{-}117)$$

The quantity on the right-hand side of (11-117) is that part of the partition-function ratio of two isotopic molecules which is influential in

* The treatment follows closely that given by J. Bigeleisen and M. Wolfsberg in I. Prigogine (ed.), "Advances in Chemical Physics," vol. I, Interscience, New York, 1958. The basic paper is M. G. Mayer and J. Bigeleisen, *J. Chem. Phys.*, **15**: 261 (1947).

the analysis of isotope exchange reactions. This is an important general remark which may be illustrated by the following example: Consider the exchange reaction between deuterium and hydrogen atoms in methyl fluoride and water. If the deuterium-to-hydrogen ratio is small, the important deuterated compounds will be H_2DCF and HOD, and we can simplify the analysis by neglecting HD_2CF, D_3CF, and D_2O. The important exchange reaction is therefore

$$H_3CF + HOD \rightleftharpoons H_2DCF + H_2O \qquad (11\text{-}118a)$$
$$q_1 \qquad q_2' \qquad q_1' \qquad q_2 \qquad (11\text{-}118b)$$
$$\sigma_1 = 3 \quad \sigma_2' = 1 \quad \sigma_1' = 1 \quad \sigma_2 = 2 \qquad (11\text{-}118c)$$

The partition functions and symmetry numbers are written below each chemical species in (11-118b) and (11-118c). The equilibrium constant is

$$\frac{[H_2DCF][H_2O]}{[H_3CF][HOD]} = \frac{q_1' q_2}{q_1 q_2'} \qquad \text{since } \Delta E_0^\circ = 0 \qquad (11\text{-}119)$$

A quantity of interest for isotope-fractionation experiments is the ratio of deuterium atoms to hydrogen atoms in the methyl fluoride. This ratio we call r_org (for organic).

$$r_\text{org} = \frac{\text{organic deuterium}}{\text{organic hydrogen}} = \frac{H_2DCF}{3[H_3CF]} \qquad (11\text{-}120)$$

because there are three equivalent hydrogens in H_3CF.

In Eq. (11-120) the assumption is made that $r_\text{org} \ll 1$, so that $[H_2DCF] \ll [H_3CF]$. In the general case,

$$r_\text{org} = \frac{3[D_3CF] + 2[D_2HCF] + [H_2DCF]}{[D_2HCF] + 2[H_2DCF] + 3[H_3CF]} \qquad (11\text{-}121)$$

The same conclusion will follow from the analysis of the general case using (11-121), but for simplicity and clarity we restrict ourselves to the case $r_\text{org} \ll 1$.

In the same way, the ratio of deuterium to hydrogen in the water molecules is

$$r_\text{aq} = \frac{[HOD]}{2[H_2O]} \qquad (11\text{-}122)$$

The isotopic-fractionation factor for deuterium vs. hydrogen between methyl fluoride and water is then

$$f = \frac{r_\text{org}}{r_\text{aq}} = \frac{[H_2DCF]/3[H_3CF]}{[HOD]/2[H_2O]} = \frac{q_1'/3q_1}{q_2'/2q_2} = \frac{q_1'}{\sigma_1 q_1} \frac{\sigma_2 q_2}{q_2'} \qquad (11\text{-}123)$$

In comparing (11-123) with (11-117), we should recall that $\sigma_1' = \sigma_2' = 1$. Thus the example illustrates that the isotopic enrichment depends on the product σq for each molecular species, partially justifying the assertion that the right-hand side of (11-117) is the important expression for isotopic equilibria. Furthermore, the other factor on the left-hand side of (11-117) is the product of the mass ratios. This will cancel in a chemical equilibrium, because the mass of each atom occurs once on each side of the equation.

In the case of the hydrogen isotopes, the differences $\Delta u_j = u_j' - u_j$ may be large, and it is best to use (11-117) directly. For heavier atoms, the fractional mass differences in isotopes are smaller, and the quantity

$$\frac{u}{u + \Delta u} e^{\Delta u/2} \frac{1 - e^{-(u+\Delta u)}}{1 - e^{-u}} \quad \text{where } \Delta u = u' - u$$

which occurs in (11-117) may be expanded as a power series in Δu,

$$\frac{u}{u + \Delta u} e^{\Delta u/2} \frac{1 - e^{-(u+\Delta u)}}{1 - e^{-u}} \approx 1 + \Delta u \left(\frac{1}{2} - \frac{1}{u} + \frac{1}{e^u - 1} \right) + \cdots$$
$$= 1 + G(u) \, \Delta u \quad (11\text{-}124)$$

where
$$G(u) = \frac{1}{2} - \frac{1}{u} + \frac{1}{e^u - 1} \quad (11\text{-}125)$$

Values of $G(u)$ are tabulated.*

The resulting expression for the effective factors in the partition-function ratio in (11-117) is

$$\frac{\sigma q}{\sigma' q'} \prod_{i=1}^{3n} \left(\frac{m_i'}{m_i} \right)^{3/2} = \prod_{j=1}^{3n-6} [1 + G(u_i) \, \Delta u_i] \approx 1 + \sum_{j=1}^{3n-6} G(u_i) \, \Delta u_i \quad (11\text{-}126)$$

This is the working equation for the analysis of isotopic-exchange reactions for most cases involving heavy isotopes. For small values of u, the expression (11-125) gives $G(u) \approx u/12$. The isotope-fractionation factor approaches unity as the temperature is raised and $u \to 0$.

11-16. The Classical Partition Function. The partition function of a nonlinear molecule with no free internal rotations is

$$q = \left(\frac{2\pi MkT}{h^2} \right)^{3/2} \frac{\pi^{1/2}}{\sigma} \left(\frac{8\pi^2 I_x kT}{h^2} \right)^{1/2} \left(\frac{8\pi^2 I_y kT}{h^2} \right)^{1/2} \left(\frac{8\pi^2 I_z kT}{h^2} \right)^{1/2}$$
$$q(u_1) q(u_2) \cdots q(u_{3n-6})$$
$$= \frac{64\pi^5}{\sigma h^6} M^{3/2} (I_x I_y I_z)^{1/2} T^3 q(u_1) q(u_2) \cdots q(u_{3n-6}) \quad (11\text{-}127)$$

At high temperatures, defined by the condition $h\nu_i/kT = u_i \ll 1$, $q(u_i) \approx 1/u_i = kT/h\nu_i$. Each vibration is then behaving classically. The classical limit for the partition function is therefore

$$q_{\text{cl}} = \frac{64\pi^5}{h^{3n}} M^{3/2} (I_x I_y I_z)^{1/2} (kT)^{3n-3} \frac{1}{\nu_1 \nu_2 \cdots \nu_{3n-6}} \quad (11\text{-}128)$$

* M. G. Mayer and J. Bigeleisen, *J. Chem. Phys.*, **15**: 261 (1947).

With the product rule in the form of Eq. (11-113), the expression for q_{cl} can be rewritten as

$$q_{cl} = \frac{64\pi^5}{h^{3n}} (kT)^{3n-3} \frac{(m_1 m_2 \cdots m_n)^{3/2}}{f(f_{11}, \ldots, f_{ij}, \ldots, f_{nn})} \qquad (11\text{-}129)$$

where, as we shall see, $f(f_{11}, \ldots, f_{nn})$ is a function of the potential-energy constants and the shape of the molecule. The significance of this result is that it suggests that the classical approximation to the partition function can be written in terms of the masses, the potential-energy constants, and the shape parameters of the molecule, without explicit calculation of the normal vibration frequencies.

We can approach this problem in another way. Take a system of cartesian coordinates fixed in space. The Hamiltonian operator of the molecule is

$$H = \sum_{i=1}^{n} \frac{p_{x_i}^2 + p_{y_i}^2 + p_{z_i}^2}{2m_i} + U(x_1, \ldots, z_n) \qquad (11\text{-}130)$$

The classical partition function is

$$q_{cl} = \frac{1}{h^{3n}} \int_{p_1} \cdots \int_{z_n} e^{-\beta H(p_1, \ldots, z_n)} dp_1 \cdots dz_n \qquad (11\text{-}131)$$

The integrals over the momenta can be performed; each one is

$$\int_{-\infty}^{+\infty} e^{-\beta p^2/2m} dp = \left(\frac{2\pi m}{\beta}\right)^{1/2} = (2\pi m kT)^{1/2}$$

Thus Eq. (11-131) becomes

$$q_{cl} = Z(T,V) \prod_{i=1}^{n} \left(\frac{2\pi m_i kT}{h^2}\right)^{3/2} \qquad (11\text{-}132)$$

where $\qquad Z(T,V) = \int_{x_1} \cdots \int_{z_n} e^{-\beta U(x_1, \ldots, z_n)} dx_1 \cdots dz_n \qquad (11\text{-}133)$

In Eqs. (11-132) and (11-133) the quantity $Z(T,V)$ is known as the configuration integral; as we shall see, it occurs repeatedly in classical n-body problems.

The potential $U(x_1, \ldots, z_n)$ is actually a function of $3n - 6$ relative or internal coordinates. If this potential function is known, the configuration integral Z can be evaluated in a detailed calculation.

We shall illustrate this procedure for one of the simplest nontrivial cases. Take an unsymmetrical bent molecule BAC with the internal coordinates illustrated in Fig. 11-9.* Let r_{1e}, r_{2e}, θ_e represent the

* The example is adapted from D. Herschbach, H. S. Johnston, and D. Rapp, *J. Chem. Phys.*, **31**: 1651 (1959).

equilibrium values for the bond distances and angles. We take a valence-force potential function

$$U(\Delta r_1, \Delta r_2, \Delta\theta) = f_1(\Delta r_1)^2 + f_2(\Delta r_2)^2 + f_\theta(\Delta\theta)^2 \quad (11\text{-}134)$$

where $\Delta r_1 = r_1 - r_{1e}$, $\Delta r_2 = r_2 - r_{2e}$, and $\Delta\theta = \theta - \theta_e$ are the changes in the bond distances and angles.

We start with cartesian coordinates for all three atoms, x_A, x_B, x_C, y_A, etc. In the integral for Z[(11-133)] we transform to relative coordinates of atom A relative to B and of atom C relative to A. The new coordinates are

$$\begin{array}{lll} x_B, y_B, z_B & & \\ x_{AB} = x_A - x_B & y_{AB} = y_A - y_B & z_{AB} = z_A - z_B \\ x_{CA} = x_C - x_A & y_{CA} = y_C - y_A & z_{CA} = z_C - z_A \end{array} \quad (11\text{-}135)$$

The Jacobian of this transformation is unity; furthermore, the potential function depends only on the relative coordinates x_{AB}, x_{CA}, etc., but it does not depend on x_B, y_B, z_B. The integral over x_B, y_B, z_B then gives a factor of V, the volume of the container.

We shall give the rest of the argument in a plausible way which it is easy to make rigorous. Instead of x_{AB}, y_{AB}, z_{AB}, we use a set of polar coordinates to specify the magnitude and direction of the vector $\mathbf{r}_{AB}$; these are r_1, θ_{AB}, ϕ_{AB}. For the relative coordinates of atom C, instead of $x_C - x_A$, etc., we use polar coordinates for the orientation of $\mathbf{r}_{AC}$ relative to $\mathbf{r}_{AB}$. If we take the direction of $\mathbf{r}_{BA}$ as the z axis for specifying $\mathbf{r}_{AC}$, then the vector $\mathbf{r}_{AC}$ is specified by r_2, θ_{BAC}, and ϕ_{BAC}. Furthermore, the bond angle θ is just θ_{BAC}. The volume element is clearly

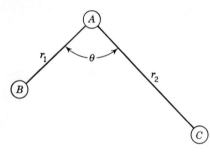

Fig. 11-9. Internal coordinates for the unsymmetrical bent molecule BAC.

$$r_1^2 \, dr_1 \, \sin\theta_{AB} \, d\theta_{AB} \, d\phi_{AB} \, r_2^2 \, dr_2 \, \sin\theta_{BAC} \, d\theta_{BAC} \, d\phi_{BAC} \quad (11\text{-}136)$$

The potential function depends only on r_1, r_2, and θ_{BAC}. For a fixed r_1, r_2, θ_{ABC}, integration over ϕ_{BAC} gives a factor of 2π (this corresponds to rotating the molecule around the AB axis); integration over θ_{BA} and ϕ_{BA} gives a factor of 4π (corresponding to all directions of AB). The potential function depends on the remaining variables. The remaining part of the volume element is

$$r_1^2 \, dr_1 \, \sin\theta \, d\theta \, r_2^2 \, dr_2 \quad (11\text{-}137)$$

where we write θ instead of θ_{BAC}. Now write $r_1 = r_{1e} + \Delta r_1$, $r_2 = r_{2e} + \Delta r_2$, $\theta = \theta_e + \Delta\theta$, and the configuration integral becomes

$$Z = 8\pi^2 V \int_{\Delta\theta = -\theta_e}^{\pi - \theta_e} \int_{\Delta r_1 = -r_{1e}}^{\infty} \int_{\Delta r_2 = -r_{2e}}^{\infty} e^{-(\beta/2)[f_1(\Delta r_1)^2 + f_2(\Delta r_2)^2 + f_\theta(\Delta\theta)^2]}$$
$$(r_{1e} + \Delta r_1)^2 \, d(\Delta r_1)(r_{2e} + \Delta r_2)^2 \, d(\Delta r_2) \, \sin(\theta_e + \Delta\theta) \, d(\Delta\theta) \quad (11\text{-}138)$$

Sec. 11-16] POLYATOMIC MOLECULES

Assume now that the potential $f_1(\Delta r_1)^2$ is sufficiently large so that $e^{-\beta f_1(\Delta r_1)^2/2}$ is negligibly small even for displacements $\Delta r_1/r_{1e} \ll 1$, and similarly for the other terms; then the integrals may be approximated as follows:

$$\int_{\Delta r_1 = -r_{1e}}^{\infty} e^{-\beta f_1(\Delta r_1)^2/2} (r_{1e} + \Delta r_1)^2 \, d(\Delta r_1)$$

$$\approx r_{1e}^2 \int_{\Delta r_1 = -\infty}^{+\infty} e^{-\beta f_1(\Delta r_1)^2/2} \, d(\Delta r_1) \quad (11\text{-}139)$$

$$= r_{12}^2 \left(\frac{2\pi kT}{f_1}\right)^{1/2}$$

(and similarly for the r_2 integral)

$$\int_{\Delta\theta = -\theta_e}^{\Delta\theta = \pi - \theta_e} e^{-\beta f_\theta (\Delta\theta)^2/2} \sin(\theta_e + \Delta\theta) \, d(\Delta\theta)$$

$$\approx \sin\theta_e \int_{-\infty}^{+\infty} e^{-\beta f_\theta (\Delta\theta^2)/2} \, d(\Delta\theta) = \sin\theta_e \left(\frac{2\pi kT}{f_\theta}\right)^{1/2} \quad (11\text{-}140)$$

The configuration integral therefore is

$$Z = 8\pi^2 V \left(\frac{2\pi kT}{f_1}\right)^{1/2} \left(\frac{2\pi kT}{f_2}\right)^{1/2} \left(\frac{2\pi kT}{f_\theta}\right)^{1/2} r_{1e}^2 r_{2e}^2 \sin\theta_e \quad (11\text{-}141)$$

The classical approximation for the partition function therefore is

$$q_{cl} = \left(\frac{2\pi m_A kT}{h^2}\right)^{3/2} \left(\frac{2\pi m_B kT}{h^2}\right)^{3/2} \left(\frac{2\pi m_C kT}{h^2}\right)^{3/2} Z \quad (11\text{-}142)$$

We see that this is of the form deduced in (11-129). It is clear that the over-all translational and rotational partition functions are, in a sense, contained in the factors involving m_A, m_B, m_C, r_{1e}, r_{2e}, and θ_e in (11-142) and (11-141).

Problem 11-7. Show that for the symmetrical molecule AB_2 of Sec. 11-7, the expression for the classical partition function [(11-142)] is equivalent to the expression (11-128), when the product of the frequencies as deduced in Sec. 11-7 is introduced into (11-128). [Note that the constant f_θ of (11-141) is identical with f_ϕ of (11-57), which is related by (11-58c) to the constant b used in the final expression for the frequencies (11-76) and (11-78).]

The procedure which was illustrated here for a simple case is practical for complicated problems, including cases where relatively free internal rotations occur. Several systematic procedures have been described in the literature.*

An outstanding advantage of this general approach is the following: The quantum-mechanical partition function for a harmonic oscillator is

---

* D. R. Herschbach, H. S. Johnston, and D. Rapp, *J. Chem. Phys.*, **31**: 1651 (1959), is particularly recommended; references to earlier papers, including the important contributions by K. S. Pitzer, will be found there.

$q(u) = (1 - e^{-u})^{-1}$. The classical limit is $q_{cl}(u) = 1/u$. The ratio of the quantum-mechanical and classical partition function is

$$R(u) = \frac{q(u)}{q_{cl}(u)} = \frac{u}{1 - e^{-u}} \qquad (11\text{-}143)$$

By taking the ratio of the expressions for the complete quantum-mechanical partition function (11-127) and the classical partition function (11-128) we therefore obtain for the complete quantum-mechanical partition function of the molecule

$$q = q_{cl} R(u_1) R(u_2) \cdots R(u_{3n-6}) \qquad (11\text{-}144)$$

If the potential function is known, q_{cl} can be calculated without doing a complete normal coordinate analysis. The correction factors $R(u_j)$ are almost unity for the low-frequency vibrations, which, as already indicated, it is difficult to estimate semiempirically or observe spectroscopically. The correction factors $R(u_j)$ are more important for the high-frequency vibrations. These high-frequency vibrations can often be recognized in an infrared or Raman spectrum even without a detailed vibrational analysis, or they can be "guesstimated" by empirical comparison with other compounds.

11-17. Computations. As already indicated, formulae suitable for the practical computation of thermodynamic functions for rotation of a polyatomic molecule and references to tables of thermodynamic functions for harmonic oscillators are given at the end of Chap. 8.

PROBLEMS

11-8. The structural parameters of CCl_3Br are given in Prob. 11-1. The vibration frequencies are given as $\nu_1 = 716.3$ cm^{-1}, $\nu_2 = 422.3$ cm^{-1}, $\nu_3 = 247.3$ cm^{-1}, $\nu_4(e) = 775.3$ cm^{-1}, $\nu_5(e) = 295.0$ cm^{-1}, $\nu_6(e) = 193.3$ cm^{-1} (cf. Fig. 11-4). Calculate the contributions of translation, rotation, and vibration to $(H - H_0^\circ)/T$, $-(F - F_0^\circ)/T$, S°, and C_P at 300°K. (*Note:* Use tables for harmonic-oscillator functions to calculate the vibrational contributions; it is not efficient to do these calculations from the formulae directly.)

(From data in the literature the teacher can add problems of this nature at will.)

11-9. Calculate the equilibrium constant of the reaction

$$CDCl_3 + HI \rightleftharpoons CHCl_3 + DI$$

from the vibration frequencies of the molecules (data in Herzberg, p. 316 [18] and [17]).

11-10. The barrier to internal rotation of C_2H_6 is believed to be 3040 cal mole^{-1} [D. Lide, *J. Chem. Phys.*, **29**: 1426 (1958)]. Use this information to calculate the contribution of the torsional motion to the entropy and heat capacity at 300°K [cf. K. S. Pitzer, *Discussions Faraday Soc.*, **10**: 66 (1951)]. Take the C—H bond distances as 1.09 Å and tetrahedral bond angles.

The observed frequency of the $1 \rightarrow 0$ transition for the torsional frequency is 290 cm^{-1}. Calculate the barrier U_0 from this result, assuming the simple harmonic-oscillator approximation, and compare with the value given above.

12

Black-body Radiation

12-1. Introduction. It is a common observation that a hot surface emits electromagnetic radiation; that the energy radiated increases with increasing temperature; and that the spectral distribution shifts toward shorter wavelengths, i.e., the surface color changes from "red" toward "blue" as the surface becomes hotter.

We are going to show in the next few paragraphs that, in an evacuated space (a *hohlraum*) enclosed by walls at a temperature T, there is an equilibrium density of radiation, $\rho(\nu,T)$, which means that the amount of energy per unit volume in the frequency range ν, $\nu + d\nu$ is $\rho(\nu,T)\,d\nu$. We shall write $\rho(\nu,T)$ for the equilibrium density of radiation and $\rho(\nu)$ for an arbitrary distribution. The equilibrium density of radiation is independent of the nature or shape of the walls of the container. Furthermore, there is an upper limit to the amount of light emitted per unit area by a surface at temperature T in the frequency interval ν, $\nu + d\nu$; this upper limit $e(\nu,T)\,d\nu$ is the emission by a black body.

The reflection coefficients of a surface are functions of the frequency of the incident light and of the temperature of the surface. We shall show that the amount of light emitted per unit area in the frequency interval ν, $\nu + d\nu$ by an opaque surface with a reflection coefficient $r(\nu,T)$ is $e(\nu,T)[1 - r(\nu,T)]\,d\nu$, where $e(\nu,T)$ is the emissivity of a black body.

We sketch the proof, which is based on the second law of thermodynamics. We consider opaque surfaces which either absorb or reflect the incident light but which do not transmit any (a semitransparent medium is considered in Sec. 12-12). We consider that, for an arbitrary surface A, the processes of light emission and absorption are entirely uncorrelated. That is, the surface has an emissivity $e_A(\nu,T)$; the amount of light of any given frequency emitted by the surface is a function of the temperature but is independent of the amount of light incident on the surface. In addition, the surface has a reflection coefficient $r_A(\nu,T)$.

Now consider the flat rectangular parallelepiped depicted in Fig. 12-1. The two large surfaces are A and B, and they are in thermal equilibrium at temperature T. The flat shape is chosen so that end effects may be neglected, and practically all the radiation that leaves B falls on A, and vice versa. For simplicity, take B as a black body; it absorbs all the light that is incident on it, and its emissivity is $e(\nu,T)$.

Then the total energy emitted per unit area of B and incident on unit area of A is $\int e(\nu,T)\,d\nu$. Of this, an amount $\int r_A(\nu,T)e(\nu,T)\,d\nu$ is reflected by A. Therefore, the amount $\int [1 - r_A(\nu,T)]e(\nu,T)\,d\nu$ is absorbed by A. But the amount absorbed by A must equal the amount radiated by A; otherwise energy would be transferred from one surface to another at the same temperature in violation of the second law. Therefore

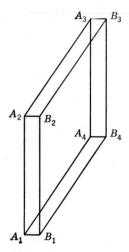

$$\int_0^\infty e_A(\nu,T)\,d\nu = \int_0^\infty [1 - r_A(\nu,T)]e(\nu,T)\,d\nu \quad (12\text{-}1a)$$

[Also note that the total light leaving A and returning to B is the amount reflected, $\int r_A(\nu,T)e(\nu,T)\,d\nu$, plus the amount emitted, $\int e_A(\nu,T)\,d\nu$; from (12-1a), this total is $\int e(\nu,T)\,d\nu$. The amount returning to B and therefore absorbed by B is equal to the amount emitted by B. Thus our calculation is self-consistent.]

Fig. 12-1. Radiative equilibrium between the faces of a narrow parallelepiped. There are two faces of large area, A and B, defined by the vertices $A_1A_2A_3A_4$ and $B_1B_2B_3B_4$, respectively.

The equality between the amount absorbed and the amount emitted by either surface must actually hold for each wavelength or frequency interval. Imagine a filter inserted in the hohlraum between A and B; the filter is a perfect reflector at most wavelengths and is transparent in one frequency interval ν, $\nu + d\nu$ (an interference filter is a fair approximation to such a device). Then the equality of energy transfer between A and B requires that

$$e_A(\nu,T) = e(\nu,T)[1 - r_A(\nu,T)] \quad (12\text{-}1b)$$

where, we repeat, $e(\nu,T)$ is the emissivity of a black body. The general relation stated above between reflectance and emission is known as Kirchhoff's law.

Imagine an evacuated space (a hohlraum) completely surrounded by a black-body surface at temperature T. Since there are emission and absorption of light by the surface, there is a density of electromagnetic radiation in the hohlraum. We have seen that, in the kinetic theory of gases, there is a definite relation between the number of molecules in the gas phase and the number hitting a surface. In the same way, if there is a definite amount of radiation being absorbed and being emitted by a black-body surface at equilibrium, there must be a definite equilibrium density of radiation, $\rho(\nu,T)$, in the hohlraum. The quantitative relation between $\rho(\nu,T)$ and $e(\nu,T)$ will be derived in Sec. 12-5.

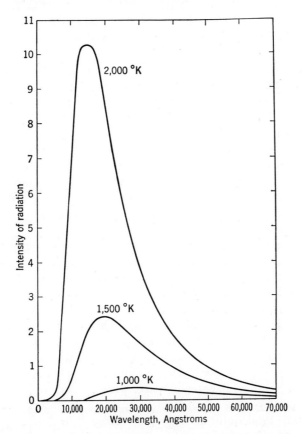

FIG. 12-2. Emission of radiation from a black body at different temperatures. The area under the curve, divided by 10,000, between specified wavelengths gives the energy in calories per second radiated from 1 cm² of a perfect radiator. Note that this is a plot of the radiation leaving a surface rather than of the density of radiation; the relation between these two quantities is given in Sec. 12-5. Furthermore, we have plotted the intensity per unit wavelength interval, not per unit frequency interval (cf. Sec. 12-6). (*Reprinted by permission from F. Daniels and R. A. Alberty, "Physical Chemistry," p. 541, Wiley, New York, 1955.*)

We have furthermore shown that the total energy leaving an opaque surface in any wavelength interval at equilibrium, including reflected light plus emitted light, is a constant, even if the surface is not a black body. It follows that in any space enclosed by an opaque surface at temperature T the equilibrium density of radiation, $\rho(\nu,T)$, is present.

The experimental results for the function $e(\nu,T)$ are plotted in Fig. 12-2, showing the increase in total radiation and the shifts to shorter wavelengths with increasing temperature.

12-2. The Number of Allowed Frequencies. In order to apply statistical mechanics to electromagnetic radiation, we need an expression for the number of normal vibrations for light waves of frequency ν, $\nu + d\nu$ in a box. The argument is really just like that for the number of translational wave functions of a perfect gas.

Consider a cubical box with sides of length L. A running wave of frequency ν and wavelength λ moving in the $+x$ direction is given by $E_0 \sin 2\pi[(x/\lambda) - \nu t]$, where E_0 is the amplitude and, *in vacuo*, $\nu\lambda = c$, the velocity of light. Set $k_x = 1/\lambda$, the wave number, and the wave is given by $E_0 \sin 2\pi(k_x x - \nu t)$.

A standing wave can be obtained by reflecting the wave at the boundary, $x = L$, and by adding the incident and reflected waves. The standing wave is given by $E_0 \sin 2\pi k_x x \sin 2\pi \nu t$. If we require as a boundary condition that the amplitude be zero at the reflecting walls, at $x = 0$ and $x = L$, then $k_x = s_x/2L$, where s_x is a positive integer. The allowed frequencies are $\nu = ck_x = cs_x/2L$.

A running wave in three dimensions is characterized by the wave-number vector (k_x, k_y, k_z) and can be written as $E_0 \sin 2\pi(k_x x + k_y y + k_z z - \nu t)$. The wave-number components add like the components of a vector, and the wavelength of this wave is $1/\lambda = (k_x^2 + k_y^2 + k_z^2)^{1/2}$.

It is by no means obvious, but it is true,* that if such a wave is allowed to reflect at the walls the standing-wave pattern is given by $E_0 \sin 2\pi k_x x \sin 2\pi k_y y \sin 2\pi k_z z \sin 2\pi \nu t$, and $\nu = c(k_x^2 + k_y^2 + k_z^2)^{1/2}$. The requirement that the wave vanish at the boundaries gives $k_x = s_x/2L$, $k_y = s_y/2L$, $k_z = s_z/2L$. When s_x, s_y, and s_z are all large numbers, we pick a radius vector s in integer space, with $s = (s_x^2 + s_y^2 + s_z^2)^{1/2}$. The number of positive sets of integers (s_x, s_y, s_z) in the interval s, $s + ds$ is approximately (and quite accurately for large s) $4\pi s^2\, ds/8$. Therefore, the number of allowed frequencies or normal vibrations between ν and $\nu + d\nu$ is obtained by setting $s^2 = 4L^2(k_x^2 + k_y^2 + k_z^2) = 4L^2\nu^2/c^2$, then $ds = (2L/c)\, d\nu$, so that $4\pi s^2\, ds/8 = 4\pi L^3 \nu^2\, d\nu/c^3$. Light waves are transverse vibrations; for any one direction of propagation, there are two possible planes of polarization; this doubles the number of possible waves. Let $L^3 = V$, the volume of the hohlraum, and we finally obtain for $g(\nu)\, d\nu$, the degeneracy of waves, or the number of allowed waves in the frequency interval ν, $\nu + d\nu$,

$$g(\nu)\, d\nu = \frac{8\pi V \nu^2\, d\nu}{c^3} \tag{12-2}$$

---

* George Joos, "Theoretical Physics," p. 572, translated from the 1st German ed. by I. M. Freeman, Stechert, New York, 1934.

This is an important and a general formula. It can be derived on the basis of a Fourier analysis of the radiation field in a box, rather than in the rather special way used here.

12-3. The Rayleigh-Jeans Formula. A theoretical derivation of the equilibrium density of radiation, $\rho(\nu,T)$, on the basis of classical statistical mechanics and classical electromagnetic theory was given by Rayleigh and by Jeans.* The contradiction between this result and experiment played a central role in the development of quantum theory.

A classical electromagnetic wave consists of an oscillating electric field $E_0 \sin 2\pi[(x/\lambda) - \nu t]$ and an oscillating magnetic field $H_0 \sin 2\pi[(x/\lambda) - \nu t]$, with E and H perpendicular to each other and perpendicular to the direction of propagation. The energy is proportional to $E_0^2 + H_0^2$. We regard each allowed vibration frequency of the hohlraum (or of the ether, as Rayleigh and Jeans would have said) as analogous to a harmonic oscillator with an energy equal to the sum of two equal terms. Then, according to classical statistical mechanics, the probability of observing amplitudes E_0 and H_0 for a given frequency of vibration is proportional to $e^{-(E_0^2+H_0^2)/kT}$. The general equipartition theorem (Sec. 10-3) then states that the average energy is $kT/2$ for the E_0^2 term and $kT/2$ for the H_0^2 term, or kT per normal vibration. In unit volume ($V = 1$), there are $g(\nu)\, d\nu = 8\pi\nu^2\, d\nu/c^3$ allowed vibrations, so the energy density is

$$\rho(\nu,T)\, d\nu = \frac{8\pi\nu^2 kT}{c^3}\, d\nu \qquad (12\text{-}3)$$

This is the Rayleigh-Jeans Law. It agrees with experiment for the low-frequency part of the spectrum (Fig. 12-2), but it contradicts both experiment and common sense in predicting an infinite density of radiation at infinitely high frequency (the "ultraviolet catastrophe"). Lord Rayleigh and James Jeans were well aware of these contradictions; they pointed out, however, that Eq. (12-3) was the necessary result of a consistent application of classical statistical mechanics.

With hindsight, one knows that the Rayleigh-Jeans law fails for frequencies such that $h\nu \geq kT$, that is, when quantum effects are important.

12-4. The Quantum Theory of Black-body Radiation. In quantum theory, radiant energy occurs in energy packets or photons or light quanta of energy $h\nu$. Photons have zero-rest mass and a spin quantum number of 1; like all particles with spin 1, they obey Bose-Einstein statistics. The fact that the angular momentum of a light quantum is 1 is not important for us; but the fact that it is a boson is. The interaction between photons *in vacuo* is negligibly small or perhaps even zero so that the perfect gas model is excellent. In a hohlraum containing a photon gas, the photons are absorbed by the walls, and new photons are emitted.

* Lord Rayleigh, *Phil. Mag.*, (5)**49**: 539 (1900), *Nature*, **72**, 54, 243 (1905); J. H. Jeans, *Phil. Mag.*, (6)**10**: 91 (1906).

If the initial photon gas was not at equilibrium with the temperature of the walls, the number and wavelength of the emitted photons will be different from the number and wavelength of the absorbed photons. This is a mechanism, and in this case the only mechanism, for the establishment of the thermal equilibrium distribution for the radiation field. We seek this equilibrium distribution.

For convenience, take unit volume, take a frequency ν_j and a small interval $d\nu$, and call the interval ν_j to $\nu_j + d\nu$ the jth region; there are $g_j = 8\pi \nu_j^2 c^{-3} d\nu$ states in this interval [by (12-2)], and the energy of any photons in these states is $h\nu_j$. Let there be N_j such photons.

The photons are bosons, and there is no restriction as to the number of light quanta in the g_j states of frequency close to ν_j. The number of ways of having the distribution $N_1, N_2, \ldots, N_j, \ldots$ with N_j photons in the g_j states of frequency ν_j was calculated in Sec. 6-4 for bosons.

$$\ln t = \sum \left(g_j \ln \frac{g_j + N_j}{g_j} + N_j \ln \frac{g_j + N_j}{N_j} \right) \tag{12-4}$$

$$d(\ln t) = \sum \frac{\partial (\ln t)}{\partial N_j} dN_j = \sum \ln \frac{g_j + N_j}{N_j} dN_j \tag{12-5}$$

We impose the constraint that the total energy in the radiation field is a constant,

$$\Sigma N_j h \nu_j = E \tag{12-6}$$

To justify this view, we can imagine that the system consists of a hohlraum surrounded by very thin walls with a negligible heat capacity. Thus, the energy content of the walls is negligible compared with the radiant energy in the hohlraum, but absorption and emission at the walls still provide the mechanism for the establishment of the equilibrium distribution.

There is no restriction on the number of particles, however, since photons can be created or destroyed at the walls.

We seek the distribution of quanta among the various frequencies, which maximizes $\ln t$, subject to the constraint that the energy is fixed [(12-6)], and identify this with the equilibrium distribution.

Use the Lagrange multiplier $-\beta$ for the constraint [(12-6)]. The condition for the most probable distribution is then

$$\frac{\partial (\ln t)}{\partial N_j} - \beta h \nu_j = 0 \quad \text{or} \quad \ln \frac{g_j + N_j}{N_j} - \beta h \nu_j = 0 \tag{12-7a}$$

$$\frac{N_j}{g_j} = \frac{1}{e^{\beta h \nu_j} - 1} \tag{12-7b}$$

If there are N_j quanta, they have energy $N_j h \nu_j$, which is, by definition, $\rho(\nu_j) \, d\nu$. Furthermore, set $g_j = 8\pi \nu_j^2 c^{-3} d\nu$, and regarding ν as a continuous variable, omit the subscript j.

$$\rho(\nu, T) \, d\nu = \frac{8\pi h \nu^3}{c^3} \frac{1}{e^{\beta h \nu} - 1} \, d\nu \tag{12-8}$$

By coupling the system to a system of ordinary particles, we find $\beta = 1/kT$ (Sec. 6-9).

Equation (12-8) is the black-body distribution law; it fits the experimental data at all frequencies. For frequencies such that $h\nu/kT \ll 1$, $e^{h\nu/kT} - 1 \approx h\nu/kT$, and

$$\rho(\nu,T)\,d\nu \approx \frac{8\pi\nu^2}{c^3} kT\,d\nu \tag{12-9}$$

which is the Rayleigh-Jeans law. When $h\nu \gg kT$,

$$\rho(\nu,T)\,d\nu \approx \frac{8\pi h\nu^3}{c^3} e^{-h\nu/kT}\,d\nu \tag{12-10}$$

An equation similar to (12-10) was proposed by Wien as an empirical law for high-frequency radiation prior to the theoretical development.

The first derivation of Eq. (12-8) by Planck constituted the birth of the quantum theory. Planck's argument* was not based on Bose-Einstein statistics. He supposed that the radiation was in equilibrium with a set of material oscillators in the walls and that these oscillators have quantized energies $vh\nu_j$ with v, the vibrational quantum number, an integer. With Boltzmann statistics, the average energy of such an oscillator is $h\nu/(e^{h\nu/kT} - 1)$ [Eq. (8-26)], and if one supposes that the average energy of each vibrational mode of the hohlraum is equal to this, formula (12-8) results. The argument is not rigorous, but this is characteristic of the beginning stages of revolutionary theoretical developments.†

The distribution law (12-8) can be rewritten in terms of the dimensionless variable $h\nu/kT$ as

$$\rho(\nu,T)\,d\nu = \frac{8\pi k^4 T^4}{h^3 c^3}\left(\frac{h\nu}{kT}\right)^3 \frac{1}{e^{h\nu/kT} - 1}\,d\left(\frac{h\nu}{kT}\right) \tag{12-11}$$

which shows that the amount of energy is a function ν/T and, for fixed ν/T, proportional to T^4.

The total energy density of radiation of all frequencies is

$$U(T) = \int_0^\infty \rho(\nu,T)\,d\nu = \frac{8\pi^5}{15} \frac{(kT)^4}{(hc)^3} \tag{12-12}‡$$

12-5. Radiation from a Surface. Consider a small element of area dS in the wall of a hohlraum at temperature T. The calculation of the

---

* M. Planck, *Ann. Physik,* (4)**4**: 553 (1901).

† Lord Rayleigh's remark, *Nature,* **72,** 55 (1905), that he has not "succeeded in following Planck's reasoning" provides an interesting insight into the intellectual climate of the time.

‡ This is obtained by integration of (12-11), since

$$\int_0^\infty \frac{x^3}{e^x - 1}\,dx = \frac{\pi^4}{15}$$

(This integral is evaluated by Mayer and Mayer, p. 372 [5].)

amount of radiation incident on dS is entirely analogous to the calculation, in Sec. 10-5, of the number of molecules striking a surface. For that case, with N molecules cc^{-1}, the number striking per unit time is $N(\bar{v}/4)\,dS$, where $\bar{v}$ is the average velocity. In the present instance, all the photons have the same velocity c. The number of quanta of frequency ν per unit volume is $\rho(\nu,T)\,d\nu/h\nu$; the number of quanta of frequency ν, $\nu + d\nu$ incident upon dS is $[\rho(\nu,T)\,d\nu/h\nu]\,(c/4)\,dS$, and the energy falling on dS is obtained by multiplying by $h\nu$. The incident energy flux (energy per unit time per unit area) is then given by

$$e(\nu,T)\,d\nu = \frac{c}{4}\,\rho(\nu,T)\,d\nu = \frac{2\pi h\nu^3}{c^2}\frac{1}{e^{h\nu/kT}-1}\,d\nu \qquad (12\text{-}13)$$

If the surface is a black body, the energy emitted is equal to the energy incident on the surface from a hohlraum; formula (12-13) therefore gives the emissivity, or rate of energy emission, from an ideal black body. As previously noted, it is assumed that a black body at temperature T will manifest its emissivity whether or not radiation is incident upon it.

An actual surface has a reflection coefficient greater than zero and, therefore, an emissivity of $e(\nu,t)[1 - r(\nu,t)]$, where $r(\nu,T)$ is the reflection coefficient. Because any actual surface has an emissivity less than that of a black body, the correct experimental way to study black-body radiation is to make a small hole dS in a hohlraum lined with reasonably "black" walls; the radiation escaping from the hole is rather accurately given by Eq. (12-13).

The total energy flux radiated by a black body is

$$e(T) = \int e(\nu,T)\,d\nu = \frac{c}{4}\,U(T)$$

where $U(T)$ is the energy density given by (12-12), or

$$e(T) = \frac{2\pi^5}{15}\frac{(kT)^4}{h^3 c^2} = \sigma T^4 \qquad (12\text{-}14)$$

where σ is the Stefan-Boltzmann constant, $\sigma = 5.669 \times 10^{-5}$ erg cm^{-2} deg^{-4} sec^{-1} (or in the units used by radio-tube engineers concerned with the radiation from hot cathodes, $\sigma = 5.7$ watts cm^{-2} kilodeg^{-4}).

Since σ is easy to look up or remember, a useful way to write (12-13) is

$$\frac{e(\nu,T)\,d\nu}{\sigma T^4} = \frac{15}{\pi^4}\left(\frac{h\nu}{kT}\right)^3 \frac{1}{e^{h\nu/kT}-1}\,d\left(\frac{h\nu}{kT}\right) \qquad (12\text{-}15)$$

In exact analogy with the molecular problem, the radiation emitted by an element of surface dS into a solid angle $d\omega$ in a direction which makes an angle θ with respect to the normal to the surface is

$$e(\nu,T)\,dS\,d\nu\,\frac{\cos\theta\,d\omega}{\pi} \qquad (12\text{-}16)$$

Problem 12-1. Show that Eq. (12-16) is properly normalized and gives the correct total amount radiated in the outward direction by a surface.

12-6. Wavelength and Frequency of Maximum Emission.

Since $\nu = c/\lambda$, $d\nu = -(c/\lambda^2)\,d\lambda$; thus unit wavelength interval $d\lambda$ corresponds to a larger frequency interval $d\nu$ at small λ than at large λ. The black-body formula can be rewritten as

$$\rho(\lambda,T)\,d\lambda = \frac{8\pi hc}{\lambda^5}\frac{1}{e^{hc/\lambda kT}-1}\,d\lambda \qquad (12\text{-}17)$$

Note that $\rho(\nu,T)$ and $\rho(\lambda,T)$ have been defined so as to have different meanings: the former is the amount of energy per unit frequency interval; the latter is the amount of energy per unit wavelength interval.

The frequency of maximum emission is defined as that frequency at which $\rho(\nu,T)$ is a maximum; the wavelength of maximum emission is conventionally defined as that wavelength at which $\rho(\lambda,T)$ is a maximum. By maximizing $\ln \rho(\nu,T)$ and $\ln \rho(\lambda,T)$, respectively, one finds

$$\frac{h\nu_{\max}}{kT} = 3(1 - e^{-h\nu_{\max}/kT})$$

and

$$\frac{hc}{\lambda_{\max}T} = 5(1 - e^{-hc/\lambda_{\max}kT})$$

or

$$\frac{h\nu_{\max}}{kT} = 2.82$$

and

$$\frac{hc}{\lambda_{\max}kT} = 4.965$$

or

$$\lambda_{\max}T = \frac{hc}{k}\frac{1}{4.965} = \frac{1.438}{4.965} \text{ cm deg} \qquad (12\text{-}18)$$

12-7. Thermodynamics of the Radiation Field.

In a volume V, the total equilibrium amount of energy in a radiation field is

$$E = U(T)V = aT^4V \qquad (12\text{-}19)$$

where a is a constant ($4\sigma/c$ actually). From the relation $(\partial S/\partial E)_V = 1/T$, and knowing that the entropy is zero at $T=0$,

$$dS \text{ (at constant } V) = 4aT^2V\,dT$$

$$S = \int_0^T 4aT^2V\,dT = \tfrac{4}{3}aT^3V = \frac{4}{3}\frac{E}{T} \qquad (12\text{-}20)$$

Then
$$A = E - TS = -\tfrac{1}{3}E = -\tfrac{1}{3}U(T)V \qquad (12\text{-}21)$$

[where $U(T)$ is the total energy density] and

$$P = -\left(\frac{\partial A}{\partial V}\right)_T = \tfrac{1}{3}U(T) \qquad (12\text{-}22)$$

The free energy is

$$F = A + PV = 0 \qquad (12\text{-}23)$$

This last result may seem less surprising if we recall that $F = N\mu$, that μ is related to the Lagrange multiplier α for the conservation of particles, and that in photon statistical mechanics this constraint is not applied.

The above relations were derived by thermodynamic arguments. The same results could of course also be obtained from the statistical formula for entropy, etc., but this would entail more work.

12-8. Absorption Coefficients. The relations between the emission and the absorption coefficients for light are an interesting illustration of a way in which equilibrium and rate considerations can be combined to obtain fruitful results. We first note the connection between the usual light-absorption coefficients and the so-called Einstein coefficients.

In an ordinary spectroscopic experiment, one has an approximately collimated beam of light of intensity $I(\nu)$; that is, the amount of light

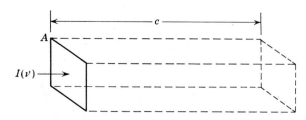

FIG. 12-3. A perfectly collimated beam of light of intensity $I(\nu)$ passes at normal incidence through the surface of unit area A. All the light that passes through the surface in 1 sec is contained in the right prism with A as a base and length c, where c is the velocity of light.

energy passing through the surface A of unit area in 1 sec in the frequency range ν, $\nu + d\nu$ is $I(\nu)\,d\nu$. Since the velocity of light is c, the light that passes through A in 1 sec extends out a distance c; the energy $I(\nu)\,d\nu$ occupies a volume c, and the density of radiation is $\rho(\nu) = I(\nu)/c$ (Fig. 12-3).

We are going to discuss the radiative equilibrium between an atom or molecule in its ground state n and in an excited state m. The excitation refers to internal energy, not translational energy. The energy difference is E_{mn}, and the frequency of light absorbed in the transition from state n to state m is $\nu_{mn} = E_{mn}/h$. The concentration of atoms in the two states is N_n and N_m.

We denote the absorption coefficient per atom, or cross section, for absorption of light of frequency ν by $\alpha_{nm}(\nu)$.* The subscripts on α_{nm} emphasize that light absorption excites an atom from state n to state m. Let $I_x(\nu)$ be the intensity at the plane x and at the frequency ν of a beam

---

* The notation σ is traditional for a cross section; however, it is also traditional for the Stefan-Boltzmann constant and as a distance parameter in describing the potential of interaction between two molecules (Chap. 15); so we use α for a cross section.

of light moving in the x direction. The number of atoms in the ground state in a unit cross section between x, $x + dx$ is $N_n\, dx$. The amount of light $-dI(\nu)\, d\nu$, absorbed per second in dx is proportional to the absorption coefficient per atom, the number of atoms, and the incident energy

$$-dI(\nu)\, d\nu = I_x(\nu)\, d\nu\, N_n \alpha_{nm}(\nu)\, dx \qquad (12\text{-}24)$$

We can omit the frequency interval $d\nu$. When integrated from $x = 0$ where the intensity is I_0, (12-24) becomes the well-known Lambert-Beer relation

$$I_x(\nu) = I_0(\nu) e^{-N_n \alpha_{nm}(\nu) x} \qquad (12\text{-}25)$$

In terms of the decadic molar extinction coefficient or molar absorbancy index, ϵ, the absorption law is

$$I_x(\nu) = I_0(\nu) 10^{-C \epsilon(\nu) x}$$

where C is concentration in moles per liter (but the light path x is still in centimeters). The relation between ϵ and α is clearly

$$\epsilon = \frac{1}{2.3}\left(\frac{N_0}{1{,}000}\right)\alpha \qquad (12\text{-}26)$$

where N_0 is Avogadro's number.

It is essential to realize that the absorption coefficient α_{nm} is a function of frequency. As illustrated in Fig. 12-4, the maximum of α is at ν_{nm}, but no absorption line is infinitely narrow. The line will be broadened by uncertainty broadening and probably by several other causes, including Doppler broadening, pressure broadening, and other perturbations which are not included in our description of the states n and m. In a relation such as (12-24), the frequency interval $d\nu$ must be a small one, such as that encompassed in the shaded area of Fig. 12-4 over which α does not change very much.

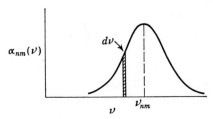

FIG. 12-4. The absorption coefficient $\alpha_{nm}(\nu)$, as a function of ν. The maximum is at ν_{nm}.

Following Einstein, we now express the absorption coefficients in terms of the density of radiation, $\rho(\nu)$, instead of the intensity. In (12-24), if $dI(\nu)\, d\nu$ is the energy absorbed, then $dI(\nu)\, d\nu/h\nu$ is the number of quanta absorbed. On the right-hand side, write $I_x(\nu)\, d\nu = c\rho(\nu)\, d\nu$, and consider absorption per unit volume, rather than in the small volume dx. Thus,

Number of quanta absorbed per unit volume per sec

$$\frac{I(\nu)\, d\nu\, N_n \alpha_{nm}(\nu)}{h\nu} = \frac{c\rho(\nu)\, d\nu\, N_n \alpha_{nm}(\nu)}{h\nu} = B_{nm}(\nu)\rho(\nu)\, d\nu\, N_n \quad (12\text{-}27)$$

where
$$B_{nm}(\nu) = \frac{c\alpha_{nm}(\nu)}{h\nu}$$

The coefficients B defined by (12-27) are called the Einstein coefficients of absorption. It seems reasonable to assume that (12-27) holds for randomly oriented absorbing atoms whether one has a collimated beam of light or uniform illumination in all directions or any other spatial distribution of the radiation. The law (12-27) holds for an arbitrary $\rho(\nu)$; it is not restricted to the equilibrium $\rho(\nu,T)$.

12-9. The Relation between Emission and Absorption Coefficients. The number of atoms absorbing light in the interval $\nu, \nu + d\nu$ and thereby going from the state n to m in unit time is $B_{nm}(\nu)N_n\rho(\nu)\, d\nu$. The total number of atoms undergoing this upward transition is obtained by integrating over the entire absorption curve,

$$-\frac{dN_n}{dt} = N_n \int_\nu B_{nm}(\nu)\rho(\nu)\, d\nu \quad (12\text{-}28)$$

There are two radiative processes whereby atoms emit light and go down from state m to n. There is spontaneous emission; and the probability that an atom in state m spontaneously emits a light quantum of frequency $\nu, \nu + d\nu$ is $A_{mn}(\nu)$; the number of such emissions in unit time is $N_m A_{mn}(\nu)\, d\nu$. The total rate of spontaneously going from m to n is

$$-\frac{dN_m}{dt} = N_m \int_\nu A_{mn}(\nu)\, d\nu \quad (12\text{-}29)$$

The integral $\int A_{mn}(\nu)\, d\nu$ is like a first-order rate constant; it has dimensions of reciprocal seconds, so that $A_{mn}(\nu)$ is dimensionless. The reciprocal of the integral $\int A_{mn}(\nu)\, d\nu$ is called the radiative lifetime of the state m.

In the presence of a radiation field, the probability of light emission, $m \to n$, is increased. This process is called induced emission, and the rate at which it occurs with the emission of light in the range $\nu, \nu + d\nu$ is $N_m B_{mn}(\nu)\rho(\nu)\, d\nu$. The process of induced emission has its classical analogue in that an oscillator of frequency ν can either absorb energy from or add energy to a radiation field, depending on the phase of the vibration with respect to the phase of the oscillating field. The quantities A_{mn} and B_{mn} are the Einstein coefficients of spontaneous and induced emission.

Suppose now the system is at equilibrium at temperature T. The radiation density is then given by the black-body formula. If the atoms are a dilute gas and obey Boltzmann statistics as regards the internal energy levels,

$$\frac{N_m}{N_n} = e^{-h\nu_{mn}/kT} \tag{12-30}$$

Note that we assume that the states m and n are single nondegenerate electronic states.

In general, the transitions $n \to m$ and $m \to n$ will be taking place both by collisions with other atoms and by light absorption and emission. We can assume that the pressure is so low that the collisional rates are negligible and the thermal equilibrium distribution of the atoms among the states m and n is maintained by interaction with the radiation field of temperature T. Thus the rate $n \to m$ by absorption of radiation must equal the rate $m \to n$ by emission,

$$N_n \int B_{nm}(\nu)\rho(\nu,T)\, d\nu = N_m \int [A_{mn}(\nu) + B_{mn}(\nu)\rho(\nu,T)]\, d\nu \tag{12-31}$$

According to the principle of microscopic reversibility or detailed balancing (Sec. 12-13), not only must the total rates be equal, but the rate of excitation by absorption of light at any one narrow frequency interval ν, $\nu + d\nu$, must equal the rate of deexcitation by emission of light in this frequency interval. That is, for any frequency

$$N_n B_{nm}(\nu)\rho(\nu,T) = N_m[A_{mn}(\nu) + B_{mn}(\nu)\rho(\nu,T)] \tag{12-32}$$

We shall explain the principle of microscopic reversibility in more detail later. At this point, we can note that the equality in (12-32) is plausible. Suppose that in a system at equilibrium the total rates of excitation and deexcitation by absorption and emission of light were equal but that the partial rates at different frequencies were different. Then, at one frequency, the rate of absorption of light would be greater than the rate of emission; at another frequency, the rate of emission would be greater. Thus the atoms in the states n and m would tend to upset the thermal equilibrium distribution of radiation. With a little ingenuity, one could arrange such a system so as to violate the second law of thermodynamics.

The principle of microscopic reversibility also guarantees that the equalities (12-31) and (12-32) hold in a system at equilibrium even if the rates of collisional excitation and deexcitation are not negligible.

From (12-32), by solving for $\rho(\nu,T)$

$$\rho(\nu,T) = \frac{N_m A_{mn}(\nu)}{N_n B_{nm}(\nu) - N_m B_{mn}(\nu)} \tag{12-33}$$

As an approximation to (12-30) we now write $N_m/N_n = e^{-h\nu/kT}$. This involves the approximation that the absorption line is narrow, and for all frequencies for which A_{mn}, B_{mn}, and B_{nm} are not zero, $\nu \approx \nu_{mn}$. Then, by substitution in (12-33),

$$\rho(\nu, T) = \frac{A_{mn}(\nu)}{B_{nm}(\nu)e^{h\nu/kT} - B_{mn}} \quad (12\text{-}34)$$

But, for thermal equilibrium,

$$\rho(\nu, T) = \frac{8\pi h \nu^3}{c^3} \frac{1}{e^{h\nu/kT} - 1} \quad (12\text{-}8)$$

If both these equations for $\rho(\nu, T)$ hold,

$$\begin{aligned} A_{mn}(\nu) &= \frac{8\pi h \nu^3}{c^3} B_{nm}(\nu) \\ B_{mn}(\nu) &= B_{nm}(\nu) \end{aligned} \quad (12\text{-}35)$$

These are the desired results. The requirements of radiative equilibrium impose the interrelations (12-35) on the several probabilities of emission and absorption.

It has been assumed that states n and m are single nondegenerate states. Suppose that state n is actually g_n-fold-degenerate, with the individual quantum states indexed by $n_1, n_2, \ldots, n_i, \ldots$ ($i = 1, \ldots, g_n$). Thus, for example, the electronic state $^2P_{3/2}$ has $g = 4$, corresponding to the individual substates $m_j = \pm 3/2, \pm 1/2$. In the absence of external electric or magnetic fields, these states all have the same energy. Similarly, let state m be g_m-fold-degenerate.

At equilibrium, the relation

$$\frac{N_{m_j}}{N_{n_j}} = e^{-h\nu_{mn}/kT} \quad \text{for all } i \text{ and all } j$$

holds, and the rates for any one transition $n_i \leftrightarrow m_j$, must be equal in the forward and backward directions, so that Eqs. (12-35) can be written

$$A_{m_j n_i}(\nu) = \frac{8\pi h \nu^3}{c^3} B_{n_i m_j}(\nu) \quad \text{for all } i = 1, \ldots, g_n; \text{ all } j = 1, \ldots, g_m \quad (12\text{-}35a)$$

$$B_{n_i m_j} = B_{m_i n_j}$$

This is often the most useful way to handle the situation, because the absorption coefficients are different for the different transitions $m_j \leftrightarrow n_i$. By summing the transition probabilities over all the substates, however, it is possible to evaluate the over-all absorption and emission coefficients for the transitions $n \leftrightarrow m$.

The total number of atoms in the state n is $N_n = g_n N_{n_i}$, since

$$N_{n_1} = N_{n_2} = N_{n_i}$$

The rate of absorption of light of frequency ν to excite atoms into the state m_j is

$$\frac{dN_{m_j}}{dt} = \sum_{n_i} B_{n_i m_j} N_{n_i} \rho(\nu) \, d\nu = \frac{1}{g_n} N_n \rho(\nu) \, d\nu \left(\sum_{n_i} B_{n_i m_j} \right)$$

The total rate of excitation by absorption of light of frequency ν to all states m_j is

$$\frac{dN_m}{dt} = \sum_{m_j} \frac{dN_{m_j}}{dt} = \frac{1}{g_n} N_n \rho(\nu)\, d\nu \left(\sum_{n_i,m_j} B_{n_i m_j} \right) \quad (12\text{-}35b)$$

But by definition of the over-all absorption coefficient B_{nm},

$$\frac{dN_m}{dt} = N_n \rho(\nu)\, d\nu\, B_{nm}(\nu) \quad (12\text{-}35c)$$

By comparison of Eqs. (12-35b) and (12-35c)

$$B_{nm}(\nu) = \frac{1}{g_n} \sum_{n_i,m_j} B_{n_i m_j} \quad (12\text{-}35d)$$

By a similar treatment of the over-all downward transitions, we obtain

$$g_n B_{nm}(\nu) = g_m B_{mn}(\nu) = \frac{g_m A_{mn}(\nu)}{(8\pi h \nu^3)/c^3} = \sum_{n_i,m_j} B_{n_i m_j}(\nu) \quad (12\text{-}35e)$$

When using the over-all absorption and emission coefficients, it should be recalled that

$$\frac{N_m}{N_n} = \frac{g_m}{g_n} e^{-h\nu_{mn}/kT}$$

12-10. Spontaneous and Induced Emission. It is interesting to compare the relative contributions of spontaneous and induced emission to the total emission in a radiating gas. Suppose that there are N_m atoms in the excited state and that the density of radiation is $\rho(\nu)$. It is not necessary, in this analysis, to assume that the number of excited atoms is in thermal equilibrium with the number of ground-state atoms. The density of radiation, $\rho(\nu)$, need not be an equilibrium distribution for any temperature. Nevertheless, the total rate of spontaneous decay is $N_m \int A_{mn}(\nu)\, d\nu = N_m/\tau$, where τ is the radiative lifetime. The total rate of induced emission is $N_m \int B_{mn}(\nu) \rho(\nu)\, d\nu$. If the absorption line is reasonably narrow and if $\rho(\nu)$ is constant over this interval, we may write, for the number of induced decays,

$$N_m \rho(\nu_{nm}) \int B_{mn}(\nu)\, d\nu = (c^3/8\pi h \nu^3) N_m \rho(\nu_{nm}) \int A_{mn}(\nu)\, d\nu$$

Then the ratio of induced to spontaneous decays is

$$\frac{\text{Induced decays}}{\text{Spontaneous decays}} = \frac{c^3}{8\pi h \nu^3} \rho(\nu_{mn})$$

Although the function $\rho(\nu)$ is not necessarily a black-body distribution, we have assumed that it does not vary very much over the width of the absorption line. It is convenient to define a "temperature" T_{mn} such that the black-body density $\rho(\nu_{mn}, T)$ in the important frequency interval around ν_{mn} is the same as the actual density of radiation, $\rho(\nu_{mn})$. Thus, T_{mn} is defined by

$$\rho(\nu_{mn}) = \frac{8\pi h \nu_{mn}^3}{c^3} \frac{1}{e^{h\nu_{mn}/kT_{mn}} - 1} \qquad (12\text{-}36)$$

Then the ratio of induced to spontaneous emission is

$$\frac{\text{Induced}}{\text{Spontaneous}} = \frac{c^3}{8\pi h \nu^3} \rho(\nu_{mn}) = \frac{1}{e^{h\nu_{mn}/kT_{mn}} - 1} \qquad (12\text{-}37)$$

If the effective "temperature" of the radiation field in the region ν_{mn} is such that $h\nu_{mn}/kT_{mn} \gg 1$, induced emission is less important than spontaneous emission. If, however, $h\nu_{mn}/kT_{mn} \ll 1$, induced emission is more important than spontaneous; in fact, the ratio of the two is approximately $kT_{mn}/h\nu_{mn}$.

From a practical point of view, for an electronic transition of an atom or a molecule with, say, $\lambda = 5{,}000$ Å, the condition $h\nu/kT = 1$ requires that the effective "temperature" of the radiation field be 30,000°K. The actual radiation density under usual experimental conditions (other than in the interior of a star or in a nuclear explosion) is much less than this. Thus, in ordinary laboratory experiments with visible and ultraviolet light, induced emission is unimportant compared with spontaneous emission. For experiments in nuclear magnetic resonance with $\nu \sim 10^7$ sec^{-1} or in microwave spectroscopy with $\nu \sim 10^{10}$ sec^{-1} the densities of radiation are such that the spontaneous emission is entirely negligible compared with induced emission. For infrared emission, both phenomena can be important.

12-11. Absorption Coefficients, Transition Moments, and f Numbers. As a matter of practical utility, and essentially in a handbook spirit, we record here several relations between the absorption and emission coefficients and molecular parameters.* The wave functions for the atom or molecule in the states n and m are $\psi_n(x_1, y_1, z_1, \ldots, x_i, y_i, z_i, \ldots)$ and $\psi_m(x_1, y_1, z_1, \ldots, x_i, y_i, z_i, \ldots)$, where x_i, y_i, z_i are the coordinates of the ith particle (electron or nucleus) in the atom or molecule. Let e_i be the charge of the ith particle. The x component of the dipole moment for the transition between states n and m is

$$\mu_{xnm} = \int \cdots \int \psi_n^* \left(\sum_i e_i x_i \right) \psi_m \, dx_1 \, dy_1 \cdots dx_i \, dy_i \, dz_i \cdots$$

Similar definitions hold for μ_y and μ_z, and the total squared dipole moment of the transition $|\mu_{nm}|^2 = |\mu_x|^2 + |\mu_y|^2 + |\mu_z|^2$. From time-

---

* Pitzer, pp. 265–266 [10].

dependent perturbation theory, it is shown that for a narrow line

$$\int B_{nm}(\nu)\, d\nu = \frac{8\pi^3}{3h^2} |\mu_{nm}|^2 \tag{12-38}$$

[In Eq. (12-38) we assume degeneracy factors $g_m = g_n = 1$. As previously remarked, it is always possible to do this by discussing the individual members of a degenerate state separately.]

The absorption cross section at any frequency is given by

$$\alpha(\nu) = B(\nu)h\nu/c$$

so for a narrow line

$$\int \alpha_{nm}(\nu)\, d\nu = \frac{8\pi^3}{3hc} \nu_{nm}|\mu_{nm}|^2 \tag{12-39}$$

It is customary also to talk about the f number, or oscillator strength, for an electronic transition; this is given by

$$\int \alpha_{nm}(\nu)\, d\nu = \frac{\pi f e^2}{mc} \tag{12-40}$$

where e and m are the electronic charge and mass. The f number is the ratio of the intensity to that expected on the basis of classical electromagnetic theory for a single, harmonically oscillating electron.

The quantity $\int \alpha(\nu)\, d\nu$ may properly be called the integrated absorption coefficient. In practical units

$$f = 4.33 \times 10^{-9} \int \epsilon\, d\omega \tag{12-41}$$

where ϵ is the decadic molar extinction coefficient (molar absorbancy index) and $d\omega$ is in wave numbers.

If the band is narrow, so that the factor ν^3 can be taken out of the integral,

$$\int A_{mn}(\nu)\, d\nu = \frac{8\pi h \nu_{mn}^3}{c^3} \int B_{nm}(\nu)\, d\nu = \frac{64\pi^4 \nu_{mn}^3}{3hc^3} |\mu_{nm}|^2 = \frac{8\pi^2 \nu_{mn}^2 f e^2}{mc^3} \tag{12-42}$$

Broad, continuous absorption bands, which are frequently encountered for polyatomic molecules, can usually be regarded as the sum of a number of reasonably sharp sub-bands, each sub-band corresponding to a transition between individual quantum states. The relations above may then be applied to each of the constituent sub-bands.

It is sometimes fruitful, however, to discuss the integrated absorption coefficient and the corresponding f number, obtained by integrating Eq. (12-41) over the entire band. For certain classes of strong transitions, the integrated absorption coefficient or the f number is constant as the temperature is changed or as various small perturbations are introduced.

12-12. The Emissivity of a Semitransparent Gas. It is common in laboratory experiments to have a radiating gas in a transparent tube. The radiation may be caused simply by heating the gas, or it may be caused by some nonthermal method of excitation such as an electric discharge or a chemiluminescent reaction. (The radiation from a flame, for example, is sometimes due to thermal excitation of the flame products and sometimes to the direct production of excited species in nonequilibrium amounts by the combustion reactions.) In any case, there are molecules (or atoms) in excited states which are radiating; there are also molecules of the same kind in the ground state which can absorb some of the emitted light. The problem we wish to investigate is the effect of this self-absorption on the light emission by the gas.

In a number of experiments (including the usual shock-tube experiments, flames, and photochemical experiments) the distribution of translational energies in the gas will be a Maxwell-Boltzmann distribution corresponding to some temperature T_r which defines the "translational temperature." This is the temperature which determines the P, V, T behavior of the gas. Each of the internal degrees of freedom (rotation, vibration, electronic excitation, and chemical reaction) for any particular molecular species may or may not be at equilibrium with the translational temperature. We consider the case that there are N_m and N_n molecules cc^{-1} in an excited state m and the ground state n with respect to one particular mode of excitation. The relation $N_m/N_n = e^{-h\nu_{mn}/kT_e}$ defines an "excitation temperature" T_e for this particular mode of excitation. If $T_e = T_{tr}$, this particular mode of excitation is in equilibrium with the translational temperature but we are not restricted to this case.

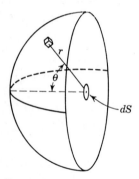

Fig. 12-5. The radiation escaping from a small hole, dS, in the center of the flat face of a hemispherical vessel.

We assume that the walls are perfectly transparent. Hence, they do not radiate back into the gas at all. If the self-absorption is not too great, a significant fraction of the radiation emitted by the gas escapes from the container. Because of the escape of the radiation, there may not be an equilibrium density of radiation, $\rho(\nu,T)$, in the tube.

We consider high-frequency radiation and assume that induced emission is negligible. Then the number of light quanta emitted per unit volume per unit time of frequency ν to $\nu + d\nu$ is $N_m A_{mn}(\nu)\, d\nu$. Let the experiment have the geometry of Fig. 12-5. There is a hemispherical transparent vessel, and we measure the radiation emerging from a small element of area, dS, at the center. Consider the element of volume, $dV = r^2 \sin\theta\, d\theta\, d\phi\, dr$, depicted in the figure. There are $N_m A_{mn}(\nu)\, d\nu\, dV$

quanta emitted from this element of volume in all directions. The fraction of these that would emerge through dS if there were no self-absorption would be $(\cos \theta \, dS)/4\pi r^2$ since $\cos \theta \, dS$ is the projected area of dS normal to the radius vector. However, some of the light is absorbed; by the Lambert-Beer law, the fraction transmitted through the distance r is $e^{-N_n\alpha(\nu)r}$.

Thus, the light leaving dV and actually passing through dS is

$$N_m A_{mn}(\nu) \frac{\cos \theta \sin \theta \, d\theta \, d\phi}{4\pi} e^{-N_n\alpha(\nu)r} \, dr \, d\nu \, dS \qquad (12\text{-}43)$$

Integrate over θ and ϕ with $0 \leq \theta \leq \pi/2$, $0 \leq \phi \leq 2\pi$, and the total emission from the spherical shell, r, $r + dr$, is

$$\tfrac{1}{4} N_m A_{mn}(\nu) e^{-N_n\alpha(\nu)r} \, dr \, d\nu \, dS \qquad (12\text{-}44)$$

The total emission through dS, obtained by integrating from $r = 0$ to $r = r_0$, where r_0 is the radius of the hemisphere, is

$$\frac{1}{4} \frac{N_m A_{mn}(\nu)}{N_n \alpha(\nu)} (1 - e^{-N_n\alpha(\nu)r_0}) \, d\nu \, dS \qquad (12\text{-}45)$$

This is the general result in a form suitable for calculation. The limiting cases are of interest. If self-absorption is small, $N_n\alpha r_0 \ll 1$, $1 - e^{-N_n\alpha r_0} \approx N_n\alpha r_0$, and the emission is

$$\tfrac{1}{4} N_m A_{mn}(\nu) r_0 \, d\nu \, dS \qquad (12\text{-}46)$$

which essentially asserts that all the radiation emitted by the gas in the direction of dS escapes.

If the self-absorption is large, $e^{-N_n\alpha r_0} \approx 0$ and Eq. (12-45) for the number of quanta emitted gives

$$\frac{1}{4} \frac{N_m A_{mn}(\nu)}{N_n \alpha(\nu)} \, d\nu \, dS$$

Now substitute $N_m/N_n = e^{-h\nu_{mn}/kT_e}$, $\alpha(\nu) = h\nu B_{nm}/c$, $A_{mn} = (8\pi h\nu^3/c^3)B_{nm}$, and the number of quanta emitted is

$$\frac{c}{4} \frac{8\pi h\nu^3}{c^3} \frac{1}{h\nu} e^{-h\nu_{mn}/kT_e} \, d\nu \, dS$$

Multiply by $h\nu$ to get the energy radiated,

$$\frac{c}{4} \frac{8\pi h\nu^3}{c^3} e^{-h\nu_{mn}/kT_e} \, d\nu \, dS \qquad (12\text{-}46a)$$

The only frequencies of interest are those for which $\nu \approx \nu_{mn}$; the density of radiation in this neighborhood in equilibrium at the excitation temperature T_e would be

$$\rho(\nu, T_e) = \frac{8\pi h \nu^3}{c^3} \frac{1}{e^{h\nu/kT_e} - 1} \approx \frac{8\pi h \nu^3}{c^3} e^{-h\nu/kT_e} \qquad (12\text{-}47)$$

(since we assume that $h\nu/kT_e \gg 1$). Therefore, by comparison of Eqs. (12-46a) and (12-47) for strong self-absorption, the energy radiated in the frequency range ν, $\nu + d\nu$ is

$$\frac{c}{4} \rho(\nu, T_e) \, d\nu \, dS \qquad (12\text{-}48)$$

This is exactly the radiation from dS if it were a black body. The general formula (12-45) reduces to

$$\text{Emission} = \frac{c}{4} \rho(\nu, T_e) \, d\nu \, dS (1 - e^{-N_n \alpha(\nu) r_0}) \qquad (12\text{-}49)$$

where $\rho(\nu, T_e)$ is the black-body density for the temperature T_e.

In summary, the situation is the following: If the self-absorption is small, the light emission, as given by (12-46), is proportional to N_m, $A_{mn}(\nu)$, and r_0. If $A_{mn}(\nu)$ is known, the light emission can be used to determine the number of atoms in the excited state. However, if we increase r_0 or at a fixed ratio N_m/N_n (that is, a fixed T_e) increase N_m by increasing the total number of atoms, then the amount of radiation increases, but self-absorption becomes important, and the radiation emitted approaches a limit which is the black-body emissivity for the excitation temperature T_e. It should be obvious on general grounds that the radiation in any frequency interval cannot exceed the black-body limit and that self-absorption is the mechanism which prevents the radiation from exceeding this limit. In view of the shape of the absorption-coefficient curve (Fig. 12-4), one can have a situation in which, in the center of a line, $N_n \alpha(\nu) r_0 \gg 1$ and the radiation is black-body-limited, but on the outside wings of the line, where $\alpha(\nu)$ is less, $N_n \alpha(\nu) r_0 \ll 1$.

For vessels with other shapes, the integration to get the total radiation is more complicated and difficult. It will, of course, still be the case that the upper limit for the amount of radiation when self-absorption is strong is the black-body emission.

12-13. Detailed Balancing or Microscopic Reversibility. This is a principle which is very useful for calculating the rate of a process in one direction from its rate in the opposite direction plus equilibrium considerations.

The Newtonian equations of motion for a system of N particles which interact with each other and with an external potential are

$$m_i \frac{d^2 x_i}{dt^2} = -\frac{\partial U(x_1, \ldots, z_N)}{\partial x_i} \qquad \text{for } x, y, z; i = 1, \ldots, N \qquad (12\text{-}50)$$

The transformation $t' = -t$ changes the signs of the velocities

$$[dx_i/dt = -dx_i/d(-t)]$$

but not of accelerations $[d^2x_i/dt^2 = d^2x_i/d(-t)^2]$; thus the equations of motion (12-50) are unaffected. Therefore, for any solution of the equations of motion, there is always another solution in which all the particles have exactly the opposite velocities, so that the system executes a reverse motion. This is the property of time reversibility in classical mechanics.

The simplest significant illustration of this phenomenon for our purposes is the collision between two atoms. Denote the velocities of one atom before and after the collision by **u** and **u'**, and let the velocities of the second atom be **v** and **v'**. The collision is then symbolically represented by

$$(\mathbf{u},\mathbf{v}) \to (\mathbf{u'},\mathbf{v'})$$

The reverse collision is

$$(-\mathbf{u'},-\mathbf{v'}) \to (-\mathbf{u},-\mathbf{v})$$

The elementary treatment of quantum mechanics in Chap. 3, which emphasized the time-independent Schrödinger equation and the properties of stationary states, is not sufficient background for a thorough discussion of the quantum-mechanical theorem of detailed balancing. Familiarity with the theory of scattering and with the general treatment of time-dependent problems in quantum mechanics is necessary for such a discussion. We shall therefore merely state and illustrate the theorem, relying on the preceding discussion of time reversal in classical mechanics to make it plausible.*

Consider a collision between two particles a and b. Let $\Psi_1(\mathbf{r}_a,\mathbf{r}_b,t)$ represent one possible state (state 1) of the pair (we shall call the system of two interacting particles a pair) and $\Psi_2(\mathbf{r}_a,\mathbf{r}_b,t)$ another state of the pair (state 2), with the same expectation value of the energy. Let $\Psi_1^-(\mathbf{r}_a,\mathbf{r}_b,t)$ and $\Psi_2^-(\mathbf{r}_a,\mathbf{r}_b,t)$ represent the reverse states, in which the velocities and the projected angular momenta of the particles have changed sign. Let $w(1 \to 2)$ be the transition probability from state 1 to state 2. Then the theorem of time reversal in quantum mechanics is

$$w(1 \to 2) = w(2^- \to 1^-) \qquad (12\text{-}51)$$

Now, both state 1 and state 2 are single quantum-mechanical states for the system of two particles (the pair) with the same energy.

It should be explained that the wave function for the pair is normalized so that

$$\int \cdots \int \Psi_1^*(\mathbf{r}_a,\mathbf{r}_b,t)\Psi_1(\mathbf{r}_a,\mathbf{r}_b,t)\, dx_a\, dy_a \cdots dy_b\, dz_b = 1$$

* The topic is discussed most ably by Landau and Lifshitz, pp. 432–436 [19].

where Ψ_1^* is the complex conjugate of Ψ_1. That is, there is a single pair in the state Ψ_1 in the container. If we take the volume as 1 cc, then $w(1 \to 2)$ is a transition probability per unit time per unit concentration, not a transition probability per collision. Thus, $w(1 \to 2)$ is analogous to a conventional chemical kinetic rate constant.

We might also remark that, in the absence of a magnetic field, the wave function for the reverse state, $\Psi^-(\mathbf{r}_a,\mathbf{r}_b,t)$ is the same as the complex conjugate $\Psi^*(\mathbf{r}_a,\mathbf{r}_b,t)$. That is, Ψ can be written as

$$\Psi(\mathbf{r}_a,\mathbf{r}_b,t) = \sum_j c_j \psi_j(\mathbf{r}_a,\mathbf{r}_b) e^{-i\epsilon_j t/\hbar}$$

where the constants c_j are real, and the functions $\psi_j(\mathbf{r}_a,\mathbf{r}_b)$ are real and are a complete set of stationary-state wave functions with energies ϵ_j for the system of two particles. The complex conjugate and the reverse state are then both given by

$$\Psi^*(\mathbf{r}_a,\mathbf{r}_b,t) = \Psi^-(\mathbf{r}_a,\mathbf{r}_b,t) = \sum_j c_j \psi_j(\mathbf{r}_a,\mathbf{r}_b) e^{+i\epsilon_j t/\hbar}$$

In the absence of an external magnetic field, state 2^- has the same energy as 2.*

Now consider a macroscopic system at thermal equilibrium. The number of pairs in state 1 is equal to the number of pairs in state 2^-, because the degeneracies of the states are both unity and the energies are the same. Since the transition probabilities are the same [(12-51)], the rate at which pairs are making the transition $1 \to 2$ is equal to the rate at which pairs are making the transition $2^- \to 1^-$. This can be taken as the principle of microscopic reversibility, or detailed balancing.†

For example, in a detailed description of a chemical reaction, we are interested in the probability that a reaction product will emerge in a particular vibrational and rotational state. Consider the particular overall reaction

$$\text{Cl} + \text{H}_2 \rightleftharpoons \text{H} + \text{HCl} \qquad (12\text{-}52)$$

---

* In the presence of an external magnetic field, time reversal does not hold in classical mechanics or in quantum mechanics. This is obvious for the classical case. The magnetic force on a charged particle is $(e/c)\mathbf{v} \times \mathbf{B}$. If the velocity is reversed, the force changes sign and the particle does not retrace its original orbit. If, however, we consider the solenoid producing the magnetic field as part of the system, then time reversal involves changing the direction of flow of the electrons in the solenoid, $\mathbf{B}$ changes sign, and $\mathbf{v} \times \mathbf{B}$ does not change. Time reversal then holds.

† Under some circumstances, with a spherically symmetric potential, it is true, not only that $w(1 \to 2) = w(2^- \to 1^-)$, but also that $w(1 \to 2) = w(2 \to 1)$. These matters are of importance for the study of the Boltzmann H theorem and the rate of approach to equilibrium; cf. Tolman, pp. 102–105, 117–120, 130–132, 521–522 [2], Landau and Lifshitz, p. 435 [19], S. Watanabe, *Revs. Modern Phys.*, **27**: 26 (1955), but they do not concern us here.

A detailed description of a single quantum-mechanical transition would be

$$\text{Cl}(\mathbf{p}) + \text{H}_2(-\mathbf{p}, J, m_J, v) \to \text{H}(\mathbf{p}') + \text{HCl}(-\mathbf{p}', J', m_J', v') \quad (12\text{-}53)$$

We are using a set of coordinates with respect to the center of mass of the system so that the translational momenta of the two particles are equal and opposite. The chlorine atom is in a state with a vector translational momentum of $\mathbf{p}$. The H_2 molecule therefore has translational momentum $-\mathbf{p}$; furthermore, it has rotational quantum numbers J and m_J and vibrational quantum number v. The component of angular momentum along a particular axis is $m_J \hbar$; we can take this axis as the direction of the vector $\mathbf{p}$. The corresponding quantities for H and HCl after the collision are denoted by $\mathbf{p}'$, J', m_J', v'. The reverse reaction is

$$\text{H}(-\mathbf{p}') + \text{HCl}(\mathbf{p}', J', -m_J', v') \to \text{Cl}(-\mathbf{p}) + \text{H}_2(\mathbf{p}, J, -m_J, v) \quad (12\text{-}54)$$

The quantities J and v are intrinsically positive and do not change sign on time reversal.

The principle of microscopic reversibility, or detailed balancing, asserts that, in a system at equilibrium, reactions (12-53) and (12-54) are proceeding at an equal rate.

In its pristine form, the principle of microscopic reversibility refers to transitions involving single, completely defined quantum-mechanical states. For complex processes, one often uses a derived version of the principle which is obtained by summing over a group of states.

For example, in the present instance, we might be interested in the rates of the reactions

$$\text{Cl} + \text{H}_2(J, v) \to \text{H} + \text{HCl}(J', v') \quad (12\text{-}55)$$
$$\text{H} + \text{HCl}(J', v') \to \text{Cl} + \text{H}_2(J, v) \quad (12\text{-}56)$$

We should first remark that the conservation of energy for reactions (12-53) and (12-54) is expressed by the equation

$$\frac{p^2}{2\mu} + \epsilon_{Jv} + \epsilon_0 = \frac{p'^2}{2\mu'} + \epsilon'_{J'v'} \quad (12\text{-}57)$$

where μ is the reduced mass of the Cl, H_2 system, μ' is the reduced mass for the H, HCl system, $\epsilon_{Jv} = \epsilon_J + \epsilon_v$ is the rotational energy plus the vibrational energy of H_2, $\epsilon'_{J'v'}$ is the same for the HCl molecule, and ϵ_0 is the energy liberated in reaction (12-52) with reactants and products in their lowest states. Thus, for a given J, v, J', v', the magnitude of $\mathbf{p}$ determines the magnitude of $\mathbf{p}'$. However, for a given $\mathbf{p}$ there are many different transitions corresponding to the different possible orientations of $\mathbf{p}'$ with respect to $\mathbf{p}$.

We can now return to the consideration of reactions (12-55) and (12-56). The first reaction (12-55) is intended to represent the total rate at which H_2 molecules in the state J, v react with chlorine atoms to give HCl molecules in the state J', v', summed over the $2J + 1$ values of m_J, the $2J' + 1$ values of m'_J, over all values of the angle between **p** and **p**′, and then over all possible directions and magnitudes for the translational momentum of the incident particle **p**. This summation can be performed in terms of the transition probabilities for the individual transitions, including all possible transitions as enumerated above, with the appropriate Boltzmann factor $e^{-p^2/2\mu kT}$ for the translational energy of the initial system. We omit this calculation because it is rather complicated, and there is not much point in doing it without a prior study of scattering theory in general.

The conclusion of the calculation is that, in a system at equilibrium, the over-all reactions (12-55) and (12-56) are taking place at an equal rate.

The equality of the rates of the reverse reactions (12-55) and (12-56) is by no means obvious. In a system at equilibrium, the total rate of destruction of H_2 molecules in the J, v state must equal their rate of formation. But it is conceivable that this constant concentration might be maintained by a cyclic process; for example,

$$\begin{aligned} \text{Cl} + \text{H}_2(J,v) &\to \text{H} + \text{HCl}(J',v') \\ \text{H} + \text{HCl}(J'v') &\to \text{Cl} + \text{H}_2(J'',v'') \\ \text{H}_2(J'',v'') &\to \text{H}_2(J,v) \quad \text{(by collision with some other molecule)} \end{aligned} \quad (12\text{-}58)$$

The principle of microscopic reversibility shows that in general equilibrium is maintained not by such cyclic processes, but by the equality of the rates of reverse processes.

The equality of the rates of Eqs. (12-55) and (12-56) can be used to derive a relation between their rate constants. Let $k(J,v;J',v')$ be the conventional chemical kinetic rate constant for reaction (12-55). The rate of reaction is then

$$\frac{d[\text{HCl}(J',v')]}{dt} = k(J,v;J',v')[\text{Cl}][\text{H}_2(J,v)] \quad (12\text{-}59)$$

where quantities in square brackets are concentrations (in units of atoms per cubic centimeter). In the same way, the rate of the reverse reaction (12-56) is

$$\frac{d[\text{H}_2(J,v)]}{dt} = k'(J'v';J,v)[\text{H}][\text{HCl}(J',v')] \quad (12\text{-}60)$$

By equating the two rates [(12-59) and (12-60)], we obtain

$$\frac{[\text{H}][\text{HCl}(J',v')]}{[\text{Cl}][\text{H}_2(J,v)]} = \frac{k(J,v;J',v)}{k'(J',v';J,v)} \quad (12\text{-}61)$$

By straightforward statistical mechanics, the equilibrium constant for the reaction (12-55) is

$$K = \left(\frac{m_\text{H} m_\text{HCl}}{m_\text{Cl} m_{\text{H}_2}}\right)^{3/2} \frac{2J'+1}{2J+1} e^{-\beta(\epsilon'_{J'v'} - \epsilon_{Jv} - \epsilon_0)} \quad (12\text{-}62)$$

The equilibrium constant may be equated to Eq. (12-61),

$$\frac{k(J,v;J',v)}{k'(J',v;J,v)} = \left(\frac{m_H m_{HCl}}{m_{Cl} m_{H_2}}\right)^{3/2} \frac{2J'+1}{2J+1} e^{-\beta(-\epsilon_0 + \epsilon'_{J'v'} - \epsilon_{Jv})} \quad (12\text{-}63)$$

If one of the rate constants is known, the other can then be calculated.

If the procedure outlined previously for calculating the rate constants for the over-all reactions (12-55) and (12-56) from the transition probabilities for the truly elementary reactions (12-53) and (12-54) were actually carried out, the same final equation (12-63) would have been obtained.

12-14. Radiative Recombination. We undertake a rather involved calculation which is an interesting illustration of the use of the principle of detailed balancing. The model is as follows: The diatomic molecule AB can dissociate by absorption of light.

$$AB + h\nu \rightleftharpoons A + B \quad (12\text{-}64)$$

There must be a corresponding radiative recombination reaction. In a system at equilibrium in the presence of an equilibrium radiation field, the rates of the forward and reverse reactions must be equal. We can calculate the equilibrium constant for the dissociation, and we can calculate the rate of dissociation in terms of the absorption coefficients of AB and the equilibrium density of radiation. Therefore, we should be able to express the rate constant for radiative recombination in terms of the absorption coefficients.

The model and approximations must be defined more completely. We are going to consider light absorption by AB in a particular vibrational and rotational state v, J. The potential curves for photodissociation are given in Fig. 12-6.

As shown in the figure, $AB(v,J)$ has a continuous absorption spectrum which may extend over a rather large frequency band. We suppose that we know the absorption coefficients $B_{vJ}(\nu)$ for molecules in this state. [One actually measures the average absorption coefficients for molecules distributed over a large number of states—but we suppose that somehow from this we extract $B_{vJ}(\nu)$.]

The reaction of interest is therefore

$$AB(v,J) + h\nu \rightarrow A + B \quad (12\text{-}65)$$

The rms angular momentum of AB is $[J(J+1)]^{1/2}\hbar$, which, for large J, we approximate as $J\hbar$. The photon may add or subtract one unit of angular momentum, but for large J we can neglect this and take $J\hbar$ as the angular momentum of the atoms A and B after dissociation.

Let ϵ_{vJ} be the sum of the vibrational energy in state v and the rotational energy in rotational state J. If D_0 is the energy of dissociation from the

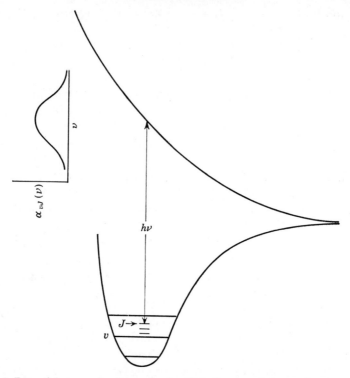

Fig. 12-6. Potential curves for the molecule AB in its ground electronic state and in a repulsive state (which dissociates to give atoms in their ground electronic state in the illustration). A molecule in the vibrational state v and rotational state J has a continuous absorption curve $\alpha_{vJ}(\nu)$, for transitions to the upper curve. The function $\alpha_{vJ}(\nu)$ is shown (with ν plotted vertically) to the left of the figure. For a further discussion of continuous absorption spectra and the Franck-Condon principle, see Herzberg, "Diatomic Molecules," pp. 390–399, especially p. 392 [17].

ground state of AB, then D_{vJ}, the energy of dissociation (to give fragments A and B with no kinetic energy) from the state vJ, is

$$D_{vJ} = D_0 - \epsilon_{vJ} \qquad (12\text{-}66a)$$

Upon absorption of light, $h\nu$, the kinetic energy K of the dissociation fragments is

$$K = h\nu - D_{vJ} \qquad (12\text{-}66b)$$

Assume that $h\nu/kT \gg 1$, so that the equilibrium density of radiation may be written

$$\rho(\nu, T)\, d\nu \approx \frac{8\pi h\nu^3}{c^3} e^{-\beta h\nu}\, d\nu \qquad (12\text{-}67)$$

In a system at equilibrium, with the equilibrium density of radiation, the rate of photodissociation of molecules $AB(v,J)$ by absorption of light ν, $\nu + d\nu$ is

$$\text{Rate} = B_{vJ}(\nu)\rho(\nu)[AB(v,J)]\,d\nu \qquad (12\text{-}68a)$$

$$\text{Rate} = \frac{8\pi h \nu^3}{c^3} B_{vJ}(\nu) e^{-\beta h \nu}[AB(v,J)]\,d\nu \qquad (12\text{-}68b)$$

The partition function for molecules in the particular state v, J is

$$q_{AB}(v,J) = (2J+1)\left[\frac{2\pi(m_A+m_B)kT}{h^2}\right]^{3/2} e^{-\beta \epsilon_{vJ}} \qquad (12\text{-}69)$$

There is a factor $2J+1$ for the $(2J+1)$ possible values of m_J, but there is otherwise no rotational or vibrational partition function because we are concerned with a molecule in the particular state v,J. Using the ordinary partition function for atoms A and B, the equilibrium constant of the reaction

$$AB(v,J) \rightleftharpoons A + B \qquad (12\text{-}70a)$$

is

$$K_e = \frac{[A][B]}{[AB(v,J)]} = \frac{q_A q_B}{q_{AB(v,J)}} e^{-\beta D_0}$$

$$= \left(\frac{2\pi\mu kT}{h^2}\right)^{3/2} \frac{e^{-\beta D_0}}{(2J+1)e^{-\beta \epsilon_{vJ}}} = \left(\frac{2\pi\mu kT}{h^2}\right)^{3/2} \frac{e^{-\beta D_{vJ}}}{2J+1} \qquad (12\text{-}70b)$$

where $\mu = m_A m_B/(m_A+m_B)$ and D_{vJ} is given by (12-66).

We now solve (12-70b) for $[AB(v,J)]$ and substitute this value in (12-68b). Recall that the kinetic energy of the escaping fragments is given by $K = h\nu - D_{vJ}[(12\text{-}66b)]$, and we obtain

$$\text{Rate} = \frac{8\pi h\nu^3}{c^3} B_{vJ}(\nu) e^{-\beta K}(2J+1)\left(\frac{h^2}{2\pi\mu kT}\right)^{3/2}[A][B]\,d\nu \qquad (12\text{-}71)$$

This is the rate of dissociation at equilibrium by absorption of light of frequency ν, $\nu + d\nu$ by molecules in the state v,J. By detailed balancing, this must also be the rate at which atoms A and B are recombining with kinetic energy K, $K + dK$ and angular momentum of about $J\hbar$ to emit light of frequency ν, $\nu + d\nu$ and thus to fall into the state $AB(v,J)$.

But expression (10-51b) gives the collision rate for collisions with kinetic energy between K and $K + dK$ and with angular momentum between L and $L + dL$ as

$$\text{Collisions} = [A][B]\frac{(8\pi)^{1/2}}{(\mu kT)^{3/2}} e^{-\beta K} L\,dL\,dK \qquad (10\text{-}51b)*$$

---

* It is obvious that in our treatment we have summed over the $2J+1$ orientations of the angular momentum of rotation. The formula (10-51b) for the number of collisions is for all possible orientations of the angular momentum L with respect to axes fixed in space. Correspondingly, the absorption coefficients $B_{vJ}(\nu)$ are average absorption coefficients, averaged over the $(2J+1)$ substates of the rotational state J.

If the emitted light is in the interval of width $d\nu$, the kinetic energy interval must be $dK = h\,d\nu$. Furthermore, to go into the quantized state J, the angular-momentum interval must be $dL = h/2\pi$; and $L = Jh/2\pi$. Then the total collision rate with the correct values of L and K to make it possible to emit light of frequency ν, $\nu + d\nu$ and go into the v, J proves to be

$$\text{Collision rate} = [A][B]\left(\frac{h^2}{2\pi\mu kT}\right)^{3/2} e^{-\beta K}(2J)\,d\nu \qquad (12\text{-}72)$$

It is consistent with our other approximations to ignore the difference between $2J + 1$ and $2J$; then the ratio of (12-71) and (12-72) is the probability per collision of a radiative recombination of the specified type. This probability is

$$\frac{8\pi h\nu^3}{c^3} B_{vJ}(\nu) \qquad (12\text{-}73)$$

This is the desired result. It is gratifying that all the temperature factors cancel in the final answer, since the probability for a process with defined energy and angular momentum should depend only on molecular parameters.

The quantity $(8\pi h\nu^3/c^3)B_{vJ}(\nu)$ is of the form of a probability of spontaneous emission per unit frequency interval, so that we could write

$$\frac{8\pi h\nu^3}{c^3} B_{vJ}(\nu) = A_{vJ}(\nu) \qquad (12\text{-}74a)$$

Then the quantity

$$\int A_{vJ}(\nu)\,d\nu \qquad (12\text{-}74b)$$

is in a sense the radiative lifetime of a molecule AB in the excited, unstable electronic state.

If the integral $\int A(\nu)\,d\nu \approx 10^8\text{ sec}^{-1}$, which is characteristic of a strong transition, and if the width of the absorption band at, say, $\lambda = 3{,}000$ Å is 300 Å wide (which is typical for such spectra), $\delta\nu \sim 10^{14}\text{ sec}^{-1}$ and so $A \sim 10^{-6}$. The probability of a radiative recombination in a collision is only of the order of 10^{-6} for a typical allowed transition.

Because of the low probability, radiative recombinations are important in practice principally in highly rarefied gases, where three-body non-radiative recombination of atoms is very improbable.

PROBLEMS

12-2. A lamp emitting 1,000 watts of radiation has a filament temperature of 3000°. Assume that the filament is a black body (it isn't quite). What is its area? How many quanta per second with wavelengths between 5,000 and 4,900 Å fall on a surface 1 cm² 10 cm away from the lamp? If all these quanta, and only these quanta, were absorbed by a sample of 1 cc of I_2 gas at $p = 0.38$ mm and $T = 300°$K, how long would it take to supply enough quanta to dissociate every molecule?

12-3. Given an equilibrium photon gas. For which frequency intervals is the number of photons greater than the number of states available so that states are multiply occupied, and for which frequencies is the number of states greater than the number of photons? Show that in the latter case Boltzmann statistics are approximately applicable.

12-4. A lamp emits a beam of monochromatic light of intensity 1 watt cm^{-2} with $\lambda = 3,000$ Å and bandwidth of $d\lambda = 1$ Å. Imagine that all the photons in 1 cc of this beam are captured in a container with walls which have zero heat capacity. The photons in the 1-cc box are then diffusely reflected without change in wavelength so that they are moving in all directions.

What is the entropy of the radiation field in the box (use the Boltzmann definition of entropy)?

The photons now interact with the walls so that the total energy in the radiation field is unchanged but the wavelengths change and the black-body distribution is established. What is the temperature of this radiation field? What is its entropy?

12-5. The radiative lifetime $[\int A(\nu)\,d\nu]^{-1}$ of the excited state of sodium, $^2P_{1/2}$, for the transition $^2P_{1/2} \to {}^2S_{1/2}$ (ground state) is 1.6×10^{-8} sec, and the wavelength is 5,896 Å.

The line shape, when determined by Doppler broadening, is given by

$$\alpha(\nu) = \alpha(\nu_0)e^{-\frac{1}{2}\{[(\nu-\nu_0)/\Delta\nu_D]^2\}}$$

where $\Delta\nu_D$, the Doppler width, is given by $\Delta\nu_D/\nu_0 = (1/c)(RT/M)^{1/2}$, R being the gas constant and M the atomic weight; and $\alpha(\nu_0)$ is the absorption coefficient at the center of the line.

What is the integrated absorption coefficient, $\int\alpha(\nu)\,d\nu$, for this transition? At $T = 3000°K$, what is the value of $\alpha(\nu_0)$? What is the corresponding molar absorption coefficient? What partial pressure of Na atoms in the ground state is needed to give an optical depth $[\alpha(\nu_0)N_{Na}l]$ of unity for a light path, l, of 1 cm?

(For further information about these topics see A. Mitchell and M. Zemansky, "Resonance Radiation and Excited Atoms," Cambridge, New York, 1934, especially pages 96 to 101.)

12-6. In a nuclear magnetic resonance (NMR) experiment, there could be a radio-frequency (rf) magnetic field of the order of 0.001 gauss, at a frequency of, say, 10^7 cycles sec^{-1}, and with a bandwidth $d\nu$ of 1 cycle sec^{-1}. The energy density in such a field is $H^2/8\pi$ ergs cc^{-1}, when H is measured in gauss.

What temperature would be necessary in order to produce a thermal-equilibrium radiation field with the same energy density per unit frequency range at 10^7 cycles sec^{-1} as in the NMR experiment?

It need hardly be added that the rf field in the NMR experiment is not a thermal-equilibrium field. In the first place there is no radiation density at other frequencies. In the second place, in an ordinary electromagnetic wave and in the equilibrium radiation field which is composed of such waves, there are oscillating electric and magnetic fields which are perpendicular. The electric field measured in cgs units (statvolts per centimeter) is equal to the magnetic field in gauss. In the NMR experiment, the electric field is much weaker.

Nevertheless, the calculation of the fictitious temperature above is useful. Students familiar with NMR can use it as a starting point to calculate the relative contributions of induced and spontaneous emission in NMR experiments and the relative populations of spin-up and spin-down states in equilibrium with the rf field in the complete absence of spin-lattice relaxation.

12-7. Here is a hard problem. Imagine the 1-cc box of photons of $\lambda = 3,000$ ($\pm\frac{1}{2}$) Å moving in all directions as described in Prob. 12-4. There is a working reservoir at a low temperature T_2 (perhaps about room temperature). What is the maximum amount of work that can be obtained from the photon gas, the heat reservoir at T_2 being used as the low-temperature heat sink and/or source?

13

Canonical and Grand Canonical Ensembles

13-1. Introduction. The statistical mechanics that we have developed so far has been derived by considering an isolated system of independent or noninteracting particles.

One of the most significant accomplishments of this theory is that it enables us to calculate the thermodynamic properties of substances in the ideal-gas state from molecular data. In several respects, however, the treatment is unsatisfactory. The most important point is that we should also like to be able to treat systems of interacting particles. Furthermore, in the laboratory, we are more likely to deal with systems that are immersed in a thermostat rather than with isolated systems. We shall now develop a statistical mechanics which is applicable to systems of interacting particles and which includes the possibility that the system can exchange energy with its surroundings.

There are some other respects in which our previous treatment is logically unsatisfactory; these are related to the approximation involved in using the properties of the most probable distribution for the properties of the system. These difficulties occur, also, in our present treatment and will not be corrected until later.

For a system of N noninteracting particles with conjugate momenta and coordinates $p_1, \ldots, p_{3N}; q_1, \ldots, q_{3N}$, the Hamiltonian of the system can be written as something like

$$H(p_1, \ldots, p_{3N}; q_1, \ldots, q_{3N})$$
$$= H(p_1, p_2, p_3, q_1, q_2, q_3) + H(p_4, p_5, p_6, q_4, q_5, q_6) + \cdots$$
$$+ H(p_{3N-2}, p_{3N-1}, p_{3N}; q_{3N-2}, q_{3N-1}, q_{3N})$$

that is, the Hamiltonian is separable, and the energy of any one molecule is independent of the energy and state of the other molecules (except, of of course, in the sense that the Pauli exclusion principle prevents two fermions from occupying the same state). There are, however, forces between molecules. In a real gas, and still more so in a liquid or solid, the potential energy of the system is, in principle, a function of the relative coordinates of all the molecules. As will be emphasized later, the kinetic-energy operator is still separable, and it may be possible to approximate the potential energy by a sum of terms which represent the interactions of pairs of molecules only. However, for the present, it is preferable to give a general treatment and to recognize that there is a Hamiltonian $H(p_1, \ldots, p_{3N}; q_1, \ldots, q_{3N})$, for the entire system. In principle, if not in practice, the Schrödinger equation

$$H(p_1, \ldots, p_{3N}; q_1, \ldots, q_{3N})\psi(q_1, \ldots, q_{3N}) = E\psi(q_1, \ldots, q_{3N}) \tag{13-1}$$

can be solved. There are then a set of stationary states ψ_i and energy levels E_i for the system as a whole.

We have emphasized the idea of a system wave function previously (in Chaps. 3 and 6, especially). For a system of noninteracting particles, any one system wave function is a suitable antisymmetrized (fermions) or symmetrized (bosons) sum over the product of the single-particle wave functions. For a system of interacting particles, the wave function cannot be expressed in terms of the product of single-particle wave functions. It must, however, still have the correct symmetry for the interchange of the coordinates of any two identical particles.

The important point at present is simply that there is a set of system wave functions $\psi_i(q_i, \ldots, q_{3N})$ and energy levels E_i for a system of interacting particles. We are considering a system in a container with a given volume V. Since the interactions between the particles are important, the allowed energy levels and wave functions are dependent on the number of particles in the system; when it is desirable to emphasize this fact, we shall write E_{Ni} and ψ_{Ni}. For example, the energy levels and wave functions for a system containing one sodium atom in 1 cc are entirely different from the energy levels and wave functions for a system containing 3×10^{22} sodium atoms in 1 cc. In the former case, the energy levels and wave functions are essentially those of a free gaseous atom; in the latter case, the interactions between the particles are such that the system has metallic properties.

It is convenient for our present purposes to index each quantum state separately so that all degeneracy factors are unity.

13-2. The Canonical Ensemble. In order to apply statistical considerations to the system of interest, with its eigenstates ψ_i, we imagine that we construct an ensemble consisting of a large number of systems, each one a replica of the system of interest, each one in an identical container with identical external electric and magnetic fields, etc., and each with the same number of particles as the actual physical system of interest. Each system in the ensemble can be a macroscopic system, with, say, 10^{24} particles.

The systems are in contact; they can exchange energy but not particles. The entire ensemble has a fixed energy; i.e., it is isolated. (As we shall discuss in Sec. 13-8, this latter assumption is really not necessary.) The ensemble described above is the canonical ensemble.

There is a very large number n of systems in the ensemble. (In fact, if one were to construct an ensemble to represent adequately a system containing 10^{24} atoms at room temperature, there would be more systems in the ensemble than there are atoms in the visible universe!) Each system in the ensemble is distinguishable (essentially because of the macroscopic nature of each system), and the systems can be labeled $s_1, s_2, \ldots, s_n$. It is meaningful to assert that system s_1 is in the quantum state of index i with wave function ψ_i and energy E_i, whereas system s_2 is in state j, s_3 in state k, ..., and s_n in state r. This specification of the quantum state for each particular system constitutes a specification of the state of the ensemble (or ensemble state). If system s_1 is in quantum state j and s_2 is in state i, this is a distinguishable and different ensemble state from the one, specified above, with s_1 in state i and s_2 in state j, etc.

Since the systems are loosely coupled and exchanging energy, we can say that any one system is constantly making transitions from one system state to another. Therefore the ensemble is undergoing transitions from one ensemble state to another.

We can examine the ensemble at any one time or at various times and determine the probability of finding systems in the different possible system quantum states. We then make the fundamental assumption that this will also give the probability that the particular physical system of interest in the laboratory will be in any given quantum state.

On examination at any one time, the ensemble will be found to be in a particular ensemble state in which there are n_1 systems in the first quantum state, n_i systems in the ith state, etc. There are a number of individual, distinguishable ensemble states which correspond to the distribution, $n_1, n_2, \ldots, n_i, \ldots, n_k, \ldots$. By the same reasoning as in Sec. 6-5, the number of ways (or number of individual ensemble states) for this particular distribution is

$$t_D = \frac{n!}{n_1! \, n_2! \cdots n_i! \cdots} = \frac{n!}{\prod_i n_i!} \tag{13-2}$$

It is to be emphasized that (13-2) contains a factor $n!$ in the numerator because the systems of the ensemble are distinguishable.

The possible distributions are subject to the two restrictions

$$\Sigma n_i = n \tag{13-3}$$
$$\Sigma n_i E_i = n\bar{E} \tag{13-4}$$

The quantity $n\bar{E}$ is the total energy of the ensemble, which is, by assumption, fixed; so $\bar{E}$ is the average energy per system.

The fundamental statistical assumption is that any one state of the ensemble which satisfies the conditions (13-3) and (13-4) is as probable as any other. We assume that the properties of the actual physical system being studied can be calculated by averaging over the properties of the systems in the ensemble. To calculate the entropy, we are interested in the total number of ensemble states, $t_{\text{total}} = \sum_D t_D$, where $\sum_D$ means summed over all possible distributions consistent with the constraints (13-3) and (13-4).

In general, if F_i is the value of some physical quantity F in the state i, then the average value $\bar{F}_D$ defined by the equation $n\bar{F}_D = \sum_i n_i F_i$ is called the ensemble average of F when the ensemble is in the distribution D, characterized by the set of numbers $n_1, n_2, \ldots, n_i, \ldots$. The observed average value of F, $\bar{F}$ for the actual system is then the average value of $\bar{F}_D$, averaged over all possible distributions,

$$\bar{F} = \frac{1}{t_{\text{total}}} \sum_D t_D F_D \tag{13-5}$$

The quantity $\bar{F}$ so defined is called the ensemble average of F.

The pressure of a system in the quantum state i is defined by

$$P_i = -\partial E_i / \partial V$$

The average value of the pressure in the particular distribution D is given by

$$n\bar{P}_D = \Sigma n_i P_i \tag{13-6a}$$

One thinks of changing the volume of each system in the ensemble by the same amount, dV; during this compression the distribution is frozen, and each system remains in the same quantum state that it started in. The quantity $\bar{P}_D$ is thus the average pressure that the ensemble exerts in resisting this change. The measured pressure is then identified with $\bar{P}$ calculated from

$$\bar{P} = \frac{1}{t_{\text{total}}} \sum_D t_D \bar{P}_D \tag{13-6b}$$

Now, just as before, some distributions correspond to a small value of t_D and are improbable. The overwhelming number of ensemble states corresponds to distributions which are almost identical with the most probable distribution. We shall find the most probable distribution and assume that $t_{\text{total}} \approx t_{\text{mp}}$ and that the pressure and other properties of the most probable distribution are the actual properties of the physical system.

To find the most probable distribution, apply the Stirling approximation to (13-2) and differentiate,

$$\ln t = n \ln n - n - \sum_i (n_i \ln n_i - n_i) \tag{13-7}$$

$$d(\ln t) = \sum_i \frac{\partial(\ln t)}{\partial n_i} dn_i = -\sum_i \ln n_i \, dn_i \tag{13-8}$$

With the usual Lagrange multipliers for the constraints (13-3) and (13-4) the condition for the most probable distribution is

$$\frac{\partial(\ln t)}{\partial n_i} + \alpha \frac{\partial n}{\partial n_i} - \beta \frac{\partial(n\bar{E})}{\partial n_i} = 0 \quad \text{for all } i \tag{13-9}$$

or
$$-\ln n_i + \alpha - \beta E_i = 0$$
or
$$n_i = e^\alpha e^{-\beta E_i} \tag{13-10}$$

We can eliminate α by using the condition (13-3),

$$\sum n_i = n = e^\alpha \sum_i e^{-\beta E_i} \tag{13-11}$$

The canonical, or system, partition function Q is defined by

$$Q(\beta, V) = \sum_i e^{-\beta E_i} \tag{13-12}$$

The energy levels are a function of the volume V, and we regard Q as a function of β and V. If there are other external parameters, such as an electric and a magnetic field which affect the energy levels, then Q is a function of these parameters, too.

This is an opportune point at which to reemphasize that there are N particles in each system. The energy levels are a function of N; therefore, the partition function depends on N. When it is desirable to emphasize this fact, we can write $Q_N(\beta, V)$.

The number of systems in the ensemble, n, is a rather arbitrary number; it is germane in the present context to discuss the probability $p_i = n_i/n$ of finding a system in the ith state. By using Eq. (13-11) for e^α, the distribution law (13-10) can be written

$$p_i = \frac{n_i}{n} = \frac{e^{-\beta E_i}}{Q} \tag{13-13}$$

The average energy of a system is related to β. Starting with (13-4),

$$n\bar{E} = \sum n_i E_i \tag{13-4}$$

or
$$\bar{E} = \sum p_i E_i = \frac{\Sigma E_i e^{-\beta E_i}}{Q} \tag{13-14a}$$

Sec. 13-3] CANONICAL AND GRAND CANONICAL ENSEMBLES 245

we obtain

$$\bar{E} = \frac{-(\partial Q/\partial \beta)_V}{Q} = -\left[\frac{\partial(\ln Q)}{\partial \beta}\right]_V \quad (13\text{-}14b)$$

It is to be emphasized that we have not yet either proved or assumed that the Lagrange multiplier β is $1/kT$ in the theory of the canonical ensemble.

The pressure of the system is taken to be the average pressure of the most probable distribution,

$$\bar{P} = \sum_i p_i \left(-\frac{\partial E_i}{\partial V}\right)$$

$$= \frac{-\Sigma(\partial E_i/\partial V)e^{-\beta E_i}}{Q}$$

$$= \frac{1}{\beta Q}\left(\frac{\partial Q}{\partial V}\right)_\beta = \frac{1}{\beta}\left[\frac{\partial(\ln Q)}{\partial V}\right]_\beta \quad (13\text{-}15)$$

13-3. The Perfect Gas in the Canonical Ensemble. Consider for simplicity a system composed of only three noninteracting atoms a, b, c with identical energy levels. Imagine, to begin with, that the particles are distinguishable so that there are no symmetry requirements for the wave functions. The single-particle wave functions and energy levels are $\phi_1, \phi_2, \ldots, \phi_i, \ldots$; $\epsilon_1, \epsilon_2, \ldots, \epsilon_i, \ldots$. The system wave function ψ_{ijk} is given by $\psi_{ijk} = \phi_i(a)\phi_j(b)\phi_k(c)$. This particular wave function ψ_{ijk} asserts that particle a is in state i, particle b in state j, and particle c in state k, with

$$E_{ijk} = \epsilon_i + \epsilon_j + \epsilon_k$$

The triple index ijk specifies a single quantum state of the system. The canonical partition function is

$$Q = \sum_{ijk} e^{-\beta(\epsilon_i+\epsilon_j+\epsilon_k)} = \sum_{ijk} e^{-\beta\epsilon_i}e^{-\beta\epsilon_j}e^{-\beta\epsilon_k}$$

$$= \sum_i e^{-\beta\epsilon_i} \sum_j e^{-\beta\epsilon_j} \sum_k e^{-\beta\epsilon_k} = \left(\sum_i e^{-\beta\epsilon_i}\right)^3 = q^3 \quad (13\text{-}16)$$

The equalities in the second line of (13-16) follow because the triple sum is unrestricted and all values of i must be combined with all values of j, etc. In the last equality in Eq. (13-16), q is the molecular partition function,

$$q = \sum_i e^{-\beta\epsilon_i}.$$

The atoms are actually fermions or bosons. For the triplet of single-particle wave functions, ϕ_i, ϕ_j, ϕ_k, for which $i \neq j \neq k$, there are six equal terms, $e^{-\beta\epsilon_i - \beta\epsilon_j - \beta\epsilon_k}$, in the unrestricted sum (13-16), corresponding to the six permutations of particles a, b, c among the states i, j, k;

$$\phi_i(a)\phi_j(b)\phi_k(c), \quad \phi_j(a)\phi_k(b)\phi_i(c), \text{ etc.}$$

For fermions or bosons, however, there should be only one term. For a set of single-particle wave functions for which $i = j \neq k$, there are three terms in the sum (13-16); there should be one term for bosons and none for fermions. For the case $i = j = k$, there is one term in the sum (13-16); there should be one term for bosons and no term for fermions. If the number of single-particle states is much greater than the number of particles (three in this example), the number of terms in (13-16) for which $i \neq j \neq k$ will greatly exceed the number of terms for which $i = j \neq k$ or $i = j = k$. For such a dilute system, the correct canonical partition function is obtained from that given in (13-16) by dividing by 3!; that is, $Q = q^3/3!$.

The argument is general; for a perfect gas containing N fermions or bosons which is dilute (number of single-particle states much greater than N)

$$Q = \frac{q^N}{N!} \qquad (13\text{-}17)$$

Therefore, using the Stirling approximation in the form $N! \approx N^N e^{-N}$,

$$Q = \left(\frac{q}{N}\right)^N e^N \qquad \ln Q = N \ln \frac{q}{N} + N \qquad (13\text{-}18)$$

As was shown in Eq. (6-35), the molecular partition function,

$$q = \Sigma e^{(-\beta \epsilon_i)}$$

for a structureless particle with no internal energy levels is

$$q = \left(\frac{2\pi m}{\beta h^2}\right)^{3/2} V$$

Then, from (13-18) and (13-15), the pressure of a perfect gas in the theory of the canonical ensemble is given by

$$\bar{P} = \frac{1}{\beta}\left[\frac{\partial (\ln Q)}{\partial V}\right]_\beta = \frac{N}{\beta}\left[\frac{\partial (\ln q)}{\partial V}\right]_\beta = \frac{N}{\beta V} \qquad (13\text{-}19)$$

which leads to the identification $\beta = 1/kT$ for the perfect gas in the canonical ensemble.

But by a proof which is entirely analogous to that used in Sec. (6-9), we can show that if $\beta = 1/kT$ for a perfect gas, $\beta = 1/kT$ for any system in a canonical ensemble. To carry out this argument, we construct a canonical ensemble consisting of a large number of replicas of a system A of interest, with a set of energy levels E_i^A, and a large number of replicas of a perfect gas B, with its own set of energy levels E_j^B. We permit the systems in the composite ensemble to exchange energy and find the most probable distribution as before. We then find that each system obeys the fundamental distribution law

$$p_i^A = \frac{e^{-\beta E_i^A}}{Q_A} \qquad p_j^B = \frac{e^{-\beta E_j^B}}{Q_B}$$

with a common β. Thus, β for an arbitrary kind of system in a canonical ensemble is the same as β for a perfect gas. In general, for the canonical ensemble

$$\beta = \frac{1}{kT} \qquad (13\text{-}20)$$

In writing equations, we shall use either β or $1/kT$ as convenient.

13-4. The Entropy and Free Energy. We are going to be interested in the quantity $(1/n) \ln t$. Recall that $\Sigma p_i = \Sigma(n_i/n) = 1$. Simplifying (13-7) and dividing by n,

$$\frac{\ln t}{n} = \ln n - \sum \frac{n_i}{n} \ln n_i = \left(\sum p_i\right) \ln n - \sum p_i \ln n_i$$

$$= -\sum p_i \ln \frac{n_i}{n} = -\sum p_i \ln p_i \qquad (13\text{-}21)$$

The final expression, in terms of probabilities, is a famous and useful way of expressing the entropy equation. As already indicated, we define the average statistical mechanical entropy $\tilde{S}_{sm}$ of a system in the ensemble by

$$n\tilde{S}_{sm} = k \ln t_{total}$$

where t_{total} is the total number of states in which the ensemble can exist, subject to the constraints (13-3) and (13-4); and we assume $\ln t_{total} \approx \ln t_{mp}$, where t_{mp} is the number of ensemble states for the most probable distribution.

Substitute for the $\ln p_i$'s in (13-21) their value for the most probable distribution,

$$\tilde{S}_{sm} = -k \sum p_i \ln \frac{e^{-\beta E_i}}{Q} = k\beta \sum p_i E_i + k \ln Q = k\beta \bar{E} + k \ln Q$$

$$\qquad (13\text{-}22)$$

$$\tilde{S}_{sm} = \frac{\bar{E}}{T} + k \ln Q \qquad (13\text{-}23)$$

It is easy to show that, if all the systems in the ensemble are subjected to an identical reversible change in volume, $d\tilde{S}_{sm} = \delta \bar{q}/T$, where $\delta \bar{q}$ is the average heat absorbed by the system. The proof will be given for the grand canonical ensemble and is omitted here. This justifies identification of the statistical mechanical entropy with the thermodynamic entropy. We have thus arrived at the following relations between thermodynamic quantities and the canonical partition function. The bars denoting ensemble averages are now superfluous and can be omitted.

$$E = kT^2 \left[\frac{\partial (\ln Q)}{\partial T} \right]_V \qquad (13\text{-}24)$$

$$P = kT \left[\frac{\partial (\ln Q)}{\partial V} \right]_T \qquad (13\text{-}25)$$

$$S = \frac{E}{T} + k \ln Q \qquad (13\text{-}26)$$

$$A = E - TS = -kT \ln Q \qquad (13\text{-}27)$$

$$\mu = \left(\frac{\partial A}{\partial N} \right)_{T,V} = -kT \left[\frac{\partial (\ln Q)}{\partial N} \right]_{T,V} \qquad (13\text{-}28)$$

These and the distribution law are the fundamental relations for a canonical ensemble.

For a dilute perfect gas, $Q = q^N/N!$, and the relations

$$F = -NkT \ln \left(\frac{q}{N} \right)$$

$$A = -NkT \left[\ln \left(\frac{q}{N} \right) + 1 \right]$$

$$S = \left(\frac{E}{T} \right) + Nk \left[\ln \left(\frac{q}{N} \right) + 1 \right]$$

follow from (13-26) through (13-28). These are the same thermodynamic relations in terms of the molecular partition function that were derived in Chap. 6.

We shall return to a discussion of the physical meaning of the canonical ensemble and its distribution function in a later section.

13-5. The Grand Canonical Ensemble. The canonical ensemble is useful for discussing a variety of problems, and a few examples will be developed later. There is a more general ensemble, and a corresponding partition function, which we shall now study and which provide a more powerful tool for statistical mechanical problems of complex systems. (There are still other general ensembles which are useful for certain problems.*) The essential point is that we are now going to consider an ensemble of systems, all replicas of the actual system of interest, which can exchange energy with each other and which can also exchange particles (molecules).

For example, we can imagine that a 1-liter container full of water at room temperature is divided by imaginary boundaries into compart-

---

* Hill, p. 72 [9a]; Fowler and Guggenheim, p. 253 [1].

For completeness, it should also be mentioned that an ensemble in which each system has the same fixed energy as well as the same number of particles is known as a microcanonical ensemble. Thus the system of independent particles considered in Chap. 6 is an example (in fact, the principal important example) of a microcanonical example.

ments, each 1 by 1 by 1 μ, or 10^{-12} cc in volume. The system of interest is a 10^{-12}-cc sample of water. The 10^{15} systems are the ensemble. They can exchange energy and water molecules with each other. There are a large number of possible quantum states for the system, consisting of 10^{-12} cc of water at room temperature (actually about $10^{10^{11}}$). By examining one member of the ensemble at any one time, we can determine the number of molecules in this particular system and the quantum state that it is in. Another member of the ensemble is likely to have a different number of particles and to be in a different energy state. Our object is to determine the probability p_{Ni} that a system will have a certain number of particles and will be in quantum state i. We recall that liquid water at its normal density contains about 3.3×10^{22} molecules per cc or 3.3×10^{10} molecules per 10^{-12} cc. We expect to find that most of the systems in the ensemble have about 3.3×10^{10} water molecules therein; but this number will fluctuate slightly from time to time for any one system in the ensemble, and, at any one time, it will be slightly different for different members of the ensemble. In a properly constructed ensemble (which would have to be much bigger than 1 liter if the basic system were 10^{-12} cc), there would even be a few of the systems (each 10^{-12} cc) containing only 10^2 H_2O molecules and some, perhaps, containing 6×10^{10} H_2O molecules.

For the general case, the number of systems in the ensemble is n; each system has the same volume V; the total number of particles is fixed at $n\bar{N}$, where $\bar{N}$ is the average number of particles in a system; and the total energy is fixed at $n\bar{E}$. The number of particles in a particular system is variable and is N. The energy levels for a system are a function of V. Furthermore, it is most important to realize that, because of the potential of interaction between the particles, the energy levels are also a function of the number of particles in the volume V, and we write $E_{Ni}(V)$ as the energy in the ith quantum state of a system which has N particles. A system with a different number N' of particles has a different set of energy levels $E_{N'i'}$. Then, if we examine the entire ensemble at any instant of time, we shall discover that there is a distribution in the sense that there are n_{Ni} systems that have N particles and are in the quantum state i for N particles.

The argument proceeds exactly as before, but the notation is more complex because N is also a variable. In a given distribution, there are n_{Ni} systems of the ensemble in the Ni quantum state, n_{Nj} in the Nj state, $n_{N'k}$ in the $N'k$ state, etc. For a given distribution, the probability p_{Ni} of finding a system in the Ni state is $p_{Ni} = n_{Ni}/n$.

The number of ways of having a given distribution is

$$t = \prod_N \prod_i \frac{n!}{n_{Ni}!} \qquad (13\text{-}29)$$

Note that the distribution ranges over all values of N (theoretically from 0 to $n\bar{N}$) and for a given N over all quantum states i (theoretically for energies from 0 to $n\bar{E}$).

Proceeding as before,

$$\ln t = n \ln n - n - \sum_N \sum_i n_{Ni}(\ln n_{Ni} - 1) \qquad (13\text{-}30)$$

We seek the most probable distribution, subject to the constraints that the number of systems in the ensemble is fixed [(13-31)], the total energy is fixed [(13-32)], and the total number of particles is fixed [(13-33)]. Formally stated, the constraints and Lagrange multipliers are

$$\sum_N \sum_i n_{Ni} = n \qquad \alpha \qquad (13\text{-}31)$$

$$\sum_N \sum_i n_{Ni} E_{Ni} = n\bar{E} \qquad -\beta \qquad (13\text{-}32)$$

$$\sum_N \sum_i n_{Ni} N = n\bar{N} \qquad \ln \lambda \qquad (13\text{-}33)$$

Note that, as a matter of future convenience, the Lagrange multiplier for (13-33) is denoted by $\ln \lambda$.

The double sum over all possible states i for a given value of N occurs frequently in the theory of the grand ensemble. For notational brevity, we shall usually denote such sums by $\sum_{N,i}$.

The student must pay close attention to the context to be aware of the sense in which the n_{Ni}'s, the number of systems of the ensemble in the ith state for N particles, are variables and the sense in which N, the number of particles in a system, is a variable.

To maximize $\ln t$, first observe that

$$d(\ln t) = \sum_{N,i} \frac{\partial (\ln t)}{\partial n_{Ni}} dn_{Ni} = - \sum_{N,i} \ln n_{Ni} \, dn_{Ni} \qquad (13\text{-}34)$$

The conditions for the most probable distribution then are

$$- \ln n_{Ni} + \alpha - \beta E_{Ni} + N \ln \lambda = 0 \qquad \text{for all } N \text{ and all } i \qquad (13\text{-}35a)$$

or

$$n_{Ni} = e^\alpha \lambda^N e^{-\beta E_{Ni}} \qquad (13\text{-}35b)$$

plus the constraints (13-31) to (13-33).

We now proceed to eliminate α and identify the thermodynamic significance of λ and β. The grand partition function is defined by

$$\Xi(\lambda,\beta,V) = \sum_{N,i} \lambda^N e^{-\beta E_{Ni}} \qquad (13\text{-}36)$$

We think of the grand partition function $\Xi(\lambda,\beta,V)$ (phonetically, "Zy") as a function of λ, β, and V (since the E_{Ni}'s are functions of V).*

By summing (13-35b) and using the condition (13-31)

$$\sum_{N,i} n_{Ni} = n = e^\alpha \sum_{N,i} \lambda^N e^{-\beta E_{Ni}} = e^\alpha \Xi(\lambda,\beta,V)$$

or
$$e^\alpha = \frac{n}{\Xi(\lambda,\beta,V)}$$

Recall that the probability of finding a system with N particles in the state i is $p_{Ni} = n_{Ni}/n$; then

$$p_{Ni} = \frac{\lambda^N e^{-\beta E_{Ni}}}{\Xi(\lambda,\beta,V)} \qquad (13\text{-}37)$$

This is the fundamental distribution law for the grand ensemble. From (13-32), the energy equation is

$$\bar{E} = \sum_{N,i} p_{Ni} E_{Ni} = \sum_{N,i} \frac{E_{Ni} \lambda^N e^{-\beta E_{Ni}}}{\Xi}$$

$$= -\frac{1}{\Xi}\left[\frac{\partial \Xi(\lambda,\beta,V)}{\partial \beta}\right]_{\lambda,V} = -\left[\frac{\partial (\ln \Xi)}{\partial \beta}\right]_{\lambda,V} \qquad (13\text{-}38)$$

From (13-33), the conservation of the total number of particles gives

$$\bar{N} = \sum_{N,i} p_{Ni} N = \sum_{N,i} \frac{N \lambda^N e^{-\beta E_{Ni}}}{\Xi}$$

$$= \frac{\lambda}{\Xi}\left[\frac{\partial \Xi(\lambda,\beta,V)}{\partial \lambda}\right]_{\beta,V} = \left[\frac{\partial (\ln \Xi)}{\partial (\ln \lambda)}\right]_{\beta,V} \qquad (13\text{-}39)$$

The pressure is taken as the ensemble average of $-\partial E_{Ni}/\partial V$,

$$\bar{P} = -\sum_{N,i} p_{Ni} \frac{\partial E_{Ni}}{\partial V} = -\sum_{N,i} \frac{\lambda^N e^{-\beta E_{Ni}}(\partial E_{Ni}/\partial V)}{\Xi}$$

$$= \frac{1}{\beta \Xi}\left[\frac{\partial \Xi(\lambda,\beta,V)}{\partial V}\right]_{\lambda,\beta} = \frac{1}{\beta}\left[\frac{\partial (\ln \Xi)}{\partial V}\right]_{\lambda,\beta} \qquad (13\text{-}40)$$

It is legitimate in the second line of (13-40) to assert that the differentiation is at constant λ and β since the energy levels at fixed i and N are a function only of V and the dependence of Ξ on V is due entirely to the dependence of the E_{Ni} on V.

* The symbol Ξ for the grand partition function is used by Fowler and Guggenheim, Hill, and many other authors. It is a little awkward to write but not difficult to pronounce. Rushbrooke uses GPF, which is awkward to write and to pronounce when it occurs repeatedly in a discussion. Kirkwood and Tolman, following Gibbs, write $\exp(\Omega/kT)$ for the grand partition function. It seems to us that, to maintain symmetry with the other partition functions, it is desirable to have a symbol for the grand partition function.

It is expedient at this point to note the relation between the grand partition function and the canonical partition function. Recall that the canonical partition function for an ensemble of systems containing N particles is given by $Q_N(\beta,V) = \sum_i e^{-\beta E_i}$. Relation (13-36) can be rewritten as

$$\Xi(\lambda,\beta,V) = \sum_N \lambda^N \left(\sum_i e^{-\beta E_{Ni}} \right) = \sum_N \lambda^N Q_N(\beta,V) \qquad (13\text{-}41)$$

As usual, we identify β by reference to the perfect-gas case. For a dilute perfect gas, the canonical partition function is given by $Q_N = q^N/N!$; therefore, the grand partition function is

$$\Xi = \sum_N \frac{\lambda^N q^N}{N!} = e^{\lambda q} \qquad (13\text{-}42)$$

Then, from the equation for the pressure [(13-40)],

$$\bar{P} = \frac{1}{\beta}\left[\frac{\partial(\ln \Xi)}{\partial V}\right]_{\lambda,\beta} = \frac{\lambda}{\beta}\left(\frac{\partial q}{\partial V}\right)_{\lambda,\beta} \qquad (13\text{-}43)$$

whereas the equation for $\bar{N}$ [(13-39)] gives

$$\bar{N} = \lambda\left[\frac{\partial(\ln \Xi)}{\partial \lambda}\right]_{\beta,V} = \lambda q \qquad (13\text{-}44)$$

Now eliminate λ from (13-43) and (13-44), and recall that

$$q = \left(\frac{2\pi m}{\beta h^2}\right)^{3/2} V$$

$$\bar{P} = \frac{\bar{N}}{\beta q}\frac{\partial q}{\partial V} = \frac{\bar{N}}{\beta V} \qquad (13\text{-}45)$$

Thus, for a dilute perfect gas in the grand ensemble $\beta = 1/kT$. Just as before, one now constructs an ensemble consisting of replicas of two types of systems; system A is a perfect gas, and system B is the system of actual interest. Energy can be exchanged between the two different kinds of systems, but particles can be exchanged only between similar systems and cannot cross from a replica of A to a replica of B. One then finds that the distribution law (13-37) holds for each system separately but that they have a common β. So, in general, $\beta = 1/kT$.

13-6. Entropy and Other Thermodynamic Functions in the Grand Ensemble. Each system in the ensemble is in a container of identical volume V. Let the volume of each container be changed in an identical way and in such a way that the ensemble remains at equilibrium, i.e., in the most probable distribution during the change. The ensemble average energy per system, which is given by (13-32) as $\bar{E} = \sum_{N,i} p_{Ni} E_{Ni}$, changes because of the changes in the energy levels for each quantum state, dE_{Ni}, and because of the changes dp_{Ni} in the probability of finding a system in a given state. Therefore,

$$d\bar{E} = \sum_{N,i} p_{Ni}\, dE_{Ni} + \sum_{N,i} E_{Ni}\, dp_{Ni} \qquad (13\text{-}46)$$

It is to be noted that we do not completely specify the physical change associated with the compression, dV. For example, the compression could be isothermal, or it could be adiabatic in the thermodynamic sense. However, we restrict our consideration to changes in which the ensemble as a whole remains closed—i.e., the total number of particles in the ensemble does not change. The definition, in thermodynamics, of the heat absorbed from the surroundings, δq, for a change in which the number of particles in the system changes is a rather subtle question which we wish to avoid.

The energy levels E_{Ni} are affected only by the volume:

$$dE_{Ni} = \frac{\partial E_{Ni}}{\partial V}\, dV$$

Therefore, for the first term on the right-hand side of (13-46),

$$\sum p_{Ni}\, dE_{Ni} = \sum p_{Ni}\, \frac{\partial E_{Ni}}{\partial V}\, dV = -\bar{P}\, dV = -\delta\bar{w}$$

where $\delta\bar{w}$ is the average thermodynamic work done per system. The second term of (13-46) is identified, therefore, with the average amount of heat, $\delta\bar{q}$, absorbed per system during the reversible compression.

$$\delta\bar{q} = \sum_{N,i} E_{Ni}\, dp_{Ni} \qquad (13\text{-}47)$$

The Boltzmann definition for the average entropy of a system in the grand ensemble is

$$n\bar{S}_{sm} = k \ln t \approx k \ln t_{mp} \qquad (13\text{-}48)$$

Just as in the case of the canonical ensemble [Eq. (13-21)], this definition leads to the equation

$$\bar{S}_{sm} = -k \sum_{N,i} p_{Ni} \ln p_{Ni} \qquad (13\text{-}49)$$

For an arbitrary change at equilibrium, we must have

$$\ln p_{Ni} = -\beta E_{Ni} + N \ln \lambda - \ln \Xi$$

Furthermore, there is the condition $\Sigma dp_{Ni} = 0$; therefore by differentiation of (13-49)

$$d\bar{S}_{sm} = -k \sum_{N,i} \ln p_{Ni}\, dp_{Ni} = k\beta \sum_{N,i} E_{Ni}\, dp_{Ni} - k \ln \lambda \sum_{N,i} N\, dp_{Ni} \qquad (13\text{-}50)$$

The above relation applies to an arbitrary change imposed on the grand canonical ensemble, provided that it remains at equilibrium. But

$$\Sigma N\, dp_{Ni} = d\bar{N}$$

the change in the average number of particles. Therefore,

$$d\bar{S}_{sm} = k\beta \Sigma E_{Ni}\, dp_{Ni} - k \ln \lambda\, d\bar{N} \qquad (13\text{-}51)$$

For a closed system ($d\bar{N} = 0$), we have, by comparison with (13-47), $d\bar{S}_{sm} = \delta\bar{q}/T$. Thus, for a closed system, the increment in the statistical mechanical entropy is the same as the increment in the thermodynamic entropy, and we may again assert that the statistical mechanical entropy, as defined by Eq. (13-48) or (13-49), may be identified with the thermodynamic entropy. We shall remark on the situation for an open system shortly.

It is interesting at this point to reconsider the meaning of the assertion that we have used repeatedly that the pressure is the ensemble average of $-\partial E_{Ni}/\partial V$. In thermodynamics,

$$dE = T\, dS - P\, dV = C_V\, dT + \left(T \frac{\partial P}{\partial T} - P\right) dV$$

and $(\partial E/\partial V)_S = -P$, but $(\partial E/\partial V)_T = T(\partial P/\partial T) - P$. The process of compression of each system of the ensemble, with each system frozen in a particular quantum state, corresponds to a compression with $dp_{Ni} = 0$ and therefore at a constant entropy for the ensemble. Thus, the statistical statement that the pressure is the ensemble average of $-\partial E_{Ni}/\partial V$ is consistent with the thermodynamic statement that the pressure is $-(\partial \bar{E}/\partial V)_S$.

Just as in the canonical ensemble, we substitute

$$\ln p_{Ni} = N \ln \lambda - \ln \Xi - \beta E_{Ni}$$

into (13-49) and recall that $\bar{N} = \sum\limits_{N,i} p_{Ni} N$, so that

$$\bar{S} = -k \ln \lambda \left(\sum_{N,i} p_{Ni} N\right) + k\beta \sum_{N,i} E_{Ni} p_{Ni} + k \ln \Xi$$

$$= -k\bar{N} \ln \lambda + \frac{\bar{E}}{T} + k \ln \Xi \qquad (13\text{-}52)$$

The Helmholtz free energy is then

$$A = \bar{E} - T\bar{S} = k\bar{N}T \ln \lambda - kT \ln \Xi \qquad (13\text{-}53)$$

The quantities $\bar{N}$, $\bar{E}$, $\bar{S}$, and $\bar{P}$ are ensemble average quantities, averaged over all systems in the ensemble. We have identified these averages with the macroscopic thermodynamic functions, but it is useful to continue to write them as ensemble averages.

To evaluate the chemical potential by the relation $\mu = (\partial A/\partial \bar{N})_{T,V}$, it is necessary to realize that λ can be regarded as an implicit function of $\bar{N}, T, V$, as given by Eq. (13-39), which can be rewritten as

$$\bar{N} = \frac{\sum_{N,i} N\lambda^N e^{-\beta E_{Ni}}}{\Sigma \lambda^N e^{-\beta E_{Ni}}} \tag{13-54a}$$

The grand partition function Ξ has heretofore been regarded as an explicit function of λ, T, V; $\Xi = \Sigma \lambda^N e^{-\beta E_{Ni}}$. Therefore, the variation of Ξ with $\bar{N}$ at constant T and V is due only to the variation of λ with $\bar{N}$ at constant T and V.

$$\left[\frac{\partial(\ln \Xi)}{\partial \bar{N}}\right]_{T,V} = \left[\frac{\partial(\ln \Xi)}{\partial(\ln \lambda)}\right]_{T,V} \left[\frac{\partial(\ln \lambda)}{\partial \bar{N}}\right]_{T,V} \tag{13-54b}$$

But, from (13-39), $\bar{N} = [\partial(\ln \Xi)/\partial(\ln \lambda)]_{T,V}$. Therefore, from (13-54b)

$$\left[\frac{\partial(\ln \Xi)}{\partial \bar{N}}\right]_{T,V} = \bar{N}\left[\frac{\partial(\ln \lambda)}{\partial \bar{N}}\right]_{T,V} \tag{13-54c}$$

We can now evaluate the chemical potential. By differentiation of (13-53),

$$\mu = \left(\frac{\partial A}{\partial \bar{N}}\right)_{T,V} = kT \ln \lambda + k\bar{N}T\left[\frac{\partial(\ln \lambda)}{\partial \bar{N}}\right]_{T,V} - kT\left[\frac{\partial(\ln \Xi)}{\partial \bar{N}}\right]_{T,V}$$

The last two terms neatly drop out because of (13-54c), and

$$\mu = kT \ln \lambda \tag{13-55}$$

Thus, all the parameters in the grand partition function now have their thermodynamic significance. The quantity λ is called the absolute activity; it is closely related to the various thermodynamic definitions of activity.*

The function $\bar{P}V$ may be evaluated as

$$\bar{P}V = F - A = \bar{N}\mu - A = kT \ln \Xi \tag{13-56}$$

This is a fundamental equation in the theory of the grand ensemble, analogous to the equations $A = -kT \ln Q$ for the canonical ensemble and $\mu = -kT \ln (q/N)$ for a system of independent particles in a microcanonical ensemble.

The quantities $\bar{P}V$ and $\bar{P}V/T$ are perfectly good thermodynamic functions, albeit rather uncommon ones. We recall the equation

---

* The quantity λ is an absolute activity in the sense that it is independent of a choice of a standard state; its value, however. is still dependent on the choice of the zero of energy.

(Prob. 4-2) for PV/T, which, for the case of a one-component system, becomes

$$d\left(\frac{PV}{T}\right) = \frac{E}{T^2}dT + \frac{P}{T}dV + N\,d\left(\frac{\mu}{T}\right) \tag{13-57}$$

The "natural" thermodynamic variables for PV/T as a thermodynamic function are T, V, and μ/T. But in the theory of the grand ensemble, $\bar{P}V/T = \ln \Xi$, and the simplest variables for expressing Ξ are T, V, and λ; but λ is directly related to μ/T [(13-55)]. In this sense then, it is not surprising that $\bar{P}V$ is closely related to the grand partition function Ξ.

It is worthwhile to return briefly to Eq. (13-51) and discuss its meaning for an open system ($d\bar{N} \neq 0$);

$$d\bar{S}_{sm} = k\beta \Sigma E_{Ni}\,dp_{Ni} - k \ln \lambda\, d\bar{N} \tag{13-51}$$

which we rewrite as

$$d\bar{S}_{sm} = k\beta(dE - \Sigma p_{Ni}\,dE_{Ni}) - k \ln \lambda\, d\bar{N}$$

or

$$d\bar{S}_{sm} = \frac{d\bar{E}}{T} - \frac{\Sigma p_{Ni}\,dE_{Ni}}{T} - k \ln \lambda\, d\bar{N} \tag{13-58a}$$

For the analogous thermodynamic equation, we start with the equation for a one-component open system,

$$dE = T\,dS - P\,dV + \mu\,dN$$

and write

$$dS = \frac{dE}{T} + \frac{P}{T}dV - \frac{\mu}{T}dN \tag{13-58b}$$

The term $\Sigma p_{Ni}\,dE_{Ni}$ in (13-58a) is $-\bar{P}\,dV$ even if $d\bar{N} \neq 0$. In view of the identification, $kT \ln \lambda = \mu$ [(13-55)] we see that the statistical equation (13-58a) and the thermodynamic equation (13-58b) are identical.

13-7. The Probability of a Given Value of N. It must first be emphasized that, in the equation $\Xi = \sum_{N,i} e^{\beta N\mu - \beta E_{Ni}}$, the chemical potential μ is not a function of N and i; it is the same for all members of the ensemble and is the chemical potential of the system of which the members of the ensemble are hypothetical replicas. We can think of μ as an independent variable, or we can think of it as determined by $\bar{N}$, the average number of particles per system, and by T and V.

We have already noted that the grand partition function can be written in terms of the canonical partition function [(13-41)],

$$\Xi = \sum_N Q_N(T,V)e^{\beta N\mu} \tag{13-59a}$$

The probability of being in a state Ni is $p_{Ni} = e^{\beta N\mu - \beta E_{Ni}}/\Xi$. The probability of finding that a member of the ensemble contains N particles irrespective of energy is

$$p_N = \sum_i p_{Ni} = \frac{\sum_i e^{N\mu/kT}e^{-E_{Ni}/kT}}{\Xi} = \frac{Q_N e^{N\mu/kT}}{\Xi} \qquad (13\text{-}59b)$$

The probability that the system will contain N particles is proportional to the canonical partition function $Q_N(T,V)$. Recall that in the canonical ensemble $A_N = -kT \ln Q_N$, where A_N is the Helmholtz free energy of a system of N particles. Then, we can also write

$$p_N = \frac{e^{(N\mu - A_N)/kT}}{\Xi} \qquad \Xi = \sum_N e^{(N\mu - A_N)/kT} \qquad (13\text{-}60)*$$

13-8. Review and Reconsideration. It is time to pause and consider just what we have accomplished in the preceding sections. Either the canonical ensemble or the grand canonical ensemble provides a logically satisfactory statistical mechanical treatment of the equilibrium properties of a system of interacting particles. If we can calculate the energy levels of such a system, we can calculate the canonical partition function or the grand partition function and then, by using the relations derived in this chapter, the thermodynamic properties of the system.

The distribution law for the canonical ensemble, $p_i = e^{-\beta E_i}/Q$, is formally very similar to the distribution law for molecular energies that we have so intensively studied in Chaps. 6, 8, and 11. For noninteracting boltzons, the probability that a molecule will be found in a given molecular state i is $p_i = e^{-\beta \epsilon_i}/q$. In the same way, the molecular partition function q is formally similar to the system partition function Q. If the system is a macroscopic one, however, there is a tremendous difference between the two distributions and between q and Q. The difference is essentially this: The distribution of molecular energies among the molecules of a system is not sharply peaked; there is a reasonable probability that a molecule will have, say, one-half or twice as much energy as the most probable amount of energy for a particular degree of freedom. How-

---

* It is interesting to note the following point: The ensemble average of the free energy A_N is

$$\bar{A} = \sum_N p_N A_N = \frac{\sum_N A_N e^{-A_N/kT} e^{N\mu/kT}}{\Xi} \qquad (13\text{-}61)$$

On the other hand, the free energy defined by $A = \bar{E} - T\bar{S}$ was shown to be (13-53),

$$A = \bar{N}\mu - kT \ln \Xi = kT \ln \frac{e^{N\mu/kT}}{\Xi} \qquad (13\text{-}62)$$

The two expressions (13-61) and (13-62) are not identically equal; we shall see in the next chapter that for all practical purposes A and $\bar{A}$ are equal.

ever, in the macroscopic ensemble, the distribution is very sharply peaked; the probability that a system will be found with an energy that differs from the average energy by a perceptible fraction, say, 10^{-6}, is negligible. In the same way, the molecular partition function

$$q = \Sigma e^{-\beta \epsilon_i}$$

receives significant contributions from a large number of states of significantly different energy ϵ_i. The only significant contributions to the canonical partition function $Q = \Sigma e^{-E_i/kT}$ are from states with energy that is extremely close to the average energy $\bar{E}$. Of course, there are many such states.

The discussion of fluctuations (Chap. 14) will provide a general justification of these statements. It is important, however, to have a good intuitive feeling for the situation at the present stage; and, to this end, we consider the canonical partition function of a perfect gas.

We have heretofore written Q as $Q = \sum_i e^{-\beta E_i}$, in which each quantum state i of the system was separately listed. It is now convenient to reintroduce the idea of degeneracy. We collect together all the states with energy very close to E_i as being degenerate; if there are $t(E_i)$ such states, we write

$$Q = \Sigma t(E_i) e^{-\beta E_i} \tag{13-63}$$

Let us now evaluate $t(E_i)$ for a perfect gas. In Chap. 6 we considered a perfect gas with a fixed energy E_i (a microcanonical ensemble). The energy E_i determines a temperature T_i according to the equation

$$E_i = \tfrac{3}{2} NkT_i \tag{13-64a}$$

The molecular partition function was $(2\pi m k T_i/h^2)^{3/2} V$; if we regard E_i as the independent variable rather than T_i, we have

$$q(E_i) = \left(\frac{4\pi m E_i}{3Nh^2}\right)^{3/2} V \tag{13-64b}$$

The entropy of the gas with energy E_i can be written

$$S(E_i) = \tfrac{5}{2} Nk + Nk \ln \frac{q(E_i)}{N} \tag{13-65}$$

This entropy was actually calculated from $\ln t_{\text{mp}}$, with t_{mp} the number of states in the most probable distribution. By assumption, the total number of states for energy E_i is approximately equal to the number of states in the most probable distribution; so we may say

$$t(E_i) = e^{S(E_i)/k} \approx e^{5N/2} \left(\frac{V}{N}\right)^N \left(\frac{4\pi m}{3Nh^2}\right)^{3N/2} E_i^{3N/2} \tag{13-66}$$

The constants are unimportant now; the important fact is that $t(E_i) \sim E_i^{3N/2}$. Thus the term in the canonical partition function for states of energy E_i, $t(E_i)e^{-E_i/kT}$, is proportional to

$$t(E_i)e^{-E_i/kT} \sim E_i^{3N/2} e^{-E_i/kT} \tag{13-67}$$

Since N is a very large number, the first term in E_i increases very rapidly with E_i and the second exponential term decreases very rapidly. The value of E_i for the maximum term in the partition function can be found from the condition

$$\frac{\partial \{\ln [t(E_i)e^{-E_i/kT}]\}}{\partial E_i} = 0 \quad \text{or} \quad \frac{3N}{2E_i} - \frac{1}{kT} = 0 \tag{13-68}$$

$$E_i = \tfrac{3}{2}NkT = \bar{E}$$

The maximum term is for an energy equal to $\tfrac{3}{2}NkT$, which we know to be the average energy of the system. Now let us see how sharply peaked the distribution is. Consider another term in the partition function for which E_i differs by a fractional amount α from $\bar{E}$, $E_i = (1 + \alpha)\bar{E}$. The ratio of the two terms

$$\frac{t(E_i)e^{-E_i/kT}}{t(\bar{E})e^{-\bar{E}/kT}}$$

is, by the distribution law, the ratio of probabilities $p(E_i)/p(\bar{E})$ of finding the system with energy E_i to the probability of finding that it has the average energy $\bar{E}$. Recalling that $\bar{E} = \tfrac{3}{2}NkT$,

$$\frac{p(E_i)}{p(\bar{E})} = \frac{t(E_i)}{t(\bar{E})} e^{-(E_i - \bar{E})/kT} = (1+\alpha)^{3N/2} e^{-\alpha \bar{E}/kT} = (1+\alpha)^{3N/2} e^{-\alpha 3N/2}$$

$$\ln \frac{p(E_i)}{p(\bar{E})} = \frac{3N}{2} \ln(1+\alpha) - \frac{3N}{2}\alpha$$

But, for small α, $\ln(1+\alpha) \approx \alpha - \alpha^2/2$; so

$$\ln \frac{p(E_i)}{p(\bar{E})} = -\frac{3N}{4}\alpha^2 \tag{13-69}$$

Suppose that $N = 10^{24}$; then, if $\alpha = 10^{-12}$, $p(E_i)/p(\bar{E}) \approx e^{-1}$, but if $\alpha = 10^{-10}$, $p(E_i)/p(\bar{E}) \approx e^{-10^4}$, which is ultranegligible.

Thus, we arrive at the view that the only states in this particular canonical ensemble which are significantly populated are those for which E_i is within 1 part in 10^{12} of the average energy $\bar{E}$ and the only terms in the partition function $Q = \sum_i t(E_i) e^{(-E_i/kT)}$ which need to be evaluated are those for which the energy is within the same small interval around $\bar{E}$.

For many practical problems, the difficulty is to enumerate all the configurations of the system for which the energy is close to $\bar{E}$. Since there may be on the order of $10^{10^{24}}$ such states, this enumeration is a formidable problem.

There is another specific remark that helps to emphasize the difference between the distribution law for the microcanonical ensemble and the distribution law for the canonical ensemble. For the former case, in the distribution law $p_i = e^{-\epsilon_i/kT}/q$, ϵ_i is a molecular energy; there are a large number of states for which ϵ_i is of the order of kT which are populated. For a canonical ensemble for a macroscopic system, $p_i = e^{-E_i/kT}/Q$. But E_i is a macroscopic energy, perhaps of the order of RT, and the exponential term $e^{-E_i/kT}$ is $10^{-2.6 \times 10^{23}}$, which is a very small number. If we pick E_i' to be greater than E_i by a factor of 1.000001, the ratio

$$\frac{e^{-E_i'/kT}}{e^{-E_i/kT}} = e^{-0.000001 E_i/kT} = 10^{-2.6 \times 10^{17}}$$

Thus, a very small fractional change in the energy produces a very large change in the exponential Boltzmann factor.

The entire discussion applies *mutatis mutandis* to the grand ensemble and the grand partition function. The distribution law $p_{Ni} = e^{(N\mu - E_{Ni})/kT}/\Xi$ is sharply peaked; only those values of N and E_{Ni} which are close to the ensemble average values, $\bar{N}$ and $\bar{E}$, represent states which are significantly populated. In the same way, in the expression for the grand partition function, $\sum_{N,i} e^{(N\mu - E_{Ni})/kT}$, the only important terms are for $N \approx \bar{N}$, $E_{Ni} \approx \bar{E}$.

The canonical ensemble is often taken as a model for a closed system in a thermostat, and the distribution law

$$p(E_i) = \frac{t(E_i) e^{-\beta E_i}}{Q} \tag{13-70}$$

may be used to calculate the probability that such a system will have an energy that is different from the most probable amount of energy.

Taken literally, the canonical ensemble is a rather unusual thermostat. We take one member of the ensemble as the system of interest. The rest of the ensemble then constitutes the thermostat. The analogy to an actual thermostat is readily improved by constructing an ensemble consisting of two different kinds of systems: the A-type systems, which are replicas of the system of interest; and B-type systems, each of which could be a large water bath or other thermostatic fluid. Energy can be exchanged between the A-type systems and the B-type systems. The distribution law for A-type systems is still given by (13-70).

In the same way, the grand ensemble is a model for an open system in a thermostat, and the distribution law enables us to calculate the probability of a fluctuation in the number of particles in a system as well as of a fluctuation in the energy.

13-9. Systems with More than One Component. In order to simplify both the presentation of the basic concepts and the notation, we have so far considered only systems with one component. There is no difficulty in generalizing to the case of several components. If, in a canonical ensemble, there are N_A particles of type A and N_B particles of type B, the energy levels are a function of both N_A and N_B and can be written as $E_{N_A N_B i}$. The partition function is $Q_{N_A N_B}(T,V) = \sum_i e^{-\beta E_{N_A N_B i}}$, and the calculation of thermodynamic properties proceeds just as before. For a grand ensemble, the number of type-A particles, N_A, and the number of type-B particles, N_B, are both variable. Let μ_A and μ_B be the respective chemical potentials of the two components. The grand partition function is

$$\Xi(\mu_A,\mu_B,T,V) = \sum_{N_A,N_B,i} e^{\beta(N_A\mu_A+N_B\mu_B-E_{N_A N_B i})}$$
$$= \sum_{N_A,N_B} e^{\beta(N_A\mu_A+N_B\mu_B)} Q_{N_A N_B} \qquad (13\text{-}71)$$

All the fundamental relations are deduced just as before. The relations

$$\bar{N}_A = kT\left[\frac{\partial(\ln \Xi)}{\partial \mu_A}\right]_{T,V,\mu_B}$$
$$\bar{N}_B = kT\left[\frac{\partial(\ln \Xi)}{\partial \mu_B}\right]_{T,V,\mu_A} \qquad (13\text{-}72)$$

are the generalization of (13-39).

13-10. Summary. The working relations for a canonical ensemble are

$$p_i = \frac{e^{-E_i/kT}}{Q} \qquad Q = \sum e^{-E_i/kT} \qquad (13\text{-}73)$$

or $\quad p_i = \dfrac{t(E_i)e^{-E_i/kT}}{Q} \qquad Q(T,V) = \sum t(E_i)e^{-E_i/kT} \qquad (13\text{-}74)$

where $t(E_i)$ is the degeneracy of the states with energy E_i and p_i is the probability of finding the system in the state i [(13-73)] or in the set of $t(E_i)$ states of energy E_i [(13-74)].

$$E = kT^2\left[\frac{\partial(\ln Q)}{\partial T}\right]_V \qquad (13\text{-}75)$$

$$A = -kT \ln Q \qquad (13\text{-}76)$$

$$P = kT\left[\frac{\partial(\ln Q)}{\partial V}\right]_T \qquad (13\text{-}77)$$

$$S = \frac{E-A}{T} = \frac{E}{T} + k \ln Q \qquad (13\text{-}78)$$

$$\mu = \left(\frac{\partial A}{\partial N}\right)_{V,T} \qquad (13\text{-}79)$$

Our summary of the relations for a grand ensemble is written for the case of a two-component system, with the notation of Sec. 13-9.

$$\Xi(\mu_A,\mu_B,T,V) = \sum_{N_A,N_B,i} e^{(N_A\mu_A+N_B\mu_B-E_{N_AN_B i})/kT} \qquad (13\text{-}80a)$$

$$\Xi(\mu_A,\mu_B,T,V) = \sum_{N_A,N_B} e^{(N_A\mu_A+N_B\mu_B)/kT} \sum_i e^{-E_{N_AN_B i}/kT}$$

$$= \sum_{N_A,N_B} e^{(N_A\mu_A+N_B\mu_B)/kT} Q_{N_AN_B}(T,V)$$

$$= \sum_{N_A,N_B} e^{(N_A\mu_A+N_B\mu_B)/kT} e^{-A_{N_AN_B}(T,V)/kT}$$

$$= \sum_{N_A,N_B} \lambda_A^{N_A} \lambda_B^{N_B} e^{-A_{N_AN_B}(T,V)/kT} \qquad (13\text{-}80b)$$

where $\ln \lambda_A = \mu_A/kT$ and λ_A is called the absolute activity of component A.

If it is desirable to group together a number $t_{N_AN_B}(E_i)$ of states with the same energy E_i (and of necessity with the same values for N_A and N_B), the formula for the grand partition function (13-80a) can be written as

$$\Xi = \sum_{N_A,N_B,i} t_{N_AN_B}(E_i) e^{(N_A\mu_A+N_B\mu_B-E_{N_AN_B i})/kT} \qquad (13\text{-}81)$$

The expression for $Q_{N_AN_B}$ can be written as

$$Q_{N_AN_B}(T,V) = \sum_i t_{N_AN_B}(E_i) e^{-E_{N_AN_B i}/kT} \qquad (13\text{-}82)$$

and the formulae (13-80b) are unaffected by the introduction of the concept of degenerate states.

The probability of finding that a system in the ensemble has N_A and N_B particles and is in the state i is

$$p_{N_AN_B i} = \frac{e^{(N_A\mu_A+N_B\mu_B-E_{N_AN_B i})/kT}}{\Xi} \qquad (13\text{-}83a)$$

or

$$p_{N_AN_B i} = \frac{t_{N_AN_B}(E_i) e^{(N_A\mu_A+N_B\mu_B-E_{N_AN_B i})/kT}}{\Xi} \qquad (13\text{-}83b)$$

The probability of finding a system with N_A and N_B particles, irrespective of energy, is

$$p_{N_AN_B} = \frac{e^{(N_A\mu_A+N_B\mu_B)/kT} Q_{N_AN_B}}{\Xi} \qquad (13\text{-}84)$$

The other important relations are

$$\bar{E} = kT^2 \left\{ \frac{\partial[\ln \Xi(\mu_A,\mu_B,T,V)]}{\partial T} \right\}_{\mu_A/T,\mu_B/T,V} \qquad (13\text{-}85)$$

$$\bar{P} = kT \left[\frac{\partial(\ln \Xi)}{\partial V} \right]_{\mu_A,\mu_B,T} \qquad (13\text{-}86)$$

$$\bar{N}_A = kT \left[\frac{\partial(\ln \Xi)}{\partial \mu_A} \right]_{\mu_B,T,V} \qquad (13\text{-}87)$$

and, similarly, for $\bar{N}_B$

$$\bar{P}V = kT \ln \Xi \tag{13-88}$$

$$\bar{S} = \frac{\bar{E}}{T} + k \ln \Xi - \frac{N_A \mu_A + N_B \mu_B}{T} \tag{13-89}$$

$$A = \bar{E} - T\bar{S} = N_A \mu_A + N_B \mu_B - kT \ln \Xi \tag{13-90}$$

We shall see in the following chapters that the grand partition function is very useful for the treatment of statistical mechanical problems of systems of interacting particles. One difficulty which arises is that the results are usually directly obtained in terms of the unfamiliar thermodynamic variables μ, T, V. There is then the additional problem of transforming the results so that they are expressed in terms of more familiar variables such as $\bar{N}$, T, V or $\bar{N}$, T, $\bar{P}$.

PROBLEMS

13-1. (a) Consider a one-component system with N_1 molecules of type A in volume V. The concentration of molecules is $N_2 = N_1/V$. Consider a second system with N_2 molecules of type A in unit volume. What is the relation between the canonical partition functions $Q_{N_1}(T,V)$ and $Q_{N_2}(T,1)$, of the two systems? (*Hint:* What function, related to Q, is an extensive function?)

(b) In the same way, what is the relation between the grand partition functions $\Xi(\mu,T,V_1)$ and $\Xi(\mu,T,V_2)$ for two one-component systems which have the same chemical potential and the same temperature, but different volumes V_1 and V_2?

13-2. In Sec. 13-8, an argument is presented to show that, for a perfect gas with N particles in the canonical ensemble, the most probable value of the energy is the average energy $\bar{E} = \frac{3}{2}NkT$ and that the probability of finding a system with energy E_i that differs from $\bar{E}$ by the fractional amount α $[E_i = \bar{E}(1 + \alpha)]$ is given by Eq. (13-69).

Carry out a similar analysis for a perfect gas in the grand ensemble with activity λ, temperature T, and volume V, to find the relation between the most probable number of particles and the ensemble average number of particles, $\bar{N}$. Find an expression for $\ln (p_N/p_{\bar{N}})$, where p_N is the probability that the system has N particles and N differs from $\bar{N}$ by the fractional amount α $[N = \bar{N}(1 + \alpha)]$.

13-3. This problem is a very simple illustration of the application of the grand partition function in adsorption problems or in titration problems for polyelectrolytes. Let each system in the grand ensemble consist of four distinguishable adsorption sites 1, 2, 3, and 4 in the shape of a square. There is a gas composed of atoms, A. The energy of adsorption at a site is $-\epsilon$. There is an energy of interaction between atoms adsorbed on neighboring sites of ω. Thus, the enumeration of the possible states of the system begins as follows:

One atom adsorbed at sites 1, 2, 3, or 4; energy $= -\epsilon$.

Two atoms adsorbed at neighboring sites, such as (1,2) (2,3), etc.; energy $= -2\epsilon + \omega$.

Two atoms adsorbed at opposite sites, such as (1,3), etc.; energy $= -2\epsilon$, and so on, for three atoms adsorbed or four atoms adsorbed.

The activity λ of the gas is prescribed by the partial pressure. The temperature is T.

Set up the grand partition function for this sytem. Use the general statistical thermodynamic formulae to calculate the average number of particles adsorbed at any given λ.

(The calculation is obviously also applicable to the addition of protons to the anion of a tetrabasic acid, with the pH of the solution prescribing the activity of the hydrogen ions.)

13-4. It is important in the study of a purely theoretical chapter, such as the present one, to be able to carry out the derivations in detail.

Without consulting your notes or the text, start with the equation

$$ t = \prod_N \prod_i \frac{n!}{n_{Ni}!} $$

and the constraints of a fixed total energy, a fixed total number of particles, and a fixed number of systems in the grand ensemble, and derive the distribution law for the grand ensemble in its usual form. Then derive the relations between $\bar{P}$, $\bar{E}$, and $\bar{N}$ and the grand partition function. Accepting the statistical mechanical definition of the average entropy per system, and for convenience accepting the identification $\beta = 1/kT$, derive the fundamental equations for A, μ, and PV in the theory of the grand ensemble.

13-5. In the theory of the grand ensemble, what are the statistical mechanical equivalents of the following thermodynamic equations?

$$ \left[\frac{\partial E}{\partial(\mu/T)}\right]_{T,V} = T^2 \left(\frac{\partial N}{\partial T}\right)_{V,\mu/T} $$

$$ \left(\frac{\partial P}{\partial N}\right)_{T,V} = -\left(\frac{\partial \mu}{\partial V}\right)_{T,N} $$

Prove these relations by direct manipulation of the statistical mechanical equations for $\bar{E}$, $\bar{P}$, $\bar{N}$, μ, etc.

14

Fluctuations, Noise, Brownian Motion

14-1. Introduction. A basic assumption in the theory developed in the preceding chapter is that if we make a large number of measurements of the value of some physical quantity—energy, density, composition, pressure, light absorption, electrode potential—on a macroscopic system, the average of these measurements can be identified with the ensemble average for this property for an appropriately constructed ensemble. The theory provides definite expressions for the probability of observing a value for a given physical quantity which is different from the average value. This is the theory of fluctuations at equilibrium.

It is a general and typical result in statistical studies of populations, whether one is considering public-opinion polls, or the number of disintegrations of a radioactive sample, or the fluctuations in density of a fluid, that the magnitude of the fractional fluctuations decreases as the size of the sample increases. Typically, the magnitude of the fractional fluctuation is proportional to $(1/\bar{N})^{1/2}$, where $\bar{N}$ is the number of particles in the sample. It will be seen that this relation holds for the fluctuations that we study.

In some cases, the main result of fluctuation theory is the comforting conclusion that the equilibrium fluctuations are so small as to be practically unmeasurable. There are other cases, most notably for light scattering, Brownian motion, and voltage noise, where the fluctuations are measurable, and the measurements provide useful information.

One of the important modern topics in fluctuation theory is the study of the frequency spectrum—or the time dependence—of a randomly fluctuating physical quantity. This is not strictly a topic in equilibrium statistical mechanics. Nevertheless, because of the practical importance and the interest of the subject, we present a brief and elementary introduction.

The chapter concludes with a discussion of the justification for using the number of states in the most probable distribution rather than the total number of states available to the system in calculating the entropy.

14-2. The Mean of a Distribution and the Mean-square Deviation.

If p_i is the probability of finding a system in the state i, and if F_i is the value of a physical quantity when the system is in this state, then the average value of F, which we denote by $\bar{F}$ or $\langle F \rangle$, is defined by

$$\bar{F} = \sum_i p_i F_i \tag{14-1}$$

(In this particular chapter, we shall ordinarily use $\bar{F}$ to represent an ensemble average and $\langle F \rangle$ for a time average for a particular system.) Recall that $\Sigma p_i = 1$. The average deviation from the mean $\overline{F - \bar{F}}$ is defined by $\overline{F - \bar{F}} = \Sigma p_i (F_i - \bar{F})$; it was shown in Sec. 5-4 that this quantity is identically zero.

$$\overline{F - \bar{F}} = \Sigma p_i (F_i - \bar{F}) = \Sigma p_i F_i - \bar{F} \Sigma p_i = \bar{F} - \bar{F} = 0 \tag{14-2}$$

A significant measure of the sharpness of the distribution is the mean-square deviation, or variance,

$$\begin{aligned}\overline{(F - \bar{F})^2} &= \Sigma p_i (F_i - \bar{F})^2 = \Sigma p_i F_i^2 - 2\bar{F} \Sigma p_i F_i + \bar{F}^2 \Sigma p_i \\ &= \overline{F^2} - 2\bar{F}^2 + \bar{F}^2 = \overline{F^2} - \bar{F}^2 \end{aligned} \tag{14-3a}$$

The final result in (14-3a) (which has already been derived in Sec. 5-4) is important: the mean-square deviation is the mean of the square minus the square of the mean. It will be convenient to write, for the mean-square deviation, $\overline{\delta F^2}$; so (14-3a) may be written as

$$\overline{\delta F^2} = \overline{(F - \bar{F})^2} = \overline{F^2} - \bar{F}^2 \tag{14-3b}$$

14-3. Fluctuation in Energy in a Canonical Ensemble.

For a canonical ensemble, the probability of finding the system in the state i is $p_i = e^{-\beta E_i}/Q$. We wish to calculate

$$\overline{\delta E^2} = \overline{E^2} - \bar{E}^2 = \Sigma p_i (E_i - \bar{E})^2$$

Recall that $\bar{E} = -(1/Q)(\partial Q/\partial \beta)_V$, so that $(\partial Q/\partial \beta)_V = -\bar{E}Q$. Then

$$\begin{aligned}\overline{E^2} &= \sum p_i E_i^2 = \frac{\Sigma E_i^2 e^{-\beta E_i}}{Q} = \frac{1}{Q} \frac{\partial^2 Q}{\partial \beta^2} = \frac{1}{Q} \frac{\partial}{\partial \beta} \left(\frac{\partial Q}{\partial \beta} \right) \\ &= \frac{1}{Q} \frac{\partial}{\partial \beta} (-\bar{E}Q) = -\frac{\bar{E}}{Q} \left(\frac{\partial Q}{\partial \beta} \right) - \frac{\partial \bar{E}}{\partial \beta} = \bar{E}^2 - \frac{\partial \bar{E}}{\partial \beta} \end{aligned} \tag{14-4a}$$

But

$$-\frac{\partial \bar{E}}{\partial \beta} = kT^2 \left(\frac{\partial \bar{E}}{\partial T} \right)_V = kT^2 C_v$$

where C_v is the heat capacity. Then, from (14-4a),

$$\overline{\delta E^2} = \overline{E^2} - \bar{E}^2 = kT^2 C_v \tag{14-4b}$$

so that

$$\left(\frac{\overline{\delta E^2}}{\bar{E}^2} \right)^{1/2} = \left(\frac{kT^2 C_v}{\bar{E}^2} \right)^{1/2} \tag{14-5}$$

The quantity $(\overline{\delta E^2}/\bar{E}^2)^{1/2}$ in (14-5) is a reasonable measure of the fractional fluctuation in energy. For a perfect monatomic gas, $\bar{E} = \tfrac{3}{2}NkT$, and $C_v = \tfrac{3}{2}Nk$, so

$$\left(\frac{\overline{\delta E^2}}{\bar{E}^2}\right)^{1/2} = \frac{2}{3N^{1/2}} \tag{14-6}$$

The fractional fluctuation in energy is of the order of the reciprocal of the square root of the number of particles. The order of magnitude of the energy fluctuation should be about the same for many other systems which are not perfect gases but which have comparable energies and heat capacities per particle.

Problem 14-1. For an arbitrary system, how does the fractional fluctuation in energy $(\overline{\delta E^2}/\bar{E}^2)^{1/2}$ vary with the number of particles in the system at constant T and constant particle density N/V?

14-4. Density Fluctuations in the Grand Ensemble. One-component System. The probability of observing N particles in a system in a grand ensemble is

$$p_N = \frac{e^{\beta N \mu} Q_N(T,V)}{\Xi} \qquad \text{where} \quad \Xi(\mu,T,V) = \sum_N e^{\beta N \mu} Q_N(T,V)$$

and

$$\bar{N} = \sum_N N p_N = \frac{1}{\beta \Xi}\left(\frac{\partial \Xi}{\partial \mu}\right)_{T,V} \tag{13-39}$$

so that

$$\left(\frac{\partial \Xi}{\partial \mu}\right)_{T,V} = \beta \bar{N} \Xi$$

Then

$$\overline{N^2} = \sum_N N^2 p_N = \frac{\Sigma N^2 e^{\beta N \mu} Q_N(T,V)}{\Xi} = \frac{1}{\beta^2 \Xi}\left(\frac{\partial^2 \Xi}{\partial \mu^2}\right)_{T,V}$$

$$= \frac{1}{\beta^2 \Xi} \frac{\partial}{\partial \mu}\left(\frac{\partial \Xi}{\partial \mu}\right) = \frac{1}{\beta^2 \Xi}\left[\frac{\partial(\beta \Xi \bar{N})}{\partial \mu}\right]_{T,V}$$

$$= \frac{\bar{N}}{\beta \Xi}\left(\frac{\partial \Xi}{\partial \mu}\right)_{T,V} + \frac{1}{\beta}\left(\frac{\partial \bar{N}}{\partial \mu}\right)_{T,V}$$

$$= \bar{N}^2 + kT\left(\frac{\partial \bar{N}}{\partial \mu}\right)_{T,V}$$

Then

$$\overline{\delta N^2} = \overline{N^2} - \bar{N}^2 = kT\left(\frac{\partial \bar{N}}{\partial \mu}\right)_{T,V} \tag{14-7}$$

Let ρ be the density. For a grand ensemble, $\rho \sim N$, since the systems are of constant volume.

$$\frac{\overline{\delta \rho^2}}{\bar{\rho}^2} = \frac{\overline{\delta N^2}}{\bar{N}^2} = \frac{kT}{\bar{N}^2 (\partial \mu/\partial \bar{N})_{T,V}} \tag{14-8}$$

Fundamentally, Eqs. (14-7) and (14-8) are acceptable solutions to the problem of calculating the density fluctuations in a grand ensemble.

However, it is desirable to express $(\partial \mu/\partial \bar{N})_{T,V}$ in terms of more familiar thermodynamic quantities for a closed system.

Consider any intensive property y as a function of T, V, and $\bar{N}$; $y = y(T,V,\bar{N})$. The intensive quantity y is a function of the intensive variable T and the two extensive variables V and $\bar{N}$. Therefore, by Euler's theorem,

$$V\left(\frac{\partial y}{\partial V}\right)_{T,\bar{N}} + \bar{N}\left(\frac{\partial y}{\partial \bar{N}}\right)_{T,V} = 0$$

or

$$\left(\frac{\partial y}{\partial \bar{N}}\right)_{T,V} = -\frac{V}{\bar{N}}\left(\frac{\partial y}{\partial V}\right)_{T,\bar{N}} \tag{14-9}$$

For the purpose at hand, we take y as the chemical potential μ and as the pressure $\bar{P}$. Then

$$\left(\frac{\partial \mu}{\partial \bar{N}}\right)_{T,V} = -\frac{V}{\bar{N}}\left(\frac{\partial \mu}{\partial V}\right)_{T,\bar{N}} \tag{14-10a}$$

$$\left(\frac{\partial \bar{P}}{\partial \bar{N}}\right)_{T,V} = -\frac{V}{\bar{N}}\left(\frac{\partial \bar{P}}{\partial V}\right)_{T,\bar{N}} \tag{14-10b}$$

For a one-component system, $dA = -\bar{S}\,dT - \bar{P}\,dV + \mu\,d\bar{N}$; and by cross differentiation

$$\left(\frac{\partial \mu}{\partial V}\right)_{T,\bar{N}} = -\left(\frac{\partial \bar{P}}{\partial \bar{N}}\right)_{T,V} = \frac{V}{\bar{N}}\left(\frac{\partial \bar{P}}{\partial V}\right)_{T,\bar{N}}$$

where the second equality follows from (14-10b). With these relations for evaluating $(\partial \mu/\partial \bar{N})$, and recalling that the compressibility is defined by $\kappa = (-1/V)(\partial V/\partial \bar{P})_{T,\bar{N}}$, we obtain from (14-8)

$$\frac{\overline{\delta\rho^2}}{\bar{\rho}^2} = \frac{\overline{\delta N^2}}{\bar{N}^2} = \frac{kT\kappa}{V} \tag{14-11}$$

This is the desired general result.

In the two-phase region for a liquid-vapor equilibrium, or at the critical point, the compressibility is infinite. The fluctuations in density are indeed large and observable. However, expression (14-10) is not correct for this case, and the fluctuations are not infinite.*

Problem 14-2. Evaluate the rms fractional fluctuation in density for a sample of a perfect gas. Comment on the meaning of the result.

14-5. Energy Fluctuations in the Grand Ensemble. For this calculation, it is expedient to write the relations for the grand ensemble in terms of the activity $\lambda = e^{\mu/kT}$, rather than μ. Recall that

---

* Cf. Hirschfelder et al., pp. 125, 373 [15]. Our treatment is not applicable, because our evaluation of $(\partial \mu/\partial \bar{N})_{T,V}$ uses Euler's theorem and therefore neglects surface tension and boundary effects.

Sec. 14-5] FLUCTUATIONS, NOISE, BROWNIAN MOTION

$$p_{Ni} = \frac{\lambda^N e^{-\beta E_{Ni}}}{\Xi} \qquad \Xi(\lambda,T,V) = \sum_{N,i} \lambda^N e^{-\beta E_{Ni}}$$

$$\bar{E} = \sum_{N,i} p_{Ni} E_{Ni} = -\frac{1}{\Xi}\left(\frac{\partial \Xi}{\partial \beta}\right)_{\lambda,V}$$

$$\left(\frac{\partial \Xi}{\partial \beta}\right)_{\lambda,V} = -\bar{E}\Xi$$

We then have

$$\overline{E^2} = \sum_{N,i} E_{Ni}^2 p_{Ni} = \frac{\Sigma E_{Ni}^2 \lambda^N e^{-\beta E_{Ni}}}{\Xi}$$

$$= \frac{1}{\Xi}\frac{\partial}{\partial \beta}\left(\frac{\partial \Xi}{\partial \beta}\right)_{\lambda,V} \tag{14-12a}$$

The advantage of using λ as a variable rather than μ is that only the second factor in the product $\lambda^N e^{-\beta E_{Ni}}$ varies with β at constant λ, whereas both factors in the term $e^{\beta N \mu} e^{-\beta E_{Ni}}$ are functions of β at constant μ.

Proceeding from (14-12a),

$$\overline{E^2} = \frac{1}{\Xi}\frac{\partial}{\partial \beta}(-\bar{E}\Xi)_{\lambda,V} = -\left(\frac{\partial \bar{E}}{\partial \beta}\right)_{\lambda,V} + \bar{E}^2$$

and so
$$\overline{\delta E^2} = kT^2 \left(\frac{\partial \bar{E}}{\partial T}\right)_{\lambda,V} \tag{14-12b}$$

The problem now is to express $(\partial \bar{E}/\partial T)_{\lambda,V}$ in more familiar terms. We are accustomed to expressing the thermodynamic energy function in terms of T, V, and $\bar{N}$; $\bar{E} = \bar{E}(T,V,\bar{N})$, but we are unaccustomed to expressing it as a function of T, λ, V. Then

$$\left(\frac{\partial \bar{E}}{\partial T}\right)_{V,\lambda} = \left(\frac{\partial \bar{E}}{\partial T}\right)_{V,\bar{N}} + \left(\frac{\partial \bar{E}}{\partial \bar{N}}\right)_{T,V}\left(\frac{\partial \bar{N}}{\partial T}\right)_{\lambda,V}$$

$$= C_v + \left(\frac{\partial \bar{E}}{\partial \bar{N}}\right)_{T,V}\left(\frac{\partial \bar{N}}{\partial T}\right)_{\lambda,V} \tag{14-13}$$

To evaluate $(\partial \bar{N}/\partial T)_{\lambda,V}$, we first remark that, when λ is constant, μ/T is constant, and we can just as well write $(\partial \bar{N}/\partial T)_{\mu/T,V}$. Now μ/T can be a function of T, V, and $\bar{N}$; $\mu/T = f(T,V,\bar{N})$. Therefore, at constant volume

$$d\left(\frac{\mu}{T}\right) = \left[\frac{\partial(\mu/T)}{\partial T}\right]_{V,\bar{N}} dT + \left[\frac{\partial(\mu/T)}{\partial \bar{N}}\right]_{T,V} d\bar{N}$$

and so
$$\left(\frac{\partial \bar{N}}{\partial T}\right)_{\mu/T,V} = \frac{-\left[\frac{\partial(\mu/T)}{\partial T}\right]_{\bar{N},V}}{\left[\frac{\partial(\mu/T)}{\partial \bar{N}}\right]_{T,V}} = -\left[\frac{\partial(\mu/T)}{\partial T}\right]_{\bar{N},V} T\left(\frac{\partial \bar{N}}{\partial \mu}\right)_{T,V}$$

Recall now the thermodynamic relation

$$d\left(\frac{A}{T}\right) = -\frac{\bar{E}}{T^2}dT + \frac{\bar{P}}{T}dV + \frac{\mu}{T}d\bar{N}$$

so that, by cross differentiation,

$$\left[\frac{\partial(\mu/T)}{\partial T}\right]_{\bar{N},V} = -\frac{1}{T^2}\left(\frac{\partial\bar{E}}{\partial\bar{N}}\right)_{T,V}$$

When all these relations are collected, Eq. (14-13) becomes

$$\left(\frac{\partial\bar{E}}{\partial T}\right)_{V,\lambda} = C_v + \frac{1}{T}\left(\frac{\partial\bar{E}}{\partial\bar{N}}\right)^2_{T,V}\left(\frac{\partial\bar{N}}{\partial\mu}\right)_{T,V} \qquad (14\text{-}14)$$

Recall that, in the grand ensemble, $\overline{\delta N^2} = kT(\partial\bar{N}/\partial\mu)_{T,V}$ [(14-7)]; then from (14-12b) and using (14-14),

$$\overline{\delta E^2} = kT^2 C_v + \left(\frac{\partial\bar{E}}{\partial\bar{N}}\right)^2_{T,V}\overline{\delta N^2} \qquad (14\text{-}15)$$

This is the desired result. It is gratifying to observe that the fluctuation in energy in a grand ensemble is the sum of two terms: the first is the fluctuation in energy in the canonical ensemble, i.e., with the number of particles fixed; and the second is the fluctuation in energy associated with the fluctuation in the number of particles.

For practical computations, one can evaluate $(\partial\bar{E}/\partial\bar{N})_{T,V}$, using Euler's theorem

$$\bar{E} = \bar{E}(T,V,\bar{N}) = V\left(\frac{\partial\bar{E}}{\partial V}\right)_{T,\bar{N}} + \bar{N}\left(\frac{\partial\bar{E}}{\partial\bar{N}}\right)_{T,V}$$

$$\left(\frac{\partial\bar{E}}{\partial\bar{N}}\right)_{T,V} = \frac{\bar{E} - V(\partial\bar{E}/\partial V)_{T,\bar{N}}}{\bar{N}} = \frac{\bar{E} + \bar{P}V - TV(\partial\bar{P}/\partial T)_{V,\bar{N}}}{\bar{N}}$$

$$(14\text{-}16)$$

14-6. Another Method for Fluctuation Problems. There is another general method of treating fluctuation problems, which works for more difficult problems and which is needed in Sec. 14-7. We illustrate it here for the problem of the density fluctuations in a one-component system in a grand ensemble which was treated in Sec. 14-4.

The probability of finding a system with N particles is given by

$$p_N = \frac{e^{\beta N\mu}e^{-\beta A(N,T,V)}}{\Xi}$$

where μ is the chemical potential of a system which has the ensemble average $\bar{N}$ particles at T and V, that is, $\mu = \mu(\bar{N},T,V)$, and A is the free energy of a system represented by a canonical ensemble which has N particles at T and V. For what value of N, $\bar{N}$ is p_N a maximum? It will

Sec. 14-6]   FLUCTUATIONS, NOISE, BROWNIAN MOTION

prove to be true that to a high degree of approximation $\tilde{N} = \bar{N}$, but this is not immediately obvious. We recall, for example, that in the Maxwell-Boltzmann velocity distribution the most probable velocity and the average velocity are different. We start with

$$\ln p_N = \beta N \mu - \beta A(N,T,V) - \ln \Xi \qquad (14\text{-}17)$$

The condition for a maximum is that $\partial(\ln p_N)/\partial N = 0$ when $N = \tilde{N}$.

$$\frac{\partial(\ln p_N)}{\partial N} = \beta \mu(\tilde{N},T,V) - \beta \left[\frac{\partial A(\tilde{N},T,V)}{\partial N}\right]_{T,V} = 0 \qquad (14\text{-}18)$$

Note that, by $\partial A(\tilde{N},T,V)/\partial N$ we mean $\partial A(N,T,V)/\partial N$, evaluated at $N = \tilde{N}$. But $\partial A(N,T,V)/\partial N$ is the chemical potential of a system in a canonical ensemble with N particles. It seems reasonable to assume that the chemical potential in the grand ensemble and in the canonical ensemble are the same when the number of particles in the canonical ensemble is the same as the ensemble average number of particles in the grand ensemble. Then, the solution to Eq. (14-18) is $\bar{N} = \tilde{N}$. In order to simplify our notation, we shall write $\mu(\tilde{N},T,V)$ for $\partial A(N,T,V)/\partial N$ evaluated at $N = \tilde{N}$; but we shall continue to distinguish between $\bar{N}$ and $\tilde{N}$ since an independent proof that they are equal will emerge from our arguments.

A function $f(x)$ can be expanded around its maximum value $f(\tilde{x})$ by $f(x) = f(\tilde{x}) + \frac{1}{2}[\partial^2 f(\tilde{x})/\partial x^2](x - \tilde{x})^2$. The notation $\partial^2 f(\tilde{x})/\partial x^2$ means the second derivative evaluated at $x = \tilde{x}$. Now

$$\frac{\partial^2(\ln p_N)}{\partial N^2} = -\beta \left[\frac{\partial^2 A(\tilde{N},T,V)}{\partial N^2}\right]_{T,V} = -\beta \left[\frac{\partial \mu(\tilde{N},T,V)}{\partial N}\right]_{T,V} \qquad (14\text{-}19)$$

and so by the Taylor expansion

$$\ln p_N \approx \ln p_{\tilde{N}} - \frac{\beta}{2}\left[\frac{\partial \mu(\tilde{N},T,V)}{\partial N}\right]_{T,V}(N - \tilde{N})^2 \qquad (14\text{-}20)$$

or

$$p_N = C p_{\tilde{N}} e^{-(\beta/2)[\partial\mu(\tilde{N},T,V)/\partial N]_{T,V}(N-\tilde{N})^2} \qquad (14\text{-}21)$$

The constant C is a renormalization constant which is chosen so that $\int p_N\, dN = 1$ and which corrects for the approximation in the series expansion (14-20). We shall not have to evaluate C, but it is important to emphasize that it need not be identically unity. Equation (14-21) shows that the distribution p_N is a Gaussian distribution around the maximum, $\tilde{N}$. The relations for a normalized Gaussian distribution are

$$p(x) = \left(\frac{1}{2\pi\sigma^2}\right)^{1/2} e^{-x^2/2\sigma^2} \qquad (14\text{-}22a)$$

$$\bar{x} = \int_{-\infty}^{+\infty} x p(x)\, dx = 0 \qquad (14\text{-}22b)$$

$$\overline{x^2} = \int_{-\infty}^{+\infty} x^2 p(x)\, dx = \sigma^2 \qquad (14\text{-}22c)$$

Therefore, from (14-21),

$$\overline{\bar{N} - \tilde{N}} = \overline{N - \bar{N}} = \int (N - \tilde{N}) p_N \, dN = 0 \qquad (14\text{-}23)$$

or
$$\bar{N} = \tilde{N}$$

which again justifies our previous assertion.

Furthermore,

$$\overline{(N - \bar{N})^2} = \frac{kT}{[\partial \mu(\bar{N}, T, V)/\partial N]_{T,V}} \qquad (14\text{-}24)$$

which is the same result as Eq. (14-7). Thus the expansion method illustrated in the present section gives the same result as that obtained previously by the straightforward evaluation of the sum for $\overline{(N - \bar{N})^2}$.

14-7. Concentration Fluctuations. Light Scattering. The results of the previous sections show that, except at the critical point, the density fluctuations in a macroscopic fluid are extremely small and hence can be observed only by a very sensitive technique. Light scattering is just such a sensitive method. It is caused by density fluctuations in a one-component system and by density and composition fluctuations in a multicomponent system. We present here a simplified version for two components of the general treatment for a multicomponent system due to Kirkwood and Goldberg.*

Our object is to outline the rigorous treatment of the density and composition fluctuations. For a more complete discussion of the actual phenomena of light scattering, the reader is referred to specialized articles.†

For a two-component system, denote the number of particles of types 1 and 2 by N_1 and N_2 and their chemical potentials by μ_1 and μ_2. In this notation, Eq. (13-84) is

$$p(N_1, N_2) = \frac{e^{\beta N_1 \mu_1} e^{\beta N_2 \mu_2} Q(N_1, N_2, T, V)}{\Xi} \qquad (14\text{-}25)$$

By using essentially the same procedure as in Sec. 14-4, it is easy to show that

$$\overline{\delta N_1^2} = kT \left(\frac{\partial N_1}{\partial \mu_1} \right)_{T,V,\mu_2} \qquad (14\text{-}26)$$

and

$$\overline{\delta N_1 \, \delta N_2} = \overline{(N_1 - \bar{N}_1)(N_2 - \bar{N}_2)} = kT \left(\frac{\partial \bar{N}_2}{\partial \mu_1} \right)_{\mu_2, T, V} = kT \left(\frac{\partial \bar{N}_1}{\partial \mu_2} \right)_{\mu_1, T, V} \qquad (14\text{-}27)$$

---

*J. G. Kirkwood and R. Goldberg, *J. Chem. Phys.*, **18**: 54 (1950). This important paper has been admirably restated, with many details filled in, by Hill, pp. 113–121 [9a].

† G. Oster, *Chem. Rev.*, **43**: 319 (1948); P. Doty and J. Edsall, *Advances in Protein Chem.*, **6**: 35 (1951); P. W. Allen (ed.), "Techniques of Polymer Characterization," chap. 5, by F. W. Peaker, Butterworth, London, 1959.

Sec. 14-7]   FLUCTUATIONS, NOISE, BROWNIAN MOTION

Note that the fluctuations in the concentrations of the two components are not uncorrelated; that is, $\overline{\delta N_1\, \delta N_2} \neq 0$, if the cross derivatives $\partial N_2/\partial \mu_1 = \partial N_1/\partial \mu_2$ are not zero. It is rather difficult to transform the thermodynamic derivatives, such as $(\partial \bar{N}_1/\partial \mu_1)_{T,V,\mu_2}$, which occur in Eqs. (14-26) and (14-27), into more familiar quantities because it is difficult to use the chemical potentials as independent variables.

It is expedient to calculate the fluctuations by using the Taylor-expansion–Gaussian-approximation method of Sec. 14-6. We therefore begin again and write (14-25) as

$$\ln p(N_1, N_2) = \beta[N_1\mu_1(\bar{N}_1,\bar{N}_2,T,V) + N_2\mu_2(\bar{N}_1,\bar{N}_2,T,V) - A(N_1,N_2,T,V)] \\ - \ln \Xi(\bar{N}_1,\bar{N}_2,T,V) \quad (14\text{-}28)$$

The notation in Eq. (14-28) reminds us that the chemical potentials μ_1 and μ_2 are parameters which characterize the grand ensemble and are functions of the ensemble average numbers of particles $\bar{N}_1$ and $\bar{N}_2$, whereas $A(N_1,N_2,T,V)$ is a function of the variables N_1 and N_2 which characterize different systems in the grand ensemble. Then

$$\frac{\partial (\ln p)}{\partial N_1} = \beta \mu_1(\bar{N}_1,\bar{N}_2,T,V) - \beta \left[\frac{\partial A(N_1,N_2,T,V)}{\partial N_1} \right]_{N_2,T,V}$$

and, of course, $\partial (\ln p)/\partial N_1 = 0$ and $\partial (\ln p)/\partial N_2 = 0$ at the point $N_1 = \bar{N}_1$, $N_2 = \bar{N}_2$.

Furthermore,

$$\frac{\partial^2 (\ln p)}{\partial N_1^2} = -\beta \frac{\partial^2 A(N_1,N_2,T,V)}{\partial N_1^2}$$

$$\frac{\partial^2 (\ln p)}{\partial N_1 \partial N_2} = -\beta \frac{\partial^2 A(N_1,N_2,T,V)}{\partial N_1 \partial N_2}$$

Let $\delta N_1 = N_1 - \bar{N}_1$, $\delta N_2 = N_2 - \bar{N}_2$. The Taylor expansion for $\ln p(N_1,N_2)$ is then

$$\ln p(N_1 N_2) = \ln p(\bar{N}_1,\bar{N}_2) - \frac{\beta}{2}\left[\frac{\partial^2 A(\bar{N}_1,\bar{N}_2,T,V)}{\partial N_1^2} \delta N_1^2 \right. \\ \left. + 2\frac{\partial^2 A(\bar{N}_1,\bar{N}_2,T,V)}{\partial N_1 \partial N_2} \delta N_1\, \delta N_2 + \frac{\partial^2 A(\bar{N}_1,\bar{N}_2,T,V)}{\partial N_2^2} \delta N_2^2 \right] \quad (14\text{-}29)$$

In the important second term on the right-hand side of (14-29),

$$\delta N_1 = N_1 - \bar{N}_1$$

is a variable, but $\partial^2 A(\bar{N}_1,\bar{N}_2,T,V)/\partial N_1^2$ means the value of $\partial^2 A/\partial N_1^2$ evaluated at the point $N_1 = \bar{N}_1$, $N_2 = \bar{N}_2$. We write

$$\left[\frac{\partial^2 A(N_1,N_2,T,V)}{\partial N_1^2} \right]_{N_2,T,V} = \left[\frac{\partial \mu_1(N_1,N_2,T,V)}{\partial N_1} \right]_{N_2,T,V}$$

$$\frac{\partial^2 A(N_1,N_2,T,V)}{\partial N_1 \partial N_2} = \left[\frac{\partial \mu_1(N_1,N_2,T,V)}{\partial N_2} \right]_{N_1,T,V} = \left[\frac{\partial \mu_2(N_1,N_2,T,V)}{\partial N_1} \right]_{N_2,T,V}$$

$$(14\text{-}30)$$

where μ_1 and μ_2 are the chemical potentials in the canonical ensemble for N_1 and N_2 particles. However, we identify the derivatives such as $\partial \mu_1 / \partial N_1$ in Eq. (14-30) at $N_1 = \bar{N}_1$, $N_2 = \bar{N}_2$ with the derivative

$$\left[\frac{\partial \mu_1(\bar{N}_1, \bar{N}_2, T, V)}{\partial \bar{N}_1} \right]_{\bar{N}_2, T, V}$$

of the chemical potential of the grand ensemble with $\bar{N}_1$ and $\bar{N}_2$ particles.

We then write

$$\ln \frac{p(N_1, N_2)}{p(\bar{N}_1, \bar{N}_2)} \approx -\frac{\beta}{2} \left[\frac{\partial \mu_1}{\partial \bar{N}_1} \delta N_1^2 + 2 \frac{\partial \mu_2}{\partial \bar{N}_1} \delta N_1 \, \delta N_2 + \frac{\partial \mu_2}{\partial \bar{N}_2} \delta N_2^2 \right] \quad (14\text{-}31a)$$

$$p(N_1, N_2) = C p(\bar{N}_1, \bar{N}_2) e^{-\frac{\beta}{2} \left[\frac{\partial \mu_1}{\partial \bar{N}_1} \delta N_1^2 + 2 \frac{\partial \mu_2}{\partial \bar{N}_1} \delta N_1 \, \delta N_2 + \frac{\partial \mu_2}{\partial \bar{N}_2} \delta N_2^2 \right]} \quad (14\text{-}31b)$$

where C is a normalizing constant which we need not evaluate.

Notice that by the device of using the Gaussian expansion we have expressed the fluctuations in terms of quantities such as $[\partial \mu_1(N_1, N_2, T, V) / \partial N_2]_{N_2, T, V}$, which are much more tractable than functions such as $(\partial N_1 / \partial \mu_1)_{\mu_2, T, V}$ which occur in Eqs. (14-26) and (14-27).

The problem now is to convert the quadratic form in the exponent of (14-31b) into an expression in which the coefficients are more readily related to experimental quantities. In particular, we want to use $(\partial \mu / \partial N)_{T,P}$ rather than $(\partial \mu / \partial N)_{T,V}$.

The rest of this section is mainly a long thermodynamic argument. We abandon our usual practice of writing $\bar{P}$ for the pressure and $\bar{N}$ for the ensemble average number of particles and instead write P and N for the equilibrium or ensemble average values. All the symbols used refer to the equilibrium or ensemble average values. Fluctuations from equilibrium are all denoted by δN_1, δV, δc (concentration fluctuation as defined later), etc., except that the variables ξ and ξ_2 introduced later also obviously represent fluctuations from equilibrium.

The standard transformation from derivatives at constant T and V to derivatives at constant T and P is now developed, for the case of the chemical potential μ_2. Think of $\mu_2 = \mu_2(T, P, N_1, N_2)$ and

$$P = P(T, V, N_1, N_2)$$

$$\left(\frac{\partial \mu_2}{\partial N_1} \right)_{T,V,N_2} = \left(\frac{\partial \mu_2}{\partial N_1} \right)_{T,P,N_2} + \left(\frac{\partial \mu_2}{\partial P} \right)_{T,N_1,N_2} \left(\frac{\partial P}{\partial N_1} \right)_{T,V,N_2} \quad (14\text{-}32)$$

But since, in thermodynamics

$$dF = -S \, dT + V \, dP + \mu_1 \, dN_1 + \mu_2 \, dN_2$$

$$\left(\frac{\partial \mu_2}{\partial P} \right)_{T,N_1,N_2} = \left(\frac{\partial V}{\partial N_2} \right)_{T,P,N_1} = \tilde{V}_2 \quad (14\text{-}33)$$

In Eq. (14-32), $\tilde{V}_2$ is the partial molecular volume and is a derivative at constant pressure. Furthermore, by a standard argument

Sec. 14-7]   FLUCTUATIONS, NOISE, BROWNIAN MOTION   275

$$\left(\frac{\partial P}{\partial N_1}\right)_{T,V,N_2} = \frac{-(\partial V/\partial N_1)_{T,P,N_2}}{(\partial V/\partial P)_{T,N_1,N_2}} = \frac{\tilde{V}_1}{V\kappa}$$

[κ being the compressibility, $-(1/V)(\partial V/\partial P)_{T,N_1,N_2}$]

and so
$$\left(\frac{\partial \mu_2}{\partial N_1}\right)_{N_2,T,V} = \left(\frac{\partial \mu_2}{\partial N_1}\right)_{N_2,T,P} + \frac{\tilde{V}_1 \tilde{V}_2}{V\kappa} \tag{14-34}$$

In general,
$$\left(\frac{\partial \mu_i}{\partial N_j}\right)_{T,V,N_k \neq j} = \left(\frac{\partial \mu_i}{\partial N_j}\right)_{T,P,N_k \neq j} + \frac{\tilde{V}_i \tilde{V}_j}{V\kappa} \tag{14-35}$$

With this substitution, the quadratic form which is the exponent of Eq. (14-31b) may be rewritten as

$$\left(\frac{\partial \mu_1}{\partial N_1}\right)_{T,P,N_2} \delta N_1^2 + 2\left(\frac{\partial \mu_2}{\partial N_1}\right)_{T,P,N_2} \delta N_1 \, \delta N_2 + \left(\frac{\partial \mu_2}{\partial N_2}\right)_{T,P,N_1} \delta N_2^2$$
$$+ \frac{\tilde{V}_1^2}{V\kappa} \delta N_1^2 + 2\frac{\tilde{V}_1 \tilde{V}_2}{V\kappa} \delta N_1 \, \delta N_2 + \frac{\tilde{V}_2^2}{V\kappa} \delta N_2^2 \tag{14-36}$$

The last half of this expression can be simplified by the substitution

$$\xi = \frac{\tilde{V}_1 \, \delta N_1 + \tilde{V}_2 \, \delta N_2}{V}$$

We now have for (14-36)

$$\left(\frac{\partial \mu_1}{\partial N_1}\right)_{T,P,N_2} \delta N_1^2 + 2\left(\frac{\partial \mu_2}{\partial N_1}\right)_{T,P,N_2} \delta N_1 \, \delta N_2 + \left(\frac{\partial \mu_2}{\partial N_2}\right)_{T,P,N_1} \delta N_2^2 + \frac{V\xi^2}{\kappa} \tag{14-37}$$

To simplify the terms involving $\partial \mu / \partial N$, use a Gibbs-Duhem relationship,

$$N_1 \left(\frac{\partial \mu_1}{\partial N_1}\right)_{T,P,N_2} + N_2 \left(\frac{\partial \mu_2}{\partial N_1}\right)_{T,P,N_1} = 0$$

and substitute for $\partial \mu_1 / \partial N_1$ in Eq. (14-37) in terms of $\partial \mu_2 / \partial N_1$. Then use the Euler-theorem relation for the intensive function μ_2,

$$N_1 \left(\frac{\partial \mu_2}{\partial N_1}\right)_{T,P,N_2} + N_2 \left(\frac{\partial \mu_2}{\partial N_2}\right)_{T,P,N_1} = 0$$

and substitute for $\partial \mu_2 / \partial N_1$ in terms of $\partial \mu_2 / \partial N_2$. The quadratic form (14-37) becomes

$$\left(\frac{\partial \mu_2}{\partial N_2}\right)_{T,P,N_1} \left[\delta N_2^2 - \frac{2N_2}{N_1} \delta N_1 \, \delta N_2 + \left(\frac{N_2}{N_1}\right)^2 \delta N_1^2\right] + \frac{V\xi^2}{\kappa}$$

or
$$N_2^2 \frac{\partial \mu_2}{\partial N_2}\left(\frac{\delta N_2}{N_2} - \frac{\delta N_1}{N_1}\right)^2 + \frac{V\xi^2}{\kappa} \tag{14-38}$$

Now introduce the new variable

$$\xi_2 = \frac{\delta N_2}{N_2} - \frac{\delta N_1}{N_1}$$

which represents the fractional fluctuation in N_2 (the solute) compared with the fractional fluctuation in N_1 (the solvent). The quadratic form is now

$$N_2^2 \frac{\partial \mu_2}{\partial N_2} \xi_2^2 + \frac{V\xi^2}{\kappa} \tag{14-39}$$

Thus by means of a transformation to the new variables ξ and ξ_2,

$$\xi = \frac{\tilde{V}_1 \,\delta N_1 + \tilde{V}_2 \,\delta N_2}{V}$$
$$\xi_2 = \frac{\delta N_2}{N_2} - \frac{\delta N_1}{N_1} \tag{14-40}$$

we have got rid of the cross term in $\delta N_1 \, \delta N_2$ in our Gaussian distribution, and we can now write

$$p(\xi, \xi_2) = C e^{-(\beta/2)[N_2^2(\partial \mu_2/\partial N_2)\xi_2^2 + V\xi^2/\kappa]} \, d\xi \, d\xi_2 \tag{14-41}$$

Without further ado and without troubling to determine C by normalization, we can assert that the mean-square fluctuations in ξ and ξ_2 are given by

$$\overline{\xi^2} = \frac{\kappa}{\beta V} \tag{14-42a}$$

$$\overline{\xi_2^2} = \frac{1}{\beta N_2^2 (\partial \mu_2/\partial N_2)_{N_1,T,P}} \tag{14-42b}$$

By comparison with (14-11), it is clear that the fluctuation in ξ is the fluctuation in density at constant composition.

We write the concentration of solute as $c = N_2/N_1$. Then

$$\frac{dc}{c} = \frac{dN_2}{N_2} - \frac{dN_1}{N_1} = \xi_2$$

Thus the quantity $\overline{\xi_2^2}$ is the mean-square fractional fluctuation in concentration. Now, since μ_2 is a function of concentration, T, and P,

$$\mu_2 = \mu_2(c, T, P)$$

$$\left(\frac{\partial \mu_2}{\partial N_2}\right)_{N_1,T,P} = \left(\frac{\partial \mu_2}{\partial c}\right)_{T,P} \left(\frac{\partial c}{\partial N_2}\right)_{N_1} = \frac{1}{N_1}\left(\frac{\partial \mu_2}{\partial c}\right)_{T,P} \tag{14-43}$$

so that
$$\overline{\xi_2^2} = \frac{kT}{c^2 N_1 (\partial \mu_2/\partial c)_{T,P}} \tag{14-44}$$

Equations (14-42a) and (14-44) are the desired result. The derivation is an interesting illustration of the way in which persistent and skillful manipulation can sometimes convert an expression into a more meaningful and tractable form.

It should be added that, for a system with three or more components, a similar derivation shows that there is a fluctuation in density which is independent (no cross term) of the fluctuations in concentration. For the fluctuations of the two solute components (N_2 and N_3) there are variables

$$\xi_2 = \frac{\delta N_2}{N_2} - \frac{\delta N_1}{N_1} \qquad \xi_3 = \frac{\delta N_3}{N_3} - \frac{\delta N_1}{N_1}$$

but in the quadratic form in the exponential of the expression which is the generalization of (14-41) there is a cross term in $\xi_2 \xi_3$. Therefore, the quantity $\overline{\xi_2 \xi_3}$ is not necessarily zero; it is in fact related to the cross derivative $\partial \mu_2 / \partial c_3$.

We shall now consider briefly how the results for the fluctuations in concentration and density are used in the theory of light scattering. The random fluctuations in concentration and density of a fluid produce corresponding random fluctuations in the refractive index of the medium. Light scattering is essentially due to the instantaneous nonuniform refractive index. One way of expressing the amount of light scattering is in terms of the attenuation of the incident beam; we write $I = I_0 e^{-\tau L}$, where I is the light intensity after traversing a distance L in the fluid and τ is the turbidity of the fluid.

The theory of light scattering gives for the turbidity τ of an otherwise homogeneous medium

$$\tau = \frac{8\pi^3}{3\lambda^4} V \overline{\delta \epsilon^2} \tag{14-45}$$

where λ is the wavelength (*in vacuo*) and $\overline{\delta \epsilon^2}$ is the mean-square fluctuation in the dielectric constant in a region of volume V. The turbidity τ is an intensive property, and we shall see that the combination $V \overline{\delta \epsilon^2}$ is independent of the size of the region. The dielectric constant that matters here is the dielectric constant for electrical vibrations at frequency $\nu = c/\lambda$; that is, $\epsilon = n^2$, where n is the refractive index for wavelength λ.

Now, write

$$\epsilon = \epsilon(T,P,c) \qquad d\epsilon = \left(\frac{\partial \epsilon}{\partial P}\right)_{T,c} dP + \left(\frac{\partial \epsilon}{\partial c}\right)_{P,T} dc$$

We are interested in the fluctuations in ϵ in a system with a fixed volume; at fixed volume [since $P = P(T,V,N_1,N_2)$]

$$dP = \left(\frac{\partial P}{\partial N_1}\right)_{T,V} dN_1 + \left(\frac{\partial P}{\partial N_2}\right)_{T,V} dN_2$$

$$= \frac{-(\partial V/\partial N_1)_{T,P,N_2} dN_1 - (\partial V/\partial N_2)_{T,P,N_1} dN_2}{(\partial V/\partial P)_{T,N_1,N_2}}$$

$$= \frac{\tilde{V}_1 dN_1 + \tilde{V}_2 dN_2}{V\kappa} = \frac{\xi}{\kappa}$$

So

$$d\epsilon = \left(\frac{\partial \epsilon}{\partial P}\right)_{T,c} \frac{\xi}{\kappa} + \left(\frac{\partial \epsilon}{\partial c}\right)_{T,P} dc \tag{14-46}$$

We have already noted that, for a general variation in c,

$$\frac{dc}{c} = \frac{dN_2}{N_2} - \frac{dN_1}{N_1} = \xi_2$$

For molecular-weight determinations, the second term in (14-46) is the important one. The first term is usually relatively small. It is possible, of course, to measure the coefficient $(\partial \epsilon/\partial P)_{T,c}$, but it is sometimes sufficient to estimate it theoretically. For this purpose, it is more convenient to deal with $(\partial \epsilon/\partial \rho)_{T,c}$, and we wish to express $(\partial \epsilon/\partial P)_{T,c}$ in terms of $(\partial \epsilon/\partial \rho)_{T,c}$. Recall that

$$\kappa = -\frac{1}{V}\left(\frac{\partial V}{\partial P}\right)_{T,c} = +\frac{1}{\rho}\left(\frac{\partial \rho}{\partial P}\right)_{T,c}$$

so that

$$\left(\frac{\partial \epsilon}{\partial P}\right)_{T,c} = \left(\frac{\partial \epsilon}{\partial \rho}\right)_{T,c}\left(\frac{\partial \rho}{\partial P}\right)_{T,c} = \left(\frac{\partial \epsilon}{\partial \rho}\right)_{T,c} \rho\kappa$$

We thus obtain

$$d\epsilon = \left(\frac{\partial \epsilon}{\partial \rho}\right)_{T,c} \rho\xi + \left(\frac{\partial \epsilon}{\partial c}\right)_{T,P} c\xi_2 \qquad (14\text{-}47)$$

Then since the fluctuations in ξ and ξ_2 are uncorrelated (that is, $\overline{\xi\xi_2} = 0$),

$$\overline{\delta\epsilon^2} = \left[\frac{\partial \epsilon}{\partial \rho}\right]_{T,c}^2 \rho^2\overline{\xi^2} + \left[\frac{\partial \epsilon}{\partial c}\right]_{T,P}^2 c^2\overline{\xi_2^2} \qquad (14\text{-}48)$$

but $\overline{\xi^2} = kT\kappa/V$ [(14-42a)], $\overline{\xi_2^2} = kT/c^2 N_1 (\partial \mu_2/\partial c)$ [(14-44)], and so

$$\overline{\delta\epsilon^2} = \left(\frac{\partial \epsilon}{\partial \rho}\right)^2 \frac{\rho^2 kT\kappa}{V} + \left[\frac{\partial \epsilon}{\partial c}\right]_{T,P}^2 \frac{kT}{N_1(\partial \mu_2/\partial c)_{T,P}} \qquad (14\text{-}49)$$

From Eq. (14-45), the turbidity is then

$$\tau = \frac{8\pi^3}{3\lambda^4}\left[\left(\frac{\partial \epsilon}{\partial \rho}\right)^2 \rho^2 kT\kappa + \left(\frac{\partial \epsilon}{\partial c}\right)_{T,P}^2 \frac{kT}{(N_1/V)(\partial \mu_2/\partial c)_{T,P}}\right] \qquad (14\text{-}50)$$

Equation (14-50) is the desired final result.* The turbidity due to density fluctuations can be computed if $\partial \epsilon/\partial \rho$ (or $\partial \epsilon/\partial P$) is measured or estimated. There are several empirical or theoretical equations which relate ϵ to ρ, but none of them is very accurate. In Chap. 18 we shall

---

* There are several important corrections to this formula which should be mentioned. In the first place, the result is derived in electromagnetic theory by assuming that the molecules are optically isotropic. It is then found that polarized light is scattered without depolarization. If the molecules are optically anisotropic, there is a depolarization of the scattered light and the intensity of the scattered light is increased by a factor of $(1 + \rho_u)/(1 - \frac{7}{6}\rho_u)$, where ρ_u is the "depolarization ratio." For a more complete discussion, see the references cited. The quantity ρ_u is often of the order of 0.5; so the correction is quite large.

Furthermore, the equations developed here are applicable only if the dimensions of the scattering molecules are small compared with the wavelength of the light, so that interference effects are unimportant.

derive the Clausius-Mosotti relation

$$\frac{\epsilon - 1}{\epsilon + 2} = A\rho \qquad A \text{ being a constant}$$

so that
$$\frac{\partial \epsilon}{\partial \rho} = \frac{(\epsilon - 1)(\epsilon + 2)}{3\rho} \tag{14-51}$$

and, for visible light, $\epsilon = n^2$.*

The light scattering due to the concentration fluctuation term is of greater practical significance. Suppose that the solution is sufficiently dilute so that $\mu_2 = \mu_2^\circ + kT \ln c$; $\partial \mu_2 / \partial c = kT/c$. The turbidity due to this term is

$$\tau(\text{ideal}) = \frac{8\pi^3}{3\lambda^4} \left(\frac{\partial \epsilon}{\partial c}\right)^2 c \left(\frac{V}{N_1}\right) \tag{14-52}$$

If the molecular weight of the solute is unknown, the concentration $N_2/N_1 = c$ (moles solute per mole solvent) is unknown. Let g be the concentration of solute in grams per cubic centimeter and m_2 the weight of one molecule of solute. Then

$$g = \frac{m_2 N_2}{V} = m_2 \left(\frac{N_2}{N_1}\right)\left(\frac{N_1}{V}\right) = m_2 c \left(\frac{N_1}{V}\right)$$

Substituting in Eq. (14-52),

$$\tau(\text{ideal}) = \frac{8\pi^3}{3\lambda^4} \left[\frac{\partial \epsilon}{\partial g}\right]_{T,P}^2 g m_2 \tag{14-53}$$

This equation illustrates the most important application of light scattering, which is the determination of the molecular weight of large molecules; $(\partial \epsilon / \partial g)_{T,P}$ is directly measurable, and so from the measured turbidity one can calculate m_2.

Equation (14-53) is of course applicable only to ideal dilute solutions. It can actually be derived by considering the scattering from independent particles. This is a much simpler argument than the one developed here, which is based on fluctuation theory. The advantage of the present treatment is that Eq. (14-50) is applicable to nonideal solutions and that furthermore the treatment is readily generalized to include three or more components. In practice, to determine molecular weights, the quantity τ/g is extrapolated to infinite dilution, where ideal behavior may be assumed.

Suppose that the preparation being studied is a mixture of a homologous series of polymers [for example, $(C_2H_4)_i$] with a varying degree of

---

* C. I. Carr and B. H. Zimm, *J. Chem. Phys.*, **18**: 1616 (1950), find that the Gladstone-Dale relation $n - 1 = A\rho$ and the Eykman equation $(n^2 - 1)/(n - 0.4) = A\rho$ are useful for some liquids. Typically $n \sim 1.4$, $\epsilon \sim 2$, and the difference between the Clausius-Mosotti equation and the Gladstone-Dale equation is appreciable, but not tremendous.

polymerization, i, and molecular weights m_{2i}. We should expect, for a homologous series, the quantity $\partial\epsilon/\partial g$ to be independent of the degree of polymerization. Let N_{2i} be the number of molecules of the ith polymer and g_i its weight concentration (grams per cubic centimeter). Note that

$$g_i = \frac{N_{2i}m_{2i}}{V}$$

The total weight concentration is given by $g = \Sigma g_i$, and the total number of molecules of solute by $N_2 = \Sigma N_{2i}$.

Equation (14-53) should then be replaced by

$$\tau(\text{ideal}) = \frac{8\pi^3}{3\lambda^4}\left(\frac{\partial\epsilon}{\partial g}\right)^2 \sum_i g_i m_{2i} \qquad (14\text{-}54)$$

But from the measurements, since we do not know the individual g_i's, we calculate an average molecular weight, which we call $\overline{m_{2w}}$, by using (14-53). Comparison of Eqs. (14-53) and (14-54) then shows that

$$\overline{gm_{2w}} = \Sigma g_i m_{2i}$$

or

$$\overline{m_{2w}} = \frac{\Sigma g_i m_{2i}}{\Sigma g_i} \qquad (14\text{-}55a)$$

Thus we see that light scattering determines a weight average molecular weight $\overline{m}_{2w}$, defined by (14-55a). The number average molecular weight is defined by

$$\overline{m}_{2n} = \frac{\Sigma N_{2i}m_{2i}}{\Sigma N_{2i}} \qquad (14\text{-}55b)$$

By substituting $g_i = N_{2i}m_{2i}/V$ in (14-55a), we obtain

$$\overline{m_{2w}} = \frac{\Sigma N_{2i}m_{2i}^2}{\Sigma N_{2i}m_{2i}} = \frac{\Sigma N_{2i}m_{2i}^2}{\overline{m_{2n}}\Sigma N_{2i}} = \frac{\overline{m_2^2}}{\overline{m}_{2n}}$$

so that $\overline{m_{2w}}\,\overline{m_{2n}} = \overline{m_2^2}$. The weight average molecular weight $\overline{m_{2w}}$ is determined by light scattering; the number average molecular weight is determined by a measurement of a colligative property, such as osmotic pressure. An estimate of the polydispersity of the polymer preparation is

$$\overline{\delta m_2^2} = \overline{m_2^2} - \overline{m_{2n}^2} = \overline{m}_{2n}(\overline{m}_{2w} - \overline{m}_{2n})$$

14-8. A Basic Problem in Probability and Statistics.* Consider N boxes, each containing Q white balls and R black balls. One ball is taken from each box. What is the probability $p_N(m)$ that m of the balls will be white and $N - m$ black? This is a famous problem in probability theory.

* The problem is discussed by Margenau and Murphy, pp. 422–425 [23].

Sec. 14-8] FLUCTUATIONS, NOISE, BROWNIAN MOTION

The probability of drawing a white ball from a box is $q = Q/(R + Q)$, and the probability of drawing a black ball is $r = R/(R + Q) = 1 - q$. The probability of drawing m white balls from m particular boxes and $N - m$ black balls from the $N - m$ other boxes is then $q^m r^{N-m}$. But there are $N!/m!(N - m)!$ ways of selecting m boxes out of N. Therefore the total probability of drawing m white balls is

$$p_N(m) = \frac{N!}{m!(N - m)!} q^m r^{N-m} \qquad (14\text{-}56)$$

Note that by the binomial expansion

$$(r + q)^N = \sum_{m=0}^{m=N} \frac{N!}{m!(N - m)!} q^m r^{N-m}$$

$$= \sum_m p_N(m) q^m r^{N-m}$$

but $r + q = 1$, $(r + q)^N = 1$; so the probability distribution (14-56) is properly normalized, that is, $\sum_m p_N(m) = 1$.

We are interested in an approximation for the case where both N and m are large. Use Stirling's approximation in its more exact form, $\ln N! = \frac{1}{2} \ln 2\pi N + N \ln N - N$, and from (14-56)

$$\ln p_N(m) = \frac{1}{2} \ln \frac{N}{2\pi m(N - m)} + N \ln N - m \ln m$$
$$- (N - m) \ln (N - m) + m \ln q + (N - m) \ln r \qquad (14\text{-}57a)$$

Now write the term $N \ln N$ as $m \ln N + (N - m) \ln N$, and (14-57a) becomes

$$\ln p_N(m) = \frac{1}{2} \ln \frac{N}{2\pi m(N - m)} - m \ln \frac{m}{Nq} - (N - m) \ln \frac{N - m}{Nr} \qquad (14\text{-}57b)$$

We surmise that the mean value of m will be $\bar{m} = Nq$ and that the rms deviation will be of the order of $N^{1/2}$. Let us therefore introduce a variable which measures the deviation from the mean in units of the rms deviation, i.e.,

$$x = \frac{m - \bar{m}}{N^{1/2}}$$

so that
$$m = \bar{m} + xN^{1/2} = Nq\left(1 + \frac{x}{qN^{1/2}}\right) \qquad (14\text{-}58)$$

$$N - m = N(1 - q) - N^{1/2}x = Nr\left(1 - \frac{x}{rN^{1/2}}\right)$$

If, as we surmise, the important values of x are of the order of unity, $x/qN^{1/2} \ll 1$; we now substitute in (14-57b) and use the expansion $\ln(1+y) \approx y - y^2/2$, taking care consistently to include all terms of the order of y^2. The three terms of (14-57b) are separately evaluated below.

$$-m \ln \frac{m}{Nq} = -Nq\left(1 + \frac{x}{qN^{1/2}}\right)\ln\left(1 + \frac{x}{qN^{1/2}}\right)$$

$$\approx -Nq\left(1 + \frac{x}{qN^{1/2}}\right)\left(\frac{x}{qN^{1/2}} - \frac{x^2}{2q^2N}\right) \approx -xN^{1/2} - \frac{x^2}{2q} \quad (14\text{-}59a)$$

$$-(N-m)\ln\frac{N-m}{Nr} = -Nr\left(1 - \frac{x}{rN^{1/2}}\right)\ln\left(1 - \frac{x}{rN^{1/2}}\right)$$

$$\approx -Nr\left(1 - \frac{x}{rN^{1/2}}\right)\left(-\frac{x}{rN^{1/2}} - \frac{x^2}{2r^2N}\right) \approx xN^{1/2} - \frac{x^2}{2r} \quad (14\text{-}59b)$$

$$\frac{1}{2}\ln\frac{N}{2\pi m(N-m)} = \frac{1}{2}\ln\frac{N}{2\pi(Nq)(Nr)} - \frac{1}{2}\ln\left(1 + \frac{x}{qN^{1/2}}\right)\left(1 - \frac{x}{rN^{1/2}}\right)$$

$$\approx \frac{1}{2}\ln\frac{1}{2\pi Nqr} - \frac{1}{2}\left(\frac{x}{qN^{1/2}} - \frac{x}{rN^{1/2}}\right). \quad (14\text{-}59c)$$

With r and q and x of the order of unity and N very large, we retain the terms $x^2/2q$ and $x^2/2r$ from Eqs. (14-59a) and (14-59b) but omit the terms like $x/qN^{1/2}$ in (14-59c). Furthermore, $x^2/2q + x^2/2r = x^2/2qr$ (since $q + r = 1$). Thus, adding up the three terms from Eqs. (14-59a), (14-59b), and (14-59c), we obtain

$$p_N(m) = \frac{1}{(2\pi Nqr)^{1/2}} e^{-x^2/2rq}$$

Substitute for x in terms of m and $\bar{m}$.

$$p_N(m) = \frac{1}{(2\pi Nqr)^{1/2}} e^{-(m-\bar{m})^2/2Nrq}$$

This is the probability (for the case that N and m are large) that the particular integral value m for the number of white balls will occur. The probability that m will lie in the interval m, $m + dm$ is therefore

$$p_N(m)\,dm = \frac{1}{(2\pi Nqr)^{1/2}} e^{-(m-\bar{m})^2/2Nrq}\,dm \quad (14\text{-}60a)$$

or, since $\bar{m} = Nq$,

$$p_N(m)\,dm = \left(\frac{1}{2\pi \bar{m}r}\right)^{1/2} e^{-(m-\bar{m})^2/2\bar{m}r}\,dm \quad (14\text{-}60b)$$

Equations (14-60) are the desired results. Observe that, by using the more exact Stirling approximation, we obtain a result which is in fact properly normalized. It would not, however, have been surprising if renormalization had been necessary to correct for the various approximations.*

Thus, in the limit of a large number of tries, the distribution resulting from the game described in the first paragraph is a Gaussian distribution. The mean-square value of the deviation from the mean is

$$\overline{(m - \bar{m})^2} = Nrq = \bar{m}r \qquad (14\text{-}61)$$

For the case of $r = q = \frac{1}{2}$, $\bar{m} = N/2$, the distribution becomes

$$p_N(m) = \left(\frac{2}{\pi N}\right)^{1/2} e^{-2(m-\bar{m})^2/N} \qquad (14\text{-}62a)$$

and

$$\overline{(m - \bar{m})^2} = \frac{N}{4} = \frac{\bar{m}}{2} \qquad (14\text{-}62b)$$

For the fractional fluctuation

$$\frac{\overline{(m - \bar{m})^2}}{\bar{m}^2} = \frac{N/4}{(N/2)^2} = \frac{1}{N} \qquad (14\text{-}62c)$$

It may be noted that, in the perfect-gas case, for the various thermodynamic fluctuations calculated in the previous sections the fractional mean-square fluctuations ($\overline{\delta E^2}/\bar{E}^2$, for example) are usually of the order of $1/N$, where N is the number of particles. This makes sense, since for a perfect gas the motion of any one particle is independent of that of all the others.

14-9. The Random Walk. Consider a random walk in one dimension. The walker (a molecule or a drunk is the usual example) makes ν' jumps per unit time, each of length l, at random, in either the $+x$ or the $-x$ direction. What is the probability that he will be a certain distance x from the origin after time t? The number of jumps is $N = \nu't$. One possible result of this random walk is that the number of jumps in the plus direction is m; then $\nu't - m = $ number of jumps in the minus direction. The mean or most probable value for m is $\bar{m} = \nu't/2$. The distance traveled for the case of m forward jumps is

$$x = ml + (\nu't - m)(-l) = l(2m - \nu't) = 2l(m - \bar{m}) \qquad (14\text{-}63)$$

For a large number of jumps ($\nu't \gg 1$), the probability of a given value of m is given by Eq. (14-62a). Since the relation between the number of jumps in the forward direction and the distance traveled is given by $x = 2l(m - \bar{m})$, an interval dm in the number of jumps corresponds to

---

* From the point of view of learning how to do this kind of problem, it is important to realize that it is necessary to be consistent about the order of terms included. The interested reader can work out for himself what would happen to the answer were we to use only the first-order term in the expansion $\ln (1 + x/qN^{1/2}) \approx x/qN^{1/2} - x^2/2q^2N$, in (14-59), leaving out the x^2 term.

an interval $dx = 2l\,dm$ in the distance traveled. Therefore the probability of the random walker being in the interval x, $x + dx$ after time t is

$$p(x)\,dx = \left(\frac{2}{\pi \nu' t}\right)^{1/2} e^{-x^2/2l^2\nu' t}\left(\frac{dx}{2l}\right)$$

$$= \left(\frac{1}{2\pi \nu' l^2 t}\right)^{1/2} e^{-x^2/2l^2\nu' t}\,dx \qquad (14\text{-}64)$$

Therefore, the mean-square distance traveled is $\overline{x^2} = l^2\nu' t$.

14-10. Diffusion and the Random Walk. For the random-walk calculation to represent a solute molecule diffusing in a solvent in three dimensions, we use the simplified model in which the solute molecule makes ν jumps per unit time, and these are at random along the x, y, or z axes. The number of jumps per unit time along the x axis is $\nu' = \nu/3$; so, from (14-64), the probability of having diffused a distance x in time t is

$$p(x)\,dx = \left(\frac{3}{2\pi \nu l^2 t}\right)^{1/2} e^{-3x^2/2l^2\nu t}\,dx \qquad (14\text{-}65)$$

There are identical expressions for the independent probabilities of having diffused a certain distance in the y and z directions, and the mean-square distances diffused are

$$\overline{x^2} = \overline{y^2} = \overline{z^2} = \frac{l^2\nu t}{3} \qquad (14\text{-}66a)$$

$$\overline{r^2} = \overline{x^2} + \overline{y^2} + \overline{z^2} = l^2\nu t \qquad (14\text{-}66b)$$

It is shown in books on diffusion that the macroscopic-diffusion equation is

$$\frac{\partial c}{\partial t} = D\left(\frac{\partial^2 c}{\partial x^2} + \frac{\partial^2 c}{\partial y^2} + \frac{\partial^2 c}{\partial z^2}\right) \qquad (14\text{-}67)$$

where D is the diffusion coefficient and c is the concentration of the diffusing species.*

For diffusion in one direction, if one starts at $t = 0$ with a finite amount of material Q in a very narrow zone in the yz plane at $x = 0$ (Fig. 14-1), the solution of the diffusion equation is

$$c(x,t) = \frac{Q}{(4\pi Dt)^{1/2}}\,e^{-x^2/4Dt} \qquad (14\text{-}67a)$$

and the mean distance diffused (in the x direction) is

$$\overline{x^2} = 2Dt \qquad (14\text{-}67b)$$

* See, for example, R. M. Barrer, "Diffusion in and through Solids," Cambridge, New York, 1934, or any one of a variety of texts.

From the probability, random-walk point of view, suppose that at $t = 0$ there are a large number of molecules in the yz plane at $x = 0$ and none elsewhere and that each one independently takes a random walk; then the probability of finding a molecule at any place x at a time t is given by (14-65); Eq. (14-65) should therefore give the concentration profile as a function of time. To make (14-65) and (14-67a) identical, we need only set

$$D = \frac{1}{6} v l^2 \tag{14-68}$$

This is an important relation, which can be derived in several ways in the kinetic theory of liquids.*

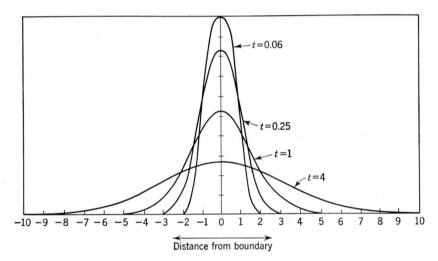

FIG. 14-1. One-dimensional diffusion from a thin layer of solution (at time $t = 0$) into a solvent. (*From R. M. Barrer, "Diffusion in and through Solids," p. 45, Cambridge, New York, 1934.*)

14-11. Brownian Motion.

In 1827 the naturalist, Robert Brown, while examining suspensions of various pollens with the aid of one of the then newly constructed achromatic objectives, discovered that the individual particles were constantly in a very animated state of motion. After his announcement of this fact, there followed the deluge of experiments and theories which sought to arrive at the true nature and cause of this motion. The explanation was first looked for in the possibility that the particles were alive. This theory was quickly disproved however, for very soon particles of glass, minerals, petrified wood, pollen known to be over 100 years old, and even stone dust from the Egyptian Sphinx were shown to behave similarly to Brown's original pollen particles. All such possibilities as convection currents

* See, for example, J. Frenkel, "The Kinetic Theory of Liquids," p. 19, Oxford, New York, 1949.

in the solution, internal motion due to uneven evaporation, hygroscopic or capillary action, mutual forces between particles, formation of small bubbles of gas, temperature effect of the illumination, etc., were carefully investigated. None of these having proved to be the true cause, the search continued.*

We now know that Brownian motion is simply an illustration of the phenomenon of thermal agitation of molecules. If we could observe smaller molecules in a liquid, we should see that they too were undergoing the same kind of agitated, random motion. In either case, the average kinetic energy per molecule is $\tfrac{1}{2}m\overline{v^2} = \tfrac{3}{2}kT$ (when classical statistical mechanics is applicable; cf. Sec. 20-3). The random-walk treatment of the preceding sections enables us to calculate how the random motions result in the phenomenon of diffusion. We have not developed a theory, however, which related the diffusion coefficient (D or $\tfrac{1}{6}\nu l^2$) to the thermal agitation. In our treatment, the mean jump frequency ν and the mean jump distance l are parameters which are introduced arbitrarily. There is a more fundamental treatment of the relation of Brownian motion and diffusion to thermal agitation and fluctuations, in which it is recognized that the particle of interest is subjected to bombardment on all sides by solute molecules. Because of fluctuations, the forces will not always cancel, and the particle will be knocked about in a random way. The calculation of the Brownian motion according to this model is rather difficult, and we shall not attempt it here.†

14-12. Diffusion Coefficients, Friction Coefficients, and Electrical Mobility. To complete the discussion of Brownian motion, we present a derivation of the relation between the diffusion coefficient and the electrical mobility. A simple kinetic-theory model for electrical conductivity which is needed in Sec. 14-16 is also developed.

We shall use v for the thermal-agitation velocity of a molecule and u for its systematic mean drift velocity in any direction. Under the influence of a force F which tends to make the molecule move in one direction, a macroscopic particle in a viscous medium attains a steady-state drift velocity

$$u = \frac{F}{f}$$

where f is the friction coefficient of the particle. We assume that the same law applies to particles of molecular dimensions. (For macroscopic spheres of radius r in a medium of viscosity η, Stokes' law is that $f = 6\pi\eta r$.) For a charged particle, with charge Ze, in an electric field E, the force is ZeE,

---

* R. B. Barnes and S. S. Silverman, *Revs. Modern Phys.*, **6**: 162 (1934).
† S. Chandresekar, *Revs. Modern Phys.*, **15**: 1 (1943).
 The important contributions to the theory of Brownian motion by A. Einstein are collected in A. Einstein, R. Fürth, (ed.), A. D. Cowper (transl.), Dutton, New York, 1926; Dover reprint, 1956.

and the drift velocity is

$$u = \frac{ZeE}{f} \tag{14-69}$$

The electrical mobility, or drift velocity in unit field, is

$$u_0 = \frac{Ze}{f} \tag{14-70}$$

For diffusion in a concentration gradient, the driving force for the diffusive drift is the negative of the gradient of the chemical potential, $\partial\mu/\partial x$, so that

$$u = -\frac{1}{f}\frac{\partial \mu}{\partial x} \tag{14-71a}$$

For an ideal dilute solution, $\mu = kT \ln c + \mu_0$, and

$$u = -\frac{kT}{cf}\frac{\partial c}{\partial x} \quad \text{or} \quad cu = -\frac{kT}{f}\frac{\partial c}{\partial x} \tag{14-71b}$$

Fick's first law for diffusion is an empirical law which asserts that the flux of material, or the amount flowing through unit area per unit time, is proportional to the diffusion coefficient and the concentration gradient, or

$$\text{Flux} = cu = -D\frac{\partial c}{\partial x} \tag{14-72}$$

Comparison of (14-71b) and (14-72) gives

$$D = \frac{kT}{f} \tag{14-73}$$

which indicates the relation of the diffusion coefficient to thermal agitation, and the friction coefficient.

Comparison of (14-73) and (14-70) gives the well-known Nernst relation between the electrical mobility and the diffusion coefficient for charged particles.

$$u_0 = \frac{DZe}{kT} \tag{14-74}$$

We now develop an equation for the electrical conductivity of an ion in solution, based on a simple model. (The same derivation was used in the classical theory of electrons in metals, but it is not really applicable because the electrons do not obey Boltzmann statistics.) The ion is a random walker. It has a thermal-agitation velocity of v, which is approximately $(3kT/M)^{1/2}$, where M is the mass of the particle. Let l be the mean free path between collisions with solvent molecules. Let the time between collisions be τ_e (e for encounters) and ν the frequency of collisions,

$$\tau_e = \frac{1}{\nu} = \frac{l}{v} \tag{14-75}$$

On each collision with solvent molecules, the ion is deflected in a random direction. Now apply an electric field E in, say, the y direction. In between collisions, while in free flight, the ion undergoes an acceleration a, in the y direction with $a = ZeE/M$. The distance it moves in time τ_e in the y direction is $\frac{1}{2}a\tau_e^2$, and the average drift velocity u in the y direction is distance divided by time, $\frac{1}{2}a\tau_e = ZeE\tau_e/2M$. On each collision, in our model, the ion loses (on the average) the small amount of energy (compared with kT) it has acquired from the electric field and moves off in a random direction. Thus, just after the collision, it has, on the average, no component of velocity in the field direction. The average drift velocity in unit field is

$$u_0 = \frac{Ze}{2M} \tau_e \tag{14-76}$$

If there are N ions per unit volume, the conductance is the current in unit field, or

$$\sigma = NZeu_0 = \frac{NZ^2e^2}{2M} \tau_e \tag{14-77}$$

The resistance of a sample of solution of area A and length L is

$$R = \frac{L}{A\sigma} = \frac{2ML}{ANZ^2e^2\tau_e} \tag{14-78}$$

This is the relation we shall need later.* It is interesting to rewrite Eq. (14-76) by using $\nu\tau_e = 1$ and then $\tau_e = l/v$,

$$u_0 = \frac{Ze}{2M} \tau_e = \frac{Ze}{2M} \nu\tau_e^2 = \frac{Zevl^2}{2Mv^2} \tag{14-79}$$

But we have assumed that $2Mv^2 = 6kT$, and our simple random-walk model gave $D = \frac{1}{6}vl^2$. Then (14-79) becomes $u_0 = ZeD/kT$. Thus our simple approximate models for diffusion and for drift in an electric field give results in agreement with the exact equation (14-74) for the relation between u_0 and D.

14-13. Fourier Series and Fourier Transforms. This section is a mathematical preparation for the next section. It is a concise review of a few of the main facts about Fourier series and integrals; readers who are not sufficiently familiar with the subjects should consult other texts. Our treatment follows Margenau and Murphy.†

---

* For the practical application of this formula, it is advisable to use either mks units or cgs units consistently (cf. Chap. 17). Thus, if R is in ohms, take $e = 1.602 \times 10^{-19}$ coulomb, M in kilograms, N in ions per cubic meter, and A and L in meters. Alternatively, with R in esu units (to obtain R in esu units, multiply R in ohms by $10^{-11}/9$) take $e = 4.803 \times 10^{-10}$ statcoulomb, M in grams, N in ions per cubic centimeter, and A and L in centimeters.

† Pages 245–246 [23].

Any function which is defined in the range $-L \le x \le L$ and which has the property

$$f(-L) = f(L)$$

can be expanded in a Fourier series,

$$f(x) = \sum_{k=1}^{k=\infty} a_k \sin \frac{k\pi x}{L} + \tfrac{1}{2} b_0 + \sum_{k=1}^{\infty} b_k \cos \frac{k\pi x}{L} \quad (14\text{-}80)$$

Thus the function $f(x)$ can be analyzed into a sum of standing waves of wavelengths $\lambda_k = 2L/k$. Because of the relations

$$\int_{-L}^{+L} \sin^2 \frac{k\pi x}{L} dx = \int_{-L}^{+L} \cos^2 \frac{k\pi x}{L} dx = L \quad (14\text{-}81a)$$

and, for $m \ne k$,

$$\int_{-L}^{+L} \sin \frac{k\pi x}{L} \sin \frac{m\pi x}{L} dx = \int_{-L}^{+L} \cos \frac{k\pi x}{L} \sin \frac{m\pi x}{L} dx = 0 \quad (14\text{-}81b)$$

$$\int_{-L}^{+L} \sin \frac{k\pi x}{L} \cos \frac{k\pi x}{L} dx = \int_{-L}^{+L} \sin \frac{k\pi x}{L} \cos \frac{m\pi x}{L} dx = 0 \quad (14\text{-}81c)$$

the coefficients a_k and b_k can be evaluated as

$$a_k = \frac{1}{L} \int_{-L}^{+L} f(u) \sin \frac{k\pi u}{L} du \qquad b_k = \frac{1}{L} \int_{-L}^{+L} f(u) \cos \frac{k\pi u}{L} du \quad (14\text{-}82)$$

The expressions are formally simplified by using imaginary exponentials.

$$f(x) = \sum_{k=-\infty}^{k=+\infty} c_k e^{ik\pi x/L} \quad (14\text{-}83a)$$

with

$$c_k = \frac{1}{2L} \int_{-L}^{+L} f(u) e^{-i\pi k u/L} du \quad (14\text{-}83b)$$

To derive Eq. (14-83b), one uses

$$\int_{-L}^{+L} e^{i\pi(m-k)u/L} du = 0 \qquad m \ne k \quad (14\text{-}84)$$

For a function which is not periodic and which is defined from $-\infty$ to $+\infty$ the analogue of the sum over all integral values of k in (14-83a) is an integral over all values of k, which is now a continuous variable

$$f(x) = \int_{-\infty}^{+\infty} c(k) e^{ikx} dk \quad (14\text{-}85)$$

where the function $c(k)$ is given by

$$c(k) = \frac{1}{2\pi} \int_{-\infty}^{+\infty} f(u) e^{-iku} du \quad (14\text{-}86)$$

Relation (14-86) is the analogue of (14-83b). The expansion (14-85) is known to be possible for functions $f(x)$ such that $\int_{-\infty}^{+\infty} |f(x)|\, dx$ exists.

The Fourier-integral relations can be written in a more symmetrical form,

$$f(x) = \left(\frac{1}{2\pi}\right)^{1/2} \int_{-\infty}^{+\infty} g(k) e^{ikx}\, dk \qquad (14\text{-}87a)$$

$$g(k) = \left(\frac{1}{2\pi}\right)^{1/2} \int_{-\infty}^{+\infty} f(u) e^{-iku}\, du \qquad (14\text{-}87b)$$

The functions $f(x)$ and $g(k)$ are said to be Fourier transforms of each other. If $f(x)$ is an even function $f(x) = f(-x)$, $g(k)$ is also an even function, and

$$f(x) = \left(\frac{2}{\pi}\right)^{1/2} \int_0^\infty g(k) \cos kx\, dk \qquad (14\text{-}88a)$$

and

$$g(k) = \left(\frac{2}{\pi}\right)^{1/2} \int_0^\infty f(u) \cos ku\, du \qquad (14\text{-}88b)$$

By substituting (14-87b) into (14-87a), one obtains the Fourier-integral theorem

$$f(x) = \frac{1}{2\pi} \int_{-\infty}^{+\infty} \int_{-\infty}^{+\infty} f(u) e^{ik(x-u)}\, dk\, du \qquad (14\text{-}89)$$

When $f(x)$ is real, the imaginary part of the integral must vanish and

$$f(x) = \frac{1}{2\pi} \int_{-\infty}^{+\infty} \int_{-\infty}^{+\infty} f(u) \cos k(x - u)\, dk\, du \qquad (14\text{-}90)$$

14-14. Fourier Analysis of a Random Function. The Wiener-Khintchine Theorem.* We can often calculate the mean-square fluctuation of a physical quantity by the arguments developed in the previous sections. In addition, however, to the magnitude of the fluctuations, their time dependence is also of interest. Suppose that one observes the pressure fluctuations in a sample of a fluid with a sensitive, rapidly responding manometer. In any system there will be positive and negative pressure fluctuations so that the average fluctuation is zero. But in one system the pressure may change rapidly; in another system the amplitude of the fluctuations may be the same, but the changes will be slower.

In general, in electromagnetic theory the energy density of the electric and magnetic fields is $(1/8\pi)(E^2 + B^2)$, with E in statvolts per centimeter and B in gauss.† In a hohlraum with an equilibrium density of radiation, there are time-varying electric and magnetic fields, with average values of zero, and with mean-square values such that $(1/8\pi)(\overline{E^2} + \overline{B^2})$ is the

---

* C. Kittel, pp. 133–136 [12]. J. L. Lawson and G. E. Uhlenbeck, "Threshold Signals," vol. 24, of MIT Radiation Laboratory Series, McGraw-Hill, New York, 1950.

† The question of electrical and magnetic units is discussed in Chap. 17.

total energy density $u(T)$ of the radiation field. These fields could be regarded as randomly fluctuating quantities. In a radiation field where there is a relatively high density of high-frequency radiation, one would expect to see the electric field fluctuating rapidly; at lower temperatures, where the energy is present in the low-frequency part of the spectrum, the random changes in the electric or magnetic field would be slower.

The energy of a sound wave in a fluid is proportional to the square of the amplitude of the pressure oscillation in the wave. We could regard the internal energy of a fluid as being composed of a set of normal modes of motion (cf. the Debye treatment of the normal vibrations of a crystal, Chap. 16). Some of the normal modes of the fluid are sound waves of different frequencies (and some are other kinds of motions, which are vortical in nature). But in any case the pressure fluctuations would then be a manifestation of the addition, with random phase and amplitude, of the pressure changes due to the various sound waves and other normal modes of the fluid. We expect that the rapidity of the pressure fluctuations would be related to the frequency spectrum of these various normal modes.

In the same way, as we shall see, there are randomly fluctuating voltages and currents in various circuit elements of an electrical circuit. One again expects that the time rate of change of the random current or voltage can in some way be related to the frequency behavior of the electric circuit.

The examples cited suggest that there should be some way of analyzing the average time history of the fluctuating quantity into a frequency spectrum. The energy of an electromagnetic wave is proportional to the square of the electric-field amplitude. The energy in a sound wave is proportional to the square of the amplitude of the pressure change. The power dissipated in a resistor is proportional to the square of the current or voltage. For a quantity, say, V, for which energy or power is proportional to V^2, we shall attempt to analyze the time dependence of V^2 into frequency components; we speak of this as the power spectrum of the fluctuations.

Suppose that there is a randomly fluctuating quantity V which varies with time. We observe its behavior for a particular system in the ensemble from time 0 to D. The time interval from 0 to D is supposed to be long so that the fluctuating quantity V has gone positive and negative many times. For convenience in the Fourier analysis, we choose the length of the interval, D, such that $V(0) = V(D)$. This particular set of measurements can be expressed as a Fourier series:

$$V(t) = \sum_{k=0}^{k=\infty} a_k \sin \frac{2\pi k t}{D} + \sum_{k=0}^{k=\infty} b_k \cos \frac{2\pi k t}{D} \qquad (14\text{-}91)$$

The time average of $V(t)^2$ is given by

$$\langle V(t)^2 \rangle = \frac{1}{D} \int_0^D \left(\sum_k a_k \sin \frac{2\pi k t}{D} + \sum_k b_k \cos \frac{2\pi k t}{D} \right)^2 dt$$

(Note that to take the time average we divide by the length of the time interval, D.) By using the several orthogonality relations in Eq. (14-81) the expression for $\langle V(t)^2 \rangle$ is easily reduced to

$$\langle V(t)^2 \rangle = \frac{1}{2} \sum_k a_k^2 + b_k^2 \tag{14-92}$$

If we now make measurements on a large number of systems, we can take the ensemble average of the time average:

$$\overline{\langle V(t)^2 \rangle} = \frac{1}{2} \sum_k (\overline{a_k^2} + \overline{b_k^2}) \tag{14-93}$$

Each positive integer k corresponds to a vibration frequency $\nu_k = k/D$. The amount of power at this frequency is $\frac{1}{2}(\overline{a_k^2} + \overline{b_k^2})$. The time period from 0 to D is really arbitrary; so the exact frequencies $\nu_k = k/D$ are rather arbitrary. It would be more realistic to talk about the amount of power $G(\nu)\,d\nu$ in the frequency interval $d\nu$ between $\nu = (k - \frac{1}{2})/D$ and $\nu = (k + \frac{1}{2})/D$, so that $d\nu = 1/D$.

For this choice of $d\nu = 1/D$, we have

$$G(\nu)\,d\nu = \frac{1}{2}(\overline{a_k^2} + \overline{b_k^2}) \qquad \text{and} \qquad G(\nu) = \frac{D}{2}(\overline{a_k^2} + \overline{b_k^2})$$

We can then write (14-93) as

$$\overline{\langle V(t)^2 \rangle} = \int_{\nu=0}^{\nu=\infty} G(\nu)\,d\nu \tag{14-94}$$

The function $G(\nu)$ is the spectral density of the randomly fluctuating quantity $V(t)^2$; it is closely analogous to the density of radiation $\rho(\nu)$ in radiation theory.

We now seek a way of measuring the "rapidity" or conversely the "persistence" of the fluctuations for a randomly fluctuating function. If the voltage is measured as $V(t)$ at one particular time, how long does it take, on the average, before the voltage has decayed to zero or changed sign? This can be quantitatively expressed by the correlation function $C(\tau)$, defined by

$$C(\tau) = \overline{\langle V(t) V(t + \tau) \rangle} \tag{14-95}$$

We look at a particular oscilloscope trace of the randomly fluctuating quantity $V(t)$ and take a time average, for different delay times τ,

$$\langle V(t)V(t+\tau)\rangle = \frac{1}{D}\int_0^D V(t)V(t+\tau)\,dt \qquad (14\text{-}96)$$

An ensemble average over a large number of such experiments then gives $C(\tau)$ as defined by Eq. (14-95). The integral in (14-96) is a function of the delay time τ. For τ very small, we expect $V(t+\tau)$ to be almost the same as $V(t)$, so that $C(\tau) = \overline{\langle V(t)V(t+\tau)\rangle}$ would be large and just slightly less than $\overline{\langle V(t)^2\rangle}$. In fact, $C(0) = \overline{\langle V(t)^2\rangle}$. As τ increases, the probability that $V(t+\tau)$ is different from $V(t)$ and possibly of the opposite sign increases; so we should expect $C(\tau)$ generally to decrease as τ increases (however, it need not decrease monotonically). For τ very large, one would expect $C(\tau)$ to approach zero because there should then be no correlation between $V(t)$ and $V(t+\tau)$. For a rapidly fluctuating random function $V(t)$, $C(\tau)$ should approach zero for smaller values of τ than would be the case for a slowly fluctuating $V(t)$. Inspection of the oscilloscope traces (Fig. 14-5) of randomly fluctuating voltages with two different correlation times may help the reader to comprehend the significance of the function $C(\tau)$.

Note that, just to be on the safe side, and to make sure that we have a statistically significant answer, $C(\tau)$ is defined as an ensemble average over a large number of measurements of the time average $\langle V(t)V(t+\tau)\rangle$, each measurement extending over the time interval D. It is intuitively plausible that the same result would be obtained by making D very large.

The correlation function can be evaluated in terms of the Fourier expansion of $V(t)$ [Eq. (14-91)],

$$C(\tau) = \overline{\langle V(t)V(t+\tau)\rangle}$$
$$= \overline{\frac{1}{D}\int_0^D \left(\sum_k a_k \sin\frac{2\pi kt}{D} + b_k \cos\frac{2\pi kt}{D}\right)}$$
$$\overline{\left(\sum_j a_j \sin 2\pi j\frac{t+\tau}{D} + b_j \cos 2\pi j\frac{t+\tau}{D}\right) dt} \qquad (14\text{-}97)$$

Use the relations

$$\sin 2\pi j\frac{t+\tau}{D} = \sin\frac{2\pi jt}{D}\cos 2\pi\frac{j\tau}{D} + \cos 2\pi j\frac{t}{D}\sin 2\pi j\frac{\tau}{D}$$

$$\cos 2\pi j\frac{t+\tau}{D} = \cos 2\pi j\frac{t}{D}\cos 2\pi j\frac{\tau}{D} - \sin 2\pi j\frac{t}{D}\sin 2\pi j\frac{\tau}{D}$$

Then, by multiplying out the various terms in (14-97) and using the orthogonality relations (14-81) (which eliminates almost all the cross terms), we get a pleasingly simple result.

$$C(\tau) = \frac{1}{2}\sum_k (\overline{a_k^2} + \overline{b_k^2})\cos\frac{2\pi k\tau}{D} \qquad (14\text{-}98)$$

As noted before, the frequency analysis in terms of the particular frequencies $\nu_k = k/D$ is arbitrary and results from the selection of a particular interval D for each experiment. For the general case, we replace $\tfrac{1}{2}(\overline{a_k^2} + \overline{b_k^2})$ by $G(\nu)\,d\nu$, to represent the power in the interval ν, $\nu + d\nu$. At the same time replace k/D by ν. The sum in (14-98) is converted into an integral, and

$$C(\tau) = \int_{\nu=0}^{\nu=\infty} G(\nu)\cos 2\pi\nu\tau\, d\nu \qquad (14\text{-}99)$$

Thus, $C(\tau)$ and $G(\nu)$ are essentially Fourier transforms of each other. By comparison with (14-88), we see that

$$G(\nu) = 4\int_{\tau=0}^{\tau=\infty} C(\tau)\cos 2\pi\nu\tau\, d\tau \qquad (14\text{-}100)$$

Equations (14-99) and (14-100) plus the definition of the correlation function and the spectral-density function are the Wiener-Khintchine

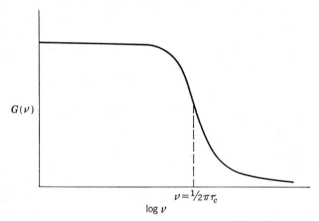

Fig. 14-2. Plot of $G(\nu)$ [Eq. (14-102)] versus $\log \nu$; the fall-off occurs at $\nu \approx \tfrac{1}{2}\pi\tau_c$. The spectral density of the fluctuations is almost constant ("white") for lower frequencies.

theorem. They satisfy our intuitive feeling that there should be a relation between the spectral-density function and the correlation function for a randomly fluctuating quantity.

We know that $C(0) = \overline{\langle V(t)^2\rangle}$, and we intuitively expect that for many cases $C(\tau)$ decreases monotonically. It is often assumed that for such random processes

$$C(\tau) = C(0)e^{-\tau/\tau_c} = \overline{\langle V(t)^2\rangle}e^{-\tau/\tau_c} \qquad (14\text{-}101)$$

where τ_c is called the correlation time for the random process. Then, since $\int_0^\infty e^{-ax}\cos mx\, dx = a/(a^2 + m^2)$,

$$G(\nu) = \frac{4\tau_c\overline{\langle V(t)^2\rangle}}{1 + 4\pi^2\nu^2\tau_c^2} \qquad (14\text{-}102)$$

Sec. 14-15]   FLUCTUATIONS, NOISE, BROWNIAN MOTION   295

The plot of this function is displayed in Fig. 14-2. The spectrum is "white," that is, $G(\nu)$ is almost constant for frequencies between zero and $\nu \approx 1/(2\pi\tau_c)$, and it rapidly goes to zero for larger frequencies. This all is intuitively reasonable.

The Wiener-Khintchine theorem is useful for a variety of problems. One important application is for the calculation of the frequency spectrum of the fluctuating magnetic field in a liquid containing magnetic molecules as the magnetic particles diffuse past each other or rotate around each other. This calculation is important in the study of nuclear magnetic relaxation times.*

In the next section we shall apply the Wiener-Khintchine theorem to the study of electrical noise.

14-15. Electrical Noise.

A tale told by an idiot, full of sound and fury, signifying nothing.
(*Macbeth*, Act V, scene 5)

As discussed in Chap. 12, any surface, other than that of a perfectly "white body," at a finite temperature emits electromagnetic radiation. The molecular cause of this emission is the random thermal motions of the charged species—electrons, ions, or heteropolar bonds—in the material. We can thus say that there are fluctuating currents and voltages in any material. These fluctuations, called thermal noise or Johnson noise, set a natural limit to the accuracy with which an electrical measurement can be made.†

The usual derivation of the magnitude and frequency dependence of thermal noise is based on consideration of a circuit, including a transmission line, which is not likely to be familiar to the readers of this book.‡ The formula can also be obtained by imposing the requirement that the fluctuating currents in the conductor should cause the emission of radiation which is exactly equal to that absorbed by the conductor from the black-body radiation field.§ This derivation requires a pretty good understanding of electromagnetic theory.

The derivation presented here is based on a more detailed and specialized model, but it is an instructive illustration of the general approach to fluctuation theory that we have been using.

Consider a resistor of area A and length L, conforming to the assumptions of Sec. 14-12. There are NAL particles in the resistor; each one has a mean thermal-agitation velocity $v = (3kT/M)^{1/2}$, and the time

---

* F. R. Andrews, "Nuclear Magnetic Resonance," app. 3, Cambridge, New York, 1954; N. Bloembergen, E. M. Purcell, and R. V. Pound, *Phys. Rev.*, **73**: 679 (1948).

† J. B. Johnson, *Phys. Rev.*, **32**: 97 (1928).

‡ H. Nyquist, *Phys. Rev.*, **32**: 110 (1928).

§ N. L. Balazs, *Phys. Rev.*, **105**: 896 (1957).

between collisions is τ_e. There is no applied electric field, and no systematic drift of the charges. At any one time, $N_x = NAL/3$ particles are moving in the plus or minus direction along the x axis, which we take as the direction of the length L of the resistor. If m is the number moving in the plus direction, $N_x - m$ are moving in the minus direction. Current density is charge density, or charges per unit volume, times velocity; so

$$\text{Current density} = Ze\left(\frac{m}{AL}v - \frac{N_x - m}{AL}v\right)$$
$$= Ze\frac{2m - N_x}{AL}v$$

and current is current density times area, or

$$I(m) = \frac{Ze(2m - N_x)}{L}v$$

Note that $I(m)$ is the random current in the x direction in the specimen when m charges are moving in the forward direction and $N_x - m$ in the $-x$ direction. The mean or most probable value of m is $\bar{m} = N_x/2$; so

$$I(m) = \frac{2Ze(m - \bar{m})v}{L} \tag{14-103}$$

But the probability of the fluctuation $m - \bar{m}$ is [cf. (14-62a)]

$$p(m) = \left(\frac{2}{\pi N_x}\right)^{1/2} e^{-2(m-\bar{m})^2/N_x} \tag{14-62a}$$

and, from (14-62b)

$$\overline{(m - \bar{m})^2} = \frac{N_x}{4} = \frac{NAL}{12} \tag{14-104}$$

The mean-square fluctuation current is then

$$\overline{\delta I^2} = \frac{4Z^2e^2v^2}{L^2}\int_m (m - \bar{m})^2 p(m)\,dm = \frac{4Z^2e^2v^2}{L^2}\overline{(m - \bar{m})^2}$$
$$= \frac{NAZ^2e^2v^2}{3L} = \frac{NAZ^2e^2kT}{ML} \tag{14-105}$$

when we set $v^2 = 3kT/M$.

What is the frequency spectrum for this fluctuating current? Since the time between collisions for a charged particle is τ_e, and hence each particle changes its direction every τ_e sec, it seems reasonable to assume that the correlation time for the fluctuating current is τ_e. Let $C_I(\tau)$ be the correlation function for the current; we assume that

$$C_I(\tau) = \overline{\delta I^2}\,e^{-\tau/\tau_e} \tag{14-106}$$

The spectral-density function that is the Fourier transform of this particular correlation function has already been calculated [cf. Eqs. (14-101) and (14-102)]. The result is

$$G_I(\nu) = \frac{4\,\overline{\delta I^2}\,\tau_e}{1 + 4\pi^2\nu^2\tau_e^2} \qquad (14\text{-}107)$$

For the usual cases of interest, $\tau_e \sim 10^{-12}$ sec; the reciprocal, $1/\tau_e \sim 10^{12}$ sec^{-1}, is much greater than the frequencies of interest in circuit analysis, that is, $\nu \ll 1/\tau_e$. Then the denominator of (14-107) is unity, and

$$G_I(\nu) = 4\,\overline{\delta I^2}\,\tau_e = \frac{4NAZ^2e^2kT\tau_e}{ML} \qquad (14\text{-}108)$$

But from (14-78) the resistance is

$$R = \frac{2ML}{NAZ^2e^2\tau_e} \qquad (14\text{-}78)$$

so that $G_I(\nu) = 8kT/R$. The voltage fluctuation is related to the resistance and the current fluctuation by $(\delta V)^2 = (\delta I)^2 R^2$. Therefore, the spectral density of the square of the voltage noise is

$$G_V(\nu) = 8RkT$$

The formula is in fact incorrect by a factor of 2; the correct result for the spectral density of the thermal voltage noise generated by a resistor is

$$G_V(\nu) = 4RkT \qquad (14\text{-}109)$$

That is, the magnitude of the mean-square voltage fluctuations in a resistor in the frequency interval ν, $\nu + d\nu$ is $4RkT\,d\nu$.

It is not surprising or disappointing that our derivation gives a result which is correct as to order of magnitude and functional form but incorrect in numerical detail, in view of the various approximate features of the model used. These include the assumptions that the motions take place along the coordinate axes only and the use of mean velocities and jump distances. In retrospect, one could obtain the correct answer by assuming that the correlation function is given by $C(\tau) = C(0)e^{-2\tau/\tau_e}$, which is just as plausible as the assumption $C(\tau) = C(0)e^{-\tau/\tau_e}$.

In most textbooks, the relation (14-109) is written in the form

$$\overline{\langle V(t)^2 \rangle} = 4RkT\,d\nu \qquad (14\text{-}110a)$$

by which it is meant that the component of the mean-square voltage fluctuation in the frequency interval ν, $\nu + d\nu$ is $4RkT\,d\nu$.

It would be more consistent with our general notation to write

$$d\overline{\langle \delta V^2 \rangle} = 4RkT\,d\nu \qquad (14\text{-}110b)$$

to represent the component of the mean-square noise voltage (voltage fluctuation), $\overline{\langle \delta V^2 \rangle}$, in the frequency interval ν, $\nu + d\nu$.

One must be careful about units when applying the above formula. With voltage expressed in practical volts and resistance in ohms, the combination V^2/R has dimensions of watts or joules per second. It is then appropriate to take $k = 1.38 \times 10^{-23}$ joule deg^{-1}, so that the combination $kT\,d\nu$ has units of watts also.

In all cases, to understand the noise voltages in an actual electrical circuit, one must know the range of frequencies to which the circuit will respond. The voltage fluctuations generated by thermal noise in other frequency bands will not be detected.

It is easy to show by a second-law argument that the detailed molecular nature of the resistor is unimportant. Thus, if we take two resistors R_A and R_B, which are at the same temperature, in the circuit of Fig. 14-3, the noise voltage $\overline{E_A^2}$ generated in resistor A will produce a current $\overline{I_A^2} = \overline{E_A^2}/(R_A + R_B)^2$. The power transferred to R_B is $P_{AB} = \overline{I_A^2} R_B = \overline{E_A^2} R_B/(R_A + R_B)^2$. The noise power generated in resistor B and transferred to R_A is similarly $P_{BA} = \overline{E_B^2} R_A/(R_A + R_B)^2$. To satisfy the second law, $\overline{E_A^2}/\overline{E_B^2} = R_A/R_B$; irrespective of the nature of R_A and R_B. Thus R_A could be an electrolyte resistor which satisfies the conditions of our derivation, whereas R_B could be a metal in which the conduction electrons obey Fermi statistics and for which it is not true that $Mv^2 = 3kT$. Nevertheless, the noise formula (14-109) still holds.

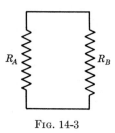

FIG. 14-3

It should be added that at very high frequencies the formula fails for a variety of reasons. It is valid, however, for the frequencies encountered in practical circuit problems.

The magnitude of the thermal noise in any frequency interval is directly proportional to the absolute temperature of the resistor. Noise thermometers for the measurement of high temperatures based on this principle have been constructed.* The most important practical application of the voltage-noise formula, however, is in the analysis of the limits of sensitivity of an electrical detecting circuit.

We spoil the poetry, but improve the aptness as a description of noise, of Macbeth's comment on life quoted at the beginning of this section, by changing the last phrase to read "signifying the mean thermal agitation of all atoms."

14-16. Fluctuations in Radioactive Disintegrations. We return to the problem treated in Sec. 14-8. If the probability that an event will occur in a single trial is q, so that $r = 1 - q$ is the probability of no event, then the expected number of events in N trials is $\bar{m} = Nq$. The

---

* J. B. Garrison and A. W. Lawson, *Rev. Sci. Instr.*, **20**: 785 (1949); E. T. Patronic, H. Marshak, C. A. Reynald, V. J. Saitor, and F. J. Shore, *Rev. Sci. Instr.*, **30**: 578 (1959).

probability that, in N trials, the number of events will be between m and $m + dm$ is [Eq. (14-60b)]

$$p(m)\, dm = \left(\frac{1}{2\pi \bar{m} r}\right)^{1/2} e^{-(m-\bar{m})^2/2\bar{m}r}\, dm$$

Suppose that there is a large number R of radioactive atoms. The decay constant or probability of decay in unit time is λ; we count the number of disintegrations in time t; the probability of disintegration in this time is $q = \lambda t$. The expected number of disintegrations is $\bar{m} = R\lambda t$. Assume that $\bar{m}$ and R are both large numbers so that the Gaussian approximation applies. However, assume that $\lambda t \ll 1$, so that $r = 1 - \lambda t \approx 1$; that is, one expects only a small fraction of the atoms to disintegrate during this time. The probability of actually having m disintegrations is then

$$p(m) = \left(\frac{1}{2\pi \bar{m}}\right)^{1/2} e^{-(m-\bar{m})^2/2\bar{m}} \tag{14-111}$$

and
$$\overline{(m - \bar{m})^2} = \bar{m} \tag{14-112}$$

$$\frac{\overline{(m - \bar{m})^2}}{\bar{m}^2} = \frac{1}{\bar{m}} \tag{14-113}$$

Thus, the expected fractional error in counting $\bar{m}$ random events is of the order of $(1/\bar{m})^{1/2}$.

14-17. Shot Noise. When a current flows through a photocell or certain kinds of vacuum tubes, the discrete nature of electricity causes a statistical fluctuation in the signal which is just like the fractional fluctuation in a radioactive counting experiment. In Fig. 14-4, P is a device in which electrons are emitted from the negative electrode and flow to the positive electrode and in which the probability of emitting any one electron is independent of when other electrons were emitted. The current is measured by the meter M. Because of the corpuscular nature of electricity, there will be a fluctuation in the current through M. This is shot noise.

A photocell is a pretty good realization of the device P. The number of light quanta falling on the photocathode is usually 10 to 100 times greater than the number of photo-

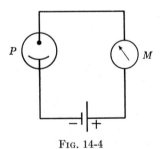

Fig. 14-4

electrons emitted; so the statistical fluctuation in the number of light quanta is smaller than the statistical fluctuation in the number of photoelectrons. If the current density is not too large, there will be no space-charge effects and the probability of emitting any one electron is independent of when other electrons were emitted.

We suppose that the meter M has a response time τ_M; it can detect slowly varying fluctuations, but if there is a very sudden change in current, it takes τ_M sec for the meter to change and read the new value. (We think actually, of course, of an exponential approach of the meter reading to a new value.) In effect, the meter is a counter which counts the number of electrons which flow through it every τ_M sec.

If the average current is $\bar{I}$, the number of electrons per second is $\bar{I}/e$. The mean number of electrons during a measuring period is

$$\bar{n} = \frac{\bar{I}}{e}\tau_M$$

The mean-square fluctuation in this quantity is

$$\overline{\delta n^2} = \bar{n}$$

and the fractional fluctuation is

$$\frac{\overline{\delta n^2}}{\bar{n}^2} = \frac{1}{\bar{n}} \qquad \frac{\overline{\delta I^2}}{\bar{I}^2} = \frac{e}{\bar{I}\tau_M} \qquad (14\text{-}114)$$

This is shot noise. It should be explicitly noted that the rms fractional fluctuation $(\overline{\delta I^2}/\bar{I}^2)^{1/2}$ represents the magnitude of the random fluctuations

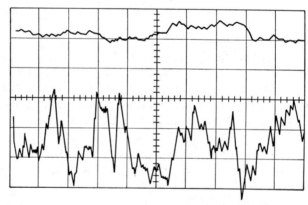

FIG. 14-5. Oscilloscope trace of the shot noise in a photocell circuit. There is a small photocurrent flowing through a resistor of 10^4 ohms; the voltage across the resistor is measured by the oscilloscope. In the lower and upper traces, the input capacitances shunting the 10^4-ohm resistor were 5×10^{-10} farad and 5×10^{-9} farad, respectively, corresponding to RC times of 5 μsec and 50 μsec. The sweep rate is 20 μsec cm^{-1}. In agreement with Eq. (14-114), we see that the amplitude of the noise decreases by a factor of about $10^{1/2}$ as the response time τ_M is increased by 10, whereas the rapidity of the fluctuations or the correlation time changes by a factor of about 10. (*Photograph courtesy of R. Stewart.*)

compared with the magnitude of the current being observed and therefore is the order of magnitude of the smallest fractional change in current which can be detected. It is therefore reasonably called the noise-to-signal ratio. The shorter the response time of the detector and the smaller the current, the greater the observed noise-to-signal ratio. Oscilloscope traces illustrating these relations are displayed in Fig. 14-5.

The derivation above applies only when the emission of each electron is independent of the emission of the other electrons. In some devices, such as a typical triode or pentode, where the current is space-charge-limited, this is not the case and the noise is less than that calculated from the above.

Shot noise and thermal-resistor noise are two of the most common sources of noise in electrical circuits. It should be emphasized that Eq. (14-114) is not exact but only an order of magnitude equation, because our definition of τ_M is not precise. However, roughly speaking we can say that the meter M responds to frequencies between zero and $\nu_M \approx 1/\tau_M$. If we change the construction of the meter and decrease τ_M, we increase the noise and we can say that the increase in bandwidth is given by $d\nu_M = d(1/\tau_M)$. Then from (14-114)

$$d\overline{\langle \delta I^2 \rangle} = G_I(\nu)\, d\nu = e\bar{I}\, d\nu \tag{14-115}$$

As remarked above, it is not to be expected that the above relation is numerically correct. The correct relation for the spectral density of the shot noise is

$$G_I(\nu)\, d\nu = 2e\bar{I}\, d\nu \tag{14-116}$$

For practical calculations, if the current is expressed in amperes or coulombs per second, we take $e = 1.602 \times 10^{-19}$ coulomb.

14-18. Brownian Motion in a Galvanometer. Fluctuations due to thermal agitation are a general phenomenon which limits the accuracy of physical measurements. Thus, suppose that there is a galvanometer suspension with a torsion constant, G. If θ is the deflection, the potential energy is $V = G\theta^2/2$. We regard the thermal vibrations of the galvanometer as one degree of freedom of a mechanical system at temperature T; by the equipartition theorem, the average energy in the degree of freedom is $kT/2$. That is, the mean-square fluctuation in θ is given by

$$\tfrac{1}{2} G \overline{\theta_N^2} = \frac{kT}{2} \quad \text{or} \quad \overline{\theta_N^2} = \frac{kT}{G}$$

The wobbling of the galvanometer can be regarded as being due to the irregular bombardment of the mirror by air molecules. But the fluctuations persist if the galvanometer suspension is *in vacuo* owing to the thermal agitation of the atoms in the suspension. The correlation time for the fluctuations is short in an atmosphere of air, and one observes rapid irregular excursions of the mirror. *In vacuo*, the galvanometer oscillates at its natural frequency with one amplitude for many vibrations; then the amplitude changes, and it carries out regular oscillations at the new amplitude for a while.

It is instructive to consider the signal-to-noise problem for a galvanometer. If I is the moment of inertia of the galvanometer suspension, the vibration period is $\tau_G = 2\pi(I/G)^{1/2}$. The galvanometer cannot respond to changes in current in times shorter than τ_G. For a given force of deflection, the angular deflection θ_S, or signal, is obviously inversely proportional to G, whereas the noise is inversely proportional to $G^{1/2}$. The signal-to-noise ratio is

$$\frac{\theta_S}{(\overline{\theta_N^2})^{1/2}} \sim \frac{1}{G^{1/2}} \sim \tau_G \tag{14-117}$$

Thus, as always, when we consider fundamental noise (as distinct from vibrations due to the street cars in the next block!), the signal-to-noise ratio is improved by increasing the effective duration of the observation time, which is equivalent to decreasing the bandwidth of the detector.

14-19. On the Approximation of Using $t_{\rm mp}$. We have now developed sufficient background so that it is profitable to consider the validity of the approximation $\ln t_{\rm total} \approx \ln t_{\rm mp}$, which was used in Chap. 6 in calculating the entropy for a system of independent particles. In the above assertion, $t_{\rm total}$ is the total number of system wave functions available to the system, and $t_{\rm mp}$ is the number of system wave functions in the most probable distribution calculated by using a particular form of the Stirling approximation for $N!$.

The same assumption was made in calculating the entropy of a canonical ensemble as the logarithm of the number of ensemble states. While our remarks will be specifically concerned with a system of independent particles with fixed energy (a microcanonical ensemble), it should be obvious that similar arguments are applicable to the more general ensembles considered in Chap. 13.

In the interests of clarity, it is well to state the conclusions that we now intend to prove before becoming involved in the details of the arguments. Our formula for the entropy, $S = k \ln t_{\rm mp}$, is based on an evaluation of $t_{\rm mp}$, the number of system states in the most probable distribution, calculated by using the simplified form $n! \approx n^n e^{-n}$ of the Stirling approximation. We shall evaluate $t_{\rm total}$, the total number of states available to the system. Both $t_{\rm total}$ and $t_{\rm mp}$ are very large numbers, of the order of N^N. Recall that N, the number of particles in the system, is of the order of 10^{20} to 10^{24}. We shall see that $t_{\rm total}$ may differ from $t_{\rm mp}$ by a very large factor, of the order of N. However, we are actually interested in $\ln t$, not t, and *we shall see that the difference between* $\ln t_{\rm total}$ *and* $\ln t_{\rm mp}$ *is negligible.*

We shall also consider the effect of using the more accurate form of Stirling's approximation, $n! \approx (2\pi n)^{1/2} n^n e^{-n}$. The number of states in the most probable distribution evaluated by using this approximation we shall denote by $\tilde{t}$. We shall see that $\tilde{t}$ is much less than $t_{\rm mp}$, which was calculated by the less accurate formula $n! \approx n^n e^{-n}$. However, again the difference in $\ln \tilde{t}$ and $\ln t_{mp}$ is negligible. Furthermore, t_{mp} is an overestimate of the number of states in the most probable distribution, and we shall discover that this overestimate approximately compensates for the underestimate involved in using the number of states in the most probable distribution rather than the total number of states available to the system.

Let us recall the argument used in arriving at the Boltzmann distribution law for the most probable distribution and in evaluating the number of states in the most probable distribution. We consider a system of independent particles with a very large number of closely spaced single-particle energy levels. We divided the energy levels, somewhat arbitrarily, into intervals ϵ_j, $\epsilon_j + d\epsilon_j$; with g_j states. The size of the interval was large enough so that g_j and the occupation number N_j were large numbers to which Stirling's approximation applies but small enough so that from the viewpoint of a physical measurement $d\epsilon$ was small.

For a given distribution D the number of system wave functions, provided that corrected Boltzmann statistics applies, is

$$t_D = \prod_k \frac{g_k^{N_k}}{N_k!} \tag{14-118}$$

Let us now use Stirling's approximation in its more exact form

$$n! = (2\pi n)^{1/2} n^n e^{-n} \tag{14-119}$$

$$\ln n! = n \ln n - n + \tfrac{1}{2} \ln 2\pi n$$

so that

$$\ln t = \sum_k \left[N_k \left(\ln \frac{g_k}{N_k} + 1 \right) - \frac{1}{2} \ln 2\pi N_k \right] \tag{14-120}$$

and

$$\frac{\partial (\ln t)}{\partial N_j} = \ln \frac{g_j}{N_j} - \frac{1}{2N_j} \tag{14-121}$$

$$\frac{\partial^2 (\ln t)}{\partial N^2} = -\frac{1}{N_j} + \frac{1}{2N_j^2} \approx \frac{1}{N_j} \qquad \frac{\partial^2 (\ln t)}{\partial N_j \, \partial N_k} = 0 \tag{14-122}$$

By imposing the usual constraints as to the constancy of the energy and the number of particles, the condition for the most probable distribution becomes

$$\frac{\partial (\ln t)}{\partial N_j} + \alpha - \beta \epsilon_j = 0 \qquad \text{for all } j$$

or

$$\frac{\tilde{N}_j}{g_j} = e^{\alpha} e^{-\beta \epsilon_j} e^{-1/2\tilde{N}_j} \tag{14-123}$$

If N_j is at all large, $e^{-1/2N_j} \approx 1$ and we obtain our usual equation

$$\frac{\tilde{N}_j}{g_j} = e^{\alpha} e^{-\beta \epsilon_j} \qquad \text{or} \qquad \frac{\tilde{N}_j}{g_j} = \frac{N}{q} e^{-\beta \epsilon_j} \tag{14-124}$$

where q is the molecular partition function and we write $\tilde{N}_j$ to denote N_j in the most probable distribution. Similarly, let $\tilde{t}$ be t for the most probable distribution, calculated using the more exact form (14-119) of the Stirling approximation. Then

$$\tilde{t} = \prod_k \left(\frac{g_k}{\tilde{N}_k}\right)^{\tilde{N}_k} e^{\tilde{N}_k} \left(\frac{1}{2\pi\tilde{N}_k}\right)^{1/2}$$

$$= \left(\frac{q}{N}\right)^N e^{E/kT} e^N \prod_k \left(\frac{1}{2\pi\tilde{N}_k}\right)^{1/2} \quad (14\text{-}125)$$

where $N = \Sigma n_i$ is the total number of particles and E is the total energy. When we use the cruder form of the Stirling approximation, $n! = n^n e^{-n}$, we obtain the expression which we have used in all our previous calculations for t in the most probable distribution; call this approximate result t_{mp}.

$$t_{\text{mp}} = \left(\frac{q}{N}\right)^N e^{E/kT} e^N \quad (14\text{-}126)$$

We may then write (14-125) as

$$\tilde{t} = t_{\text{mp}} \prod_k \left(\frac{1}{2\pi N_k}\right)^{1/2} \quad (14\text{-}127)$$

For the moment, we shall be content to observe that $\tilde{t}$ is a good deal less than t_{mp}. We shall return to the significance of this comparison later.

Right now, let us look at the problem in a different way. There is a straightforward method of calculating the total number of translational states available to a gas by using a classical phase-space argument. How does this number compare with t_{mp}?

For a system of N atoms, each with mass m, the total kinetic energy is given by

$$2mE = \sum_{i=1}^{3N} p_i^2 \quad (14\text{-}128)$$

where p_1, p_2, p_3 are the three components (in cartesian coordinates) of the momentum of the first particle, p_4, p_5, p_6 refer to the second particle, etc. The volume in phase space $\phi(E)$, for all possible classical motions with energy less than E, is

$$\phi(E) = \int \cdots \int dx_1\, dy_1\, dz_1 \cdots dx_N\, dy_N\, dz_N\, dp_1 \cdots dp_{3N} \quad (14\text{-}129)$$

The configuration coordinates range over the volume of the apparatus and contribute a factor of V^N to the integral (14-129). The limits of integration for the momentum coordinates are all values of p_i between zero and the upper limit defined by Eq. (14-128). This is the volume of a $3N$-dimensional hypersphere of radius $(2mE)^{1/2}$.

The volume of an n-dimensional hypersphere of radius r (for n even)* is

$$V_n = \frac{\pi^{n/2} r^n}{(n/2)!} \qquad (14\text{-}130a)$$

Upon using the Stirling approximation in its more exact form,

$$n! = n^n e^{-n} (2\pi n)^{1/2}$$

(14-130a) becomes

$$V_n \approx \left(\frac{2\pi e r^2}{n}\right)^{n/2} \left\{\frac{1}{(\pi n)^{1/2}}\right\} \qquad (14\text{-}130b)$$

In the above and the next few equations, the factor or term enclosed in braces { } arises from the $(2\pi n)^{1/2}$ term in the Stirling approximation.

For our problem, $n = 3N$, and $r^2 = 2mE$. Thus, the integral (14-129) becomes

$$\phi(E) = V^N \left(\frac{4\pi e m E}{3N}\right)^{3N/2} \left\{\frac{1}{(3\pi N)^{1/2}}\right\} \qquad (14\text{-}131)$$

If we write $\phi(E) = aE^{3N/2}$, we see that, for large values of N, the volume in phase space increases very rapidly with increasing E. By differentiation, we have

$$\phi(E + \delta E) - \phi(E) \approx \frac{3N}{2} aE^{(3N/2)-1} \delta E = \frac{3N}{2} aE^{3N/2} \left(\frac{\delta E}{E}\right)$$

$$= \phi(E) \left(\frac{3N}{2}\right)\left(\frac{\delta E}{E}\right) \qquad (14\text{-}132a)$$

Although we glossed over this fact in Chap. 6, it is clear, and is shown in (14-132a), that the volume in phase space and the number of states available to the system are proportional to the allowed uncertainty δE in the total energy. If the system were a member of a canonical ensemble, $\delta E/E \approx 1/N^{1/2}$ and

$$\phi(E + \delta E) - \phi(E) \approx \phi(E) \frac{3N^{1/2}}{2} \qquad (14\text{-}133)$$

Thus the volume in phase space between E and $E + \delta E$ is greater than the volume from 0 to E, for this very small δE. The differential formulae (14-132a) and (14-133) are of course not literally applicable, but by the mean-value theorem it will always be all right to write

$$\phi(E + \delta E) - \phi(E) = \phi(\bar{E}) \left(\frac{3N}{2}\right)\left(\frac{\delta E}{E}\right) \qquad (14\text{-}132b)$$

* Mayer and Mayer, p. 433 [5]; for n odd, the formula is

$$V_n = \frac{2^n \pi^{(n-1)/2} \left[\left(\frac{n-1}{2}\right)!\right] r^n}{n!}$$

which reduces to essentially the same result when the Stirling approximation is used.

where $\bar{E}$ is some mean value between E and $E + \delta E$. The total number of system quantum states available to the system if its energy is in the interval E, $E + \delta E$ is obtained by dividing the volume in phase space by $N!$ to correct for the identity of the particles and by h^{3N},

$$t_{\text{total}} = \frac{1}{N!} \frac{1}{h^{3N}} \phi(E) \left(\frac{3N}{2}\right) \left(\frac{\delta E}{E}\right) \qquad (14\text{-}134a)$$

or, by using Stirling's approximation for $N!$ and Eq. (14-131) for $\phi(E)$,

$$t_{\text{total}} = \left(\frac{4\pi mE}{3Nh^2}\right)^{3N/2} \left(\frac{V}{N}\right)^N e^{5N/2} \left(\frac{3N}{2}\right) \left(\frac{\delta E}{E}\right) \left\{\frac{1}{6^{1/2}\pi N}\right\} \qquad (14\text{-}134b)$$

We make the substitution $E = \tfrac{3}{2}NkT$ and find

$$\ln t_{\text{total}} = N \ln \left(\frac{2\pi mkT}{h^2}\right)^{3/2} \frac{V}{N} + \frac{5N}{2} + \ln \left(\frac{3N}{2}\right)\left(\frac{\delta E}{E}\right) - \{\ln 6^{1/2}\pi N\}$$
$$(14\text{-}135a)$$

$$\ln t_{\text{total}} = N \ln \left(\frac{2\pi mkT}{h^2}\right)^{3/2} \frac{V}{N} + \frac{5N}{2} + \ln \frac{3}{2\pi 6^{1/2}}\left(\frac{\delta E}{E}\right) \qquad (14\text{-}135b)$$

However, the formula (14-126) for t_{mp} gives

$$\ln t_{\text{mp}} = N \ln \left(\frac{2\pi mkT}{h^2}\right)^{3/2} \frac{V}{N} + \tfrac{5}{2}N \qquad (14\text{-}136)$$

Thus our approximate formula for $\ln t_{\text{mp}}$ gives a number which is slightly greater than our classical phase-space estimate of $\ln t_{\text{total}}$, because in the latter we used the more accurate Stirling approximation

$$N! = (2\pi N)^{1/2} N^N e^{-N}$$

The difference between $\ln t_{\text{total}}$ and $\ln t_{\text{mp}}$ depends on the size of $\delta E/E$. The important point, however, is that the difference between the two values of $\ln t$ is negligible. For atomic weight of unity and 1 mole of gas at 273°K, 1 atm pressure, the main terms are

$$\ln t_{\text{mp}} = N \ln \left(\frac{2\pi mkT}{h^2}\right)^{3/2} \frac{V}{N} + \frac{5N}{2} \approx 10^{26} \qquad (14\text{-}137)$$

The difference between the two expressions (14-135) and (14-136) is essentially $\ln (\delta E/E)$. If we take $\delta E/E = 1/N$, $\ln (\delta E/E) \approx -55$; if we take $\delta E/E = 1/N^{1/2}$, $\ln (\delta E/E) \approx -27.5$. Thus, we arrive at the conclusion that our errors in estimating t may be of the order of a factor of N; this contributes a factor of $\ln N$ to $\ln t$, which is negligible compared with the terms of order N which are the main part of the answer.

There is another way of treating the problem which is interesting. We shall take the formula for $\bar{t}$, which is an estimate of the number of states in the most probable distribution, and then calculate the total number of states available to the system, using the Taylor-expansion–Gaussian-distribution method to estimate the probability of other distributions.

We start with the formula (14-127) for $\tilde{t}$, the number of states in the most probable distribution, evaluated by using the more accurate version of the Stirling approximation.

$$\tilde{t} = t_{mp} \prod_k \left(\frac{1}{2\pi N_k}\right)^{1/2}$$

Let us now evaluate the probability of other distributions, using the Taylor-expansion–Gaussian-distribution method.

$$\ln t(\ldots, N_i, \ldots, N_k, \ldots)$$
$$= \ln \tilde{t} + \frac{1}{2} \sum_j \frac{\partial^2(\ln t)}{\partial N_j^2}(\ldots, N_i, \ldots, N_k, \ldots)(\delta N_j)^2 \quad (14\text{-}138)$$

where
$$\delta N_j = N_j - \tilde{N}_j$$

But, by (14-122), $\partial^2(\ln t)/\partial N_j^2 = -N_j$; so

$$t(\ldots, N_i, \ldots, N_k, \ldots) \approx \tilde{t} \prod_j e^{-(\delta N_j^2/2\tilde{N}_j)}$$

or, using the relation (14-127) between $\tilde{t}$ and t_{mp},

$$t(\ldots, N_i, \ldots, N_k, \ldots) \approx t_{mp} \prod_j \left(\frac{1}{2\pi \tilde{N}_j}\right)^{1/2} e^{-\delta N_j^2/2\tilde{N}_j} \quad (14\text{-}139)$$

The total number of states available to the system is approximately

$$t_{\text{total}} = \int \cdots \int t(\ldots, N_i, \ldots, N_k, \ldots)\, dN_1 \cdots dN_i \cdots dN_k \cdots \quad (14\text{-}140)$$

The volume in N space for the multiple integration in (14-140) should be constrained by the conditions $\Sigma N_i \epsilon_i = E$, $\Sigma N_i = N$; but if we ignore these restrictions, we can only get too large a value for t_{total}. Using (14-139),

$$t_{\text{total}} \approx t_{mp} \prod_j \left(\frac{1}{2\pi \tilde{N}_j}\right)^{1/2} \left(\int_0^{+\infty} e^{-(N_j - \tilde{N}_j)^2/2\tilde{N}_j}\, dN_j\right) \quad (14\text{-}141a)$$

Each integral is the integral over a Gaussian distribution,

$$\int_0^{+\infty} e^{-(N_j - \tilde{N}_j)^2/2\tilde{N}_j}\, dN_j \approx (2\pi \tilde{N}_j)^{1/2} \quad \text{for large } \tilde{N}_j$$

which cancels the factor $(1/2\pi \tilde{N}_j)^{1/2}$ in front of the integral. Thus

$$t_{\text{total}} \approx t_{mp} \quad (14\text{-}141b)$$

We thus arrive at a surprising and pleasant conclusion. The quantity $\tilde{t}$ is less than t_{mp} by a factor $\prod_j (1/2\pi \tilde{N}_j)^{1/2}$; the integration over all possible distributions approximately cancels this error.

Our general conclusion then is that the formulae derived by using $\ln t_{mp}$ for the entropy are satisfactory. In part this is due to a fortuitous cancellation of errors owing to the fact that the use of the less accurate version of Stirling's approximation makes $\ln t_{mp}$ too large. The main reason, however, why our approximate calculations give the correct answer is because $\ln N$ is utterly negligible compared with N for large values of N.

PROBLEMS

14-3. The compressibility of toluene is $\kappa = 92.7 \times 10^{-12}$ dyne^{-1} cm^2 at $T = 25°$C. Estimate the fractional fluctuations in density if observations are made on a 10^{-12}-cc sample.

14-4. Discuss the question of the fluctuation in radiant energy in a hohlraum of volume 1 cc at different temperatures.

14-5. Obtain an expression for the mean-square fluctuation in pressure in the canonical ensemble, using the methods of Secs. 14-3 and 14-4.

Hint: The answer contains a term

$$\overline{\frac{\partial^2 E}{\partial V^2}} = -\left(\overline{\frac{\partial P}{\partial V}}\right) = \frac{1}{Q}\sum \frac{\partial^2 E_i}{\partial V^2} e^{-\beta E_i}$$

In general, this term cannot be identified with a macroscopic thermodynamic function (cf. Hill, p. 101 [9a]). Note that $\overline{(\partial P/\partial V)}$ is not the same as $(\partial \bar{P}/\partial V)_T$. However, for a perfect monatomic gas, where the energy levels are proportional to $V^{-\frac{2}{3}}$, you can show that

$$\overline{\frac{\partial^2 E}{\partial V^2}} = -\left(\frac{\partial \bar{P}}{\partial V}\right)_S = -\frac{5}{3}\left(\frac{\partial \bar{P}}{\partial V}\right)_T$$

That is, it is related to the isentropic compressibility.

14-6(a) Estimate the turbidity of liquid toluene at $T = 298°$K and at $\lambda = 5{,}461$ Å, given its compressibility (Prob. 14-3), its refractive index ($n = 1.494$), density ($\rho = 0.867$), and depolarization ratio ($\rho_u = 0.48$). Estimate $dn^2/d\rho$ from the Clausius-Mosotti relation.

The experimental value for the turbidity is 3.43×10^{-4} cm^{-1} [S. H. Maron and R. I. H. Lou, *J. Polymer Sci.*, **14**: 276 (1954)]. Incidentally, if you look at this or other references, the Rayleigh ratio $R_{90°}$, which is usually reported in light-scattering investigations, is related to the turbidity by $\tau = (16\pi/3)R_{90°}$.

(b) The refractive-index increment of polystyrene in toluene at $\lambda = 5{,}461$ Å is $\partial n/\partial g = 0.11$, where g is concentration of solute in grams per cubic centimeter. Calculate the contribution by the solute to the turbidity of a 1 per cent solution of polystyrene in toluene for a molecular weight of the polymer of 10^4 and for a molecular weight of 10^5.

14-7. Consider the circuit depicted in Fig. 14-6. The photocurrent $\bar{I}$ which flows through the photocell is measured by measuring the voltage across the resistance R, and, of course, $V = IR$. If $\overline{\delta I^2_{sh}}$ is the mean-square fluctuation in the photocurrent due to the shot effect, there is a corresponding voltage fluctuation $\overline{\delta V^2_{sh}} = R^2 \overline{\delta I^2_{sh}}$.

Suppose that $I = 10^{-6}$ amp, $R = 10^4$ ohm, the detecting circuit which measures the voltage across the resistor, has a uniform response for all frequencies from 0 to 10^6 sec^{-1}, and its response drops abruptly to zero for higher frequencies. Calculate the magnitudes of the mean-square voltage fluctuations due to (a) the shot effect and (b) Johnson noise in the resistor. What is the approximate noise-to-signal ratio for this experiment?

Suppose that the detecting circuit responds only to frequencies of 0 to 10^2 sec^{-1}; what are the magnitudes of the several noise voltages?

Given the general setup of Fig. 14-6. For arbitrary I and R, what is the ratio of the Johnson noise to the shot noise in any frequency interval?

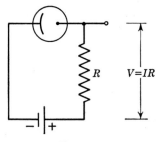

FIG. 14-6

14-8. Suppose that the spectral-density function for the fluctuations of a given physical quantity is

$$G(\nu) = G(0)e^{-\alpha\nu}$$

What is the correlation function for this quantity?

14-9. Use the results of the previous problem to calculate the correlation function when the spectral-density function is

$$G(\nu) = A\nu^3 e^{-\alpha\nu}$$

Use this result to calculate the correlation function for the fluctuating electric field in a hohlraum, neglecting the low-frequency part of the spectrum and using the approximate form of the black-body law,

$$\rho(\nu) = \frac{8\pi h\nu^3}{c^3} e^{-h\nu/kT}$$

It is worthwhile, albeit laborious, to plot the correlation function $C(\tau)$, as a function of the dimensionless parameter, $2\pi\tau/\alpha$, with $\alpha = h/kT$. In pondering the significance of the results, it may be helpful to note that the Fourier transform of a monochromatic power spectrum, that is of the delta function $\delta(\nu - \nu_0)$, is $\cos 2\pi\nu_0\tau$.

$$\int \delta(\nu - \nu_0) \cos 2\pi\nu\tau \, d\nu = C(\tau) = \cos 2\pi\nu_0\tau$$

This correlation function is alternately positive and negative, as the monochromatic radiation propagates past the observer. What qualitative feature of a power spectrum will lead to negative values of $C(\tau)$ for a certain range of τ?

14-10. Consider liquid water, which is slightly ionized into H$^+$ and OH$^-$ ions.

$$H_2O \rightleftharpoons H^+ + OH^- \qquad K = 10^{-14} \text{ mole}^2 \text{ liter}^{-2}$$

Suppose that a measurement is made of the conductivity of a 1-μ^3 (10^{-12}-cc) sample of H$_2$O and that the conductivity is proportional to the number of ions in the sample. Estimate the rms fractional fluctuation in the conductivity or in the concentration of ions.

Hint: For the purposes of this problem it is appropriate to regard liquid water as consisting of two components, un-ionized water and H$^+$ plus OH$^-$ ions. Because of electrostatic repulsions, the only important configurations are ones in which $N_{H^+} = N_{OH^-}$. Let $c = N_{H^+}/N_{H_2O}$. The chemical potential of the ions is $\mu_2 = \mu_2^\circ + 2kT \ln c$.

15

Real Gases

15-1. Introduction. It is an experimental fact that real gases do not obey the ideal or perfect-gas law $PV/NkT = 1$. The fractional deviation of (for example) the PV product for the real gas from that for an ideal gas at any given temperature and pressures varies from substance to substance; however, in all cases, the deviations approach zero as the pressure approaches zero or the volume per molecule approaches infinity.

The perfect-gas law was derived theoretically in Chaps. 6 and 10 on the basis of the assumption that there is no potential energy of interaction between molecules. The behavior of a real gas shows that this assumption is false.

In the present chapter we shall see that the statistical mechanical apparatus developed in Chap. 13 for systems of interacting particles can be put to use to obtain practical numerical results for the P, V, T behavior of dilute gases, where the deviations from ideality are small.[*]

For dense gases and liquids, the problem is much more difficult, and there is no successful statistical mechanical theory for calculating the behavior of dense gases and liquids from molecular data. Some features of this general problem are treated here; Chap. 20 is devoted to this subject also.

The curves of Fig. 15-1 illustrate the behavior of the compressibility factor $z = PV/NkT$ as a function of P for different temperatures, for a typical substance. For a perfect gas $z = 1$ for all P and all T. For a real gas, as $P \to 0$, $z \to 1$. At temperatures not too much above the critical temperature (curve A), z first decreases with increasing P, and then, at higher P, it increases. At somewhat higher temperatures (curve B), z is always greater than unity. Below the critical temperature (curve C), as P increases, z usually decreases; at the liquefaction pressure, there is a vertical decrease in PV/NkT in the two-phase region, as the gas condenses to the liquid at constant pressure. The region C_1 to C_2 repre-

---

[*] Probably the most thorough and detailed treatment of the whole field of real gases and liquids is Hirschfelder, Curtiss, and Bird, "Molecular Theory of Gases and Liquids" [15]. In the present chapter, we shall refer to this as HCB.

sents the P, V, T behavior of the liquid. If the liquid were truly incompressible, C_1C_2 would be a straight line with intercept at $z = 0$. At the critical temperature (curve D), the vertical discontinuity becomes a point singularity.

The condensation of a gas to liquid is striking evidence for the existence of attractive forces between molecules. The negative deviations of z from unity are also qualitative evidence for these forces.

Liquids and solids have a finite volume and are relatively incompressible. Thus there must also be short-range repulsive forces between molecules.* The positive value of dz/dP at high densities is further evidence for short-range repulsive forces.

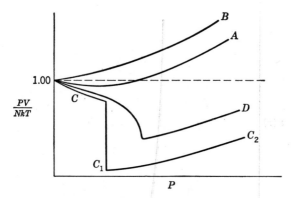

FIG. 15-1. Schematic illustration of the compressibility factor $z = PV/NkT$ as a function of P for different constant temperatures for a real fluid. A is an isotherm at a temperature somewhat greater than the critical temperature, B is at a still greater temperature, C is somewhat below the critical temperature, and D is at the critical temperature.

We conclude, therefore, that a typical intermolecular potential should be as depicted in Fig. 15-2. There is an attractive region, $du/dr > 0$, at large distances, and a steeper repulsive region, $du/dr < 0$, at smaller r. The depth of the minimum is denoted by ϵ, and the value of r at which $u(r) = 0$ is called σ.

Presumably the value of r at which the potential is a minimum should be approximately equal to the interatomic distance in the crystal. In a close-packed crystal structure of spherical atoms, each atom has 12 nearest neighbors. We might expect that the energy of sublimation of the crystal would be about six times the depth ϵ of the minimum in the potential curve. These relations are not at all accurate (especially the state-

---

* The finite volume and the small compressibility for condensed phases could conceivably be due to zero-point vibration effects; elementary quantitative considerations show that this is not the main factor for most substances, although it is significant for light atoms.

ment about the sublimation energy) and are mentioned only because they indicate in a simple and direct way the expected order of magnitude of the quantities ϵ and r_{min}.

That even the attractive forces are reasonably short-range is suggested by the fact that gases such as nitrogen or argon at 1 atm pressure at their boiling points obey the ideal-gas law to within a few per cent, even though the average intermolecular distance in such a gas at this pressure is only about ten times greater than in the liquid.

We shall discuss the quantitative representation of the intermolecular potential in a later section. These weak intermolecular forces are frequently called van der Waals forces—and the term refers to both the

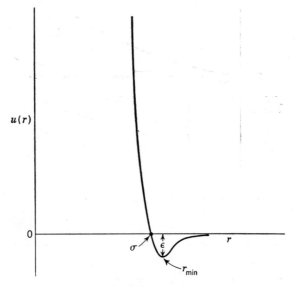

Fig. 15-2. Qualitative graph of the variation of the intermolecular potential with distance r between centers for two atoms or molecules.

attractive and repulsive interactions. Both experiment and theory indicate that they are nonspecific in that they operate between any pair of atoms and molecules. Unlike valence forces, they do not saturate—that is, the interaction between an argon atom and each of its 12 nearest neighbors in a crystal is at least approximately the same as the interaction on collision of a pair of argon atoms in a gas.

15-2. The Virial Expansions. Since the compressibility factor approaches the ideal-gas value of unity as $P \to 0$ or $V \to \infty$, the variation of PV/NkT with P or V for any particular substance can be represented as a power series in P or in $1/V$.

Let v ($= N_0 V/N$) be the volume per mole. The virial expansion in $1/v$ is then

$$\frac{PV}{NkT} = 1 + \frac{B(T)}{\text{v}} + \frac{C(T)}{\text{v}^2} + \cdots \qquad (15\text{-}1a)$$

or
$$\frac{PV}{NkT} = 1 + \frac{NB(T)}{N_0 V} + \frac{N^2 C(T)}{N_0^2 V^2} + \cdots \qquad (15\text{-}1b)$$

The virial equation of state with pressure as a variable is

$$\frac{PV}{NkT} = 1 + B^P(T)P + C^P(T)P^2 + \cdots \qquad (15\text{-}2)$$

The quantities $B(T)$ and $B^P(T)$ are known as second virial coefficients; and $C(T)$ and $C^P(T)$ are third virial coefficients; etc.

Problem 15-1. Show by comparison of Eqs. (15-1) and (15-2) that

$$B(T) = N_0 k T B^P(T) \qquad (15\text{-}3)$$
$$C(T) = (N_0 k T)^2 \{C^P(T) + [B^P(T)]^2\} \qquad (15\text{-}4)$$

15-3. The Configuration Integral and the Canonical Partition Function.

Consider a fluid consisting of N identical molecules or atoms. The coordinates of the N particles in space are specified by the N vectors $\mathbf{r}_1, \mathbf{r}_2, \ldots, \mathbf{r}_N$ (where $\mathbf{r}_j$ has components x_j, y_j, z_j). In integrations, we represent the volume element in configuration space, $dx_1\, dy_1 \cdots dy_N\, dz_N$, symbolically as $d^3\mathbf{r}_1 \cdots d^3\mathbf{r}_N$. Let the vector momentum of the jth particle be $\mathbf{p}_j$ (components $p_{x_j}, p_{y_j}, p_{z_j}$); we represent the volume element in phase space symbolically as $d^3\mathbf{r}_1 \cdots d^3\mathbf{r}_N\, d^3\mathbf{p}_1 \cdots d^3\mathbf{p}_N$. That is, the symbol $d^3\mathbf{r}_1$ stands for the volume element $dx_1\, dy_1\, dz_1$, and $d^3\mathbf{p}_1$ represents $dp_{x_1}\, dp_{y_1}\, dp_{z_1}$. Let the potential energy of interaction between the N molecules be represented by $U(\mathbf{r}_1, \ldots, \mathbf{r}_N)$. In writing this potential, we limit our consideration to molecules which are effectively spherically symmetrical in the sense that the potential of interaction, as written above, depends only on the relative positions of the molecules and not on their relative orientation. The extension to the more general case, however, offers no difficulties in principle.

The molecules or atoms may have a set of internal energy levels; if so, we assume that these are unaffected by the other molecules. This is usually a very good assumption for dilute gases; it may or may not be satisfactory in dense gases or liquids.

The internal energy levels are denoted by ϵ_k, and $q_\text{int} = \sum_k e^{-\beta \epsilon_k}$.

Assume that corrected Boltzmann statistics apply. In setting up the partition function, we shall integrate over the entire volume in phase space, thus effectively regarding the molecules as distinguishable, and then divide the partition function by $N!$ since the molecules are actually indistinguishable.

The canonical partition function is

$$Q = \sum_t e^{-E_t/kT} \qquad (15\text{-}5)$$

The energy E_t of the tth system quantum state may be expressed as

$$E_t = \sum_{i=1}^{N} \epsilon_{k,\text{int}} + \sum_{i=1}^{3N} \frac{p_{x_i}^2 + p_{y_i}^2 + p_{z_i}^2}{2m} + U(\mathbf{r}_1, \ldots, \mathbf{r}_N) \quad (15\text{-}6)$$

The first term refers to the internal energy state, denoted by the quantum number k, for each of the N molecules in the system; the second and third terms are treated classically and represent the translational kinetic energy and the potential energy of the atoms at the point in phase space, $\mathbf{p}_1, \ldots, \mathbf{p}_N, \mathbf{r}_1, \ldots, \mathbf{r}_N$. That is, the tth quantum state is characterized by having one particle in the internal state k_1, with translational momentum p_{x_1}, p_{y_1}, p_{z_1}, and with position vector $\mathbf{r}_1$; a second particle in the internal state k_2, with translational momentum p_{x_2}, p_{y_2}, p_{z_2}, and with position vector $\mathbf{r}_2$; etc.

The several degrees of freedom are separable, and the partition function is

$$Q = (q_{\text{int}})^N \frac{1}{N! h^{3N}} \int \cdots \int e^{-\beta \sum_{i=1}^{N}(p_{x_i}^2 + p_{y_i}^2 + p_{z_i}^2)/2m} e^{-\beta U(\mathbf{r}_1, \ldots, \mathbf{r}_N)} d^3\mathbf{p}_1 \cdots d^3\mathbf{p}_N \, d^3\mathbf{r}_1 \cdots d^3\mathbf{r}_N \quad (15\text{-}7)$$

The factor h^{3N} arises as usual because the partition function is approximated by an integral in classical phase space.

The integral over the momenta can be performed; each of the $3N$ factors is

$$\int_{-\infty}^{+\infty} e^{-p^2/2mkT} \, dp = (2\pi mkT)^{1/2}$$

Set
$$\Lambda = \left(\frac{h^2}{2\pi mkT}\right)^{1/2} \quad (15\text{-}8)$$

As previously remarked (Prob. 6-2), Λ is some sort of an average De Broglie wavelength for particles at temperature T.

The partition function therefore is

$$Q = \frac{(q_{\text{int}})^N}{N!} \left(\frac{2\pi mkT}{h^2}\right)^{3N/2} Z$$

$$= \frac{(q_{\text{int}})^N}{N!} (\Lambda)^{-3N} Z \quad (15\text{-}9)$$

where $\quad Z(T,V) = \int \cdots \int e^{-\beta U(\mathbf{r}_1, \ldots, \mathbf{r}_N)} d^3\mathbf{r}_1 \cdots d^3\mathbf{r}_N \quad (15\text{-}10)$

The function $Z(T,V)$ is known as the configuration integral.

The free energy is

$$A = -kT \ln Q = -NkT \ln q_{\text{int}} + NkT \ln N - NkT + 3NkT \ln \Lambda - kT \ln Z(T,V) \quad (15\text{-}11)$$

The only term in (15-11) which depends on the volume is the last one; the other terms contribute to the thermodynamic functions of the substance at infinite dilution, i.e., to the perfect-gas thermodynamic functions, but the real-gas effects are all due to the configuration integral $Z(T,V)$. In particular, the pressure is given by

$$P = -\left(\frac{\partial A}{\partial V}\right)_T = kT\frac{\partial[\ln Z(T,V)]}{\partial V} \qquad (15\text{-}12)$$

15-4. The Second Virial Coefficient. We assume: (*a*) that the potential energy of interaction between a pair of molecules is a function of their distance apart, and not of their relative orientation; (*b*) that the potential energy of the gas is a sum of pair potentials,

$$U(\mathbf{r}_1, \ldots, \mathbf{r}_N) = \sum_{1 \le i < j \le N} u(r_{ij}) \qquad (15\text{-}13)$$

where r_{ij} is the scalar magnitude of the distance between particles i and j, $r_{ij} = |\mathbf{r}_j - \mathbf{r}_i|$.

Assumptions (*a*) and (*b*) are quite different in character. It is not too difficult to extend the treatment we shall give for spherically symmetrical potentials [assumption (*a*)] to include angle-dependent potentials. This latter, more general treatment is necessary for molecules which have permanent dipole moments or have very nonspherical shapes. Assumption (*b*) asserts that the forces between molecules are additive and do not show any saturation characteristics, so that the interaction between molecules i and j, for example, is not changed by the fact that they are both interacting with a nearby molecule k.

If (15-13) holds,

$$e^{-\beta U(\mathbf{r}_1,\ldots,\mathbf{r}_N)} = \prod_{1 \le i < j \le N} e^{-\beta u(r_{ij})} \qquad (15\text{-}14)$$

If the interaction is typically small, that is, if $u(r)/kT \ll 1$ for most of the important values of r, it is expedient to introduce a symbol for the deviation from unity; we set

$$\begin{aligned} f_{ij} &= e^{-\beta u(r_{ij})} - 1 \\ e^{-\beta u(r_{ij})} &= 1 + f_{ij} \end{aligned} \qquad (15\text{-}15)$$

The f_{ij} have the property that $f_{ij} \to 0$ as $r_{ij} \to \infty$. Then,

$$e^{-\beta U(\mathbf{r}_1,\ldots,\mathbf{r}_N)} = \prod_{1 \le i < j \le N} (1 + f_{ij}) \qquad (15\text{-}16)$$

Expand the infinite product,

$$e^{-\beta U} = 1 + \sum_{1 \le i < j \le N} f_{ij} + \sum f_{ij} f_{i'j'} + \cdots \qquad (15\text{-}17)$$

[We shall be more explicit later about the allowed values of i, j, i', and j' in the double sum $\Sigma f_{ij} f_{i'j'}$ of (15-17).]

The f_{ij} are by assumption small; so we begin our analysis of (15-17) by assuming that only the first term Σf_{ij} is important. Then

$$Z(T,V) = \int \cdots \int e^{-\beta U(\mathbf{r}_1,\ldots,\mathbf{r}_N)} \, d^3\mathbf{r}_1 \cdots d^3\mathbf{r}_N$$
$$\approx \int \cdots \int (1 + \Sigma f_{ij}) \, d^3\mathbf{r}_1 \cdots d^3\mathbf{r}_N \quad (15\text{-}18)$$

The integral $\int \cdots \int d^3\mathbf{r}_1 \cdots d^3\mathbf{r}_N = V^N$. The second term in the integral in (15-18) consists of $N(N-1)/2$ identical terms, since each of the position vectors $\mathbf{r}_1, \ldots, \mathbf{r}_N$ ranges over the entire volume of the system. A typical term is

$$\int \cdots \int f_{12} \, d^3\mathbf{r}_1 \cdots d^3\mathbf{r}_N = \int \cdots \int (e^{-\beta u(r_{12})} - 1) \, d^3\mathbf{r}_1 \cdots d^3\mathbf{r}_N$$

We can integrate over $\mathbf{r}_3, \mathbf{r}_4, \ldots, \mathbf{r}_N$,

$$\int \cdots \int f_{12} \, d^3\mathbf{r}_1 \cdots d^3\mathbf{r}_N = V^{N-2} \int\int (e^{-\beta u(r_{12})} - 1) \, d^3\mathbf{r}_1 \, d^3\mathbf{r}_2 \quad (15\text{-}19)$$

Introduce the definition

$$b_2 = \frac{1}{2V} \int\int (e^{-\beta u(r_{12})} - 1) \, d^3\mathbf{r}_1 \, d^3\mathbf{r}_2 \quad (15\text{-}20)$$

Then
$$\int \cdots \int f_{12} \, d^3\mathbf{r}_1 \cdots d^3\mathbf{r}_N = 2V^{N-1} b_2 \quad (15\text{-}21)$$

We now transform coordinates to introduce the relative separation $\mathbf{r}_{12} = \mathbf{r}_1 - \mathbf{r}_2$ explicitly. Instead of cartesian coordinates $x_1, y_1, z_1, x_2, y_2, z_2$, we use x_1, y_1, z_1 and $x_{21} = x_2 - x_1$; $y_{21} = y_2 - y_1$, $z_{21} = z_2 - z_1$. The Jacobian of the transformation is readily seen to be unity, and the integral

$$b_2 = \frac{1}{2V} \int\int (e^{-\beta u(r_{12})} - 1) \, d^3\mathbf{r}_1 \, d^3\mathbf{r}_2$$

becomes

$$b_2 = \frac{1}{2V} \int \cdots \int (e^{-\beta u(r_{12})} - 1) \, dx_1 \, dy_1 \, dz_1 \, dx_2 \, dy_2 \, dz_2$$
$$= \frac{1}{2V} \int \cdots \int (e^{-\beta u(r_{12})} - 1) \, dx_1 \, dy_1 \, dz_1 \, dx_{21} \, dy_{21} \, dz_{21}$$

In principle, the limits of integration for x_{21}, y_{21}, z_{21} are dependent on the coordinates x_1, y_1, z_1 in the sense that, for values of x_1, y_1, z_1 close to a wall of the container, x_{21}, y_{21}, z_{21} are limited so that particle 2 remains in the container. Since the range of the intermolecular forces is small, this restriction is unimportant; furthermore, as we shall see, as $r_{ij} \to \infty$, $f_{ij} \to 0$ sufficiently rapidly so that we can take the limits of integration for x_{21}, y_{21}, z_{21} as $-\infty$ to $+\infty$ and the limits of integration for x_1, y_1, z_1 as the boundaries of the container. The integral over x_1, y_1, z_1 then gives V and

$$b_2 = \frac{1}{2V} V \int_{-\infty}^{+\infty} \cdots \int (e^{-\beta u(r_{12})} - 1) \, dx_{21} \, dy_{21} \, dz_{21}$$

Transform the spherical coordinates; $x_{21} = r_{12} \sin \theta \cos \phi$, $y_{21} = r_{12} \sin \theta \sin \phi$, $z_{21} = r_{12} \cos \theta$; and integrate over the angle variables.

$$b_2 = 2\pi \int_{r_{12}=0}^{r_{12}=\infty} (e^{-\beta u(r_{12})} - 1) r_{12}^2 \, dr_{12} = 2\pi \int_{r=0}^{r=\infty} (e^{-\beta u(r)} - 1) r^2 \, dr \quad (15\text{-}22)$$

In the approximate evaluation of the configuration integral in (15-18), there are $N(N-1)/2 \approx N^2/2$ identical terms, $\int \cdots \int f_{ij} \, d^3\mathbf{r}_1 \cdots d^3\mathbf{r}_N$, and each one is equal to $2V^{N-1} b_2$. Therefore,

$$Z(T,V) \approx V^N + V^{N-1} N^2 b_2 = V^N \left(1 + \frac{N^2 b_2}{V}\right) \quad (15\text{-}23)$$

The pressure is to be calculated from $P = kT \, \partial[\ln Z(T,V)]/\partial V$.

$$\ln Z(T,V) = N \ln V + \ln\left(1 + \frac{N^2 b_2}{V}\right) \quad (15\text{-}24)$$

If b_2 is sufficiently small, so that $N^2 b_2/V \ll 1$, the second term on the right-hand side of (15-24) may be expanded to give

$$\ln\left(1 + \frac{N^2 b_2}{V}\right) \approx \frac{N^2 b_2}{V} \quad (15\text{-}25a)$$

Then
$$\ln Z(T,V) = N \ln V + \frac{N^2 b_2}{V} \quad (15\text{-}25b)$$

$$P = kT \frac{\partial (\ln Z)}{\partial V} = \frac{NkT}{V} - kT \frac{N^2 b_2}{V^2} = \frac{NkT}{V}\left(1 - \frac{Nb_2}{V}\right) \quad (15\text{-}26)$$

We rearrange (15-26) and then introduce the molar volume, $\mathrm{v} = VN_0/N$,

$$\frac{PV}{NkT} = 1 - \frac{Nb_2}{V} = 1 - \frac{N_0 b_2}{\mathrm{v}} \quad (15\text{-}27)$$

By comparison with (15-11a), the second virial coefficient is

$$B(T) = -N_0 b_2 = 2\pi N_0 \int_0^\infty (1 - e^{-\beta u(r)}) r^2 \, dr \quad (15\text{-}28)$$

The result in (15-28) is correct and important. The derivation given is a fraud and a hoax. The error can be recognized by considering the expression $\ln(1 + N^2 b_2/V)$, which was expanded on the assumption that $N^2 b_2/V \ll 1$. But, as we shall see in detail later, the quantity b_2, which has the dimensions of volume, is of the order of magnitude of the volume of an atom. Then, for a mole of substance, Nb_2 is of the order of the molar volume of the liquid; at 1 atm pressure, molar volume of the gas is 200 to 1,000 times greater. Then $Nb_2/V \sim 10^{-3}$; but $N^2 b_2/V \sim 10^{21}$, and the approximation used to obtain (15-25) is thoroughly unjustified.

The resolution of this dilemma is the following: In the expansion of the product,

$$\prod_{1 \le i < j \le N}^{N} (1 + f_{ij}) = 1 + \sum_{1 \le i < j \le N} f_{ij} + \sum f_{ij} f_{i'j'} + \cdots \quad (15\text{-}29)$$

[cf. (15-16) and (15-17)] we retained only the terms Σf_{ij} and neglected the further terms such as $\Sigma f_{ij}f_{i'j'}$. This is not justified. When proper account is taken of the higher terms, it is found that, at low densities, the configuration integral can be written as

$$Z(T,V) = V^N\left(1 + \frac{Nb_2}{V}\right)^N + \text{"other terms"} \tag{15-30}$$

where the contribution of the "other terms" becomes negligible as the gas becomes dilute, that is, when $N_2b_2/V \ll 1$.

If we write

$$Z(T,V) = V^N\left(1 + \frac{Nb_2}{V}\right)^N \tag{15-31}$$

then, by the binomial expansion,

$$Z(T,V) = V^N\left[1 + \frac{N^2b_2}{V} + \frac{N^3(N-1)}{2}\left(\frac{b_2}{V}\right)^2 + \cdots\right] \tag{15-32}$$

The first two terms are identical with (15-23), but of course the further terms, starting with $\frac{1}{2}N^3(N-1)(b_2/V)^2$, are by no means negligible.

Proceeding from the correct approximate expression (15-31),

$$\ln Z(T,V) = N \ln V + N \ln\left(1 + \frac{Nb_2}{V}\right) \tag{15-33}$$

Since $Nb_2/V \ll 1$, the expansion $\ln(1 + Nb_2/V) \approx Nb_2/V$ is justified and

$$\ln Z(T,V) = N \ln V + \frac{N^2b_2}{V} \tag{15-34}$$

This is the equation previously derived in an unjustified way [(15-25b)]. Then, proceeding as before, since $P = kT\, \partial(\ln Z)/\partial V$, we arrive again at Eq. (15-27) for the second virial coefficient.

The detailed analysis of the terms in Eq. (15-17) is known as the theory of the cluster integrals.* This is a very complicated and slightly difficult subject, which we shall not treat.

However, the general nature of the derivation of Eq. (15-30) can be indicated in a rather simple way. We shall consider the terms in the sum $\Sigma f_{ij}f_{i'j'}$.

Consider the case of four particles. The product is

$$(1 + f_{12})(1 + f_{13})(1 + f_{14})(1 + f_{23})(1 + f_{24})(1 + f_{34})$$

The terms which are the product of two factors $f_{ij}f_{i'j'}$ are $f_{12}f_{13} + f_{12}f_{14} + f_{12}f_{23} + f_{12}f_{24} + \underline{f_{12}f_{34}} + f_{13}f_{14} + f_{13}f_{23} + \underline{f_{13}f_{24}} + f_{13}f_{34} + f_{14}f_{23} + f_{14}f_{24} + f_{14}f_{34} + f_{23}f_{24} + \underline{f_{23}f_{34}}$. There are 15 terms; for the 3 underscored, $f_{12}f_{34}$,

---

* Mayer and Mayer, chap. 13 [5]; Hill, "Statistical Mechanics," pp. 122–129, 136–144 [9a].

$f_{13}f_{34}$, and $f_{14}f_{23}$, the two f's have no indices in common; for the 12 other terms, the two f's have one index number in common.

In the general case, the number of f_{ij} terms is $\frac{1}{2}N(N-1)$. The total number of product terms $f_{ij}f_{i'j'}$ is

$$\frac{\frac{N(N-1)}{2}\left[\frac{N(N-1)}{2}-1\right]}{2} \tag{15-35}$$

For large N, this is approximately $N^4/8$.

There are $\frac{1}{2}N(N-1)$ f_{ij}'s; for any choice of f_{ij}, there are $\frac{1}{2}(N-2)(N-3)$ ways of choosing two indices $i'j'$ for $f_{i'j'}$ excluding i and j; we divide by 2 again so that each product $f_{ij}f_{i'j'}$ will occur only once. Therefore the total number of two product terms with no indices in common is

$$\frac{N(N-1)(N-2)(N-3)}{2 \cdot 2 \cdot 2} \tag{15-36}$$

For large N, this is also approximately $N^4/8$.

The number of ways of having the product $f_{ij}f_{i'j'}$ with one index in common is the difference between (15-35) and (15-36),

$$\frac{N(N-1)}{8}[N(N-1)-2-(N-2)(N-3)] = \frac{N(N-1)(N-2)}{2}$$

$$\approx \frac{N^3}{2} \quad \text{for large } N \tag{15-37}$$

Therefore, for large N, there are many more terms with no indices in common. It is then plausible to believe that, at high dilution, the product terms with no indices in common make the principal contribution to the sum,

$$\Sigma \int \cdots \int f_{ij}f_{i'j'}\, d^3\mathbf{r}_1 \cdots d^3\mathbf{r}_N$$

For no indices in common, a typical term is

$$\int \cdots \int f_{12}f_{34}\, d^3\mathbf{r}_1\, d^3\mathbf{r}_2\, d^3\mathbf{r}_3\, d^3\mathbf{r}_4 \cdots d^3\mathbf{r}_N \tag{15-38}$$

Integration over $d^3\mathbf{r}_5 \cdots d^3\mathbf{r}_N$ gives a factor of V^{N-4}. The integral over $d^3\mathbf{r}_1\, d^3\mathbf{r}_2$ is completely independent of the integrals over $d^3\mathbf{r}_3\, d^3\mathbf{r}_4$; each such integral gives a factor of $2b_2V$. Thus

$$\int \cdots \int f_{ij}f_{i'j'}\, d^3\mathbf{r}_1 \cdots d^3\mathbf{r}_N = 4V^{N-2}b_2^2 \tag{15-39}$$

There are approximately $N^4/8$ such terms; so the contribution to the configuration integral is $\frac{1}{2}N^4V^{N-2}b_2^2$.

In the same way, the number of triple product terms $f_{ij}f_{i'j'}f_{i''j''}$ with no indices in common is

$$\frac{N(N-1)}{2} \cdot \frac{(N-2)(N-3)}{2} \cdot \frac{(N-4)(N-5)}{2} \cdot \frac{1}{6} \approx \frac{N^6}{48} \tag{15-40}$$

(The factor $\frac{1}{6}$ ensures that no term is counted twice.)

The value of each term

$$\int \cdots \int f_{12} f_{34} f_{56} \, d^3\mathbf{r}_1 \cdots d^3\mathbf{r}_N = (2Vb_2)^3 V^{N-6}$$
$$= 8 b_2^3 V^{N-3} \tag{15-41}$$

The contribution of these terms to the configuration integral is $\tfrac{1}{6} N^6 V^{N-3} b_2^3$.

Adding up all the contributions to the configuration integral from the expansion of $\Pi(1 + f_{ij})$ up through the triple product terms, we have

$$Z(T,V) \approx V^N + N^2 V^{N-1} b_2 + \tfrac{1}{2} N^4 V^{N-2} b_2^2 + \tfrac{1}{6} N^6 V^{N-3} b_2^3 + \cdots \tag{15-42a}$$

$$Z(T,V) = V^N \left[1 + N \frac{Nb_2}{V} + \tfrac{1}{2} N^2 \left(\frac{Nb_2}{V}\right)^2 + \tfrac{1}{6} N^3 \left(\frac{Nb_2}{V}\right)^3 + \cdots \right] \tag{15-42b}$$

If the first approximation to the configuration integral for a dilute gas is really

$$Z(T,V) \approx V^N \left(1 + \frac{Nb_2}{V}\right)^N \tag{15-31}$$

as asserted previously, then the binomial expansion of the right-hand side gives

$$Z(T,V) = V^N \left[1 + N \frac{Nb_2}{V} + \frac{N(N-1)}{2} \left(\frac{Nb_2}{V}\right)^2 \right.$$
$$\left. + \frac{N(N-1)(N-2)}{6} \left(\frac{Nb_2}{V}\right)^3 + \cdots \right] \tag{15-43}$$

For large N, Eqs. (15-42b) and (15-43) are identical up to terms of the order of $(Nb_2/V)^3$. Thus, we have made it plausible that, by a consistent look at the dominant terms in the expansion of $\Pi(1 + f_{ij})$, Eq. (15-30) or (15-31) will result and the formula (15-28) for the second virial coefficient is justified.

An alternative derivation of the equations for the virial coefficient is presented in Sec. 15-14.

15-5. The Second Virial Coefficient for the Rigid-sphere Model. If atoms were rigid spheres of diameter σ, then the distance of closest approach between the centers of two atoms would be σ and the potential function is

$$u(r) = 0 \quad r > \sigma \qquad u(r) = \infty \quad r \leq \sigma \tag{15-44}$$

so that $\quad 1 - e^{-\beta u(r)} = 0 \quad r > \sigma \qquad 1 - e^{-\beta u(r)} = 1 \quad r \leq \sigma$
$$\tag{15-45}$$

The second virial coefficient is given by

$$B(T) = 2\pi N_0 \int_0^\infty (1 - e^{-\beta u(r)}) r^2 \, dr \tag{15-28}$$

In the present instance, this is

$$B(T) = 2\pi N_0 \int_0^\sigma r^2 \, dr = \frac{2\pi N_0 \sigma^3}{3} \tag{15-46}$$

The volume of a molecule is $(4\pi/3)(\sigma/2)^3 = \frac{1}{4}(2\pi/3)\sigma^3$; for the rigid-sphere model, the virial coefficient is four times the molar volume of the molecules. It is positive and independent of temperature.

Problem 15-2. Suppose that the potential of interaction, $u(r)$, is represented by

$$u(r) = -\frac{A}{r^m} \quad \text{for } r > \sigma \qquad u(r) = \infty \quad r \leq \sigma \tag{15-47}$$

where m is a positive integer and A is a positive constant. For sufficiently large r, $|\beta u(r)| \ll 1$, and the exponential in the integral for the virial coefficient [(15-28)] can be expanded. What values of m are acceptable if the integral is to converge as $r \to \infty$?

Problem 15-3. Show that, provided that the conditions of Prob. 15-2 are satisfied, the integral for the second virial coefficient [(15-28)] can be transformed (by integration by parts) to

$$B(T) = -\frac{2\pi N_0}{3} \int_0^\infty \beta e^{-\beta u(r)} \left(\frac{du}{dr}\right) r^3 \, dr \tag{15-48}$$

15-6. More Realistic Intermolecular Potentials. We shall give only a brief descriptive résumé of this subject.

It can be shown on quite general grounds that the leading term in the attractive potential between a pair of nonpolar molecules is due to the so-called London dispersion forces, which arise from the distortion of the charge distribution of the isolated molecule due to the electrostatic interaction with the electrons of the neighboring molecule.*

An approximate perturbation-theory calculation gives for the potential of interaction between two atoms A and B (in S states, with no orbital angular momentum)

$$u(r) = \frac{-3\alpha_A \alpha_B}{r^6} \frac{I_A I_B}{I_A + I_B} \tag{15-49}$$

where α_A and α_B are the electrostatic polarizabilities (cf. Chap. 18) of the two atoms and I_A and I_B are energy quantities of the order of magnitude of the ionization potentials of the atoms. Another approximate formula for the potential is

$$u(r) = \frac{-3e\hbar}{2r^6} \alpha_A \alpha_B \left[\left(\frac{\alpha_A}{n_A}\right)^{1/2} + \left(\frac{\alpha_B}{n_B}\right)^{1/2}\right] \tag{15-50}$$

where n_A is the number of electrons in the outer shell of atom A.

These formulae are not sufficiently accurate for practical computations;

---

* F. London, *Z. physik. Chem.*, **B11**: 222 (1931); *Z. Physik*, **63**: 245 (1930). The present status of the subject is reviewed by K. S. Pitzer in I. Prigogine (ed.), "Advances in Chemical Physics," vol. II, pp. 59–82, Interscience, New York, 1959.

they do serve to emphasize that the attractive potential varies as $1/r^6$ and that atoms of high polarizability and low ionization potential show the strongest intermolecular potential.

In a vertical column in the periodic table, the attractive forces increase with increasing atomic number.

At shorter distances, when the electron clouds of the two atoms begin to interpenetrate, there is a steep repulsive interaction. Neither theory nor experiment has yielded an accurate general functional representation for this potential. One approximate representation is to take the potential as varying as $1/r^n$ (with $n > 6$); this leads to the generalized Lennard-Jones potential

$$u(r) = \frac{\lambda}{r^n} - \frac{\nu}{r^6} \qquad n > 6 \tag{15-51}$$

The most commonly used potential is the Lennard-Jones 6–12 potential

$$u(r) = 4\epsilon \left[\left(\frac{\sigma}{r}\right)^{12} - \left(\frac{\sigma}{r}\right)^{6} \right] \tag{15-52}$$

The potential is zero owing to the balancing of the attractive and repulsive terms at $r = \sigma$; by differentiation of (15-52) it can be seen that the minimum in the potential occurs at $r = r_m = 2^{1/6}\sigma \ (= 1.122\sigma)$ and that the depth of the minimum is ϵ. If we use r_m rather than σ as a parameter, the equation for the potential is

$$u(r) = \epsilon \left[\left(\frac{r_m}{r}\right)^{12} - 2\left(\frac{r_m}{r}\right)^{6} \right] \tag{15-53}$$

The potential is positive for $r < \sigma$; it is negative for $r > \sigma$, and for large r,

$$u(r) \approx -4\epsilon \frac{\sigma}{r^6} \tag{15-54}$$

The curve of Fig. 15-2 is actually a Lennard-Jones 6–12 potential.

The repulsive contribution to the potential is sometimes represented by an exponential, $u(r) \sim e^{-\alpha r}$. There are a number of potentials of this type which have been used; one of the more popular ones is the modified Buckingham potential

$$u(r) = \frac{\epsilon}{1 - (6/\alpha)} \left[\frac{6}{\alpha} e^{\alpha(1-r/r_m)} - \left(\frac{r_m}{r}\right)^{6} \right] \quad \text{for } r \geq r_{\max}$$
$$= \infty \qquad\qquad\qquad\qquad\qquad\qquad\qquad\qquad \text{for } r < r_{\max} \tag{15-55}$$

This potential has a minimum of depth ϵ at $r = r_m$; the point $r = \sigma$ at which $u(r) = 0$ depends on α, for $\alpha = 12$, which is close to the values encountered in practice, $\sigma/r_m = 0.876$, whereas, for the Lennard-Jones

6–12 potential, $\sigma/r_m = 0.891$. For $r < \sigma$, the repulsive potential is essentially $u(r) = A\, e^{-\alpha r/r_m}$; thus the larger α, the steeper the potential.

There is a maximum in the analytical expression for $u(r)$ of (15-55) at $r = r_{\max}$. This is physically unimportant because it occurs at small values of r and for large values of $u(r)$. For example, for $\alpha = 12$, $r_{\max}/r_m = 0.30$, and $u(r_{\max}) = 1{,}705\epsilon$.[*] To avoid mathematical difficulties, we take $u(r) = \infty$ for $r < r_{\max}$.

In principle, the potential between two molecules that are not spherically symmetric (or, for that matter, between such a molecule and an atom) is not spherically symmetric but depends on the relative orientation of the two molecules with respect to each other and with respect to the vector between the centers of the two molecules. The anisotropic nature of the interaction will to a very considerable extent be averaged out because of the rotational motions of the molecules. In any case, it is customary to use the spherically symmetric potentials discussed in this section for simple molecules with shapes that do not differ too much from spheres. Presumably, the more unsymmetrical the molecule, the less useful this approximation.[†]

A molecule such as CCl_4 is approximately spherically symmetric. As two molecules approach, however, the dominant interaction is between those chlorine atoms of the two molecules which are closest to each other. If the chlorine-atom–chlorine-atom interactions are approximately described by a Lennard-Jones 6–12 potential, the functional form for the potential when expressed in terms of the distance between the centers of the molecules would not be a 6–12 potential. Nevertheless, it is customary to interpret experimental data for such substances with the same potential that is used for atoms.[‡]

15-7. The Second Virial Coefficient and Other Properties of a Dilute Gas for the Lennard-Jones 6–12 Potential. For the potential

$$u(r) = 4\epsilon\left[\left(\frac{\sigma}{r}\right)^{12} - \left(\frac{\sigma}{r}\right)^{6}\right]$$

---

[*] HCB, p. 234.

[†] For a calculation of the effective potential around an N_2 molecule taking these effects into account, see I. Amdur, E. Mason, and J. Jordan, *J. Chem. Phys.*, **27**: 527 (1957).

[‡] S. D. Hamann and J. A. Lambert, *Australian J. Chem.*, **7**: 1 (1954), have made a quantitative calculation for the interaction between a pair of quasi-spherical molecules (such as a pair of CF_4, CCl_4, SiF_4, or SF_6 molecules), assuming that the potential energy of interaction of the molecules is the sum of the potentials for the interactions between the constituent atoms and that the latter are given by Lennard-Jones 6–12 potentials. The conclusion is that, by using as a variable the distance between the centers of molecules, a Lennard-Jones 7–28 potential gives a good representation for the potential energy of interaction. It is claimed that this potential explains the properties of such fluids more satisfactorily than the 6–12 potential.

we introduce dimensionless variables and functions

$$r^* = \frac{r}{\sigma} \qquad T^* = \frac{kT}{\epsilon} \qquad B^* = \frac{B}{\frac{2}{3}\pi N_0 \sigma^3} \qquad (15\text{-}56)$$

The potential can now be written as

$$\frac{u(r)}{kT} = \frac{4}{T^*}\left(\frac{1}{r^{*12}} - \frac{1}{r^{*6}}\right) \qquad (15\text{-}57)$$

so that

$$\frac{1}{kT}\frac{du(r)}{dr} = \frac{4}{T^*}\left(-\frac{12}{r^{*13}} + \frac{6}{r^{*7}}\right)\frac{1}{\sigma} \qquad (15\text{-}58)$$

The second-virial-coefficient formula

$$B(T) = 2\pi N_0 \int (1 - e^{-u(r)/kT}) r^2 \, dr \qquad (15\text{-}28)$$

or

$$B(T) = -\frac{2\pi N_0}{3} \int \frac{e^{[-u(r)/kT]}}{kT}\left(\frac{du}{dr}\right) r^3 \, dr \qquad (15\text{-}48)$$

becomes

$$B^*(T^*) = \frac{B(T)}{(2\pi/3)N_0\sigma^3} = 3\int_0^\infty (1 - e^{-(4/T^*)(1/r^{*12} - 1/r^{*6})}) r^{*2} \, dr^* \qquad (15\text{-}59)$$

or

$$B^*(T^*) = -\int_0^\infty e^{-(4/T^*)(1/r^{*12} - 1/r^{*6})} \frac{4}{T^*}\left(-\frac{12}{r^{*12}} + \frac{6}{r^{*6}}\right) r^{*3} \, dr^* \qquad (15\text{-}60)$$

The function $B^*(T^*)$ is a function of the single parameter T^*. Values are tabulated.† If the molecular parameter ϵ is known for a substance, T^* can be calculated for any given T. From the value of B^* and the value of σ for the particular substance, the actual second virial coefficient can be calculated. The function $B^*(T^*)$ is plotted in Fig. 15-3.

It can be seen from the figure that parameters ϵ and σ can be chosen for a group of atoms and more or less spherical, nonpolar molecules such that the Lennard-Jones 6–12 potential gives a good representation of the virial-coefficient data over a wide temperature range. This does not constitute a critical test of the Lennard-Jones potential since equally good fits can be obtained with other two-parameter or three-parameter potentials. Any generalized Lennard-Jones potential (15-51) with a repulsive exponent n in the range 10 to 14 does just about as well as the 6–12 potential. To some extent, the popularity of the $1/r^{12}$ repulsive potential is based on its mathematical convenience.

Inspection of Fig. 15-3 reveals that, at low temperatures, $B(T)$ is negative $(pV/NkT < 1)$, whereas, at high temperatures, $B(T)$ is positive. In the former case, the dominant contribution to the integral (15-28) for $B(T)$ comes from the region where $u(r) < 0$, so that $1 - e^{-\beta u(R)} < 0$; at high temperatures, the typical collisions are more energetic, and the important contribution to the second virial coefficient arises from the region of small r, where $1 - e^{-\beta u(r)} > 0$.

† HCB, table I, pp. 1114–1115.

The Boyle point, where the second virial coefficient is zero because of the balance between repulsive and attractive forces, occurs at $T^* = 3.40$.

Representative values of the Lennard-Jones (6–12) parameters as deduced from second virial coefficients are given in Table 15-1.

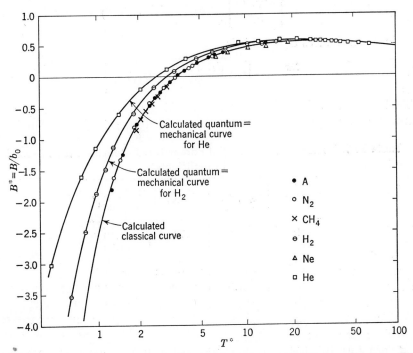

FIG. 15-3. The reduced second virial coefficient for the Lennard-Jones potential. The classical curve of $B^*(T^*)$ is shown, along with experimental points for several gases. Also shown are the curves for hydrogen and helium, which have been calculated by the quantum-mechanical theory of virial coefficients (cf. Sec. 15-13). (*From J. O. Hirschfelder, C. F. Curtiss, and R. B. Bird, "Molecular Theory of Gases and Liquids,"* p. 164, Wiley, New York, 1954; these authors took the figure from the doctoral dissertation by R. J. Lunbeck, Amsterdam, 1950.)

The Joule-Thompson coefficient, $\mu = (\partial T/\partial P)_H$, is related to the equation of state as follows:

$$\mu = \frac{1}{C_p}\left[T\left(\frac{\partial V}{\partial T}\right)_P - V\right] = \frac{T^2}{C_p}\left[\frac{\partial(V/T)}{\partial T}\right]_P \quad (15\text{-}61)$$

For a perfect gas, $\mu = 0$. For a real gas, the molal heat capacity is itself a function of pressure, but, for the limiting low-pressure value of μ, we can use c_p°, the molal heat capacity at zero pressure (which is the same as the molal heat capacity for the hypothetical ideal gas). To evaluate $[\partial(v/T)/\partial T]_P$, write the virial expansion in pressure units [cf. Eq. (15-2)].

$$\frac{Pv}{RT} = 1 + \frac{B(T)P}{RT} + \cdots$$

$$\frac{v}{T} = \frac{R}{P} + \frac{B(T)}{T} + \cdots \qquad (15\text{-}62)$$

$$T^2\left[\frac{\partial(v/T)}{\partial T}\right]_P = T\frac{dB(T)}{dT} - B(T) \qquad (15\text{-}63)$$

In dimensionless units, this is

$$\frac{2\pi}{3} N_0 \sigma^3 \left[T^* \frac{dB^*(T^*)}{dT^*} - B^*(T^*) \right]$$

Tables are provided in HCB for calculating the Joule-Thompson coefficient at zero pressure.* At low temperatures, μ is positive, and the

Table 15-1. Lennard-Jones Potential Parameters Determined from Second Virial Coefficients*

Gas	ϵ/k, °K	σ, Å	$(2\pi/3)N_0\sigma^3$, cc mole^{-1}
Ne	34.9	2.78	27.10
A	119.8	3.405	49.80
Kr	171	3.60	58.86
Xe	221	4.10	86.94
N_2	95.05	3.698	63.78
O_2	118	3.46	52.26
CH_4	148.2	3.817	70.16
CO_2	189	4.486	113.9

* From J. O. Hirschfelder, C. F. Curtiss, and R. B. Bird, "Molecular Theory of Gases and Liquids," p. 165, Wiley, New York, 1954, where the references to experimental data are given; a more complete table is given by these authors on pp. 1110–1112.

gas cools in an isoenthalpic expansion; at high temperatures, μ is negative, and the reverse is the case. The inversion temperature occurs at $T^* = 6.47$. An understanding of these phenomena is of course essential for the design of gas liquefiers based on Joule-Thompson expansion cooling.

We have already indicated the straightforward way in which tables of $B^*(T^*)$ as a function of T^* can be used to calculate second virial coefficients from Lennard-Jones parameters. The inverse problem of determining ϵ and σ when $B(T)$ is given over a range of temperature is a somewhat less straightforward process, which we shall not study.†

15-8. Second Virial Coefficients of Mixtures. We consider a mixture of N_A molecules of type A and N_B molecules of type B. The mole

* HCB, pp. 1114–1115.
† HCB, pp. 166–167.

fractions are

$$X_A = \frac{N_A}{N_A + N_B} \qquad X_B = \frac{N_B}{N_A + N_B} \qquad (15\text{-}64)$$

We assume that the internal energy levels are unaffected by the interactions between molecules, and we omit the internal partition functions. We denote the coordinates of A-type particles by $\mathbf{r}_1, \ldots, \mathbf{r}_{N_A}$ and of B-type particles by $\mathbf{s}_1, \ldots, \mathbf{s}_{N_B}$.

If we proceed exactly as in Sec. 15-3, the canonical partition function for the mixture is

$$Q = \frac{(\Lambda_A)^{-3N_A}(\Lambda_B)^{-3N_B}}{N_A! N_B!} Z \qquad (15\text{-}65)$$

where

$$Z = \int \cdots \int e^{-\beta U(\mathbf{r}_1, \ldots, \mathbf{r}_{N_A}; \mathbf{s}_1, \ldots, \mathbf{s}_{N_B})} d^3\mathbf{r}_1 \cdots d^3\mathbf{r}_{N_A} d^3\mathbf{s}_1 \cdots d^3\mathbf{s}_{N_B} \qquad (15\text{-}66a)$$

and

$$\Lambda_A = \left(\frac{h^2}{2\pi m_A kT}\right)^{1/2} \qquad \Lambda_B = \left(\frac{h^2}{2\pi m_B kT}\right)^{1/2} \qquad (15\text{-}66b)$$

Let r_{ij} be the distance between two A particles, s_{kp} the distance between two B particles, and t_{ik} the distance between an A particle and a B particle: $r_{ij} = |\mathbf{r}_j - \mathbf{r}_i|$; $s_{kp} = |\mathbf{s}_p - \mathbf{s}_k|$; $t_{ik} = |\mathbf{s}_k - \mathbf{r}_i|$. If the total potential energy is the sum of pair potentials, it may be written as

$$U(\mathbf{r}_1, \ldots, \mathbf{r}_{N_A}; \mathbf{s}_1, \ldots, \mathbf{s}_{N_B}) = \sum_{1 \le i < j \le N_A} u_{AA}(r_{ij})$$
$$+ \sum_{1 \le k < p \le N_B} u_{BB}(s_{kp}) + \sum_{\substack{1 \le i \le N_A \\ 1 \le k \le N_B}} u_{AB}(t_{ik}) \qquad (15\text{-}67)$$

where u_{AA}, u_{BB}, and u_{AB} are the potentials of interaction between two A particles, two B particles, and an A and a B particle, respectively.

Then define the f's as before:

$$f_{ij}^A = e^{-\beta u_{AA}(r_{ij})} - 1 \qquad f_{kp}^B = e^{-\beta u_{BB}(s_{kp})} - 1 \qquad f_{ki}^{AB} = e^{-\beta u_{AB}(t_{ki})} - 1 \qquad (15\text{-}68)$$

Then, just as for Eq. (15-16),

$$e^{-\beta U(\mathbf{r}_1, \ldots, \mathbf{s}_{N_B})} = \Pi (1 + f_{ij}^A)(1 + f_{kp}^B)(1 + f_{ki}^{AB}) \qquad (15\text{-}69)$$

By expanding the product in (15-69) and again using the unjustified procedure of retaining only the first term, we obtain

$$e^{-\beta U(\mathbf{r}_1, \ldots, \mathbf{s}_{N_B})} \approx 1 + \sum_{1 \le i < j \le N_A} f_{ij}^A + \sum_{1 \le k < p \le N_b} f_{kp}^B + \sum_{\substack{1 \le k \le N_B \\ 1 \le i \le N_A}} f_{ki}^{AB} \qquad (15\text{-}70)$$

There are $N_A(N_A - 1)/2 \approx N_A^2/2$ terms in the sum Σf_{ij}^A; there are $N_B(N_B - 1)/2 \approx N_B^2/2$ terms in the sum Σf_{kp}^B; there are $N_A N_B$ terms in the sum Σf_{ki}^{AB}.

We define the quantities

$$b_{2AA} = \frac{1}{2V} \iint (e^{-\beta u_{AA}(r_{12})} - 1)\, d^3\mathbf{r}_1\, d^3\mathbf{r}_2 = 2\pi \int_0^\infty (e^{-\beta u_{AA}(r)} - 1) r^2\, dr \tag{15-71a}$$

$$b_{2BB} = \frac{1}{2V} \iint (e^{-\beta u_{BB}(s_{12})} - 1)\, d^3\mathbf{s}_1\, d^3\mathbf{s}_2 = 2\pi \int_0^\infty (e^{-\beta u_{BB}(s)} - 1) s^2\, ds \tag{15-71b}$$

$$b_{2AB} = \frac{1}{2V} \iint (e^{-\beta u_{AB}(t_{12})} - 1)\, d^3\mathbf{r}_2\, d^3\mathbf{s}_1 = 2\pi \int_0^\infty (e^{-\beta u_{AB}(t)} - 1) t^2\, dt \tag{15-71c}$$

Proceeding just as before, we integrate Eq. (15-70) over the configuration space of the $N_A + N_B$ particles and obtain

$$Z(T,V,N_A,N_B) = V^{N_A+N_B} + V^{N_A+N_B-1}(N_A^2 b_{2AA} + N_B^2 b_{2BB} + 2N_A N_B b_{2AB}) + \cdots$$

$$\approx V^{N_A+N_B}\left(1 + \frac{N_A^2 b_{2AA} + N_B^2 b_{2BB} + 2N_A N_B b_{2AB}}{V} + \cdots\right) \tag{15-72}$$

To calculate $\ln Z$, we again make the unjustified assumption that the several quantities $N^2 b_2/V$ are all small compared with unity; thus we obtain

$$\ln Z = (N_A + N_B)\ln V + \frac{N_A^2 b_{2AA} + 2N_A N_B b_{2AB} + N_B^2 b_{2BB}}{V} \tag{15-73}$$

But $P = kT[\partial(\ln Z)/\partial V]_{T,N_A,N_B}$; by differentiation of (15-73),

$$\frac{P}{kT} = \frac{N_A + N_B}{V} - \frac{N_A^2 b_{2A} + 2N_A N_B b_{2AB} + N_B^2 b_{2B}}{V^2}$$

or

$$\frac{PV}{(N_A + N_B)kT} = 1 - \frac{1}{V}\left(\frac{N_A^2 b_{2AA}}{N_A + N_B} + \frac{2N_A N_B b_{2AB}}{N_A + N_B} + \frac{N_B^2 b_{2BB}}{N_A + N_B}\right) \tag{15-74}$$

The molar volume is

$$\mathrm{v} = \frac{VN_0}{N_A + N_B} \tag{15-75}$$

When this is introduced into the right-hand side of (15-74), the desired final equation is

$$\frac{PV}{(N_A + N_B)kT} = 1 - \frac{1}{\mathrm{v}}\left[\left(\frac{N_A}{N_A + N_B}\right)^2 N_0 b_{2AA} + 2\frac{N_A}{N_A + N_B}\frac{N_B}{N_A + N_B} N_0 b_{2AB} + \left(\frac{N_B}{N_A + N_B}\right)^2 N_0 b_{2BB}\right] \tag{15-76}$$

The virial coefficients are

$$B_{AA}(T) = -N_0 b_{2AA} = 2\pi N_0 \int_0^\infty (1 - e^{-\beta u_{AA}(r)}) r^2 \, dr \quad (15\text{-}77a)$$

$$B_{BB}(T) = -N_0 b_{2BB} = 2\pi N_0 \int_0^\infty (1 - e^{-\beta u_{BB}(r)}) r^2 \, dr \quad (15\text{-}77b)$$

$$B_{AB}(T) = -N_0 b_{2AB} = 2\pi N_0 \int_0^\infty (1 - e^{-\beta u_{AB}(r)}) r^2 \, dr \quad (15\text{-}77c)$$

Note that $B_{AA}(T)$ and $B_{BB}(T)$ are the second virial coefficients of pure A and pure B, respectively. With these definitions and the definitions of the mole fractions (15-64), Eq. (15-76) becomes

$$\frac{PV}{(N_A + N_B)kT} = 1 + \frac{1}{V}[X_A^2 B_{AA}(T) + 2X_A X_B B_{AB}(T) + X_B^2 B_{BB}(T)] \quad (15\text{-}78)$$

For a mixture containing C components ($i = 1, 2, \ldots, C$) at mole fractions $X_1, \ldots, X_i, \ldots, X_C$, the mean molar virial coefficient is the obvious generalization of (15-78),

$$\overline{B(T)} = \sum_{i=1}^{C} \sum_{j=1}^{C} X_i B_{ij}(T) X_j \quad (15\text{-}79)$$

where the virial coefficients $B_{ij}(T)$ are defined analogously to Eqs. (15-77).

Equations (15-76) and (15-77) were derived on the basis of the unjustified assumption that the terms of the type $N^2 b_2/V \ll 1$. The equations are correct, although the derivation is not, for the same reasons as were discussed in Sec. 15-4 for a one-component system.

In principle, by an accurate study of the mean second virial coefficient of mixtures and of the pure components (or, more particularly, by measuring the volume change on mixing), one can determine experimental values of the interaction term $B_{AB}(T)$, as well as of $B_{AA}(T)$ and $B_{BB}(T)$.

It has been found for the few cases that have been carefully studied that, if the virial coefficients are interpreted by means of Lennard-Jones potential parameters, the combining rules

$$\sigma_{AB} = \tfrac{1}{2}(\sigma_A + \sigma_B) \qquad \epsilon_{AB} = (\epsilon_A \epsilon_B)^{1/2} \quad (15\text{-}80)$$

give a reasonably satisfactory prediction for the potential of interaction between an A atom and a B atom,

$$u_{AB}(r) = 4\epsilon_{AB}\left[\left(\frac{\sigma_{AB}}{r}\right)^{12} - \left(\frac{\sigma_{AB}}{r}\right)^{6}\right] \quad (15\text{-}81)$$

and for the interaction term for the virial coefficient of a mixture.*

15-9. Thermodynamics of Imperfect Gases. We start with the fundamental equation for a one-component open system.

$$dF = -S \, dT + V \, dP + \mu \, dN$$

* HCB, table 3.6-3, p. 169.

(where μ is again the chemical potential, not the Joule-Thomson coefficient), from which it follows that

$$\left(\frac{\partial \mu}{\partial P}\right)_{T,N} = \left(\frac{\partial V}{\partial N}\right)_{T,P} = \tilde{V} \qquad (15\text{-}82)$$

The integral of this equation may be written

$$\mu(T,P) = \mu^\circ(T) + kT \ln \frac{P}{P_0} + \int_0^P \left(\tilde{V} - \frac{kT}{P}\right) dP \qquad (15\text{-}83)$$

where P_0 is the pressure of the standard state. By direct differentiation, we can verify that (15-83) satisfies the differential equation (15-82). We shall show below that

$$\lim_{P \to 0} \int_0^P \left(\tilde{V} - \frac{kT}{P}\right) dP = 0 \qquad (15\text{-}84)$$

Therefore, from (15-83),

$$\lim_{P \to 0} \mu(T,P) = \mu^\circ(T) + kT \ln \frac{P}{P_0} \qquad (15\text{-}85)$$

This is the equation for the chemical potential of a perfect gas, and we see that $\mu^\circ(T)$ is the chemical potential of the hypothetical perfect gas in the standard state $P = P_0$. It is the quantity calculated in terms of the molecular partition function by the formulae of Sec. 8-10.

Since P is being used as the independent variable, we write the virial expansion in the form

$$\frac{PV}{NkT} = 1 + B^P P + C^P P^2 + \cdots$$

$$V = N\left(\frac{kT}{P} + B^P kT + C^P kTP + \cdots\right)$$

so that $\quad \tilde{V} = \left(\frac{\partial V}{\partial N}\right)_{T,P} = \frac{kT}{P} + B^P kT + C^P kTP + \cdots \qquad (15\text{-}86)$

The integral in Eq. (15-83) is then

$$\int_0^P \left(\tilde{V} - \frac{kT}{P}\right) dP = kT \int_0^P (B^P + C^P P + \cdots) dP$$

$$= kT\left(B^P P + \frac{C^P P^2}{2} + \cdots\right) \qquad (15\text{-}87)$$

The assertion (15-84) is clearly justified; furthermore we have, as a formula for the chemical potential,

$$\mu(T,P) = \mu^\circ(T) + kT \ln \frac{P}{P_0} + kT\left(B^P P + \frac{C^P P^2}{2} + \cdots\right) \qquad (15\text{-}88)$$

The fugacity f is defined by the equation

$$\mu(T,P) = \mu^\circ(T) + kT \ln f \qquad (15\text{-}89)$$

For an ideal gas, $f = P/P_0$; it is effectively the pressure in units of P_0. In general [by comparison of (15-83) and (15-89)],

$$f = \frac{P}{P_0} e^{(1/kT)\int_0^P (\tilde{V} - kT/P)\,dP} \qquad (15\text{-}90)$$

and, in terms of the virial expansion, from (15-87),

$$f = \frac{P}{P_0} e^{B^P P + C^P P^2/2 + \cdots} \qquad (15\text{-}91)$$

For a two-component system, the fundamental relations are

$$dF = -S\,dT + V\,dP + \mu_A\,dN_A + \mu_B\,dN_B$$

$$\left(\frac{\partial \mu_A}{\partial P}\right)_{T,N_A,N_B} = \left(\frac{\partial V}{\partial N_A}\right)_{T,P,N_B} = \tilde{V}_A \qquad (15\text{-}92)$$

The appropriate integral of this equation is

$$\mu_A(T,P) = \mu_A^\circ(T) + kT \ln \frac{PX_A}{P_0} + \int_0^P \left(\tilde{V}_A - \frac{kT}{P}\right) dP \qquad (15\text{-}93)$$

The integral (15-93) satisfies the differential equation (15-92). The first approximation to the virial equation of state is

$$\frac{PV}{(N_A + N_B)kT} = 1 + P\left[\frac{N_A^2}{(N_A+N_B)^2} B_{AA}^P(T) \right.$$
$$\left. + \frac{2N_A N_B}{(N_A+N_B)^2} B_{AB}^P + \frac{N_B^2}{(N_A+N_B)^2} B_{BB}^P(T)\right]$$

or
$$V = kT\frac{(N_A + N_B)}{P}$$
$$+ kT\frac{N_A^2 B_{AA}^P + 2N_A N_B B_{AB}^P + N_B^2 B_{BB}^P}{N_A + N_B} \qquad (15\text{-}94)$$

so that

$$\tilde{V}_A = \frac{\partial V}{\partial N_A} = \frac{kT}{P} + kT\frac{2N_A B_{AA}^P + 2N_B B_{AB}^P}{N_A + N_B}$$
$$- \frac{N_A^2 B_{AA}^P + 2N_A N_B B_{AB}^P + N_B^2 B_{BB}^P}{(N_A + N_B)^2} \qquad (15\text{-}95)$$

With a little manipulation, and by introducing the mole fractions, this becomes

$$\tilde{V}_A - \frac{kT}{P} = kT[X_A(2 - X_A)B_{AA}^P + 2X_B(1 - X_A)B_{AB}^P - X_B^2 B_{BB}^P]$$
$$\qquad (15\text{-}96)$$

With the help of this relation, the chemical potential can be expressed as a function of the virial coefficients.

The fugacity of component A is defined by the equation

$$\mu_A(T,P,X_A) = \mu_A^\circ(T) + kT \ln f_A \qquad (15\text{-}97)$$

so that, by comparison with (15-93),

$$f_A = \frac{PX_A}{P_0} e^{(1/kT)\int_0^P (\bar{V}_A - kT/P)\, dP} \qquad (15\text{-}98)$$

By using (15-96), this can be expressed in terms of the virial coefficients.

It is also instructive to consider the energy of a real gas at a finite concentration. The appropriate thermodynamic equation for a one-component closed system is

$$\left(\frac{\partial E}{\partial V}\right)_T = T\left(\frac{\partial P}{\partial T}\right)_V - P = T^2 \left[\frac{\partial (P/T)}{\partial T}\right]_V \qquad (15\text{-}99)$$

But, from the virial equation,

$$\frac{PV}{NkT} = 1 + \frac{NB(T)}{N_0 V} + \frac{N^2 C(T)}{N_0^2 V^2} + \cdots$$
$$\frac{P}{T} = \frac{Nk}{V} + \frac{N^2 k B(T)}{N_0 V^2} + \frac{N^3 k C(T)}{N_0^2 V^3} + \cdots \qquad (15\text{-}100)$$

Let $B'(T) = dB/dT$, $C'(T) = dC/dT$; then

$$\left[\frac{\partial (P/T)}{\partial T}\right]_V = \frac{N^2 k B'(T)}{N_0 V^2} + \frac{N^3 k C'(T)}{N_0^2 V^3} + \cdots \qquad (15\text{-}101)$$

The general integral of (15-99) is

$$E(T,V) = E^\circ(T) - \int_\infty^V T^2 \frac{\partial (P/T)}{\partial T}\, dV \qquad (15\text{-}102)$$

where $E^\circ(T)$ is the thermodynamic-energy function of the gas at infinite dilution or in the hypothetical ideal-gas state.

Using (15-101), we have

$$E(T,V) = E^\circ(T) + NkT^2 \left[\frac{B'(T)}{\mathrm{v}} + \frac{C'(T)}{2\mathrm{v}^2} + \cdots\right] \qquad (15\text{-}103a)$$

The contribution of gas imperfection to the heat capacity can be calculated from the relation*

$$\left[\frac{\partial C_V(T,V)}{\partial V}\right]_T = T\left(\frac{\partial^2 P}{\partial T^2}\right)_V \qquad (15\text{-}103b)$$

---

*cf. HCB, p. 231.

Tables of the derivatives $B'(T)$, $B''(T)$, $C'(T)$, $C''(T)$ for the Lennard-Jones 6–12 potential are listed by HCB.*

15-10. The Principle of Corresponding States. The critical point of a fluid is defined by the equations

$$\left(\frac{\partial P}{\partial V}\right)_T = 0 \qquad \left(\frac{\partial^2 P}{\partial V^2}\right)_T = 0$$

Let P_c, V_c, T_c represent values at the critical point; the reduced dimensionless variables P_r, V_r, and T_r are then defined by

$$P_r = \frac{P}{P_c} \qquad V_r = \frac{V}{V_c} \qquad T_r = \frac{T}{T_c} \qquad (15\text{-}104)$$

The empirical principle of corresponding states is the assertion that, for pressures, volumes, and temperatures measured in terms of P_r, V_r, and T_r, there is a universal equation of state which applies to all fluids. For example, the reduced compressibility factor is $P_r \mathrm{v}_r/T_r$; the principle of corresponding states asserts that

$$\frac{P_r \mathrm{v}_r}{T_r} = f(\mathrm{v}_r, T_r) \qquad (15\text{-}105)$$

where f is a universal function, i.e., the same for all substances. At the critical point, $P_r = \mathrm{v}_r = T_r = 1$, and $f(1,1) = 1$. The ordinary compressibility factor is

$$z = \frac{P\mathrm{v}}{RT} = \frac{P_r \mathrm{v}_r}{T_r} \frac{P_c \mathrm{v}_c}{RT_c} = f(\mathrm{v}_r, T_r) \frac{P_c \mathrm{v}_c}{RT_c} \qquad (15\text{-}106)$$

As $\mathrm{v} \to \infty$ (and $\mathrm{v}_r \to \infty$), for any T, the fluid approaches perfect-gas behavior and $z = P\mathrm{v}/RT \to 1$. Since $f(\mathrm{v}_r, T_r)$ is a universal function, $f(\infty, T_r)$ must be the same for all substances. From (15-106), we therefore have, for all substances,

$$\frac{P_c \mathrm{v}_c}{RT_c} = \frac{1}{f(\infty, T_r)} = \text{a universal constant} \qquad (15\text{-}107)$$

Thus, the principle of corresponding states plus the perfect-gas law predicts that $P_c \mathrm{v}_c / RT_c$ should be the same for all substances.

Table 15-2 displays the values of the compressibility factors at the critical point for a series of polar and nonpolar molecules. We see that, for simple, almost spherical molecules, z_c is 0.29 to 0.30. For hydrocarbons, it is slightly less. For molecules with permanent dipole moments, z_c is somewhat smaller.

(We shall not discuss the forces between molecules with permanent dipole moments. In the lower part of the table, μ is the permanent dipole moment, and the parameter $t^* = \mu^2/\epsilon\sigma^3 \sqrt{8}$ measures the relative importance of dipole-dipole interactions and the London interactions.)

* Appendix, table I-B, pp. 1114–1117.

The principle of corresponding states is useful for predicting, with greater or less accuracy, the P, V, T behavior of substances over a wide range of conditions in the gaseous and liquid states, and not merely for predicting the compressibility factor at the critical point.* It is of limited accuracy and generality. It applies most accurately to a series of

Table 15-2. The Compressibility Factor $P_c v_c/RT_c$ at the Critical Point*
Simple, Almost Spherical Nonpolar Molecules

Substance	$P_c v_c/RT_c$	Substance	$P_c v_c/RT_c$
He	0.300	Xe	0.293
H$_2$	0.304	N$_2$	0.292
Ne	0.307	O$_2$	0.292
A	0.291	CH$_4$	0.290
		CO$_2$	0.287

Hydrocarbons

Substance	$P_c v_c/RT_c$	Substance	$P_c v_c/RT_c$
Ethane	0.267	Benzene	0.265
Propane	0.270	Cyclohexane	0.276
Isobutane	0.276	Diisopropyl	0.266
n-Butane	0.257	Diisobutyl	0.262
Isopentane	0.268	Ethyl ether	0.262
n-Pentane	0.266	Ethylene	0.291
n-Hexane	0.260	Propylene	0.273
n-Heptane	0.258	Acetylene	0.275
n-Octane	0.258		

Polar Molecules

Substance	$P_c v_c/RT_c$	$t^* = \mu^2/(\epsilon \sigma^3 \sqrt{8})$
CH$_3$CN	0.181	1.2
H$_2$O	0.224	1.2
NH$_3$	0.238	1.0
CH$_3$OH	0.220	0.8
CH$_3$Cl	0.258	0.6
C$_2$H$_5$Cl	0.269	0.2

* From J. O. Hirschfelder, C. F. Curtiss, and R. B. Bird, "Molecular Theory of Gases and Liquids," p. 237, Wiley, New York, 1954.

related substances. It should also be mentioned that empirical two-parameter equations of state such as the van der Waals equation, the Berthelot equation, and the Dieterici equation are all consistent with the principle of corresponding states.

* For examples, see HCB, pp. 239–244; and O. A. Hougen and K. M. Watson, "Chemical Process Principles," pt. II, chap. XII, Wiley, New York, 1948.

Sec. 15-10]   REAL GASES

The theoretical principle of corresponding states is based on the assumption that the potential energy of interaction of the N molecules can be expressed as a sum of pair potentials and the pair potential $u(r)$ can be represented by a universal function ϕ and two scale factors ϵ and σ which are characteristic of the particular substance.

$$u(r) = \epsilon\phi\left(\frac{r}{\sigma}\right) \tag{15-108}$$

The Lennard-Jones potential $u(r) = 4\epsilon[(\sigma/r)^{12} - (\sigma/r)^6]$ is of this nature; the modified Buckingham potential (15-55) is not.

The real-gas behavior of the substance depends on the configuration integral,

$$Z(T,V) = \int \cdots \int e^{-\beta U(\mathbf{r}_1,\ldots,\mathbf{r}_N)}\, d^3\mathbf{r}_1 \cdots d^3\mathbf{r}_N \tag{15-109}$$

We introduce reduced-distance variables

$$\mathbf{r}_i^* = \frac{\mathbf{r}_i}{\sigma}$$

that is, $\quad x_i^* = \dfrac{x_i}{\sigma} \quad y_i^* = \dfrac{y_i}{\sigma} \quad z_i^* = \dfrac{z_i}{\sigma} \tag{15-110a}$

and the volume element $d^3\mathbf{r}_1 \cdots d^3\mathbf{r}_N$ becomes $\sigma^{3N}\, d^3\mathbf{r}_1^* \cdots d^3\mathbf{r}_{3N}^*$ (recall that $d^3\mathbf{r}_i$ stands for $dx_i\, dy_i\, dz_i$). The reduced temperature T^* is defined by

$$T^* = \frac{kT}{\epsilon} = \frac{1}{\beta\epsilon} \tag{15-110b}$$

The pair potential function is $u(r) = \epsilon\phi(r^*)$. The total potential function is

$$U(\mathbf{r}_1,\ldots,\mathbf{r}_N) = \epsilon\Phi(\mathbf{r}_1^*,\ldots,\mathbf{r}_N^*) \tag{15-111}$$

where Φ is a universal function and is the appropriate sum over all the pairs $\phi(r_{ij}^*)$.

In a cubical container, each position variable x_i ranges over the values $0 \le x_i \le V^{1/3}$; the corresponding range for the reduced variables is $0 \le x_i^* \le V^{1/3}/\sigma = (V/\sigma^3)^{1/3}$. The integral (15-109) becomes

$$Z(T,V) = \sigma^{3N}\int_0^{V^{1/3}/\sigma} \cdots \int_0^{V^{1/3}/\sigma} e^{-\beta\epsilon\Phi(\mathbf{r}_1^*,\ldots,\mathbf{r}_N^*)}\, dx_1^*\, dy_1^* \cdots dz_N^* \tag{15-112}$$

Note in the integrand that $\beta\epsilon = 1/T^*$. Then

$$Z(T,V) = \sigma^{3N}Z^*\left(T^*, \frac{V}{\sigma^3}\right) \tag{15-113}$$

where $Z^*(T^*, V/\sigma^3)$ is a reduced universal configuration integral, and, as indicated, it is a function of the reduced variable T^* and the reduced limits of integration, V/σ^3. The pressure is given by

$$P = kT\frac{\partial[\ln Z(T,V)]}{\partial V} = \epsilon T^*\frac{\partial[\ln Z(T,V)]}{\partial V} = \epsilon T^*\frac{\partial[\ln Z^*(T^*, V/\sigma^3)]}{\partial V}$$
$$= \frac{\epsilon T^*}{\sigma^3}\frac{\partial[\ln Z^*(T^*, V/\sigma^3)]}{\partial(V/\sigma^3)} \qquad (15\text{-}114)$$

The quantity ϵ/σ^3 has the dimensions of energy divided by volume or pressure. The reduced dimensionless pressure is defined by

$$P^* = \frac{P\sigma^3}{\epsilon} \qquad (15\text{-}115)$$

Equation (15-114) can then be written

$$P^* = T^*\frac{\partial[\ln Z^*(T^*, V/\sigma^3)]}{\partial(V/\sigma^3)} \qquad (15\text{-}116)$$

Thus the theoretical principle of corresponding states indicates that the reduced pressure $P^* = P\sigma^3/\epsilon$, temperature $T^* = kT/\epsilon$, and volume

Table 15-3. Critical Constants for Some Almost Spherical Nonpolar Molecules Reduced by Means of Lennard-Jones (6–12) Parameters[†]

Gas	T_c, °K	v_c, cc	P_c, atm	T_c^*	$v_c/N_0\sigma^3$	P_c^*
He	5.3	57.8	2.26	0.52	5.75	0.027
H_2	33.3	65.0	12.8	0.92	4.30	0.064
Ne	44.5	41.7	26.9	1.25	3.33	0.111
A	151	75.2	48	1.26	3.16	0.116
Xe	289.8	120.2	57.9	1.31	2.90	0.132
N_2	126.1	90.1	33.5	1.33	2.96	0.131
O_2	154.5	74.4	49.7	1.31	2.69	0.142
CH_4	190.7	99.0	45.8	1.29	2.96	0.126

[†] J. O. Hirschfelder, C. F. Curtiss, and R. B. Bird, "Molecular Theory of Gases and Liquids," p. 245, Wiley, New York, 1954.

V/σ^3, defined by reference to Lennard-Jones molecular parameters, obey a universal equation of state. The deviations from ideal thermodynamic functions all depend on $\ln Z^*(T^*, V/\sigma^3)$; suitable dimensionless combinations for the heat capacity, the energy, and the entropy, will also be universal functions of V/σ^3 and T^*.

The principal significance of the treatment just given is that it provides a theoretical justification for the empirical principle of corresponding states. It is also useful for predicting the P, V, T behavior of a substance if its Lennard-Jones parameters are known.

In Table 15-3, the critical-constant data for some almost spherical nonpolar molecules are reduced by means of the Lennard-Jones parameters of the substances.

Except for H_2 and He, where quantum effects are important, the quantities T_c^*, $v_c/N_0\sigma^3$, and P_c^* are reasonably constant; we have

$$T_c^* = \frac{kT_c}{\epsilon} = 1.28 \ (\pm 0.03)$$

$$\frac{v_c}{N_0\sigma^3} = 3.0 \ (\pm 0.3) \qquad (15\text{-}117)$$

$$P_c^* = \frac{P_c\sigma^3}{\epsilon} = 0.127 \ (\pm 0.015)$$

There are additional semiempirical rules for predicting the boiling point, the molar volume of the liquid at its boiling point, and other properties, from the Lennard-Jones parameters.

This section completes the presentation of material which is commonly accepted as basic for an introduction to the study of real-gas behavior. In the following sections, we shall discuss several additional interesting topics.

15-11. The Virial Coefficient and Pair Formation. There is an alternative point of view toward virial coefficients and intermolecular interactions which is interesting and is useful for some problems. We shall develop this in an intuitive, nonrigorous way.

Consider a monatomic gas. If there are attractive forces between atoms, there is an increased probability of finding two atoms close together; we can call such a pair of atoms a molecule which is held together by the weak intermolecular forces.

We then have the chemical equilibrium

$$A + A \rightleftharpoons A_2 \qquad (15\text{-}118)$$

There are various definitions for when a pair of atoms is to be identified as a molecule. The simplest one is that any two atoms which are within a defined distance r_{mol} of each other constitute a molecule. We take the case of a dilute gas where the degree of association to form A_2 molecules will be small and the probability of forming A_3 molecules is, by assumption, sufficiently small so as to be negligible.

We now calculate the equilibrium constant for reaction (15-118). For a standard state of 1 atom cc^{-1}, the translational partition function of the atoms is $(2\pi m_A kT/h^2)^{3/2}$. The translational partition function of A_2 is $[2\pi(2m_A)kT/h^2]^{3/2}$.

Since the bonding is weak, the vibration frequency of A_2 is low and we can calculate the internal partition function for A_2 as a classical phase integral. The internal Hamiltonian is

$$H = \frac{p_r^2}{2\mu} + \frac{p_\theta^2}{2\mu r^2} + \frac{p_\phi^2}{2\mu r^2 \sin^2 \theta} + u(r) \qquad (15\text{-}119)$$

where $\mu = m_A/2$ is the reduced mass and the symbols otherwise have their usual meaning.

The internal partition function (with a symmetry number of 2) is

$$q_{int} = \frac{1}{2h^3} \int_{p_r=-\infty}^{+\infty} \int_{p_\theta=-\infty}^{+\infty} \int_{p_\phi=-\infty}^{+\infty} \int_{r=0}^{r_{mol}} \int_{\theta=0}^{\pi} \int_{\phi=0}^{2\pi} e^{-\beta p_r^2/2\mu} e^{-\beta p_\theta^2/2\mu r^2} e^{-\beta p_\phi^2/2\mu r^2 \sin^2\theta} \, dp_r \, dp_\theta \, dp_\phi \, e^{-\beta u(r)} \, dr \, d\theta \, d\phi \quad (15\text{-}120)$$

By integration over the momenta,

$$q_{int} = \frac{1}{2h^3}(2\pi\mu kT)^{3/2} \int_{r=0}^{r_{mol}} \int_{\theta=0}^{\pi} \int_{\phi=0}^{2\pi} e^{-\beta u(r)} r^2 \sin\theta \, d\theta \, d\phi \, dr$$

$$= 2\pi \left(\frac{2\pi\mu kT}{h^2}\right)^{3/2} \int_0^{r_{mol}} e^{-\beta u(r)} r^2 \, dr \quad (15\text{-}121)$$

Combining this with the translational partition function, we obtain for the equilibrium constant for reaction (15-118)

$$K_a \text{ (atom}^{-1}\text{ cc)} = \frac{[A_2]}{[A]^2} = 2\pi \int_0^{r_{mol}} e^{-\beta u(r)} r^2 \, dr \quad (15\text{-}122a)$$

or

$$K_c \text{ (mole}^{-1}\text{ cc)} = 2\pi N_0 \int_0^{r_{mol}} e^{-\beta u(r)} r^2 \, dr \quad (15\text{-}122b)$$

Let us now inquire as to the effect of this association on the P, V, T behavior of the gas.

We take one formal mole of atoms, N_0, in a volume V. If there are N_2 molecules and N_1 atoms at equilibrium,

$$N_1 + 2N_2 = N_0 \quad (15\text{-}123)$$

The equilibrium condition is

$$\frac{N_2/V}{(N_1/V)^2} = K_a \quad \text{or} \quad \frac{N_2}{N_1^2} = \frac{K_a}{V} \quad \text{or} \quad N_2 = \frac{K_a}{V} N_1^2 \quad (15\text{-}124)$$

By assumption, the ratio $N_2/N_1 \ll 1$, or $K_a N_1/V \ll 1$. From (15-123)

$$N_0 = N_1 + 2\frac{K_a}{V} N_1^2 = N_1\left(1 + 2\frac{K_A}{V} N_1\right) \quad (15\text{-}125)$$

The first approximate solution of this equation for N_1 is $N_1 = N_0$. The second approximation is

$$N_1 \approx \frac{N_0}{1 + 2(K_a/V)N_0} \approx N_0\left(1 - 2\frac{K_a N_0}{V}\right) \quad (15\text{-}126)$$

To the same order of approximation

$$N_2 = \frac{KN_1^2}{V} = N_0 \frac{K_a N_0}{V} \quad (15\text{-}127)$$

The total number of free particles at equilibrium is

$$N_1 + N_2 = N_0\left(1 - \frac{K_a N_0}{V}\right) \quad (15\text{-}128)$$

If the two species obey the perfect-gas law,

$$\frac{PV}{kT} = N_1 + N_2 = N_0\left(1 - \frac{K_a N_0}{V}\right)$$

or
$$\frac{PV}{N_0 kT} = 1 - \frac{K_a N_0}{V} \qquad (15\text{-}129)$$

In terms of the formal number of atoms, N_0, the compressibility factor is PV/N_0kT; we see that a weak association in an otherwise perfect gas gives rise to a second virial coefficient

$$B(T) = -K_a N_0 = -K_c \qquad (15\text{-}130)$$

There is one awkward feature of the treatment so far. Referring to Eq. (15-122), since $u(r) \to 0$ as $r \to \infty$, $K_a \to \infty$ as $r_{\text{mol}} \to \infty$. The forces are negligible at large distances, and so the distribution of atoms is random. The larger the value of r_{mol}, the greater the number of pairs of atoms which will be, by definition, molecules. It is therefore necessary, in this treatment, to choose r_{mol} carefully.

This difficulty can be overcome as follows: The second virial coefficient for rigid spheres of diameter r_{mol} is

$$B(T) = \frac{2\pi r_{\text{mol}}^3}{3} N_0$$

But no two atoms can approach to a distance closer than r_{mol}; otherwise they are not two atoms but an A_2 molecule. Adding on this excluded volume correction to the perfect gas with association [Eq. (15-129)], we obtain

$$\frac{PV}{N_0 kT} = 1 + \left(\frac{2\pi r_{\text{mol}}^3}{3} N_0 - K_a N_0\right)\frac{1}{V}$$
$$= 1 + \left[2\pi N_0 \int_0^{r_{\text{mol}}} (1 - e^{-\beta u(r)}) r^2\, dr\right]\frac{1}{V} \qquad (15\text{-}131)$$

The value of the integral in (15-131) is not significantly changed by letting $r_{\text{mol}} \to \infty$; so (15-131) may be revised to read

$$\frac{PV}{N_0 kT} = 1 + \left[2\pi N_0 \int_0^\infty (1 - e^{-\beta u(r)}) r^2\, dr\right]\frac{1}{V} \qquad (15\text{-}132)$$

which is the usual second virial coefficient equation.

This discussion can be extended in several ways. In the first place, one can, in the same intuitive way, consider formation of n-mers as well as dimers.

$$nA \rightleftharpoons A_n \qquad K_n = \frac{[A_n]}{[A]^n}$$

The equilibrium formation of such physical clusters as the pressure increases gives insight into the condensation of a gas to a liquid.*

* Pitzer, pp. 331–344 [10].

It is possible to define those pairs of atoms which constitute molecules in more sophisticated and interesting ways. One can, for example, say that a molecule exists only if a pair of atoms are in that region of their internal phase space corresponding to a negative total energy

$$\frac{p_r^2}{2\mu} + \frac{p_\theta^2}{2\mu r^2} + \frac{p_\phi^2}{2\mu r^2 \sin^2 \theta} + u(r) \leq 0$$

That is, the pair must have less kinetic energy than its potential energy $u(r)$.* These are molecules in bound states.

There are also molecules in metastable states. The total kinetic energy of a pair in the center-of-mass system is

$$K = \frac{p_r^2}{2\mu} + \frac{L^2}{2\mu r^2}$$

where L is the angular momentum, which is a constant of the motion. Thus the total energy (ϵ) equation is

$$\frac{p_r^2}{2\mu} + \frac{L^2}{2\mu r^2} + u(r) = \epsilon$$

The linear momentum in the radial direction then varies with r according to the relation

$$\frac{p_r^2}{2\mu} = \epsilon - \left[u(r) + \frac{L^2}{2\mu r^2} \right]$$
$$= \epsilon - u_{\text{ef}}(r) \quad (15\text{-}133)$$

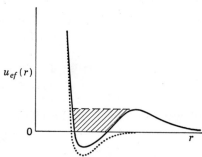

Fig. 15-4. Variation of the effective potential $u_{\text{ef}}(r) = u(r) + L^2/2\mu r^2$, with r for one value of the angular momentum L. Molecules in the shaded area of phase space have sufficient energy to dissociate, but they cannot do so in classical mechanics because they cannot penetrate the "rotational barrier." The potential curve $u(r)$ for $L = 0$ is shown as a dotted line.

where $u_{\text{ef}}(r) = u(r) + L^2/2\mu r^2$ is the effective potential. The effective potential is the sum of the real potential plus the term $L^2/2\mu r^2$, which represents the effect of centrifugal force and is always positive and repulsive. The variation of $u_{\text{ef}}(r)$ with r for a particular value of L is shown in Fig. 15-4. It is positive for large values of r but is negative for a certain region of r. Molecules which are in a region of phase space corresponding to the shaded area are in a metastable state. Although they have sufficient kinetic energy, they cannot dissociate because they cannot penetrate the region where $\epsilon - u_{\text{ef}}(r) < 0$ [cf. Eq. (15-133)]. For a calculation of the equilibrium constant for the formation of metastable molecules, see Stogryn and Hirschfelder.†

* T. Hill, *J. Chem. Phys.*, **23**: 617 (1955); D. Bunker and N. Davidson, *J. ACS*, **80**: 5090 (1958).

† D. E. Stogryn and J. O. Hirschfelder, *J. Chem. Phys.*, **31**: 1543 (1959); the centrifugal potential and the "rotational maximum" for diatomic molecules is explained in more detail by Herzberg, "Diatomic Molecules," pp. 425–427 [17].

The treatment which we have presented of association as related to virial coefficients was intuitive and approximate. Hill* gives a precise statistical mechanical treatment of the problem, in which the contributions to the virial coefficient due to bound molecules and due to the interactions of pairs of atoms that we may wish not to call molecules are clearly defined. The problem is essentially one of dividing up phase space for the pair into regions which correspond to molecules and regions which do not.

15-12. Further Remarks about the Intermolecular Potential. The potential of interaction between atoms or molecules can, to some extent, be deduced from experimental data on the second virial coefficient, as described in the preceding sections. It is intuitively obvious that the trajectory in a collision between two atoms depends on the potential of interaction between them. There is a detailed kinetic theory of the transport properties of gases (viscosity, heat conduction, and diffusion), in which the potential of interaction plays a central role.† Conversely, the parameters in the potential of interaction can be deduced from the values of the viscosity coefficient, the diffusion coefficient, or the thermal-conductivity coefficient over a range of temperatures.

The interatomic distance and the heat of sublimation at 0°K for a solid such as any one of the crystalline rare gases which is held together by van der Waals forces provide rather direct information about the depth of the potential and the position of the minimum of the potential-energy curve.‡

The region of the potential curve which has the predominant effect on the properties of the gas varies with the temperature. At low temperatures, the attractive region of the curve has the most significant effect on the virial coefficients and the transport properties. At high temperatures, where the kinetic energy per atom is large compared with the depth of the minimum ($\epsilon/kT \ll 1$), the repulsive inner portion of the curve is most important; as the temperature is raised, the parts of the curve for small r and large $u(r)$ become progressively more important.

A direct method of measuring the intermolecular potential is the study of the scattering of atoms by atoms in molecular-beam experiments. The experiments that have been performed so far are concerned with the scattering of a beam of high-velocity rare-gas atoms by rare gases or simple molecules such as N_2. The conditions of the experiments are such that they provide information about the repulsive potential in the regions where it has a value of about 10^{-12} erg, whereas, for example, the depth of

---

* T. Hill, "Statistical Mechanics," pp. 122–129, 136–144, 152–160 [9a].

† HCB, pt. II; S. Chapman and T. G. Cowling, "The Mathematical Theory of Nonuniform Gases," Cambridge, New York, 1939.

‡ HCB, pp. 1035–1041; J. Corner, *Trans. Faraday Soc.*, **44**: 914 (1948); E. A. Mason and W. E. Rice, *J. Chem. Phys.*, **22**: 843 (1954).

the potential minimum for argon is 1.7×10^{-14} erg.* The general conclusion from these experiments is that the repulsive-potential curve for small r is "softer" than the potential functions deduced from the properties of the crystal and from virial coefficients and transport properties of gases at moderate temperatures. In general, the beam data can be interpreted in terms of a repulsive potential of the form $u(r) = K/r^s$; and for argon for

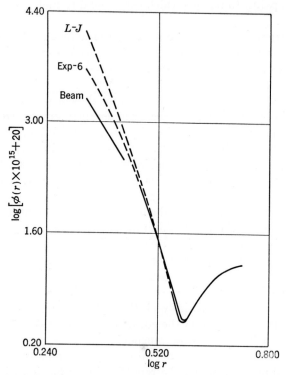

Fig. 15-5. Various intermolecular potentials for the argon-argon interaction. Dashed portions of the curves represent extrapolations outside of the ranges for which the curves fit the experimental data. [*From I. Amdur and J. Ross, Combustion and Flame*, **2**: 416 (1958).]

$2.2 < r < 2.7$ A, $s = 8.33$, whereas, in the Lennard-Jones 6–12 potential, the repulsive exponent is $s = 12.0$.

Figure 15-5 and Table 15-4 illustrate these statements for the argon-argon interaction.

This information about the repulsive potential at short distances has been used to compute thermodynamic and transport properties of gases at high temperatures.†

* I. Amdur, E. A. Mason, and J. E. Jordan, *J. Chem. Phys.*, **27**: 527 (1957), and other papers by Amdur and collaborators.

† I. Amdur and E. A. Mason, *Phys. of Fluids*, **1**: 370 (1958).

Table 15-4. Magnitudes of Various Potentials for Argon System*

Lennard-Jones 6–12		Modified Buckingham (exp-6)		Beam K/r^s	
r, Å	$u(r) \times 10^{15}$, ergs	r, Å	$u(r) \times 10^{15}$, ergs	r, Å	$u(r) \times 10^{15}$, ergs
(2.18)	(14,110)	(2.18)	(4,819)	2.18	2,058
(2.60)	(1,473)	(2.60)	(935)	2.30	1,317
(3.10)	(98.2)	3.10	93.6	2.43	833
3.60	−13.4	3.60	−12.0	2.55	558
3.837	−17.1	3.866	−17.0	2.69	357
4.30	−12.9	4.30	−13.1		
4.80	−7.76	4.80	−7.69		
		5.30	−4.41		

The potential parameters used are

L.J. 6–12 potential:

$$u(r) = \epsilon \left[\left(\frac{r_m}{r}\right)^{12} - 2\left(\frac{r_m}{r}\right)^6 \right] \qquad \frac{\epsilon}{k} = 124° \qquad r_m = 3.837 \text{ Å } (\sigma = 3.41 \text{ Å})$$

Modified Buckingham:

$$u(r) = \frac{\epsilon}{1 - 6/\alpha} \left[\frac{6}{\alpha} \exp \alpha \left(1 - \frac{r}{r_m}\right) - \left(\frac{r_m}{r}\right)^6 \right] \qquad \frac{\epsilon}{k} = 123.2° \qquad r_m = 3.866$$
$$\alpha = 14.0$$

Beam repulsive:

$$u(r) = \frac{K}{r^s} \qquad K = 9.855 \times 10^6 (°\text{K} - \text{Å}^s) \qquad s = 8.33$$

The values in parentheses in the table are extrapolated beyond the estimated range of validity of the functions used.

* I. Amdur and J. Ross, *Combustion and Flame*, **2**: 412 (1958).

For our purposes, the main point of this section is simply to emphasize the restricted range of validity of the potential curves used in the previous sections.

15-13. The Virial Coefficients of a Quantum Gas. The calculations of the preceding sections are based on the assumption that corrected Boltzmann statistics apply and on the use of the classical phase integral in evaluating the partition function. These assumptions are satisfactory for gases such as argon or nitrogen, but they are not sufficiently accurate for the light gases, especially H_2 and He, at low temperatures.

A general treatment of this topic usually begins with the theory of the quantum-mechanical density matrix. We shall not discuss the general theory. It is possible, however, to comment, in an elementary way, on one simple and important aspect of the properties of a quantum gas.

As shown in Chap. 6, a system of noninteracting particles does not satisfy the perfect-gas equation $PV/NkT = 1$ unless the system is sufficiently dilute so that corrected Boltzmann statistics apply. The general condition for the applicability of this assumption is that $N/q \ll 1$, where q is the molecular partition function.

We shall give a simple derivation of the second virial coefficient for a system of noninteracting atoms which are sufficiently dilute so that the deviations from corrected Boltzmann statistics are small, but not negligible. As shown in Chap. 6, the distribution law for bosons or fermions is

$$N_i = \frac{g_i e^{\beta\mu} e^{-\beta\epsilon_i}}{1 \mp e^{\beta\mu} e^{-\beta\epsilon_i}} = \frac{g_i y_i}{1 \mp y_i} \quad \begin{cases} \text{(upper sign, BE)} \\ \text{(lower sign, FD)} \end{cases} \quad (15\text{-}134a)$$

where $\quad y_i = e^{\beta\mu} e^{-\beta\epsilon_i} \quad\quad\quad\quad\quad\quad\quad\quad\quad\quad (15\text{-}134b)$

and μ is the chemical potential.

The condition for the applicability of corrected Boltzmann statistics is $N_i/g_i \ll 1$, or $y_i \ll 1$ (for all i). In this approximation

$$N = \Sigma N_i \approx \Sigma g_i e^{\beta\mu} e^{-\beta\epsilon_i} = q e^{\beta\mu} \quad (15\text{-}135)$$

as usual.

In general, μ is to be determined by the condition

$$N = \sum N_i = \sum \frac{g_i y_i}{1 \mp y_i} \quad (15\text{-}136)$$

The general equation for P is

$$\frac{PV}{kT} = \mp \sum g_i \ln(1 \mp y_i) \quad (15\text{-}137)$$

For the first deviations from Boltzmann statistics as the gas is compressed, we take y_i as small and expand each term of Eqs. (15-136) and (15-137) as a power series in y_i.

$$N \approx \sum g_i y_i \pm \sum g_i y_i^2 = \sum g_i y_i \left(1 \pm \frac{\Sigma g_i y_i^2}{\Sigma g_i y_i}\right) \quad (15\text{-}138a)$$

$$\frac{PV}{kT} \approx \sum g_i y_i \pm \frac{1}{2} \sum g_i y_i^2 = \sum g_i y_i \left(1 \pm \frac{1}{2} \frac{\Sigma g_i y_i^2}{\Sigma g_i y_i}\right) \quad (15\text{-}138b)$$

Now divide (15-138b) by (15-138a),

$$\frac{PV}{NkT} = \frac{1 \pm \frac{1}{2}(\Sigma g_i y_i^2/\Sigma g_i y_i)}{1 \pm (\Sigma g_i y_i^2/\Sigma g_i y_i)} \quad (15\text{-}139a)$$

We treat the quantity $\Sigma g_i y_i^2 / \Sigma g_i y_i$ as a small number and expand the denominator of (15-139a),

$$\frac{PV}{NkT} = 1 \mp \frac{1}{2} \frac{\Sigma g_i y_i^2}{\Sigma g_i y_i} \quad (15\text{-}139b)$$

The second term in (15-139b), $\mp \frac{1}{2}(\Sigma g_i y_i^2/\Sigma g_i y_i)$, is a small correction term to the corrected-boltzon ideal-gas law. We can therefore evaluate the correction term by using the ideal-gas values for y_i,

REAL GASES

$$y_i = \frac{N}{q} e^{-\beta \epsilon_i} \qquad \sum g_i y_i = \frac{N \Sigma g_i e^{-\beta \epsilon_i}}{q} = N$$

$$\sum g_i y_i^2 = \left(\frac{N}{q}\right)^2 \sum g_i e^{-2\beta \epsilon_i} = \left[\frac{N}{q(\beta)}\right]^2 q(2\beta) \tag{15-140}$$

The partition functions $q(\beta)$ and $q(2\beta)$ are

$$q(\beta) = \left(\frac{2\pi m kT}{h^2}\right)^{3/2} V = \frac{V}{\Lambda^3} \qquad q(2\beta) = \left(\frac{2\pi m kT}{2h^2}\right)^{3/2} V = \frac{V}{2^{3/2}\Lambda^3} \tag{15-141}$$

By substitution of (15-141) and (15-140) into (15-139b),

$$\frac{PV}{NkT} = 1 \mp \frac{\Lambda^3}{2^{5/2}} \frac{N}{V} = 1 \mp \frac{\Lambda^3}{2^{5/2}} \frac{N_0}{\mathrm{v}} \tag{15-142}$$

The second virial coefficient for a system of noninteracting particles due to the fact that the particles are bosons or fermions is

$$B(T) = \mp \frac{\Lambda^3 N_0}{2^{5/2}} \qquad \begin{cases} \text{(upper, BE)} \\ \text{(lower, FD)} \end{cases} \tag{15-143}$$

Bosons have a negative virial coefficient as though they attract each other; fermions have a positive virial coefficient as though they repel each other.

The deviations from corrected Boltzmann behavior make a significant contribution to the virial coefficients of the light gases at low temperature, but the effects of intermolecular forces are typically more important. Thus, the second virial coefficient of He^4 gas at its boiling point (4.2°K) is -75.5 cc mole^{-1}.* The virial coefficient for an ideal Bose-Einstein gas of atomic weight 4.0 at 4.2°K according to Eq. (15-143) is -8.0 cc mole^{-1}. The rest of the effect is due to intermolecular forces. We shall not treat this rather advanced topic of the statistical mechanics of a quantum gas with intermolecular interactions.† We may mention, however, that the essential point is that the classical phase-integral approximation for the interaction of a pair of atoms, $2\pi \int e^{-\beta u(r)} r^2 \, dr$, is not a satisfactory approximation and the pair interaction must be treated in a proper quantum-mechanical way.

15-14. A Derivation of the Virial-coefficient Formulae. We conclude this chapter with an alternative derivation of the formulae for the virial coefficients. The derivation is interesting for several reasons. It appears to avoid the cumbersome apparatus of the cluster integrals; it provides a simple method for obtaining expressions for the higher virial coefficients; it is an instructive example of the use of the grand partition function; and the formulae derived are applicable even for fluids where the quantal form

---

* W. H. Keesom and W. K. Walstra, *Physica*, **6**: 1146 (1939).

† See HCB, chap. 6; also, for a relatively simple discussion, see C. F. Squire, "Low Temperature Physics," chap. 3, McGraw-Hill, New York, 1953.

of the partition function must be used. The derivation applies to molecules of arbitrary complexity and for potentials which are not a sum of pair potentials.*

Let $Q_N(T,V)$ be the canonical partition function for the system in volume V at temperature T with N particles therein. Note that $Q_1(T,V)$ is the single-particle partition function

$$Q_1(T,V) = q_{\text{int}} \left(\frac{2\pi mkT}{h^2}\right)^{3/2} V = q_{\text{int}} \frac{V}{\Lambda^3} \qquad (15\text{-}144)$$

$Q_2, \ldots, Q_N$, etc., are more complicated and contain contributions due to the interactions of the particles.

The grand partition function is

$$\Xi(\lambda,T,V) = \sum_N Q_N(T,V)\lambda^N \qquad (15\text{-}145)$$

In Eq. (15-145), $\lambda = e^{(\mu/kT)}$. We recall that the ensemble average number of particles in the system is

$$\bar{N} = \lambda \left\{\frac{\partial \ln [\Xi(\lambda,T,V)]}{\partial \lambda}\right\}_{T,V} \qquad (15\text{-}146)$$

We begin by assuming that the system is such that the molecules may be treated as corrected boltzons so that (15-145) may be rewritten as

$$\Xi(\lambda,T,V) = \sum_N Q_N(T,V)\lambda^N = \sum_N \frac{Z_N(T,V)z^N}{N!} \qquad (15\text{-}147)$$

where Z_N is the configuration integral;

$$Z_N(T,V) = \int \cdots \int e^{-\beta u(\mathbf{r}_1, \ldots, \mathbf{r}_N)} d^3\mathbf{r}_1 \cdots d^3\mathbf{r}_N$$

For this case,

$$Q_N(T,V) = \frac{q_{\text{int}}^N}{\Lambda^{3N}} \frac{Z_N}{N!} \qquad (15\text{-}148)$$

so that, for (15-147) to be true, the definition of z must be

$$z = \frac{\lambda q_{\text{int}}}{\Lambda^3} \qquad (15\text{-}149)$$

and we note that

$$\bar{N} = \lambda \left\{\frac{\partial[\ln \Xi(T,V,\lambda)]}{\partial \lambda}\right\}_{T,V} = z \left\{\frac{\partial[\ln \Xi(T,V,z)]}{\partial z}\right\}_{T,V} \qquad (15\text{-}150)$$

In the limit of low density, where the perfect gas holds,

$$\mu/kT = -\ln (q/\bar{N})$$

---

*S. Ono, *J. Chem. Phys.*, **19**: 504 (1951); J. E. Kilpatrick, *J. Chem. Phys.*, **21**: 274 (1953); Hill, pp. 134–136 [9a].

Therefore
$$\lim_{\bar{N}/V \to 0} \lambda = \frac{\bar{N}}{q} = \frac{\Lambda^3}{q_{int}} \frac{\bar{N}}{V} \qquad (15\text{-}151)$$

so that, in this limit,

$$z = \frac{\bar{N}}{V} \quad \text{average number density of particles} \qquad (15\text{-}152)$$

The fundamental equation for the grand ensemble is

$$\frac{PV}{kT} = \ln \Xi = \ln\left(1 + Z_1 z + \frac{Z_2 z^2}{2} + \frac{Z_3 z^3}{6} + \cdots\right) \qquad (15\text{-}153)$$

In the limit as $z \to 0$, this can be expanded

$$[\ln(1+x) = x - x^2/2 + x^3/3 \cdots]$$

to give

$$\frac{PV}{kT} = \ln \Xi = zZ_1 + z^2\frac{Z_2 - Z_1^2}{2} + z^3\left(\frac{Z_3}{6} - \frac{Z_1 Z_2}{2} + \frac{Z_1^3}{3}\right) + \cdots \qquad (15\text{-}154)$$

If we write

$$\frac{PV}{kT} = \ln \Xi = V \sum_{j=1}^{j=\infty} b_j z^j \qquad (15\text{-}155)$$

or

$$\frac{P}{kT} = \sum b_j z^j \qquad (15\text{-}156)$$

the coefficients b_j are given [by comparison of (15-156) and (15-154)] as

$$Vb_1 = Z_1 = V \quad \text{or} \quad b_1 = 1 \quad Vb_2 = \tfrac{1}{2}(Z_2 - Z_1^2)$$
$$Vb_3 = \tfrac{1}{6}(Z_3 - 3Z_1 Z_2 + 2Z_1^3) \qquad (15\text{-}157)$$

We want to express PV/kT in terms of $\bar{N}$, the actual number of particles in the system, rather than in terms of z. Now

$$\bar{N} = \lambda\left[\frac{\partial(\ln \Xi)}{\partial \lambda}\right]_{T,V} = z\left[\frac{\partial(\ln \Xi)}{\partial z}\right]_{T,V}$$

Therefore, by differentiating (15-155),

$$\bar{N} = V\Sigma j b_j z^j = V(z + 2b_2 z^2 + 3b_3 z^3 + \cdots) \qquad (15\text{-}158a)$$
$$\frac{\bar{N}}{V} = z + 2b_2 z^2 + \cdots + j b_j z^j + \cdots \qquad (15\text{-}158b)$$

We wish to invert this relation and express z, which is related to the activity, as a power series in the concentration $\bar{N}/V$. A standard trick for doing this is the following: Write z as a power series in $\bar{N}/V$,

$$z = A_1 \frac{\bar{N}}{V} + A_2\left(\frac{\bar{N}}{V}\right)^2 + A_3\left(\frac{\bar{N}}{V}\right)^3 + \cdots \qquad (15\text{-}159)$$

In order to evaluate the coefficients A_1, A_2, A_3, etc., substitute the expression for z [(15-159)] into the right hand side of (15-158b), and collect coefficients of the various powers of $\bar{N}/V$. The result is

$$A_1 = 1$$

[which we already knew; cf. (15-152)].

$$A_2 = -2b_2 \qquad A_3 = 8b_2^2 - 3b_3 \qquad (15\text{-}160)$$

so that
$$z = \frac{\bar{N}}{V} - \left(\frac{\bar{N}}{V}\right)^2 2b_2 + \left(\frac{\bar{N}}{V}\right)^3 (8b_2^2 - 3b_3) + \cdots \qquad (15\text{-}161)$$

Now substitute this expression for z into Eq. (15-156) for P/kT; after collecting coefficients of $\bar{N}/V$ this becomes

$$\frac{P}{kT} = \frac{\bar{N}}{V} - \left(\frac{\bar{N}}{V}\right)^2 b_2 + \left(\frac{\bar{N}}{V}\right)^3 (4b_2^2 - 2b_3) + \cdots \qquad (15\text{-}162)$$

or
$$\frac{PV}{\bar{N}kT} = 1 - \frac{\bar{N}}{V} b_2 + \left(\frac{\bar{N}}{V}\right)^2 (4b_2^2 - 2b_3) + \cdots \qquad (15\text{-}163)$$

Then the virial coefficients are

$$B(T) = -N_0 b_2 = -\frac{N_0}{2V}(Z_2 - V^2) = \frac{N_0}{2V}\left(V^2 - \iint e^{-\beta u(\mathbf{r}_1,\mathbf{r}_2)}\, d^3\mathbf{r}_1\, d^3\mathbf{r}_2\right) \qquad (15\text{-}164)$$

$$C(T) = N_0^2(4b_2^2 - 2b_3) = \frac{Z_2(Z_2 - V^2)}{V^2} + \frac{V^3 - Z_3}{3V} \qquad (15\text{-}165)$$

For a spherically symmetrical potential of intermolecular interaction, $u(\mathbf{r}_1,\mathbf{r}_2) = u(r_{12})$, and the equation for $B(T)$ reduces to the formula previously derived.

We thus have a neat, general derivation for all the virial coefficients. The derivation is applicable to nonspherically symmetrical potentials. In the references cited at the beginning of this section, general formulae for the coefficients b_j are given.

The derivation is also applicable when quantum effects are important. In this case, the variable z is defined by

$$z = \frac{Q_1}{V}\lambda$$

and Q_1 is still the molecular partition function for a single particle. However, the quantities Z_N are no longer the configuration integrals; they are simply formally defined by

$$Z_N = \frac{V^N}{Q_1^N} N!\, Q_N \qquad (15\text{-}166)$$

which makes Eqs. (15-145) and (15-147) equivalent. In principle, the functions $Q_N(T,V)$ are calculable for a quantum gas; and the rest of the derivation is unchanged.

The derivation of this section is generally accepted as rigorous. However, it is subject to one criticism. The expansion $\ln(1 + Z_1 z + Z_2 z^2/2 + \cdots) \approx zZ_1 + z^2(Z_2 - Z_1^2)/2 + \cdots$ requires that $Z_1 z + Z_2 z^2/2 + \cdots \ll 1$; but, for example, for a dilute gas, $z \approx \bar{N}/V$, $Z_1 z \approx \bar{N}$, and, under ordinary circumstances, this is not small. The same is true for the entire expression $Z_1 z + Z_2 z^2/2 + \cdots$. We can take two points of view toward this difficulty. In one point of view, we derive the formulae by taking such an ultradilute gas that $z \to 0$ and $Z_1 z \ll 1$. [We could also take a very small volume, so that, even for ordinary values of $z \approx \bar{N}/V$, $Z_1 z = V(\bar{N}/V) = \bar{N} \ll 1$.] The relations derived must be applicable for this case. But we know, for example, that for the ordinary permanent gases at room temperature and ordinary pressures, the relations derived here are applicable, and in fact, only the second virial coefficient is needed for a reasonably satisfactory equation of state in this range. Yet under these conditions it is most emphatically not true that $Z_1 z \ll 1$ or that $Z_2 z^2 \ll 1$. The truth of the matter is that, if there are no intermolecular forces, $Z_2 = V^2$, $Z_N = V^N$. Under these circumstances, as we already know, $\Xi = \sum_N (V^N z^N/N!) = e^{Vz}$; $\ln \Xi = Vz$, and the perfect-gas law $PV/\bar{N}kT = 1$ results. If the intermolecular interactions are not very important, the quantities Z_N do not differ very much from V^N and the principal contributions to the difference between Z_N and V^N are from the interactions of pairs of molecules. This situation, which was hinted at in Sec. 15-4, is not revealed by the formal analysis of the present section, which does, however, succeed in deriving the correct formulae.

PROBLEMS

15-4. Use the tables of $B^*(T^*)$ in HCB to calculate the second virial coefficient of argon at its boiling point, 84°K, assuming a Lennard-Jones 6–12 potential. Estimate the per cent deviation from the perfect-gas law at 1 atm pressure.

15-5. Show by substitution for P in (15-88), using the virial expansion (15-1b), that the chemical potential may be expressed in terms of concentrations N/V by the equation

$$\mu(T) = \mu°(T) + kT \ln \frac{kT}{P_0} + kT \ln \frac{N}{V} + 2kT \frac{B(T)}{N_0} \frac{N}{V}$$

The correction term to ideal-gas behavior is $2kT[B(T)/N_0](N/V)$. Note that the correction term involves twice the second virial coefficient, whereas the correction term in (15-88) involves unity times the second virial coefficient. How does this come to be?

The activity coefficient γ may be defined by

$$\mu(T) = \mu°(T) + kT \ln \frac{kT}{P_0} + kT \ln \gamma \frac{N}{V}$$

What is γ in terms of the second virial coefficient?

15-6. A container with a fixed volume $V_A + V_B$ is divided into two parts, with volumes V_A and V_B, containing N_A and N_B molecules of types A and B, respectively. The pressure and the temperature are the same in the two compartments. The gases are now allowed to mix isothermally. Obtain an expression for the pressure change on mixing, assuming that the gases are sufficiently dilute so that only the second virial coefficients need be considered. (*Hint:* Use the virial expansions in powers of P and assume that the changes in P are small.)

15-7. Calculate the heat change during the mixing process of Prob. 15-6.

15-8. By a process analogous to that of Secs. 15-3 and 15-4, obtain an expression for the film pressure of a two-dimensional gas as a function of the area a if the atoms interact according to a potential $u(r)$ that depends only on the distance apart (cf. Prob. 6-5). Evaluate the second virial coefficient of the two-dimensional gas.

15-9. From the Lennard-Jones parameters (Table 15-1) and the tables in HCB of $B^*(T^*)$ and its derivatives estimate the second virial coefficients and the Joule-Thompson coefficients for argon and xenon at 273°K. [The measured Joule-Thompson coefficients of argon are given by J. R. Roebuck and H. Osterberg, *Phys. Rev.*, **46**: 785 (1934); for references to other substances, see J. R. Roebuck, T. A. Murrell, and E. E. Miller, *J. ACS*, **64**: 400 (1942).]

15-10. Derive a thermodynamic equation for $(\partial c_P/\partial P)_T$. Use Lennard-Jones parameters and the second-virial-coefficient data to estimate the change in c_P for argon and xenon between zero pressure and 2.00 atm at 273°K.

16

Lattice Vibrations and the Thermodynamic Properties of Crystals. Order-Disorder Phenomena

16-1. Introduction. Consider a monatomic crystal containing N atoms. Regarded as a giant molecule, the crystal has three degrees of translational freedom and three degrees of rotational freedom. There are therefore $3N - 6 \approx 3N$ degrees of vibrational freedom. A classical harmonic vibration has a mean kinetic energy of $kT/2$ and a mean potential energy of $kT/2$; the total vibrational energy is therefore $3NkT$, and the molal heat capacity at constant volume is predicted to be $3R$ and to be constant. This is, of course, the law of Dulong and Petit. For most monatomic solids there is a range of temperatures for which this relation is approximately true. The point of main interest to us, however, is the observation that at low temperatures the heat capacity decreases and approaches zero as $T \to 0$. The situation is schematically illustrated in Fig. 16-1. This result, which is inexplicable on the basis of

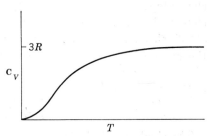

FIG. 16-1. Typical plot of the molal heat capacity (c_V) as a function of temperature.

classical statistical mechanics, is readily understandable in terms of quantum statistical mechanics. In fact, the curve is qualitatively similar to the heat-capacity function of a quantum mechanical oscillator (Fig. 8-4). Our principal task in the next few sections is to develop a quantitative theory for the heat-capacity curve.

The quantity which is most readily determined experimentally is c_P, the molal heat capacity at constant pressure, but the quantity of theoretical interest is c_V. The thermodynamic relation between these is

$$c_P - c_V = \frac{\alpha^2 \mathrm{v} T}{\kappa} \qquad (16\text{-}1)$$

351

where $\alpha = (1/V)(\partial V/\partial T)_P$ is the volume coefficient of thermal expansion and $\kappa = -(1/V)(\partial V/\partial P)_T$ is the compressibility.

It is an empirical observation that α does vary considerably with temperature but that κ does not. There is a semiempirical rule, due to Grüneisen,* that

$$\alpha = \frac{\gamma \kappa c_V}{V} \qquad (16\text{-}2)$$

where the value of γ is approximately constant for a substance and is of the magnitude $\gamma \approx 2.0$ (dimensionless).

If (16-2) is used, a dimensionless equation for $C_P - C_V$ is

$$\frac{C_P - C_V}{C_V} = \frac{\gamma^2 \kappa C_V T}{V} \qquad (16\text{-}3)$$

Problem 16-1. Estimate $(C_P - C_V)/C_V$ and α for aluminum with $\gamma = 2.0$, density $= 2.71$, $\kappa = 1.36 \times 10^{-12}$ dyne^{-1} cm^2, and, at 73°K, $c_V/R = 0.9$.†

The quantity $(C_P - C_V)/C_V$ increases with increasing T [in agreement with Eq. (16-3) and our comments on the temperature dependence of α and κ]; thus the corrections to be applied to calculate C_V from C_P are rather large at high temperatures.

16-2. The Canonical Partition Function for a System of Independent Distinguishable Elements. Let there be $3N$ distinguishable independent elements denoted by letters $a, b, c, d, \ldots, 3N$ in the system. Each element has a set of quantized energy levels; the levels of the a system are $\epsilon_1^a, \epsilon_2^a, \ldots, \epsilon_{v_a}^a, \ldots$, where v_a is the general quantum number for the energy levels of element a. A system quantum state is specified by the $3N$-component vector $v_a, v_b, \ldots, v_i, \ldots, v_{3N}$. The energy of the system in this state is

$$E = \epsilon_{v_a}^a + \epsilon_{v_b}^b + \cdots + \epsilon_{v_{3N}}^{3N}$$

The single-particle partition function for the ath element is

$$q_a = \sum_{v_a=1}^{\infty} e^{-\beta \epsilon_{v_a}^a}$$

The canonical partition function is

$$Q = \sum_{v_a, v_b, \ldots, v_{3N}} e^{-\beta(\epsilon_{v_a}^a + \cdots + \epsilon_{v_{3N}}^{3N})} \qquad (16\text{-}4a)$$

$$Q = q_a q_b \cdots q_{3N} \qquad (16\text{-}4b)$$

Equation (16-4b) follows from (16-4a) because the sums over $v_a, \ldots,$

---

* This is discussed and illustrated by Wilson, p. 159 [4].

† The values of κ are usually given in cgs units of dyne^{-1} cm^2. If the heat capacity is given in the dimensionless form of c_V/R, one need only recall that $R = 8.31 \times 10^7$ ergs mole^{-1} deg^{-1} to proceed with the calculation.

$v_d, \ldots$ are unrestricted and each value of v_a must occur with each value of v_b, etc.

All the system thermodynamic functions are sums of the thermodynamic functions for the independent elements.

$$-\frac{A}{kT} = \ln Q = \ln q_a + \ln q_b + \cdots + \ln q_{3N} \tag{16-5}$$

$$E = -T^2 \left[\frac{\partial(A/T)}{\partial T}\right]_V = kT^2 \left[\frac{\partial(\ln Q)}{\partial T}\right]_V = kT^2 \left\{\left[\frac{\partial(\ln q_a)}{\partial T}\right]_V \right.$$
$$\left. + \left[\frac{\partial(\ln q_b)}{\partial T}\right]_V + \cdots \right\} \tag{16-6a}$$

$$E = -\frac{\partial(\ln q_a)}{\partial \beta} - \frac{\partial(\ln q_b)}{\partial \beta} - \cdots - \frac{\partial(\ln q_{3N})}{\partial \beta} \tag{16-6b}$$

$$E = \bar{\epsilon}_a + \bar{\epsilon}_b + \cdots + \bar{\epsilon}_{3N} \tag{16-6c}$$

where $\bar{\epsilon}_a[= -\partial(\ln q_a)/\partial\beta]$ is the average energy in the ath element.

The heat capacity is given by

$$C_V = \left(\frac{\partial E}{\partial T}\right)_V = c_{Va} + c_{Vb} + c_{V3N} \tag{16-7}$$

where $c_{Va} = (\partial \bar{\epsilon}_a/\partial T)_V$ is the heat capacity of the ath element.

The entropy is

$$S = \frac{E - A}{T} = \frac{\epsilon_a}{T} + k \ln q_a + \frac{\epsilon_b}{T} + k \ln q_b + \cdots \tag{16-8}$$

It is to be noted that the elements are distinguishable; there is no division by $N!$ in the partition function (16-4); even the entropy is the sum of the entropies of the individual elements.

16-3. The Einstein Theory. The first great step in understanding the low-temperature heat capacity of crystal was taken by Einstein,[*] who applied the Planck theory of quantized oscillators to the vibrations of a crystal.

Consider for conceptual simplicity N atoms at the points of a simple cubic lattice. We can assume that each atom is harmonically bound to its equilibrium position in the lattice. We make the obviously absurd assumption that the motion of any one atom is not influenced by the motions of the neighboring atoms. Then each atom is a three-dimensional harmonic oscillator. In a cubic lattice, the force constant and the vibration frequency are the same in the three orthogonal directions; if this frequency is ν and if $u = h\nu/kT$, then the partition function for any one vibration is the harmonic-oscillation partition function

$$q(u) = \frac{1}{1 - e^{-u}} \tag{16-9}$$

---

[*] A. Einstein, *Ann. Physik*, **22**: 180 (1907).

Unlike the atoms in a gas, the atoms are localized at definite lattice points. In principle, we can look at the crystal and observe that the atom at one particular site is vibrating with certain quantum numbers α, β, γ, in the x, y, and z direction. This is a recognizably different state of the crystal (system) from that in which an atom at a different site is vibrating with the quantum numbers α, β, γ.* The system therefore consists of $3N$ independent harmonic oscillators, and the formulae of the preceding section are applicable.

We recall the formula for the thermodynamic functions of a harmonic oscillator, with $u = h\nu/kT$,

$$q(u) = \frac{1}{1 - e^{-u}} \tag{16-10}$$

$$\bar{\epsilon}(u) = \frac{h\nu}{e^u - 1} \quad \text{or} \quad \frac{\bar{\epsilon}}{kT} = \frac{u}{e^u - 1} \tag{16-11}$$

$$c_V(u) = \left(\frac{\partial \bar{\epsilon}}{\partial T}\right)_V = \frac{ku^2 e^u}{(e^u - 1)^2} \tag{16-12}$$

$$\frac{s}{k} = \frac{\bar{\epsilon}}{kT} + \ln q = \frac{u}{e^u - 1} - \ln(1 - e^{-u}) \tag{16-13}$$

where s is the entropy of a single harmonic oscillator.

In the Einstein theory, the thermodynamic functions for the crystals are $3N$ times the thermodynamic functions for a single oscillator. In particular, the molal heat capacity is given by

$$c_V = 3N_0 k \frac{u^2 e^u}{(e^u - 1)^2} \tag{16-14a}$$

$$c_V = 3R \left(\frac{\Theta}{T}\right)^2 \frac{e^{\Theta/T}}{[e^{\Theta/T} - 1]^2} \tag{16-14b}$$

with

$$\Theta = \frac{h\nu}{k} \tag{16-14c}$$

At high temperatures ($u \to 0$)

$$\frac{u^2 e^u}{(e^u - 1)^2} = 1 - \tfrac{1}{12} u^2 + \cdots \tag{16-15a}$$

so that

$$c_V = 3R \left[1 - \frac{1}{12}\left(\frac{\Theta}{T}\right)^2\right] \tag{16-15b}$$

We thus obtain the law of Dulong and Petit, $c_V = 3R$, at high temperatures.

At low temperatures, $u \to \infty$, $e^u \to \infty$, and the limiting equation for c_V is

$$c_V = 3R \left(\frac{\Theta}{T}\right)^2 e^{-\Theta/T} \tag{16-16}$$

Thus, the heat capacity approaches zero as $T \to 0$.

* The atoms are not really distinguishable, but the lattice sites are.

Equation (16-14) with Θ or ν chosen to fit the heat-capacity data as well as possible* gives a good qualitative description of the heat-capacity curve for monatomic solids. However, it does not give a good quantitative fit. For example, if Θ is determined by the point where the heat capacity is one-half its limiting high temperature value ($\Theta/T = 2.98$), then the observed values of the heat capacity at low temperatures are greater than the values calculated from Eq. (16-14). Nevertheless, the Einstein theory was a great contribution and pointed the way toward a further understanding of the vibrations of crystal lattices.

16-4. The Normal Vibrations of a Crystal. The vibrational contributions to the thermodynamic functions of a polyatomic molecule can be calculated if the frequencies of the normal vibrations are known (Chap. 11). Viewed as a giant molecule, a crystal containing N atoms has $3N - 6 \approx 3N$ normal vibrations. If the potential energy of the crystal were known as a function of the $3N$ displacement coordinates of the atoms from their equilibrium positions at lattice sites, a normal-coordinate analysis could, in principle, be made; the normal vibrations would be $\nu_1, \ldots, \nu_{3N}$. Each of the normal vibrations is in principle physically distinguishable from the other normal vibrations; in the harmonic-oscillator approximation, the normal vibrations do not interact, and the total energy is the sum of the energies in the different normal vibrations. Thus each vibration is an independent, distinguishable statistical element, and the considerations of Sec. 16-2 apply. The canonical partition function of the crystal would be

$$Q = q(u_1) q(u_2) \cdots q(u_{3N}) \qquad (16\text{-}17)$$

and the other thermodynamic functions are obtained by summing over the values for the $3N$ oscillators. (It should be emphasized that the independent, distinguishable statistical elements are now the normal vibrations, not the atoms; any one normal vibration involves the motion of a large number of atoms.)

We can say that the vibrational spectrum of the crystal consists of the frequencies $\nu_1, \nu_2, \ldots, \nu_{3N}$. Because of the large number of normal vibrations, we expect that their frequencies will be closely spaced. It is then sensible to introduce a continuous function $g(\nu)$, the spectral density function, such that $g(\nu)\, d\nu$ is the number of normal vibrations in the interval $\nu, \nu + d\nu$.

Clearly

$$\int_0^\infty g(\nu)\, d\nu = 3N \qquad (16\text{-}18)$$

We say that the normal-vibration spectrum of the crystal is represented

---

* It is also possible, as pointed out by Einstein, to calculate ν from the density and compressibility of the crystal.

by the function $g(\nu)$. Then (writing $u = h\nu/kT$ or ν as convenient)

$$-\frac{A}{kT} = -\int \ln(1 - e^{-u})g(\nu)\,d\nu \tag{16-19}$$

$$\frac{E}{kT} = \int \frac{u}{e^u - 1} g(\nu)\,d\nu \tag{16-20}$$

$$\frac{C_V}{k} = \int \frac{u^2 e^u}{(e^u - 1)^2} g(\nu)\,d\nu \tag{16-21}$$

$$S = \frac{E - A}{T} \tag{16-22}$$

These are the fundamental equations for an improved theory of the thermodynamic properties of crystals.

16-5. The Debye Theory.* The difficult feature of the program outlined in the preceding section is the normal coordinate analysis. A simple method for short-cutting this analysis and yet obtaining a function $g(\nu)$, which gives a surprisingly good agreement with the experimental data, was discovered by Debye.

Some of the normal modes of the crystal are low-frequency long-wavelength vibrations, in which neighboring atoms move together, with equal amplitudes and equal phases. These vibrations do not depend on the discrete atomic nature of the crystal. They are the macroscopic elastic or acoustic waves of the crystal, and their frequencies can be calculated from the macroscopic elastic constants and the density of the crystal. There are also high-frequency vibrations in which neighboring atoms are vibrating against each other; these are more like the normal vibrations of a small molecule and depend on the details of the interatomic interactions. However, in the Debye theory we treat all the frequencies from the viewpoint of elastic or acoustic waves.

---

* In most respects, the topics discussed in Secs. 16-1 through 16-7 are treated in almost exactly the same way in a number of books, including the statistical mechanics texts of the general references. An authoritative modern version, which we have in part followed, is M. Blackman, in S. Flügge (ed.), "Handbuch der Physik," vol. 7, pt. 1, p. 324, Springer, Berlin, 1955; Blackman gives a comprehensive list of references to the fundamental contributions to the theory (p. 385).

It is not our purpose to present a critical comparison of theory and experiment. Various aspects of the data, in different temperature ranges, or reduced in different ways, are presented in a number of sources. The early data were collected and analyzed by E. Schrödinger (of all people!), *Physik. Z.*, **20**: 420, 450, 474, 523 (1919), and in H. Geiger and K. Scheel, "Handbuch der Physik," vol. 10, p. 304, Springer, Berlin, 1926; the data are discussed in some detail by Fowler and Guggenheim, chap. IV [1]. There are many references to experimental data in J. R. Partington, "An Advanced Treatise in Physical Chemistry," vol. III, pp. 329–340, Longmans, New York, 1952. See also G. T. Furukawa and T. B. Douglas, in D. E. Gray (ed.), "American Institute of Physics Handbook," pp. 4–39 to 4–43, McGraw-Hill, New York, 1957; H. M. Rosenberg, in B. Chalmers and R. King (eds.), "Progress in Metal Physics," vol. 7, p. 385, Pergamon, New York, 1958; P. H. Keesom and N. Pearlman, "Handbuch der Physik," vol. 14, p. 282 Springer, Berlin, 1956.

Consider a crystal in the shape of a cube of edge L. Let the sides be clamped, so that the boundary condition for the vibrations is that the displacements be zero at the walls.* The analysis of the number of allowed standing waves is identical with that for the number of allowed waves in a hohlraum (Chap. 12) or for the translational wave functions of a perfect gas. Each standing wave must be of the form

$$\sin \frac{\pi s_x x}{L} \sin \frac{\pi s_y y}{L} \sin \frac{\pi s_z z}{L}$$

where s_x, s_y, and s_z are positive integers.

Let $s^2 = s_x^2 + s_y^2 + s_z^2$; the wavelength is $\lambda = 2L/s$. If c is the velocity of the wave, the frequency is

$$\nu = \frac{c}{\lambda} = \frac{cs}{2L} \tag{16-23}$$

The number of allowed sets of positive integers (s_x, s_y, s_z) between s and $s + ds$ is $(\pi s^2/2)\, ds$. Therefore, from Eq. (16-23), the number of allowed frequencies for standing waves between ν and $\nu + d\nu$ is

$$\frac{\pi}{2}\left(\frac{2L}{c}\right)^3 \nu^2\, d\nu = \frac{4\pi V}{c^3} \nu^2\, d\nu \tag{16-24}$$

where $V = L^3$.

There are two kinds of waves in rigid materials. There are transverse waves with a velocity c_t and with two directions of polarization for any given direction of propagation (i.e., two possible waves for any given s_x, s_y, and s_z), and there are longitudinal waves, with a velocity c_l. The total number of modes of vibration between ν and $\nu + d\nu$ is then, from (16-24),

$$g(\nu)\, d\nu = \left(\frac{2}{c_t^3} + \frac{1}{c_l^3}\right) 4\pi V \nu^2\, d\nu \tag{16-25}$$

It is convenient to define an average velocity c by

$$\frac{3}{c^3} = \frac{2}{c_t^3} + \frac{1}{c_l^3} \tag{16-26}$$

and to rewrite (16-25) as

$$g(\nu)\, d\nu = \frac{12\pi V}{c^3} \nu^2\, d\nu \tag{16-27}$$

As already remarked, these relations, which are based on the theory of elastic vibrations of a continuous medium, would not be expected to apply to the high-frequency short-wavelength vibrations, where neighboring atoms are vibrating against each other. Nevertheless, we assume

* The boundary conditions are different for an unclamped crystal; but, for a large crystal, the function $g(\nu)$ proves to be the same. The boundary conditions considered by Debye were more general than the simple ones used here.

that they do apply. We assume there is an upper limit for the allowed vibrations, ν_D, which is determined by the requirement that the total number of normal vibrations be $3N$. That is, we require that

$$\int_0^{\nu_D} g(\nu)\, d\nu = 3N \tag{16-28a}$$

or

$$\int_0^{\nu_D} \frac{12\pi V}{c^3} \nu^2\, d\nu = 3N \tag{16-28b}$$

so that

$$\nu_D^3 = \frac{3}{4\pi} \frac{N}{V} c^3 \tag{16-28c}$$

In terms of ν_D, the distribution function can be written

$$g(\nu)\, d\nu = \frac{9N}{\nu_D^3} \nu^2\, d\nu \qquad 0 \leq \nu \leq \nu_D \tag{16-29a}$$

$$g(\nu) = 0 \qquad \nu > \nu_D \tag{16-29b}$$

This distribution is known as the Debye spectrum of frequencies. The frequency ν_D is called the Debye frequency; the corresponding characteristic temperature Θ_D is defined by

$$\Theta_D = \frac{h\nu_D}{k} \tag{16-30a}$$

When $\nu = \nu_D$, the variable $u = h\nu/kT$ becomes

$$u_D = \frac{h\nu_D}{kT} = \frac{\Theta_D}{T} \tag{16-30b}$$

The distribution function $g(\nu)\, d\nu$ can now be written [since

$$\nu^2\, d\nu = \left(\frac{kT}{h}\right)^3 u^2\, du\bigg]$$

$$g(\nu)\, d\nu = 9N\left(\frac{T}{\Theta_D}\right)^3 u^2\, du = \frac{9N}{u_D^3} u^2\, du \tag{16-31}$$

With this expression for $g(\nu)\, d\nu$, the molal heat-capacity expression becomes

$$\frac{C_V}{3N_0 k} = \frac{C_V}{3R} = \frac{3}{u_D^3} \int_0^{u_D} \frac{u^4 e^u}{(e^u - 1)^2}\, du = D(u_D) \tag{16-32}$$

The Debye function $D(u_D)$ [or, as is more commonly written, $D(T/\Theta_D)$] is defined by (16-32).

The other thermodynamic functions are given by

$$-\frac{A}{3RT} = -\frac{3}{u_D^3} \int_0^{u_D} [\ln(1 - e^{-u})] u^2\, du \tag{16-33}$$

$$\frac{E}{3RT} = \frac{3}{u_D^3} \int_0^{u_D} \frac{u^3\, du}{e^u - 1} = D_e(u_D) \left[\text{or } D_e\left(\frac{T}{\Theta_D}\right)\right] \tag{16-34}$$

The function $D_e(u_D)$ is defined in (16-34).*

Tables of the Debye heat-capacity function and of the entropy and energy functions are available.† An abbreviated table is reproduced here. The general behavior of the heat-capacity function is illustrated in Fig. 16-2.

Table 16-1. The Debye Heat-capacity Function

T/θ	$c_V/3R$	T/θ	$c_V/3R$	T/θ	$c_V/3R$
0.01	7.8×10^{-5}	0.20	0.369	0.75	0.916
0.03	0.00210	0.24	0.476	0.90	0.941
0.05	0.00974	0.30	0.608	1.0	0.952
0.070	0.0267	0.40	0.746	2.0	0.988
0.10	0.0758	0.50	0.825	3.0	0.994
0.14	0.182	0.60	0.872	∞	1.000

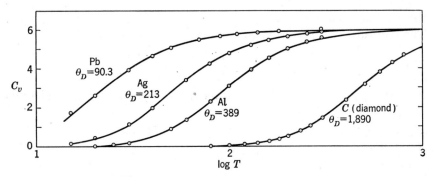

FIG. 16-2. The heat capacity in calories per degree per mole for several solid elements. The curves are from the Debye function with the Θ_D values given. (*From G. N. Lewis and M. Randall, "Thermodynamics," 2d ed., revised by K. S. Pitzer and L. Brewer, p. 56, McGraw-Hill, New York, 1961.*)

As $T \to \infty$, $u_D \to 0$ and the integral in Eq. (16-32) can be expanded to give

$$D(u_D) = 1 - \frac{u_D^2}{20} + \frac{u_D^4}{560} - \frac{u_D^6}{18,144} \qquad (16\text{-}35)$$

which, it is said (M. Blackman, in S. Flügge (ed.), "Handbuck der Physik," vol. 7, pt. 1, p. 328, Springer, Berlin, 1955), is a good approximation even for $u_D = 2$. The main point is that $D(u_L) \to 1$ and $c_V \to 3R$ as $T \to \infty$.

* The function $D_e(u_D)$ was once called the Debye function; it is now recommended that this name be reserved for the heat-capacity function (16-32) (see M. Blackman, in S. Flügge (ed.), "Handbuch der Physik," vol. 7, pt. 1 p. 328, Springer, Berlin, 1955).

† F. Simon, in H. Geiger and K. Scheel, "Handbuch der Physik," vol. 10, pp. 367–369, Springer, Berlin, 1926; G. T. Furukawa and T. B. Douglas, in D. E. Gray (ed.), "American Institute of Physics Handbook," pp. 4-44 to 4-46, McGraw-Hill, New York, 1957; Pitzer, pp. 502–503 (heat capacity and energy functions only) [10].

As $T \to 0$, $u_D \to \infty$ and the Debye function approaches

$$D\left(\frac{T}{\Theta_D}\right) = 3\left(\frac{T}{\Theta_D}\right)^3 \int_0^\infty \frac{u^4 e^u}{(e^u - 1)^2} du = \frac{4\pi^5}{5}\left(\frac{T}{\Theta_D}\right)^3 \qquad \frac{C_V}{3R} = 77.93\left(\frac{T}{\Theta_D}\right)^3 \tag{16-36}$$

The T^3 law, as expressed in Eq. (16-36), is one of the important results of the Debye theory. Equation (16-36) is accurate to 1 per cent or better for $T/\Theta_D < \frac{1}{12}$.

Another important feature of the theory is the prediction that the heat capacity and related thermodynamic functions are universal functions for all monatomic solids, determined by one parameter Θ_D.

The agreement of the Debye theory with experimental data for four substances of quite different Θ_D values is displayed in Fig. 16-2.

According to Eqs. (16-26), (16-28c), and (16-30a), the Debye temperature Θ_D can be calculated from the velocities of the elastic waves in a solid. These in turn can be calculated from the compressibility

$$\left[\kappa = -\left(\frac{1}{V}\right)\left(\frac{\partial V}{\partial P}\right)\right]$$

the density ρ, and Poisson's ratio σ for an isotropic material.* The two wave velocities are given by

$$c_t^2 = \frac{3(1 - 2\sigma)}{2(1 + \sigma)\kappa\rho} \qquad c_l^2 = \frac{3(1 - \sigma)}{(1 + \sigma)\kappa\rho} \tag{16-37a}$$

We therefore have

$$\frac{2}{c_t^3} + \frac{1}{c_l^3} = \rho^{3/2}\kappa^{3/2}\left\{2\left[\frac{2(1 + \sigma)}{3(1 - 2\sigma)}\right]^{3/2} + \left[\frac{1 + \sigma}{3(1 - \sigma)}\right]^{3/2}\right\} \tag{16-37b}$$

$$\frac{2}{c_t^3} + \frac{1}{c_l^3} = \rho^{3/2}\kappa^{3/2}f(\sigma) \tag{16-37c}$$

A comparison of Θ (elastic) and Θ (heat capacity) is given in Table 16-2. In view of the marked variations of κ, $f(\sigma)$, and Θ from substance to substance, the agreement between the two sets of values for Θ in Table 16-2 is a tribute to the general accuracy of the Debye theory, at least as regards its ability to predict low-temperature heat capacities.

The Debye theory does not fit the heat-capacity data perfectly. We shall comment on these discrepancies later. There is one point which should be mentioned here, however. There are two important contributions to the heat capacity of a metal, the lattice vibrations and the electronic heat capacity.

---

* Poisson's ratio may be defined as the ratio of the relative linear contraction of the diameter of a cylindrical specimen to its relative axial elongation when elastically stressed along the axis. For an incompressible substance it would be 0.5; for actual substances it is usually in the range of 0.33 to 0.4. Young's modulus, the ordinary linear elastic constant, is given by $Y = (3/\kappa)(1 - 2\sigma)$.

We shall not study the theory of the wave functions and energy levels of electrons in metals.* To a rather crude approximation, however, the conduction electrons of a metal can be regarded as constituting a free-electron gas. Our introductory discussion of a Fermi-Dirac gas (Sec. 6-16) shows that, at $T = 0$, the electrons are in the lowest levels available to them, consistent with the requirement that there be only two electrons in each translational level. As the temperature is raised to moderate temperatures, a relatively small fraction of all the electrons are excited from the highest filled levels to the lower unfilled levels. The corresponding contribution to the heat capacity is small. A simple and

Table 16-2. Comparison of Θ (elastic) and Θ (heat capacity)*

Substance	ρ, cc^{-1}	$\kappa \times 10^{12}$ dyne^{-1} cm^2	$f(\sigma)$	Θ (elastic), °K	Θ (heat capacity), °K
Al	2.71	1.36	10.2	399	396
Cu	8.96	0.74	10.5	329	313
Ag	10.53	0.92	15.4	212	220
Au	19.21	0.6	24.7	166	186
Cd	8.63	2.4	7.89	168	164
Sn	7.28	1.9	8.50	185	165
Pb	11.32	2.0	61.0	72	86
Bi	9.78	3.2	8.90	111	111
Pt	21.39	0.40	17.1	226	220

* From M. Blackman, in S. Flügge (ed.), "Handbuch der Physik," vol. 7, pt. 1, p. 329, Springer, Berlin, 1955.

straightforward application of Fermi-Dirac statistics to the low-temperature heat capacity of a free-electron gas gives

$$c_V = \gamma T \tag{16-38}$$

where γ is a constant.†

Values of the Debye Θ and γ for several metals are displayed in Table 16-3. Since the heat capacity at low temperature is given by

$$c_V = 465 \left(\frac{T}{\Theta}\right)^3 + \gamma T \qquad \text{cal mole}^{-1} \text{ deg}^{-1}$$

* See C. Kittel, "Introduction to Solid State Physics," 2d ed., chaps. 6, 10–12, Wiley, New York, 1956; A. H. Wilson, "Theory of Metals," 2d ed., Cambridge, New York, 1953.

† C. Kittel, "Introduction to Solid State Physics," 2d ed., p. 258, Wiley, New York, 1956; this simple and interesting calculation is recommended to the student. In the free-electron theory, $\gamma = \frac{1}{2} \pi^2 z R T / T_f$, where z is the valence and T_f is the "Fermi temperature," defined by $kT_f = \mu(0)$, where $\mu(0)$ is the energy of the highest filled level at $T = 0$, or the chemical potential at $T = 0$ (Sec. 6-16). The free-electron value for γ is not accurately in agreement with the facts but is approximately correct.

the contributions at the electronic specific heat and the lattice vibrations are equal at

$$T = \left(\frac{\gamma \Theta^3}{465}\right)^{1/2} \qquad (16\text{-}39)$$

Above this temperature, the lattice contributions are predominant; below it, the electronic contributions are most important. For $\gamma = 10^{-4}$

Table 16-3. Debye Θ's and Electronic Constants for Metals*

Metal	Ag	Al	Be	Na	Ni	Pt	Ti	Zn
Θ, °K	229	375	1160	160	413	233	430	235
γ, 10^{-4} cal mole^{-1} deg^{-2}	1.45–1.60	3.27–3.48	0.53	4.3	17.4	16.1–16.5	8.0–8.5	1.25–1.50

* G. T. Furukawa and T. B. Douglas, in D. E. Gray (ed.), "American Institute of Physics Handbook," p. 4-48, McGraw-Hill, New York. 1957.

cal mole^{-1} deg^{-2} and $\Theta = 200°K$, the temperature defined by Eq. (16-39) is 1.3°K.

16-6. Normal Coordinate Analysis for One-dimensional Crystals. In order to improve upon the Debye theory of the heat capacity due to lattice vibrations, a more accurate theory of the spectrum of normal vibrations is needed. We shall not study the difficult three-dimensional problem. We can, however, gain a feeling for the methods of treating the vibration spectrum and the nature of the results by treating the much simpler one-dimensional problem.

Consider a one-dimensional lattice with a lattice spacing of a containing N identical atoms at the points along the x axis, a, $2a$, ..., na, ..., Na. Assume that the only interaction is a harmonic interaction between neighboring atoms, with a force constant f. Denote the *displacement* of the nth atom from its equilibrium position by x_n. The kinetic and potential energies are then

$$2K = \sum_{n=1}^{N} m\dot{x}_n^2 \qquad (16\text{-}40a)$$

$$2U = \sum_{n=2}^{N} f(x_n - x_{n-1})^2 \qquad (16\text{-}40b)$$

The Lagrangian equations of motion therefore are

$$m\ddot{x}_n = f(x_{n+1} + x_{n-1} - 2x_n) \qquad n = 1, 2, \ldots, N \qquad (16\text{-}40c)$$

and we wish to find functions $x_n(t)$ which are solutions of the equations of motion.

The easiest boundary conditions to use are the so-called "cyclic" boundary conditions*

$$x_n(t) = x_{n+N}(t) \tag{16-41}$$

We then look for solutions of the form

$$x_n(t) = A e^{i(\omega t + n\phi)} \tag{16-42}$$

where A is a constant and ω is the angular frequency $\omega = 2\pi\nu$. Values of ϕ that satisfy the boundary conditions are

$$\phi_j = \frac{2\pi j}{N} \quad j = 1, 2, 3, \ldots, N \tag{16-43}$$

We can then rewrite Eq. (16-42) as

$$x_n(t) = A e^{i\omega t} e^{i 2\pi j n / N} \tag{16-44}$$

Since $e^{2\pi i} = 1$, it is clear that $x_n(t) = x_{n+N}(t)$, as required. Furthermore, for $k = j + N$, $e^{in\phi_k} = e^{in\phi_j}$; the range of values of $j = 1, \ldots, N$ in (16-44) covers all the physically different values of the function $e^{in\phi_j}$.

To each of the allowed values of ϕ there corresponds an allowed frequency or normal vibration ω. To determine these, substitute (16-42) into (16-40c); then cancel the factor $A e^{i(\omega t + n\phi)}$, and obtain

$$-m\omega^2 = f(e^{i\phi} + e^{-i\phi} - 2) = f(e^{i\phi/2} - e^{-i\phi/2})^2 = -4f \sin^2 \frac{\phi}{2}$$

The allowed values of ϕ are $2\pi j/N$; the allowed values of ω are therefore

$$\omega_j = 2 \left(\frac{f}{m}\right)^{1/2} \sin \frac{\pi j}{N} \quad j = 1, 2, \ldots, N \tag{16-45a}$$

or

$$\nu_j = \frac{1}{\pi} \left(\frac{f}{m}\right)^{1/2} \sin \frac{\pi j}{N} \tag{16-45b}$$

By differentiation of (16-45b)

$$d\nu_j = \frac{1}{N} \left(\frac{f}{m}\right)^{1/2} \cos \frac{\pi j}{N} dj \tag{16-46}$$

Since $\sin \phi = \sin (\pi - \phi)$, for each value of $j < N/2$, there is a corresponding value $k = N - j$ which gives the same value of ν_j as is given by (16-45b). Thus, to determine the spectral-density function $g(\nu)$, we can restrict our attention to values of j such that $1 \le j \le N/2$. For each integral value of j in the interval $1 \le j \le N/2$ there are two physically

---

* One can imagine that the linear crystal is bent around into a large circle, so that the Nth atom is next to the first atom. It is known that the distribution of frequencies is not seriously affected by this choice of boundary conditions. (M. Blackman, in S. Flügge (ed.), "Handbuch der Physik," vol. 7, pt. 1, p. 330, Springer, Berlin, 1955.) With cyclic boundary conditions, the sum in (16-40b) should logically extend from $n = 1$ to $n = N$, and we set $x_0 = x_N$.

distinguishable normal vibrations, corresponding to $\phi_j = 2\pi j/N$ and $\phi_{N-j} = 2\pi(N-j)/N$; both have the frequency ν_j as given by (16-45b). Then from (16-46), the number of frequencies between ν and $\nu + d\nu$ is

$$g(\nu)\,d\nu = \frac{2N(m/f)^{1/2}}{\cos(\pi j/N)}\,d\nu \qquad (16\text{-}47)$$

We rewrite (16-45b) as

$$\nu = \nu_0 \sin\frac{\pi j}{N} \quad \text{with } \nu_0 = \left(\frac{f}{\pi^2 m}\right)^{1/2} \qquad (16\text{-}48)$$

and

$$\cos\frac{\pi j}{N} = (\nu_0^2 - \nu^2)^{1/2}\,\frac{1}{\nu_0}$$

so that

$$g(\nu)\,d\nu = \frac{2N}{\pi}\,\frac{d\nu}{(\nu_0^2 - \nu^2)^{1/2}} \qquad (16\text{-}49)$$

This is the resulting density function or spectrum of frequencies for a linear lattice, with the assumed potential function.

Problem 16-2. Show that the spectral-density function (16-49) is properly normalized.

It should be noted that the Debye spectrum for a one-dimensional lattice is a constant,

$$g(\nu)\,d\nu = N\,d\nu$$

The spectrum (16-49) is a constant at very low frequencies, $\nu \ll \nu_0$, but it increases to an infinite discontinuity as ν approaches its upper limit of ν_0.

A general equation for a propagating one-dimensional wave in a continuous medium is

$$\psi(x) = \psi(0)e^{i(\omega t + kx)}$$

This represents a wave of wavelength $\lambda = 2\pi/k$ propagating with a velocity of ω/k in the $-x$ direction.

For a disturbance in a discrete lattice with a spacing a, the equation

$$x_n(t) = A e^{i(\omega t + n\phi)}$$

corresponds to a wave which repeats every $\delta n = 2\pi/\phi$ atoms, so that the wavelength is $2\pi a/\phi$ or, for $\phi_j = 2\pi j/N$,

$$\lambda_j = \frac{Na}{j} \qquad (16\text{-}50)$$

For $j = 1$, the wavelength is Na, or the length of the crystal. For $j = N/2$, the wavelength is $2a$, which means that alternate atoms are vibrating out of phase with respect to each other. This is the shortest-wavelength highest-frequency (ν_0) vibration possible in the lattice. For $j > N/2$, consider the number $k = N - j$; and, of course, $k < N/2$. Now $e^{in\phi_j} = e^{2\pi i n j/N} = e^{2\pi i n(N-k)/N} = e^{-2\pi i n k/N} = e^{-in\phi_k}$. Thus, for $j >$

$N/2$, the wavelength is the same as that for $k = N - j$; but one wave travels in the forward direction and the other in the backward direction. It was shown previously that the two waves have the same frequency.

Thus for this simple case we can get a true picture of the normal vibrations of the lattice and recognize that the spectrum differs considerably from the Debye spectrum.

It is instructive to make a calculation for a linear chain consisting of an alternating sequence of two different kinds of atoms. This is the one-dimensional analogue of substances such as the alkali halide salts.

Let the interatomic distance be a so that the lattice spacing is $2a$. There are atoms of type 1 and mass m with their equilibrium positions at $2a, 4a, \ldots, 2na, \ldots, 2Na$. There are atoms of the second type of mass μ, with their equilibrium positions at $a, 3a, \ldots, (2n-1)a, \ldots, (2N-1)a$. Let $x_{2n}(t)$ be the displacement coordinates for the atom of the first kind, which is at $2na$; $x_{2n+1}(t)$ is the displacement coordinates for the atom of the second kind, at $(2n+1)a$. We again assume only nearest-neighbor forces, so that the potential energy is

$$2U = \sum_{n=1}^{N} f[(x_{2n} - x_{2n-1})^2 + (x_{2n+1} - x_{2n})^2] \quad (16\text{-}51)$$

The kinetic energy is

$$2K = \sum_{n=1}^{N} (m\dot{x}_{2n}^2 + \mu\dot{x}_{2n-1}^2) \quad (16\text{-}52)$$

The equations of motion are therefore

$$m\ddot{x}_{2n} - f(x_{2n+1} + x_{2n-1} - 2x_{2n}) = 0 \quad (16\text{-}53a)$$
$$\mu\ddot{x}_{2n+1} - f(x_{2n+2} + x_{2n} - 2x_{2n+1}) = 0 \quad (16\text{-}53b)$$

We seek a solution of the form

$$x_{2n}(t) = A e^{i(\omega t + 2n\phi)} \quad (16\text{-}54a)$$
$$x_{2n+1}(t) = B e^{i[\omega t + (2n+1)\phi]} \quad (16\text{-}54b)$$

In Eqs. (16-54), the amplitudes A and B of the vibrations of the two different kinds of atoms can be different. We again use cyclic boundary conditions, which in this case are

$$\begin{aligned} x_{2n}(t) &= x_{2n+2N}(t) \\ x_{2n+1}(t) &= x_{2n+1+2N}(t) \end{aligned} \quad (16\text{-}55)$$

The allowed values of the phase angle ϕ, consistent with the boundary conditions, are

$$\phi_j = \frac{\pi j}{N} \quad j = 1, 2, \ldots, N \quad (16\text{-}56)$$

By straightforward substitution of Eqs. (16-54) into the equations of

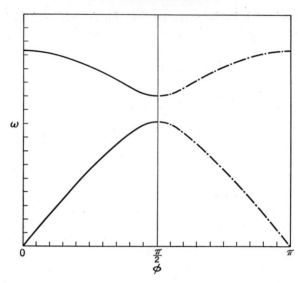

FIG. 16-3. The normal vibration frequencies ω of a linear diatomic chain as a function of the phase angle ϕ. (*From M. Blackman, in S. Flügge (ed.), "Handbuch der Physik," vol. 7, pt. 1, p. 332, Springer, Berlin.*)

motion (16-53) and canceling the factor $e^{i(\omega t + 2n\phi)}$ or $e^{i[\omega t + (2n+1)\phi]}$ from (16-53a) or (16-53b), respectively, we obtain

$$(2f - m\omega^2)A - f(e^{i\phi} + e^{-i\phi})B = 0$$
$$-f(e^{i\phi} + e^{-i\phi})A + (2f - \mu\omega^2)B = 0 \quad (16\text{-}57)$$

Solutions for the amplitudes A and B are possible if and only if the secular determinant is zero.

$$\begin{vmatrix} 2f - m\omega^2 & -2f\cos\phi \\ -2f\cos\phi & 2f - \mu\omega^2 \end{vmatrix} = 0 \quad (16\text{-}58)$$

Let ω_0 be the angular frequency given by

$$\omega_0^2 = \frac{f(m + \mu)}{m\mu} \quad (16\text{-}59)$$

The solutions of the secular equation are then

$$\omega^2 = \omega_0^2 \left\{ 1 \pm \left[1 - \frac{4\mu m \sin^2 \phi}{(\mu + m)^2} \right]^{\frac{1}{2}} \right\} \quad (16\text{-}60)$$

with the allowed values of ϕ given by $\phi_j = \pi j/N$.

There are two branches to the curve corresponding to the $\pm$ choice for the square root in (16-60). Values of ω as a function of ϕ are displayed in Fig. 16-3. The upper high-frequency branch of the curve is called the optical branch. The low-frequency branch is called the acoustical branch.

For $j = 0$ or N, the frequency of the optical branch is a maximum, $\omega = 2^{1/2}\omega_0$; the frequency of the acoustical branch is zero. For $j = N/2$, the frequency of the optical branch is a minimum, and the frequency of the acoustical branch is a maximum. For $j = N/2$, the frequencies are

$$\omega^2 = \omega_0^2 \left(1 \pm \left|\frac{\mu - m}{\mu + m}\right|\right) \qquad (16\text{-}61)$$

The lowest optical frequency equals the highest acoustical frequency for $\mu = m$; otherwise it is greater.

On the optical branch, A and B of Eq. (16-54) are of opposite sign; neighboring atoms vibrate in opposite directions. On the acoustical branch, the signs of A and B are alike.

The density function $g(\nu)$ can be calculated from the relation between ν and j as given in (16-60) and (16-56). The results of two such calculations are displayed in Fig. 16-4.

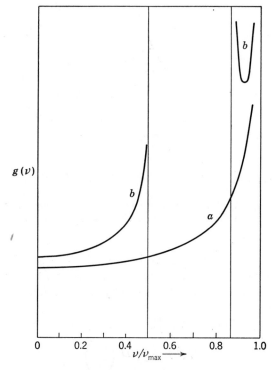

FIG. 16-4. Density of normal vibrations, $g(\nu)$, of a linear diatomic chain. (a) $m/\mu = 1$. (b) $m/\mu = 3$. The maximum frequency is $\nu_{\max}$ in each case. (*From M. Blackman, in S. Flügge (ed.), "Handbuch der Physik," vol. 7, pt. 1, p. 332, Springer, Berlin, 1955.*)

16-7. More about Lattice Vibrations and Heat Capacities. The application of the method illustrated in the previous section for a one-dimensional lattice to the analysis of the vibration spectrum of a three-dimensional lattice was initiated by Born and Von Kármán and has been carried on by a number of investigators.* Approximate calculations of the vibrational spectra for many of the simple crystal structures of the

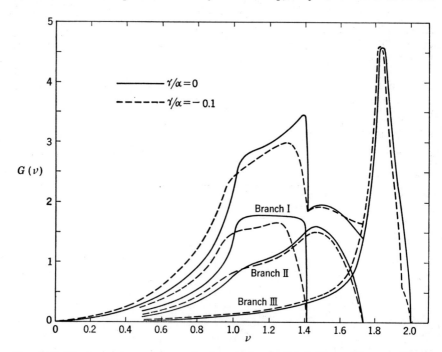

Fig. 16-5. The frequency spectrum of a face-centered cubic lattice. The quantities γ and α are the force constants for the interactions between next-nearest and nearest neighbors, respectively. There are three branches to the curve, and the sum of these three is the total density of normal vibrations. [*From* R. B. Leighton, *Revs. Modern Phys.*, **20**: 168 (1948).]

elements and for binary salts have been made. As an illustration, Fig. 16-5 is a calculation of the vibrational spectrum of a face-centered cubic lattice.† The spectrum resembles a Debye spectrum at low frequencies but differs at high frequencies.

The Debye expression for the heat capacity is

$$c_V(T) = 3RD\left(\frac{\Theta_D}{T}\right) \tag{16-62}$$

---

* For references, see M. Blackman, in "Handbuch der Physik," vol. 7, pt. 1, pp. 330–374, Springer, Berlin, 1955.

† R. B. Leighton, *Revs. Modern Phys.*, **20**: 168 (1948).

The measure of agreement between the Debye theory and experiment can be displayed by determining a value of Θ_D to satisfy (16-62) at each temperature. (This procedure is useful only for $c_V/3R$ less than about 0.6, because as $T \to \infty$, $C_V \to 3R$, irrespective of the value of Θ_D.) If the Debye theory fitted the facts very accurately, Θ_D would be constant. Figure 16-6 displays such a plot for silver. It can be seen that the "experimental" values of Θ_D vary by about 20° out of 200°; the theoretical curve based on the spectral distribution shown in Fig. 16-4 gives a better agreement with experiment than does the Debye theory. The results as shown in Fig. 16-6 for silver are typical for many cases.

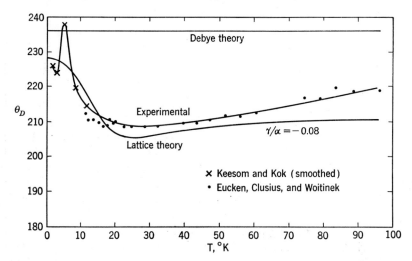

FIG. 16-6. The heat capacity of silver expressed in terms of the equivalent Debye temperature Θ_D. The value of Θ_D calculated from the Debye theory and the elastic constants is also shown. [From R. B. Leighton, Revs. Modern Phys., **20**: 172 (1948).]

For a binary compound such as NaCl or ZnS which does not contain individual molecules but in which the crystal can be considered as one giant molecule, a fairly good empirical fit to the molal heat-capacity curve can be achieved by regarding the two different kinds of atoms as identical, so that 1 mole of the crystal contains $2N_0$ atoms, and

$$c_V = 6RD\left(\frac{\Theta_D}{T}\right) \tag{16-63}$$

The vibrational spectrum of rock salt as deduced from a theoretical Born–Von Kármán treatment is shown in Fig. 16-7. It should be said that the high-frequency parts of this spectrum are an "optical branch," in which the chloride ions and the sodium ions vibrate in opposite directions. These vibrations emit and absorb infrared radiation strongly. In spite of the great deviation of the vibrational spectrum from that of the Debye theory, the specific-heat curve is approximately represented by a Debye curve with $\Theta_D = 275$ to $300°K$ (M. Blackman, in S. Flügge (ed.), "Handbuch der Physik," vol. 7, pt. 1, p. 363, Springer, Berlin, 1955).

For a crystal such as $CaCO_3$, the following argument is used: There are six internal vibrations $\nu_1, \ldots, \nu_6$ of a carbonate anion; these will not be greatly influenced by the other vibrations in the lattice and will be reasonably sharp. If we denote by $c_V(u_i)$ the harmonic-oscillator heat-capacity function $u_i^2 e^{u_i}/(e^{u_i} - 1)^2$ for an oscillator, the contribution of the vibrations of the CO_3 molecule to the molal heat capacity is given by

$$N_0 k \sum_{i=1}^{6} c_V(u_i)$$

[The function $c_V(u_i)$ is sometimes called the Einstein function.]

We now regard each Ca^{++} ion and each $CO_3^=$ ion as an individual particle; the molal heat capacity due to the lattice vibrations of these particles would be a Debye function for $2N_0$ particles. In addition, there are three motions which in zero-th approximation would be described as torsional vibrations of the $CO_3^=$ anion, corresponding to the three rotations of a free molecule. The torsional vibrations will be of low frequency, and it is intuitively clear that they will interact considerably with the lattice vibrations. We can therefore approximate the situation by saying that there are $9N_0$ degrees of freedom which can be treated as acoustical vibrations and approximated by a Debye function and $6N_0$ degrees of freedom for the internal vibrations of $CO_3^=$ which can be represented as harmonic-oscillator frequencies. Therefore

$$c_V = 9RD\left(\frac{\Theta_D}{T}\right) + R \sum_{i=1}^{6} c_V(u_i) \tag{16-64}$$

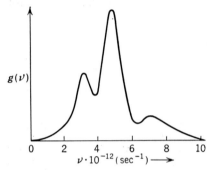

FIG. 16-7. The vibrational spectrum of sodium chloride. (*From M. Blackman, in S. Flügge (ed.), "Handbuch der Physik," vol. 7, pt. 1, p. 332, Springer, Berlin, 1955.*)

This same approach has been applied to solid benzene with considerable success.* With 12 atoms in the molecule, there are 30 internal vibrations which are known from the gas-phase spectrum, and there are six degrees of translation and torsion which are treated as acoustical modes by the Debye theory.

For general chemical purposes, an important application of low-temperature heat-capacity measurements is for the calculation of the

---

* R. C. Lord, J. E. Ahlberg, and D. H. Andrews, *J. Chem. Phys.*, **5**: 649 (1937).

entropy change in going from 0°K to ordinary temperatures. In much of this work, the heat capacity will be measured down to a temperature of either 20°K or 4°K. The heat-capacity curve and the entropy change from this temperature to 0°K are estimated by assuming the relation

$$c_P = aT^3 \tag{16-65}$$

and determining the constant a from the measured value at 20°K or 4°K. In the absence of any of the magnetic effects discussed in Chap. 19, and if there is reason for believing that there are no order-disorder transitions as discussed in Sec. 16-9, this is a useful extrapolation, even for rather complex crystals.

One other topic may be mentioned here. Because of anharmonicity effects, the vibrational heat capacity of a diatomic molecule is usually greater than the classical value of k at high temperatures. A question of some interest is whether there is a similar increase in c_v for crystals above the classical value of $3k$ per atom at high temperatures, where the amplitude of the lattice vibrations is relatively large. There are not many pertinent data (cf. M. Blackman, in "Handbuch der Physik," vol. 7, pt. 1, pp. 369–370, Springer, Berlin, 1955). In the case of NaCl (Fig. 16-8), the high-temperature value of c_v does not increase above the theoretical value of $6R$ (per mole of NaCl), but the values of c_P are much greater.

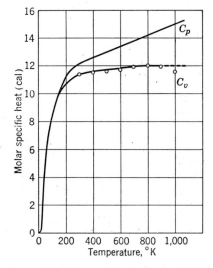

Fig. 16-8. The molar heat capacities c_P and c_V of NaCl as a function of temperature. [From L. Hunter and S. Siegel, Phys. Rev., **61**: 90 (1942).]

16-8. Entropy and Disorder in Crystals. Suppose that a crystal contains N molecules at definite lattice sites. Let each molecule have n physically distinguishable orientations with respect to the crystallographic axes. Then the number of physically distinguishable configurations of the system is n^N. This statement is based on the fact that the different lattice sites are distinguishable;* the state with the molecule at site i up and the molecule at site j down is physically distinguishable from the state with the molecule at site j up and the molecule at site i down. The entropy associated with the n^N configurations is

$$S = k \ln n^N = Nk \ln n \tag{16-66}$$

The entropy of random orientation or disorder in a crystal is most

* The molecules are not distinguishable, but the lattice sites are distinguishable.

readily recognized if it occurs at low temperatures, where the entropy due to lattice vibrations and molecular vibrations is small. Solid carbon monoxide provides a simple and interesting example. The entropy of carbon monoxide in the hypothetical ideal-gas state at 1 atm pressure at 298.16°K is 47.301 cal deg^{-1} mole^{-1}.* This is the absolute entropy of the gas molecules, as calculated from spectroscopic data, with the statistical mechanical formulae developed in Chaps. 6 and 8.

The entropy of the gas can also be determined "calorimetrically," i.e., by starting with the solid at 0°K and measuring the heat capacity as a function of temperature and the entropies for the various phase transitions leading to the gas at the desired temperature. Clayton and Giauque† have made such measurements for the entropy change on heating solid carbon monoxide from 11.70°K to the gaseous state at the boiling point of 81.61°K. The entropy change in heating the ideal gas from 81.60 to 298.16°K was calculated from the molecular constants of CO gas (but of course the same answer would have been obtained from experimental measurements of the heat capacity of the gas). Their data are reproduced in Table 16-4.

Table 16-4. Calculation of Entropy of Carbon Monoxide

0–11.70°K, Debye extrapolation $h\nu_D/k = 79.5$	0.458
11.70–61.55, graphical	9.632
Transition, 151.3/61.55	2.457
61.55–68.09, graphical	1.228
Fusion, 199.7/68.09	2.933
68.09–81.61, graphical	2.611
Vaporization, 1443.6/81.61	17.689
Entropy of carbon monoxide gas at boiling point	37.01 ± 0.1
Correction for gas imperfection, assuming Berthelot gas	0.21
Entropy corrected to the ideal state at 81.61°K	37.2
Entropy change for ideal gas, 81.61–298.16°	9.0
"Calorimetric" entropy of ideal gas at 298.16°	46.2 cal mole^{-1} deg^{-1}

The calorimetric entropy change between 0 and 298.16°K is 1.1 cal mole^{-1} deg^{-1} less than the spectroscopic entropy at 298.16°K. This discrepancy is explained by assuming that the crystal has a residual entropy of 1.1 e.u. at 0°K. The residual entropy is due to some disorder in the crystal which is still present at 0°K. In the present instance, it is reasonable to believe that in the crystal, at any one lattice site, there is an axis of orientation for a CO molecule. However, the molecules in the real crystal are oriented at random as either CO or OC units. The molar entropy for two possible configurations per molecule is $R \ln 2 = 1.38$ e.u. The actual residual entropy is slightly less than this, suggesting a slight amount of ordering.

* Rossini et al., p. 99 [24].
† J. O. Clayton and W. F. Giauque, *J. ACS*, **54:** 2610 (1932).

The CO molecule is relatively nonpolar, and the van der Waals radii of C and O are similar. Thus the molecule is only slightly unsymmetrical. Presumably at 0°K the crystal should (in the thermodynamic sense) go into a more stable configuration with the carbons and oxygens in nonequivalent positions. However, the molecules cannot rotate at this low temperature, and the disorder is frozen in.

In the same way, there is a residual entropy of about $R \ln 2$ for N_2O (which is linear, with the structure NNO) because the actual crystal at 0°K fails to distinguish between the orientations NNO and ONN.* Crystalline NO has a residual entropy of $R \ln 2$ per mole of N_2O_2.† Crystalline NO contains dimeric N_2O_2 molecules with a rectangular structure, with a short edge of 1.12 Å and a long edge of 2.40 Å:‡

$$\begin{array}{c} N{=}O \\ \vdots \quad \vdots \\ O{=}N \end{array}$$

The residual entropy is presumably due to the two possible orientations of the rectangle.

In the tetrahedral molecule $FClO_3$, there is a residual entropy of 2.42 cal deg^{-1} mole^{-1}. This is approximately $R \ln 4 = 2.75$ cal deg^{-1} mole^{-1}; evidently the crystal fails to distinguish between oxygen and fluorine atoms, and the Cl—F bond is oriented in any one of the four tetrahedral directions at random.§

One of the most interesting problems of this nature is that of the entropy of ice.∥ Calorimetric measurements give residual entropies of 0.82 and 0.77 cal mole^{-1} deg^{-1} for H_2O and D_2O crystals, respectively.¶ X-ray diffraction shows that the oxygen atoms of ice are in the wurtzite structure (hexagonal ZnS), that is, a hexagonal structure with each oxygen surrounded tetrahedrally by four other oxygen atoms at a distance of 2.76 Å. A structure which explains the entropy results and is consistent with general structural knowledge about hydrogen bonds was proposed by Pauling.** It was proposed that: (a) there is one hydrogen atom on the line joining any two neighboring oxygen atoms; (b) each oxygen has two near hydrogens at about 0.95 Å (the H—O distance in gaseous H_2O); (c) each oxygen has two distant hydrogens at about 1.81 Å.

* R. W. Blue and W. F. Giauque, *J. ACS*, **57**: 991 (1935).
† H. L. Johnston and W. F. Giauque, *J. ACS*, **51**: 3194 (1926).
‡ W. J. Dulmage, E. A. Meyers, and W. N. Lipscomb, *Acta Cryst.*, **6**: 760 (1953).
§ J. K. Koehler and W. F. Giauque, *J. ACS*, **80**: 2659 (1958).
∥ The subject is discussed by L. Pauling, "The Nature of the Chemical Bond," pp. 464–468, Cornell University Press, Ithaca, New York, 1960, and he gives a more complete set of references.
¶ W. F. Giauque and J. W. Stout, *J. ACS*, **58**: 1144 (1936); E. A. Long and J. D. Kemp, *J. ACS*, **58**: 1829 (1936).
** L. Pauling, *J. ACS*, **57**: 2680 (1935).

The exact calculation of the number of possible configurations for this structure is difficult, because if one starts with one permissible structure and moves one H atom from one side of an O—H $\cdots$ O bond to the other (O $\cdots$ H—O), then at least several other H atoms must also be moved to satisfy the rules stated above.

Pauling has given two approximate calculations which lead to the same answer and which are generally accepted as good approximations.

If there are N oxygen atoms, there are $2N$ hydrogen atoms and there would be 2^{2N} possible configurations of the H atoms according to condition (a) above, ignoring (b) and (c). Some of these configurations are eliminated by the requirements (b) and (c) that there be two near H's and two distant H's around each O atom. Consider one O atom and its four surrounding H's. There are $2^4 = 16$ possible configurations for the four H atoms. Of these, $4 \times 3/2 = 6$ satisfy conditions (b) and (c). Thus the total number of possible configurations is

$$2^{2N}(6/16)^N = (3/2)^N$$

A similar argument starts by imagining individual HOH molecules with short OH distances that maintain their identity but can rotate around to satisfy the structural requirements. Any one molecule can be oriented in six ways ($4 \times 3/2$) so that the hydrogens are pointed along two of the four O—O directions. For each O—O direction, there is a chance of $1/2$ that the other H$_2$O molecule will be using the direction for its O—H bond (since there are two H's on this O and four possible directions). Thus there is a chance of $(1/2)^2$ that a given orientation will be possible for a molecule. There are thus

$$6^N(1/4)^N = (3/2)^N$$

possible orientations in agreement with the previous argument. The calculated entropy of $R \ln 3/2 = 0.806$ cal mole^{-1} deg^{-1} is in good agreement with the experimental entropies.*

These calculations of the entropy of disorder are not rigorous, since they do not properly allow for the restrictions on the hydrogen-bond configuration around one oxygen, owing to the particular configurations

---

*   "If you are a man who wants to know
      How to find the entropy of H$_2$O

      If you care for atoms and what's beyond
      And you want to know the nature of the chemical bond

      Then the only place in the world to be
      Is the Gates and Crellin labs of chemistry."

(T. Harrold, J. Dunitz, 1954; from a skit presented at the California Institute of Technology in honor of Professor Pauling on the occasion of his Nobel Prize award; the tune is *The Eddystone Light*.)

around the neighboring oxygens. However, they are apparently approximately correct. This general interpretation of the structure of ice is confirmed by neutron diffraction investigations, by which the positions of the hydrogen and deuterium atoms in the crystal can be determined. Incidentally, the O—D distance was accurately determined as 1.01 Å,* showing that the short bond distance in the crystal is slightly larger than that of the gas-phase molecule.

It may be mentioned that there are other examples of residual entropy that may be due to the same kind of randomness of orientation of H_2O molecules. One such case is $Na_2SO_4 \cdot 10H_2O$.†

16-9. Order-Disorder Transitions. The Heat Capacity of Solid Ortho Hydrogen. We remind the reader that ortho hydrogen is the J-odd nuclear-spin-one form of hydrogen, whereas para hydrogen has J-even and nuclear spin zero. At low temperatures, the stable form is para hydrogen; but in the absence of paramagnetic catalysts, the rate of conversion is slow. By cooling of normal high-temperature hydrogen to low temperatures, one obtains a mixture containing 75 per cent ortho hydrogen. Below about 20 or 30°K, para hydrogen is exclusively in the $J = 0$ state, and ortho hydrogen is in the $J = 1$ state.

The heat capacity of solid hydrogen containing varying percentages of ortho hydrogen is shown in Fig. 16-9. The heat-capacity curve of pure para hydrogen resembles a Debye curve and is essentially due to the lattice vibrations of H_2 particles. As the amount of ortho hydrogen increases, there is an increase in the low-temperature heat capacity and at least for the rich mixtures (77 and 66 per cent) there is a sharp spike in the curve.‡ The spikes are shown in greater detail in Fig. 16-10. Heat-capacity anomalies of this nature are known as "λ-point" transitions.

The explanation of this heat-capacity anomaly is as follows: The level $J = 1$ of a free rotor is threefold-degenerate, corresponding to the states $m_J = 0, \pm 1$. Solid ortho hydrogen at 10 to 20° retains this entropy; the molecules are evidently freely rotating in the crystal. However, the forces between H_2 molecules must be somewhat anisotropic and must depend on the relative orientations of the axes of the two molecules. The crystal can therefore go to a state of lower energy, but lower entropy, by losing its free rotation and going into a state where each molecule has a definite orientation with respect to its neighbors.

The temperature of this transition is given by $T = \Delta H/\Delta S$, where $-\Delta H$ is the energy gain on ordering and $-\Delta S$ is the entropy loss. It is plausible to expect the quenching of rotation to be a cooperative phe-

---

* E. O. Wollan, W. L. Davidson, and C. G. Shull, *Phys. Rev.*, **75**: 1348 (1949); S. Peterson and H. Levy, *Acta Cryst.*, **10**: 70 (1957).

† K. S. Pitzer and L. V. Coulter, *J. ACS*, **60**: 1310 (1938); G. Brodale and W. F. Giauque, *J. ACS*, **80**: 2042 (1958).

‡ Conceivably, there are similar λ points at lower temperatures for the more dilute mixtures.

nomenon; if one molecule stops rotating, the neighboring molecules will tend to stop also. This accounts for the narrow temperature range of the transition. On the other hand, the evidence indicates that it is possible for some molecules to be rotating while others are not; thus

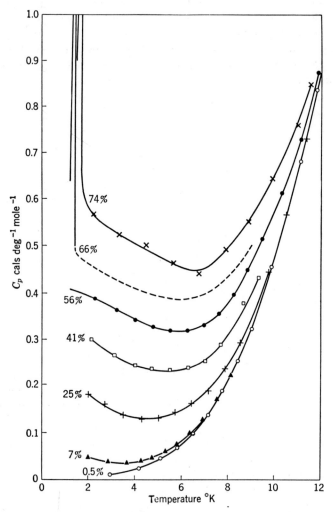

FIG. 16-9. The specific heat of solid hydrogen, as a function of the per cent ortho hydrogen. [*From R. W. Hill and B. W. A. Ricketson, Phil. Mag.*, **45**: 280 (1954).]

the transition is not a discontinuous phase transition like melting or a transition from one crystal structure to another.*

By subtracting out a small lattice contribution, one can estimate the

---

* It is now possible to prepare pure ortho hydrogen [C. M. Cunningham, D. S. Chapin, and H. L. Johnston, *J. ACS*, **80**: 2382 (1958)]. It would be interesting to study the transition in pure solid ortho hydrogen.

entropy loss due to the heat-capacity anomaly for the ortho-hydrogen-rich samples. This is about $R \ln 3$, in agreement with the above interpretation.

The exact nature of the ordered state of ortho hydrogen below the transition is not known. It should be pointed out that the quantum numbers $m_J = 0, \pm 1$ are appropriate for a rotor in a spherically symmetrical potential; it is by no means certain that these designations of the states apply to molecules in the crystal below the transition temperature.

Nuclear magnetic-resonance experiments* confirm the general description of the transition which was first discovered by heat-capacity measurements.

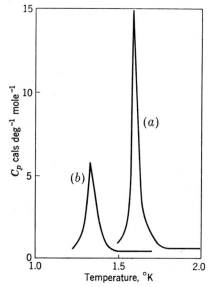

FIG. 16-10. Heat-capacity λ anomalies in solid hydrogen. (a) 74 per cent. (b) 66 per cent ortho hydrogen. [From R. W. Hill and B. W. A. Ricketson, Phil. Mag., **45**: 281 (1954).]

The transition in ortho hydrogen is a particularly interesting and simple example of a λ-point transition. The heat-capacity curves of a large number of molecular solids show λ points or at least some sort of hump in the heat-capacity curves. It is plausible that in all these cases there is a cooperative increase in freedom of the molecules to orient above the transition temperatures. It is usually a difficult task to determine the detailed nature of the motions above the transition temperature—whether they be free rotation, free rotation about one axis, an increased freedom of torsional oscillation, or sometimes the possibility of several possible orientations of a molecule. (For example, in CO or N_2O, a molecule at a particular lattice site is presumably oriented along a definite axis, but with an equal probability of pointing

* E. Reif and E. M. Purcell, Phys. Rev., **91**: 631 (1957).

in one direction or the other. One could imagine a transition on cooling, with a heat-capacity hump, in which the molecules all swing into an ordered structure, with the carbon and oxygen atoms in nonequivalent positions.) In some cases, for polar molecules, the transitions are accompanied by a large increase in the dielectric constant of the crystal, indicating that the molecules have an increased freedom of orientation in an electric field above the transition temperature.[*,†]

Order-disorder magnetic transitions are discussed in Chap. 19. For completeness' sake, we also mention the existence of order-disorder transitions in metallic alloys such as β-brass, where at high temperatures, there is a random distribution of Cu and Zn atoms over the body-centered cubic lattice points, whereas, below the λ point, there is an ordered arrangement of the atoms.[‡]

16-10. Order-Disorder Transitions in One-dimensional Systems. The Matrix Method for the Ising Model. There are a number of cooperative order-disorder transitions in three-dimensional crystals, such as λ-point transitions due to the onset of molecular rotation, which were qualitatively described in the previous section, the β-brass type of transition, and the transitions between the paramagnetic state and either the ferromagnetic or the antiferromagnetic state, which will be described in Chap. 19. Some of these transitions occur at a single temperature, while others take place over a finite temperature range.

It is not feasible at present to give a general, rigorous statistical mechanical treatment of these phenomena. However, such a treatment is feasible for one-dimensional crystals with interactions between near neighbors only.[§] This theory is of some limited interest to students of the three-dimensional phenomena because there may be some analogies

---

[*] For HBr, see C. P. Smyth and C. S. Hitchcock, *J. ACS*, **55**: 1296 (1933), **56**: 1084 (1934); W. F. Giauque and R. Wiebe, *J. ACS*, **50**: 2193 (1928). For more extensive references to the use of electric polarization for the study of internal motions in a solid, see C. J. F. Böttcher, "Theory of Electric Polarization," chap. XI, Elsevier, Houston, Tex., 1952; there is a discussion and a rather complete tabular summary of results in chap. V of C. P. Smyth, "Dielectric Behavior and Structure," McGraw-Hill, New York, 1955.

[†] References to transitions in molecular crystals are too numerous to be given in detail. A few cases with key references (these examples have been collected almost at random) are AsH_3, R. H. Sherman and W. F. Giauque, *J. ACS*, **77**: 2154 (1955); PH_3, C. C. Stephenson and W. F. Giauque, *J. Chem. Phys.*, **5**: 149 (1937); NH_4Cl, NH_4Br, NH_4I, A. W. Lawson, *Phys. Rev.*, **57**: 417 (1940); E. L. Wagner and D. F. Hornig, *J. Chem. Phys.*, **18**: 296, 304 (1950); R. C. Plumb and D. F. Hornig, *J. Chem. Phys.*, **20**: 1044 (1952).

[‡] For a discussion of the facts and statistical mechanical theories concerning these transitions, see Fowler and Guggenheim, pp. 563–607 [1]; C. Kittel "Introduction to Solid State Physics," 2d ed., pp. 337–348, Wiley, New York, 1956; F. C. Nix and W. Shockley, *Revs. Modern Phys.*, **10**: 1 (1938).

[§] E. Ising, *Z. Physik*, **31**: 253 (1925).

between the properties of one-dimensional and of three-dimensional crystals. The one-dimensional theory does have direct physical significance for problems such as the adsorption of gases onto a linear polymer* or the titration curves of linear polyelectrolytes.†

The one-dimensional theory has been most notably useful for the study of helix (ordered-state)—random-coil (disordered-state) transitions in polypeptides and nucleic acids.‡ We propose to illustrate the theoretical treatment of order-disorder transitions in linear systems by considering a simplified version of the Zimm theory for polypeptides.

There are several well-known methods for evaluating the partition functions that occur in one-dimensional systems of interacting particles.§ Of these, the matrix method is particularly well adapted for treating moderately complicated and realistic helix-coil transition models, and we shall therefore endeavor to expound this method. Although we shall review some of the mathematical operations, it is doubtful that readers who are unfamiliar with the elementary aspects of matrix algebra can profit from this section.

The Ising Model for a System of Spins. To illustrate the general features of the formalism, we study the classical Ising model, before introducing the special features of the helix-coil problem. Consider a linear chain of N elements, such as N spins, each of which has an up state $(+)$ and a down state $(-)$.‖ The energy of an isolated spin is e_+ or e_- (for example, $\pm \mu B$ in a magnetic field). In addition, there is an interaction between neighboring spins; let A be the interaction energy for an antiparallel pair and D the interaction energy for a parallel pair.

In any particular configuration of the chain, the parameters which determine the energy are the number of elements in up states, n_+, and the number of antiparallel pairs, a; correspondingly there are $N - n_+$ elements in down states and $N - 1 - a$ parallel pairs.

According to our assumptions, the energy of this configuration is

$$E(n_+,a) = n_+ e_+ + (N - n_+)e_- + aA + (N - 1 - a)D$$
$$= n_+(e_+ - e_-) + a(A - D) + Ne_- + (N - 1)D \quad (16\text{-}67)$$

Let $g(n_+,a)$ be the number of chain configurations with n_+ elements in

---

* Hill, pp. 235–241 [9b].
† J. Mazur, A. Silverburg, and A. Katchalsky, *J. Polymer Sci.*, **35**: 43 (1959).
‡ B. H. Zimm and J. K. Bragg, *J. Chem. Phys.*, **31**: 526 (1959) (polypeptides); B. H. Zimm, *J. Chem. Phys.*, **33**: 1349 (1960) (nucleic acids). The nucleic acid problem is more difficult than the polypeptide problem. References to the important contributions by Schellman, Hill, Peller, Gibbs and Di Marzio, and Rice and Wada are given in Zimm's papers.
§ Hill, pp. 318–327 [9a].
‖ For a more detailed discussion see Hill, pp. 318–327 [9a], or Wilson, pp. 476–482 [4]. Only a very elementary background in magnetism (which is otherwise not taken up until Chap. 19) is necessary for the present section.

up states and a antiparallel pairs. The canonical partition function for the chain is then

$$Q = \sum_{acc} g(n_+, a) e^{-\beta E(n_+, a)}$$

$$= \sum_{acc} g(n_+, a) e^{-\beta n_+(e_+ - e_-)} e^{-\beta a(A-D)} e^{-\beta N e_-} e^{-\beta(n-1)D}$$

where $\beta = 1/kT$ and $\sum_{acc}$ means the sum over all chain configurations.

The factor $e^{-\beta N e_-} e^{-\beta(N-1)D}$ can be factored out of the sum. For our present purpose of computing the probability of different configurations, this factor is unimportant and will be omitted from our subsequent formulae.

We define

$$\phi = e^{-\beta(e_+ - e_-)} \qquad \psi = e^{-\beta(A-D)} \qquad (16\text{-}68)$$

The quantity ϕ is the ratio of the Boltzmann factors for spin-up and spin-down states and ψ is the ratio of Boltzmann factors for antiparallel and parallel pairs.

With these definitions and omissions, the partition function becomes

$$Q = \sum_{acc} g(n_+, a) \phi^{n_+} \psi^a \qquad (16\text{-}69)$$

The probability of the group of configurations which have the parameters n_+ and a is

$$p(n_+, a) = \frac{g(n_+, a) \phi^{n_+} \psi^a}{Q} \qquad (16\text{-}70)$$

and the ensemble average values of n_+ and a are

$$\langle n_+ \rangle = \sum_{acc} n_+ p(n_+, a) = \left[\frac{\partial (\ln Q)}{\partial (\ln \phi)} \right]_\psi \qquad (16\text{-}71)$$

$$\langle a \rangle = \sum_{acc} a p(n_+, a) = \left[\frac{\partial (\ln Q)}{\partial (\ln \psi)} \right]_\phi \qquad (16\text{-}72)$$

Some Matrix Algebra. As already remarked, we shall evaluate the partition function Q of Eq. (16-69) by a matrix method. We first review some elementary facts of matrix algebra for the particularly simple case of 2×2 matrices.*

A 2×2 matrix $\mathbf{A}$ is the ordered array

$$\mathbf{A} = \begin{bmatrix} a_{11} & a_{12} \\ a_{21} & a_{22} \end{bmatrix} \quad \text{or} \quad \mathbf{A} = [a_{ij}] \quad i = 1, 2; j = 1, 2 \qquad (16\text{-}73)$$

* For further study, see Margenau and Murphy, pp. 291–316 [23].

The product of two matrices **A** and **B** is a matrix **C**,

$$\mathbf{C} = \mathbf{AB} = \begin{bmatrix} a_{11} & a_{12} \\ a_{21} & a_{22} \end{bmatrix} \begin{bmatrix} b_{11} & b_{12} \\ b_{21} & b_{22} \end{bmatrix} = \sum_{j=1,2} a_{ij}b_{jk} \qquad i = 1, 2;\ k = 1, 2$$

$$= \begin{bmatrix} a_{11}b_{11} + a_{12}b_{21} & a_{11}b_{12} + a_{12}b_{22} \\ a_{21}b_{11} + a_{22}b_{21} & a_{21}b_{12} + a_{22}b_{22} \end{bmatrix} \qquad (16\text{-}74)$$

A row vector **r** and a column vector **c** are, respectively, 1×2 and 2×1 matrices, written as

$$\mathbf{r} = [r_1, r_2] \qquad \mathbf{c} = \begin{bmatrix} c_1 \\ c_2 \end{bmatrix}$$

but, where appropriate to save space, we shall write **c** horizontally in braces as $\{c_1, c_2\}$.

The product **rA** is another row vector,

$$[r_1, r_2] \begin{bmatrix} a_{11} & a_{21} \\ a_{12} & a_{22} \end{bmatrix} = [r_1 a_{11} + r_2 a_{12},\ r_1 a_{21} + r_2 a_{22}]$$

or
$$\mathbf{rA} = \sum_i r_i a_{ij} \qquad j = 1, 2 \qquad (16\text{-}75a)$$

Similarly, the product **Ac** is a column vector,

$$\mathbf{Ac} = \begin{bmatrix} a_{11} & a_{12} \\ a_{21} & a_{22} \end{bmatrix} \begin{bmatrix} c_1 \\ c_2 \end{bmatrix} = \begin{bmatrix} a_{11}c_1 + a_{12}c_2 \\ a_{21}c_1 + a_{22}c_2 \end{bmatrix} \qquad (16\text{-}75b)$$

Note that the product $\mathbf{A}\{1,1\}$ is a column vector $\{a_{11} + a_{12},\ a_{21} + a_{22}\}$ with each component the sum of the rows of **A**.

The product **rc** is simply the vector dot product,

$$[r_1, r_2] \begin{bmatrix} c_1 \\ c_2 \end{bmatrix} = r_1 c_1 + r_2 c_2 \qquad (16\text{-}75c)$$

The reciprocal of a matrix **T** is $\mathbf{T}^{-1}$ such that their product is the unit matrix,

$$\mathbf{T}^{-1}\mathbf{T} = \mathbf{TT}^{-1} = \begin{bmatrix} 1 & 0 \\ 0 & 1 \end{bmatrix} \qquad (16\text{-}76)$$

Except in certain special cases,* given a square matrix **A** there is a matrix **T** such that the similarity transformation $\mathbf{TAT}^{-1}$ gives a diagonal matrix.

$$\mathbf{TAT}^{-1} = \mathbf{\Lambda} = \begin{bmatrix} \lambda_0 & 0 \\ 0 & \lambda_1 \end{bmatrix} \qquad (16\text{-}77a)$$

The diagonal elements λ_0 and λ_1 of $\mathbf{\Lambda}$ are the two roots of the deter-

* Cf. Margenau and Murphy, pp. 305–307 [23].

minantal (secular) equation

$$\begin{vmatrix} a_{11} - \lambda & a_{12} \\ a_{21} & a_{22} - \lambda \end{vmatrix} = 0 \qquad (16\text{-}77b)$$

By straightforward matrix multiplication, we see that the Nth power of a diagonal matrix is the diagonal matrix with elements λ_0^N, λ_1^N.

$$\mathbf{\Lambda}^N = \begin{bmatrix} \lambda_0 & 0 \\ 0 & \lambda_1 \end{bmatrix}^N = \begin{bmatrix} \lambda_0^N & 0 \\ 0 & \lambda_1^N \end{bmatrix} \qquad (16\text{-}78)$$

The Matrix Method for the Partition Function. Consider a chain of seven elements. One possible configuration of the chain is: $- + + - - + -$. This configuration contains three $+$ spins (and four $-$ spins) and four antiparallel pairs (therefore two parallel pairs). For each $+$ spin, there is a Boltzmann factor of ϕ, and for each antiparallel pair, there is a Boltzmann factor of ψ, according to our previous discussion. Therefore, the above configuration contributes a term $\psi^4 \phi^3$ to the chain partition function.

The formal rules for computing the partition function term of a given configuration are:

(a) $+$ after $+$ gives ϕ.
(b) $+$ after $-$ gives $\psi\phi$.
(c) $-$ after $-$ gives 1.
(d) $-$ after $+$ gives ψ.
(e) For the first element in the chain, $-$ and $+$ give 1 and ϕ, respectively.

The total partition function for the seven-element chain is the sum of terms for all configurations.

For the general case of a chain consisting of N elements, let $\nu_1, \nu_2, \ldots, \nu_i, \ldots, \nu_N$ be variables, with ν_i specifying the state of the ith element; $\nu_i = -$ or $+$ $(i = 1, 2, \ldots, N)$.

We now construct some vectors and matrices. We agree to list the $-$ state before the $+$ state. Let $\mathbf{r}_{\nu_1}$ be the row vector $[1,\phi]$ (that is, $r_- = 1$, $r_+ = \phi$, which says that the first element contributes 1 or ϕ to the partition function, depending on whether it is in the $-$ or $+$ state).

Let $\mathbf{c}_{\nu_N}$ be the column vector $\{1,1\}$ (that is, $c_- = 1$, $c_+ = 1$). For the matrix $\mathbf{A}_{\nu_{i-1},\nu_i}$,

$$\mathbf{A}_{\nu_{i-1},\nu_i} = \begin{matrix} - \\ + \end{matrix}\begin{bmatrix} 1 & \psi\phi \\ \psi & \phi \end{bmatrix} \quad \text{or} \quad \begin{matrix} A_{--} = 1 & A_{-+} = \psi\phi \\ A_{+-} = \psi & A_{++} = \phi \end{matrix} \qquad (16\text{-}79)$$

With the aid of this formalism, the partition function of the chain of N elements can be written

$$Q(\psi,\phi) = \sum_{\substack{\nu_1=-,+ \\ \nu_2=-,+ \\ \cdots\cdots \\ \nu_N=-,+}} r_{\nu_1} A_{\nu_1,\nu_2} A_{\nu_2,\nu_3} \cdots A_{\nu_{N-1},\nu_N} c_{\nu_N} \qquad (16\text{-}80a)$$

A moment's reflection should make it clear that the sum in (16-80a) is a sum over all configurations and that with the definitions of $\mathbf{r}$, $\mathbf{A}$, and $\mathbf{c}$ given it satisfies the rules for calculating the contribution of each configuration to the partition function. Furthermore, the partition function sum in (16-80a) can be seen to be the product of the row vector $\mathbf{r}$, the matrix $\mathbf{A}^{N-1}$, and the column vector $\mathbf{c}$.

$$\begin{aligned} Q(\psi,\phi) &= \mathbf{r} A_{\nu_1,\nu_2} A_{\nu_2,\nu_3} \cdots A_{\nu_{N-1},\nu_N} \mathbf{c} \\ &= \mathbf{r} \mathbf{A}^{N-1} \mathbf{c} \end{aligned} \qquad (16\text{-}80b)$$

This is the desired final result. Before considering the problem of the practical evaluation of the partition function, we construct a few simple examples.

For a system of two elements, the allowed states and partition function terms are: $++$, ϕ^2; $+-$, $\psi\phi$; $-+$, $\psi\phi$; $--$, 1; or

$$Q = \phi^2 + 2\phi\psi + 1$$

According to the formal equation (16-80b)

$$Q = [1,\phi]\begin{bmatrix} 1 & \psi\phi \\ \psi & \phi \end{bmatrix}\begin{bmatrix} 1 \\ 1 \end{bmatrix} = \phi^2 + 2\psi\phi + 1$$

in agreement with the above.

Problem 16-3. Evaluate the partition function for a system of three elements by direct summation of the contributions from all eight possible configurations and by application of the formal equations (16-80).

We now return to the problem of evaluating the partition function from Eq. (16-80). Let $\mathbf{T}$ be the matrix of the similarity transformation which diagonalizes $\mathbf{A}$,

$$\mathbf{T}^{-1}\mathbf{A}\mathbf{T} = \begin{bmatrix} \lambda_0 & 0 \\ 0 & \lambda_1 \end{bmatrix} \qquad (16\text{-}81)$$

Now, $\mathbf{T}\mathbf{T}^{-1}$ is the unit matrix. We write out the partition function as

$$Q = [1,\phi]\mathbf{A}\mathbf{A} \cdots \mathbf{A}\{1,1\}$$

and insert the factor $\mathbf{T}\mathbf{T}^{-1}$ in front of and behind every matrix $\mathbf{A}$ in the product. Then

$$\begin{aligned} Q(\psi,\phi) &= [1,\phi][\mathbf{T}\mathbf{T}^{-1}\mathbf{A}\mathbf{T}\mathbf{T}^{-1}]^{N-1}\{1,1\} = [1,\phi]\mathbf{T}[\mathbf{T}^{-1}\mathbf{A}\mathbf{T}]^{N-1}\mathbf{T}^{-1}\{1,1\} \\ &= [1,\phi]\mathbf{T}\begin{bmatrix} \lambda_0^{N-1} & 0 \\ 0 & \lambda_1^{N-1} \end{bmatrix}\mathbf{T}^{-1}\begin{bmatrix} 1 \\ 1 \end{bmatrix} \end{aligned} \qquad (16\text{-}82)$$

Thus the evaluation of the partition function is reduced to the straightforward and not very tedious problem of finding the roots of the secular equation for the matrix **A**, of finding the matrix **T**, and then of performing the operations indicated in Eq. (16-82).*

Consider, for example, the case for $\phi = 1$, that is, for equal probability of spin-up and spin-down states. This would apply to a system of paramagnetic spins with zero external magnetic field. In this case the matrix **A** is

$$\mathbf{A} = \begin{bmatrix} 1 & \psi \\ \psi & 1 \end{bmatrix}$$

Since **A** is symmetrical, it is a theorem in matrix algebra that the matrices **T** and **T**$^{-1}$ are transposes of each other. It is readily found that

$$\begin{matrix} \lambda_0 = 1 - \psi \\ \lambda_1 = 1 + \psi \end{matrix} \quad \mathbf{T} = \frac{1}{2^{1/2}} \begin{bmatrix} 1 & 1 \\ -1 & 1 \end{bmatrix} \quad \mathbf{T}^{-1} = \frac{1}{2^{1/2}} \begin{bmatrix} 1 & -1 \\ 1 & 1 \end{bmatrix} \quad (16\text{-}83a)$$

and that

$$Q = 2(1 + \psi)^{N-1} \tag{16-83b}$$

$$\langle a \rangle = \frac{\partial (\ln Q)}{\partial (\ln \psi)} \quad \frac{\langle a \rangle}{N - 1} = \frac{\psi}{1 + \psi} \tag{16-83c}$$

The important result of this problem is (16-83c), which asserts that the fractional number of antiparallel pairs $\langle a \rangle /(N - 1)$ is given by $\psi/(1 + \psi)$. We recall the definition of ψ as the Boltzmann factor for the energy difference between antiparallel pairs and parallel pairs. If this energy difference is negative (antiparallel pairs more stable), then ψ varies from ∞ to 1 as T goes from 0 to ∞. The fractional number of antiparallel pairs varies from 1 in the completely ordered state ($T = 0$, $\psi = \infty$) to $\frac{1}{2}$ in the random state ($T = \infty$, $\psi = 1$). However, our analysis shows that this transition takes place gradually over a rather broad temperature range.†

Problem 16-4. Verify the assertions in Eqs. (16-83).

* It may be remarked that the two components of the row vector

$$[1, \phi] \begin{bmatrix} 1 & \psi\phi \\ \psi & \phi \end{bmatrix} = [1 + \psi\phi, \psi\phi + \phi^2]$$

are the partition-function contributions (or relative probabilities) for the $-$ and $+$ states of the second element, averaged over all allowed configurations of the first element. In the same way, the two components of the row vector $[1, \phi]\mathbf{A}^{i-1}$ are the partition-function contributions, or relative probabilities, for the two possible states of the ith element, averaged over all possible configurations of the preceding $i - 1$ elements.

† Cf. Hill, fig. 14-1, p. 239 [9b].

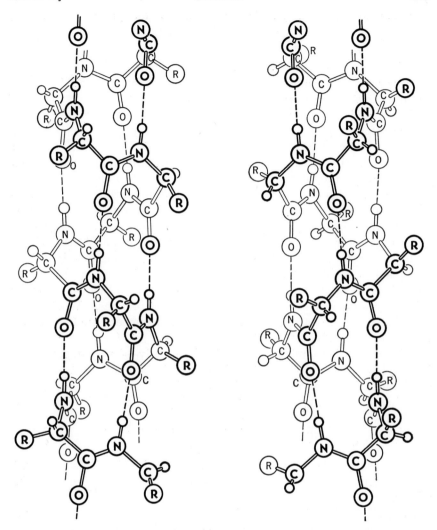

FIG. 16-11. The Pauling-Corey alpha helix. (*From L. Pauling, "The Nature of the Chemical Bond," p. 500, Cornell University Press, Ithaca, N.Y., 1960.*) Both a right-handed helix and a left-handed helix are shown.

The Helix-Coil Transition in Polypeptides.* The Pauling-Corey alpha-helix structure, which is known to occur in some synthetic polypeptides and proteins, is depicted in Fig. 16-11. In this ordered, helical structure, each amide (CONH) hydrogen is hydrogen-bonded to the

---

* Our treatment is a simplified version of the Zimm-Bragg theory [B. H. Zimm and J. K. Bragg, *J. Chem. Phys.*, **31**: 526 (1959), where other references are given]. We should mention that the pioneering experimental work in this field is largely due to P. Doty and E. Blout and their coworkers.

carbonyl oxygen of the third following amide group. Under certain conditions, the ordered structure is disrupted, and the polymer assumes a disordered, random-coil structure. There is evidence that the helix-coil transition occurs for some proteins and for some synthetic polypeptides. However, we shall confine our descriptive comments to the latter case, where the phenomena are somewhat simpler.

For example, a high-molecular-weight ($\sim$1,500 monomer units) preparation of poly-γ-benzyl-L-glutamate in solution in ethylene dichloride has the helical configuration, but in dichloroacetic acid as a solvent it is a random coil. In mixed solvents at 29°C, the transition between the two forms occurs as the solvent composition is changed by only a few per cent in the neighborhood of 80 weight per cent dichloroacetic acid. Furthermore in a solvent mixture with a fixed composition the helix is stable at high temperatures and the coil at low temperatures. The transition takes place over a narrow temperature range; for example, from about 24 to 34° for 80 per cent DCA (dichloracetic acid) (cf. Fig. 16-14).* The transition is rapid and reversible for the several kinds of changes of conditions mentioned. The same phenomena are observed in other solvent systems, with other polypeptides, and with proteins. Whether the helical structure is stable as the high or the low temperature form varies from case to case.†

It is reasonable to interpret the behavior of poly-γ-benzyl-L-glutamate in ethylene dichloride–dichloracetic acid solutions by assuming that dichloracetic acid can itself form hydrogen bonds with the amide groups and thus disrupt the hydrogen bonds which hold the helical structure together.

We wish to construct a statistical mechanical theory to describe these transitions. The narrow range of conditions over which the transitions occur would not be expected on the basis of the simple Ising model previously discussed, and some new feature must be introduced into the theory.

We shall simplify the statistical mechanical problem by assuming that the NH of an amide-group hydrogen bonds to the carbonyl oxygen of the next amide group, instead of the third following amide group. Thus we have to deal only with nearest-neighbor interactions.‡

A schematic diagram of our simplified model of the ordered (helical)

---

*P. Doty and J. T. Yang, *J. ACS*, **78**: 498 (1956).

† For proteins and nucleic acids, the random-coil form is always the stable form at high temperatures. For poly-L-glutamic acid in aqueous dioxane, the helix is stable at low pH and low T; the coil is formed at higher pH's and/or at higher temperatures [P. Doty, A. Wada, J. T. Yang, and E. R. Blout, *J. Polymer Sci.*, **23**: 851 (1957)].

‡ The problem for the actual alpha helix which is treated by B. H. Zimm and J. K. Bragg, *J. Chem. Phys.*, **31**: 526 (1959), is not really much more difficult. It involves 8×8 matrices instead of 2×2 matrices. The results are not very different.

structure is depicted in Fig. 16-12. We think of each amide residue

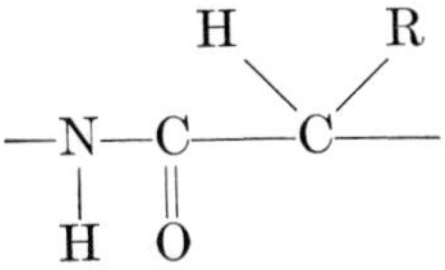

$$-N-C\underset{H}{\overset{H}{|}}\underset{O}{\overset{R}{|}}C-$$

as the elements or segments of the chain.* A segment is in a $+$ state if its oxygen is bonded to the N—H group of the preceding segment; a segment is in a $-$ state if its carbonyl oxygen is not so bonded.

A state or configuration of the chain is then specified by a sequence such as $-+++--+++-+-+---$, etc. According to our conventions, the first amide group must of necessity be in the unbonded $(-)$ state. (Furthermore, of course, we completely ignore the

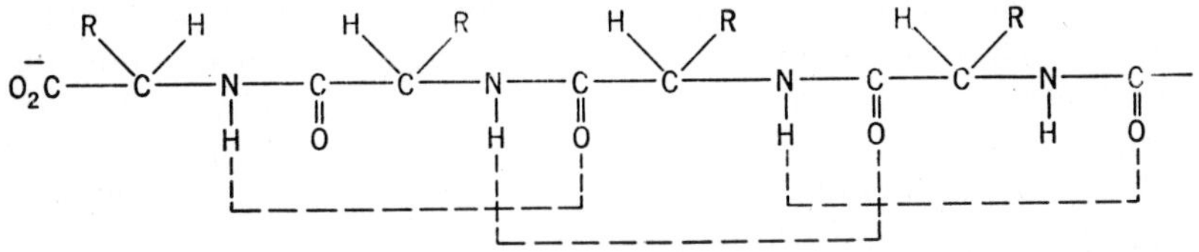

FIG. 16-12. Simplified model of an ordered chain with each N—H bonded to the C=O group of the next,

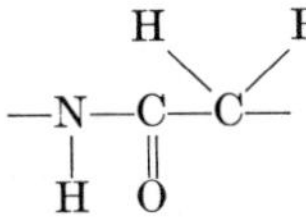

$$-N-C\underset{H}{\overset{H}{|}}\underset{O}{\overset{R}{|}}C-$$

segment, as indicated by the dotted lines. For our statistical mechanical theory, we ignore the carboxyl group $^-O_2C-$ on the far left.

carboxyl group $^-O_2C-$ at the beginning of the chain on the far left of Fig. 16-12.) We take as our reference state an unbonded segment and therefore assert that each unbonded segment contributes a factor of 1 to the partition function for a given configuration of the chain. We assume that a bonded segment that is in an ordered part of the chain contributes a factor s to the partition function. The factor s is analogous to the factor ψ in our previous discussion.

There is an important extension in our point of view, however. The

* For ordinary chemical purposes in protein and polypeptide chemistry one thinks of the polymer as composed of amino acid residues,

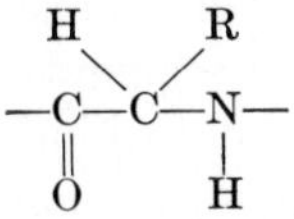

$$-C\underset{O}{\overset{H}{|}}\underset{H}{\overset{R}{|}}-N-$$

rather than of amide groups, but the latter point of view is more convenient here.

segments of the chain are not single, structureless particles but molecular units with a variety of possible configurations and possible rotational and vibrational motions. Our previous point of view was that ψ was the ratio of Boltzmann factors for antiparallel and parallel pairs. Our present point of view is that s is the ratio of the partition functions for bonded and unbonded segments. The partition function for an unbonded segment, for example, is a sum of Boltzmann factors for a large number of possible orientations and vibrations of an unbonded segment. The ratio of the two partition functions is, however, the contribution that a bonded segment makes to the partition function of the chain, with the reference state chosen as an unbonded segment.

In a naïve view of the matter, we should expect that in an inert solvent the helix would be the stable form at low temperatures and the coil at high temperatures. That is, s would be of the form $s = s_0 e^{-\epsilon/kT}$, where s_0 is less than 1 and is related to the entropy loss per segment on going from the relatively free configuration of a segment in a random coil to the more restricted configuration of a segment in a helix. The energy quantity ϵ is negative and is the energy gained on going from the unbonded coil state to the hydrogen-bonded helix state. Thus, at low T, $s > 1$, and at high T, $s \approx s_0 < 1$.

Since poly-γ-benzyl-L-glutamate in ethylene dichloride–dichloracetic acid (DCA) mixtures is stable as a helix at high temperature and as a coil at low temperature, we must assume that $s > 1$ at high T and $s < 1$ at low T and that $s = s_0 e^{-\epsilon/kT}$ with $s_0 > 1$ and ϵ positive. This may be interpreted by saying that the reaction is really*

$$\text{Coil (DCA)}_x \rightleftharpoons \text{helix} + x\text{DCA} \tag{16-84}$$

This reaction is endothermic because of the hydrogen bonds formed between DCA and the amide groups in the compound on the left-hand side of the equation; the entropy change is positive (and so $s_0 > 1$) because of the release of the DCA molecules into the solvent on forming the helix.

The main point for our present purposes is that there is a parameter s which is a function of temperature or solvent composition.

There is one extremely important new feature which makes our present problem different from the typical Ising-model problem considered in the previous subsection. In the real alpha helix, each segment is bonded to the fourth subsequent segment. This makes it relatively easy for the second segment to bind to the fifth, the third to the sixth, etc. Stated in another way, the formation of a helical region is a cooperative phenomenon; it is difficult for one segment to bind, but when this occurs, it is easier for the next segment to bind. For our simplified model of the situation, we introduce a nucleation parameter σ with $\sigma \ll 1$ and

---

* P. Doty and J. T. Yang, *J. ACS*, **78**: 498 (1956).

assume that the partition-function contribution for a bonded segment that follows an unbonded segment is σs, whereas the partition function contribution by a bonded segment that follows a bonded segment is s. We shall assume that σ is independent of temperature.

Therefore, our formal rules for writing the contribution of a configuration (specified by a sequence $-+++-+$, etc.) to the partition function are the following:

(a) Every unbonded $(-)$ segment contributes a factor 1.

(b) Every bonded $(+)$ segment that follows a bonded $(+)$ segment contributes a factor s.

(c) A bonded $(+)$ segment that follows an unbonded $(-)$ segment contributes a factor σs.

(d) The first segment must be unbonded $(-)$.

Thus, for the configuration $-+++--++-++$, the contribution to the partition function is $\sigma^3 s^7$.

In general for a configuration with m +'s that follow $-$'s and n +'s all told, the contribution to the partition function is $\sigma^m s^n$. The total partition function is the sum of these terms for all configurations. The probability of any one configuration is $\sigma^m s^n/Q$; and, as before, the ensemble average values of m and n are

$$\langle m \rangle = \frac{\partial(\ln Q)}{\partial(\ln \sigma)} \qquad \langle n \rangle = \frac{\partial(\ln Q)}{\partial(\ln s)} \qquad (16\text{-}85)$$

We recognize that $\langle m \rangle$ is the number of helical regions in the chain and $\langle n \rangle$ is the number of segments in helical regions.

The matrix formulation of rules (a) to (d), is $\mathbf{r} = [1,0]$ ($\mathbf{r}$ is the initial row vector, and the first segment is in the minus state). $A_{--} = A_{+-} = 1$ [rule (a)]; $A_{++} = s$ [rule (b)]; $A_{-+} = \sigma s$ [rule (c)]; therefore

$$\mathbf{A} = \begin{matrix} - \\ + \end{matrix} \begin{bmatrix} \overset{-}{1} & \overset{+}{\sigma s} \\ 1 & s \end{bmatrix} \qquad (16\text{-}86)$$

The partition function for N amide segments is therefore

$$Q(\sigma, s) = [1,0] \mathbf{A}^{N-1} \begin{bmatrix} 1 \\ 1 \end{bmatrix} \qquad (16\text{-}87)$$

Problem 16-5. Show that Eq. (16-87) gives the correct partition function for a system of three amide segments (which is, as previously explained, four amino acid residues).

The characteristic equation and roots for $\mathbf{A}$ are

$$\begin{vmatrix} 1-\lambda & \sigma s \\ 1 & s-\lambda \end{vmatrix} = 0 \qquad \lambda = \frac{(1+s) \pm [(1-s)^2 + 4\sigma s]^{1/2}}{2} \qquad (16\text{-}88)$$

Denote the larger of the two roots of Eq. (16-88) by λ_0 and the smaller by λ_1. The transformation matrices such that

$$\mathbf{T}^{-1}\mathbf{AT} = \begin{bmatrix} \lambda_0 & 0 \\ 0 & \lambda_1 \end{bmatrix} \qquad (16\text{-}89a)$$

are

$$\mathbf{T} = \frac{1}{\lambda_0 - \lambda_1} \begin{bmatrix} 1 - \lambda_1 & \lambda_0 - 1 \\ 1 & -1 \end{bmatrix} \qquad (16\text{-}89b)$$

$$\mathbf{T}^{-1} = \begin{bmatrix} 1 & \lambda_0 - 1 \\ 1 & \lambda_1 - 1 \end{bmatrix} \qquad (16\text{-}89c)$$

The partition function therefore is

$$Q = [1,0]\mathbf{T} \begin{bmatrix} \lambda_0^{N-1} & 0 \\ 0 & \lambda_1^{N-1} \end{bmatrix} \mathbf{T}^{-1} \begin{bmatrix} 1 \\ 1 \end{bmatrix} \qquad (16\text{-}90a)$$

or

$$Q = \frac{\lambda_0^N(1 - \lambda_1) + \lambda_1^N(\lambda_0 - 1)}{\lambda_0 - \lambda_1} \qquad (16\text{-}90b)$$

Problem 16-6. Prove the statements in Eqs. (16-88) to (16-90).

Our problem is now in principle solved, and by straightforward calculation we can obtain the properties of a chain of arbitrary length as a function of the parameters σ and s. We shall treat only the limiting case of $N \gg 1$. The cases of practical interest are for quite small values of σ, $\sigma \sim 10^{-4}$ to 10^{-2}. The values of s where interesting things happen are all close to 1. We first inquire as to the values of λ_0 and λ_1 in the neighborhood of $s = 1$, for $\sigma \ll 1$. By analysis of Eq. (16-88), it is fairly easy to see that

For $s < 1$ and $(1 - s)^2 \gg 4\sigma s$: $\lambda_0 \approx 1 + \dfrac{2\sigma s}{1 - s}$ $\lambda_1 \approx s - \dfrac{2\sigma s}{1 - s}$

For $s = 1$: $\lambda_0 = 1 + \sigma^{1/2}$ $\lambda_1 = 1 - \sigma^{1/2}$ (16-91)

For $s > 1$ and $(1 - s)^2 \gg 4\sigma s$: $\lambda_0 \approx s + \dfrac{2\sigma s}{1 - s}$ $\lambda_1 \approx 1 - \dfrac{2\sigma s}{1 - s}$

Thus, the difference between the two roots, $\lambda_0 - \lambda_1$, is a minimum of $2\sigma^{1/2}$ for $s = 1$. If N is sufficiently large so that $N\sigma^{1/2} \gg 1$, then $\lambda_0^N \gg \lambda_1^N$. The partition-function equation

$$Q = \frac{\lambda_0^N(1 - \lambda_1) + \lambda_1^N(\lambda_0 - 1)}{\lambda_0 - \lambda_1} \qquad (16\text{-}90b)$$

can then be approximated by

$$Q \approx \frac{\lambda_0^N(1 - \lambda_1)}{\lambda_0 - \lambda_1} \qquad \ln Q \approx N \ln \lambda_0 \qquad (16\text{-}92)$$

With this approximation, the equation for the number of helical segments, $\langle n \rangle = \partial(\ln Q)/\partial(\ln s)$, becomes

$$\langle n \rangle = \frac{Ns}{\lambda_0} \frac{\partial \lambda_0}{\partial s} \qquad \frac{\langle n \rangle}{N} \approx \frac{\partial \lambda_0}{\partial s} \qquad (16\text{-}93)$$

(since, in the interesting region, both s and $\lambda_0 \approx 1$).
Similarly, the equation for the number of helical regions,

$$\langle m \rangle = \frac{\partial(\ln Q)}{\partial(\ln \sigma)}$$

becomes

$$\frac{\langle m \rangle}{N} \approx \frac{\sigma}{\lambda} \frac{\partial \lambda}{\partial \sigma} \approx \sigma \frac{\partial \lambda}{\partial \sigma} \qquad (16\text{-}94)$$

We first deal with Eq. (16-93). By direct differentiation of Eq. (16-88),

$$\frac{d\lambda_0}{ds} = \frac{1}{2}\left\{1 + \frac{(s-1) + 2\sigma}{[(1-s)^2 + 4\sigma s]^{1/2}}\right\} \qquad (16\text{-}95)$$

from which we see that

$$\text{For } s < 1 \text{ and } (1-s)^2 > 4\sigma s;\ \frac{\langle n \rangle}{N} = \frac{d\lambda_0}{ds} \approx 0$$
$$\text{For } s > 1 \text{ and } (1-s)^2 > 4\sigma s;\ \frac{\langle n \rangle}{N} = \frac{d\lambda_0}{ds} \approx 1 \qquad (16\text{-}96)$$

The transition between the completely disordered and the completely ordered state occurs in the narrow zone of s for which

$$1 - 2\sigma^{1/2} < s < 1 + 2\sigma^{1/2} \qquad (16\text{-}97)$$

Figure (16-13) is a plot of the calculated values of $\langle n \rangle/N$ as a function of s for large N for three different values of σ. Since s is a parameter

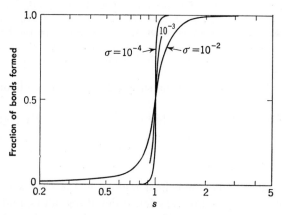

FIG. 16-13. Fraction of ordered (hydrogen-bonded) segments, $\langle n \rangle/N$, as a function of the equilibrium constant s for various values of the initiation parameter σ. [From B. H. Zimm and J. K. Bragg, J. Chem. Phys., **31**: 526 (1959).]

which changes with temperature, we see from the figure that the theory predicts a transition between a random coil to an alpha helix over a narrow temperature range for large N and small σ, in agreement with experiment. The sharpness of the transition is due to the small value of σ, that is, to the difficulty in starting an ordered region. If the partition function is evaluated for the case $\sigma = 1$, the result is $\lambda_0 = 1 + s$, $\lambda_1 = 0$, $Q = (1 + s)^{N-1}$, $\langle n \rangle / N = (1/N) \, d(\ln Q)/d(\ln s) = s/(1 + s)$. This is

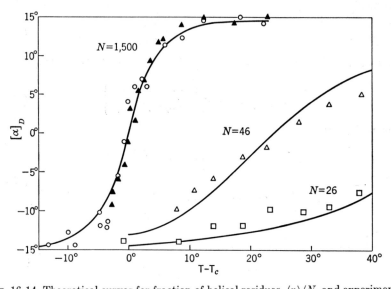

FIG. 16-14. Theoretical curves for fraction of helical residues, $\langle n \rangle /N$, and experimental results (points) for the optical rotation, $[\alpha]_D$, as a function of temperature T minus critical temperature T_c for samples of poly-γ-benzyl-L-glutamate of various degrees of polymerization N. ○, solvent 80 per cent DCA (dichloroacetic acid), 20 per cent ethylene dichloride, $T_c = 28.7°$. [P. Doty and J. T. Yang, J. Am. Chem. Soc., **78**: 498 (1956).] ▲, △, □, 70 per cent DCA, $T_c = 11.8°C$. In the latter case, $s = 1.57 e^{-890/RT}$. [From B. Zimm, P. Doty, and K. Iso, Proc. Natl. Acad. Sci., **45**: 160A (1959).] The theoretical curves were calculated for $\sigma = 2 \times 10^{-4}$.

identical with Eq. (16-83c) for the simple Ising model and corresponds to a broad transition.

Equation (16-94) for the number of helical regions gives (for large N)

$$\frac{\langle m \rangle}{N} = \frac{\sigma s}{[(1 - s)^2 + 4\sigma s]^{1/2}} \tag{16-98}$$

The maximum number of helical regions is $\tfrac{1}{2}\sigma^{1/2}$ for $s = 1$; it is less for small s, because there are very few ordered segments, and it is less for $s > 1$, because in the limit there is only one ordered region which includes the whole chain.

Both theory and experiment agree that (for the practical case of small σ)

the transition should be broader and should occur at a larger value of s for low-molecular-weight polymers. Some aspects of the comparison between theory and experiment are displayed in Fig. 16-14.

PROBLEMS

16-7. The values of Θ for Al, Be, C(diamond), and MgO are given as 375, 1160, 1860, and 800°K (D. E. Gray (ed.), "American Institute of Physics Handbook," pp. 4–47, 4–48, McGraw-Hill, New York, 1957). Estimate the values of c_V at $T = 298.16°$, and, neglecting the difference between c_P and c_V, estimate the standard entropy at 298.16°K. (Use tables of appropriate Debye functions.) Compare with the values of c_P and $s°$ at 298.16° in NBS Series I tables [24].

16-8. Given a two-dimensional crystal of HCl molecules with the chlorine atoms in hexagonal packing. Each chlorine atom is at the center of a regular hexagon with six nearest-neighbor chlorine atoms at the corners. Suppose the rules are that each Cl atom has one near H atom (an HCl molecule), that each H must lie on a line between two Cl's (a hydrogen bond), and that there can be no more than one H atom between any two Cl atoms. Estimate the entropy of a random structure conforming to these rules.

17

A Digression on Electricity, Magnetism, and Units

17-1. Introduction. Before applying statistical mechanics to the interaction of molecules with electric and magnetic fields, we shall review some of the basic laws of electricity and magnetism and consider some simple examples.

A formidable obstacle to communication on this subject is the diversity of electrical and magnetic units. Thus, many undergraduates are now educated exclusively in the rationalized mks system. Most of the contemporary as well as the older literature of atomic and molecular physics and chemistry uses electrostatic, electromagnetic, or Gaussian units. Electrostatic and electromagnetic units are, at least to some extent, readily interrelated. Gaussian units, an appropriate combination of electrostatic and electromagnetic units, are fairly easy to use, and we shall adopt them.

There is no point in debating the relative merits of the several systems in detail. It is certainly easier to keep units straight when one is making calculations in the mks system. To reap this advantage, it is necessary to express masses in kilograms and distances in meters—and these units are not customarily adopted in tabulations of physicochemical data. In order to make calculations in the Gaussian system, it is necessary to become familiar with a few important conversion factors between practical units and Gaussian units.

In any case, it is clearly necessary for students to be conversant with both systems of units. This is not a course in electricity and magnetism; we simply state some of the important results without proof.*

---

* For further study, see N. H. Frank, "Introduction to Electricity and Optics," 2d ed., McGraw-Hill, New York, 1950, especially pp. 423–429; W. K. H. Panofsky and M. Phillips, "Classical Electricity and Magnetism," Addison-Wesley, Reading, Mass., 1955, especially pp. 375–384; E. A. Guggenheim, "Thermodynamics: An Advanced Treatment," chaps. 12, 13, Interscience, New York, 1950.

17-2. Some Fundamental Electrical Relations.

The relations on the left-hand side are in cgs-Gaussian units and on the right in rationalized mks units.

In cgs units, velocities are in centimeters per second (cm sec^{-1}), masses in grams (g), force in dynes (g cm sec^{-2}), and energies in ergs (g cm^2 sec^{-2}).

In the mks system, velocities are in meters per second (m sec^{-1}), masses in kilograms (kg), forces in newtons (kg-m sec^{-2}), and energies in joules (kg-m^2 sec^{-2}).

$$10^7 \text{ ergs} = 1 \text{ joule}$$

The force on a moving particle of charge q is

$$\mathbf{F} = q\mathbf{E} + \frac{q}{c}\mathbf{v} \times \mathbf{B} \qquad \mathbf{F} = q\mathbf{E} + q\mathbf{v} \times \mathbf{B} \qquad (17\text{-}1)$$

$\mathbf{E}$ = electric field	statvolts cm^{-1}	$\mathbf{E}$	volts m^{-1}
q = charge	statcoulombs	q	coulombs
$\mathbf{B}$ = magnetic induction	gauss	$\mathbf{B}$	webers m^{-2}

c = velocity of light

1 statvolt = 300 volts*,†   1 statvolt cm^{-1} = 3 × 10^4 volts m^{-1}*

$$3 \times 10^9 \text{ statcoulombs} = 1 \text{ coulomb}*$$
$$3 \times 10^{10} \text{ statcoulombs} = 1 \text{ emu (abcoulomb)}*$$

Charge on electron = $|e^-|$ = 4.80 × 10^{-10} statcoulomb

$$= 1.60 \times 10^{-19} \text{ coulomb} = 1.6 \times 10^{-20} \text{ emu}*$$
$$10^4 \text{ gauss} = 1 \text{ weber m}^{-2}$$

The scalar magnitude of the force between two charges *in vacuo* is

$$F = \frac{q_1 q_2}{r^2} \text{ dynes} \qquad F = \frac{q_1 q_2}{4\pi\epsilon_0 r^2} \text{ newtons} \qquad (17\text{-}2)$$

ϵ_0 = permittivity of free space
$$= 8.85 \times 10^{-12} \text{ farad m}^{-1}$$

$$4\pi\epsilon_0 = \frac{1}{9.00} \times 10^{-9} \text{ farad m}^{-1}$$

---

* In the starred relations, the number 3 is derived from the velocity of light. For highly accurate calculations, the number 2.99793 should be used. The factor c or c^2 occurs in some of the other conversion factors quoted here. We have not troubled consistently to indicate this fact in the text, but all such instances are noted in Table 17-1. (Also see W. K. H. Panofsky and M. Phillips, "Classical Electricity and Magnetism," p. 384, Addison-Wesley, Reading, Mass., 1955.)

† From a practical point of view in making electrostatic calculations in the cgs system for atomic problems, it is easy for most persons to remember that the charge on the electron is 4.80 × 10^{-10} statcoulomb; the tricky point, which is very important to remember, is that 1 statvolt = 300 practical volts, or that 1 practical volt is $\frac{1}{300}$ statvolt.

The magnitude of the electric field at a distance r from a charge *in vacuo* is

$$E = \frac{q}{r^2} \quad \text{statvolts cm}^{-1} \qquad E = \frac{q}{4\pi\epsilon_0 r^2} \quad \text{volts m}^{-1} \qquad (17\text{-}3)$$

The potential due to a charge *in vacuo* is

$$\phi = \frac{q}{r} \quad \text{statvolts} \qquad\qquad \phi = \frac{q}{4\pi\epsilon_0 r} \quad \text{volts} \qquad (17\text{-}4)$$

The potential and field are related by the equation

$$\mathbf{E} = -\nabla \phi \qquad (17\text{-}5)$$

The potential energy of interaction of two charges

$$u = \frac{q_1 q_2}{r} \quad \text{ergs} \qquad\qquad u = \frac{q_1 q_2}{4\pi\epsilon_0 r} \quad \text{joules}$$

The cgs dimensions of charge ($g^{1/2}\,cm^{3/2}\,sec^{-1}$), field, and potential are not worth remembering. For dimensional checking of formulae, it is easier to remember and use the facts that q^2/r and $qEr = q\phi$ have dimensions of energy and that E^2 has dimensions of energy per unit volume.

The force between two point charges in a homogeneous isotropic material medium is

$$F = \frac{q_1 q_2}{\epsilon r^2} \qquad\qquad F = \frac{q_1 q_2}{4\pi\kappa\epsilon_0 r^2} \qquad (17\text{-}6)$$

ϵ = dielectric constant (dimensionless)

$\kappa\epsilon_0$ = permittivity of medium farads m^{-1}

κ = dielectric coefficient of medium

(κ in mks is the same as ϵ, the dielectric constant in electrostatic units.) The field and potential due to a charge immersed in a medium are

$$E = \frac{q}{\epsilon r^2} \quad \phi = \frac{q}{\epsilon r} \qquad E = \frac{q}{4\pi\kappa\epsilon_0 r^2} \quad \phi = \frac{q}{4\pi\kappa\epsilon_0 r} \qquad (17\text{-}7)$$

In a medium as well as *in vacuo*, E and ϕ are related by the equation

$$\mathbf{E} = -\nabla \phi \qquad (17\text{-}5)$$

The electric-displacement vector **D** is related to the electric-field vector **E** by

$$D = \epsilon E \qquad D = \kappa\epsilon_0 E \qquad (17\text{-}8)$$
$$D = E \quad \text{in vacuo} \qquad D = \epsilon_0 E \quad \text{in vacuo}$$
$$\textbf{D and E have the same} \qquad D, \text{ coulombs m}^{-2}$$
dimensions.

For an arbitrary charge distribution $\rho(x,y,z)$ in an arbitrary medium

$$\nabla \cdot \mathbf{D} = 4\pi\rho \qquad \nabla \cdot \mathbf{D} = \rho \qquad (17\text{-}9)$$

Charges $-q$ and $+q$ separated by the vector **r** have an electric-dipole moment:

$$\boldsymbol{\mu} = q\mathbf{r} \quad \text{statcoulomb cm} \qquad \boldsymbol{\mu} = q\mathbf{r} \quad \text{coulomb m}$$

The dipole moment per unit volume is the polarization **P**:

P, statcoulombs cm^{-2} (cgs)
P, coulombs m^{-2} (mks)

The fundamental relation between **D**, **E**, and **P** is

$$\mathbf{D} = \mathbf{E} + 4\pi\mathbf{P} \quad \text{(cgs)}$$
$$\mathbf{D} = \epsilon_0\mathbf{E} + \mathbf{P} \quad \text{(mks)} \qquad (17\text{-}10)$$

In an isotropic medium, **D**, **E**, and **P** are all in the same direction; then since $D = \epsilon E$ (cgs) or $D = \kappa\epsilon_0 E$ (mks)

$$P = \frac{(\epsilon - 1)E}{4\pi} = \frac{D}{4\pi}\frac{\epsilon - 1}{\epsilon} \quad \text{(cgs)}$$

$$P = \epsilon_0(\kappa - 1)E = \frac{\kappa - 1}{\kappa}D \quad \text{(mks)} \qquad (17\text{-}11)$$

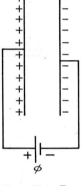

FIG. 17-1. Parallel-plate capacitor *in vacuo*.

17-3. The Parallel-plate Capacitor. Consider a parallel-plate capacitor *in vacuo*, with the area of the plates, A, and their separation, L; the capacitor is charged by a battery to a voltage ϕ. There is a surface charge density σ and $-\sigma$ (charge per unit area) on the $+$ and $-$ plates (Fig. 17-1). Assume that end effects are negligible; then it can be shown that the uniform fields E and D between the plates are given by

$$D = E = 4\pi\sigma \qquad D = \sigma \qquad E = \frac{\sigma}{\epsilon_0}$$

$$\phi = EL = 4\pi\sigma L \qquad \phi = \frac{\sigma L}{\epsilon_0}$$

The capacitance of a capacitor is the ratio of the stored charge to the applied potential. The units of capacitance are statfarads or centimeters (cgs) and farads (mks).

$$9 \times 10^{11} \text{ statfarads} = 1 \text{ farad}$$

For the parallel-plate capacitor *in vacuo*,

$$\text{Capacitance} = \frac{\sigma A}{\phi} = \frac{A}{4\pi L} \qquad \text{Capacitance} = \frac{\sigma A}{\phi} = \frac{\epsilon_0 A}{L}$$
$$\text{statfards (cm)} \hspace{4cm} \text{farads} \quad (17\text{-}12)$$

If the same parallel-plate capacitor with the same applied potential ϕ is now immersed in a medium of dielectric constant ϵ (cgs) or dielectric coefficient κ (mks—but remember that numerically ϵ and κ are the same), the charge density of the plates is increased to σ_1; there is an induced charge $-\sigma_i$ of opposite sign on the surface of the dielectric medium (Fig. 17-2).

There is a uniform polarization vector in the medium, pointing in the direction from $-\sigma_i$ to $+\sigma_i$, and of magnitude

$$P = \sigma_i$$

The magnitude of D is given by

$$D = 4\pi\sigma_1 \text{ (cgs)}$$
$$D = \sigma_1 \text{ (mks)} \qquad (17\text{-}13)$$

(that is, D is determined by the free charge σ_1).

From the relations between D, E, and P [(17-10)], we then have

$$E = D - 4\pi P = 4\pi(\sigma_1 - \sigma_i) \text{ (cgs)}$$
$$\epsilon_0 E = D - P = \sigma_1 - \sigma_i \text{ (mks)} \qquad (17\text{-}14)$$

(that is, E is determined by the "effective charge" on the plates, $\sigma_1 - \sigma_i$).

From the relations above and the equation $D = \epsilon E$ or $D = \kappa\epsilon_0 E$, we have

$$\sigma_1 - \sigma_i = \frac{\sigma_1}{\epsilon} \text{ (cgs)}$$
$$\sigma_1 - \sigma_i = \frac{\sigma_1}{\kappa} \text{ (mks)} \qquad (17\text{-}15)$$

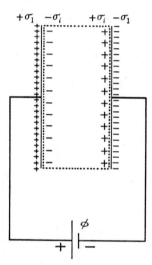

FIG. 17-2. A parallel-plate capacitor containing a medium of dielectric constant ϵ (cgs). The dielectric may be imagined to be contained within the dotted lines. There are charges $\pm\sigma_1$ on the plates of the capacitor; there are smaller induced charges of opposite sign, $\mp\sigma_i$, on the surface of the dielectric.

The potential is given by $\phi = EL$; from the relation between $\sigma_1 - \sigma_i$ and E we then have

$$\text{Capacitance} = \frac{\sigma_1 A}{\phi} = \frac{\epsilon A}{4\pi L} \qquad \text{Capacitance} = \frac{\sigma_1 A}{\phi} = \frac{\kappa\epsilon_0 A}{L}$$
$$(17\text{-}16)$$

The polarization in the medium is seen to be

$$P = \sigma_i = \sigma_1 \frac{\epsilon - 1}{\epsilon} \qquad P = \sigma_i = \sigma_1 \frac{\kappa - 1}{\kappa} \qquad (17\text{-}17)$$

17-4. Some Magnetic Relations.

According to Eq. (17-1) the force on a moving charge is determined by the magnetic induction **B**. Consider a long solenoid containing n turns per unit length *in vacuo*, with a steady current I. There is a uniform magnetic-induction vector **B** inside the solenoid, directed along the axis.

$$B = \frac{4\pi nI}{c} \qquad I \text{ in esu or statcoulombs sec}^{-1} \tag{17-18a}$$

$$B = 4\pi nI \qquad I \text{ in emu or abamp} \tag{17-18c}$$

$$B = \frac{4\pi nI}{10} \qquad I \text{ in amp or coulombs sec}^{-1} \tag{17-18d}$$

$$B = \mu_{0,\text{mks}}\, nI \qquad I \text{ in amp or coulombs sec}^{-1} \tag{17-18b}$$

$\mu_{0,\text{mks}}$ = magnetic permeability of free space
$= 4\pi \times 10^{-7}$ kg-m coulomb^{-2}

As noted in the top equation, there is a reasonably simple equation for the magnetic field in terms of the current in esu, but it is more customary to use emu [(17-18c)] or practical units [(17-18d)]. One of the disadvantages of this system of units is painfully evident here.

B is in gauss $\qquad\qquad B$ is in webers m^{-2}

$$10^4 \text{ gauss} = 1 \text{ weber m}^{-2}$$

For the magnetic-field vector **H**

$\mathbf{H} = \mathbf{B}$ (H in oersteds) $\qquad \mathbf{H} = \mathbf{B}/\mu_{0,\text{mks}}$ (H in coulombs sec^{-1} m^{-1})

Let the solenoid now be immersed in a homogeneous isotropic medium; the current is still I.

Let μ_P be the magnetic permeability of the medium; then

$$B = \mu_P H = 4\pi n \mu_P \frac{I}{c}$$

$$H = 4\pi n \frac{I}{c}$$

I in esu or statcoulombs sec^{-1}; cf. Eq. (17-18a).

Let μ_{mks} be the magnetic permeability, of the medium; then

$$B = \mu_{\text{mks}} H = \mu_{\text{mks}}\, nI \tag{17-19}$$

$$H = nI$$

The important point to note is that the field vector **H** is unaffected by the insertion of the solenoid in the medium, but the magnetic induction **B** increases or decreases by the factor μ_P (cgs-Gaussian) or $\mu_{\text{mks}}/\mu_{0,\text{mks}}$ (mks).

We shall not use the quantity, the magnetic permeability, very much. We wish to use the symbol μ to represent both electric and magnetic dipoles; therefore we use the awkward symbols μ_P, μ_{mks}, $\mu_{0,\text{mks}}$ for the several magnetic permeabilities. In the mks system, the ratio $\mu_{\text{mks}}/\mu_{0,\text{mks}}$ is the relative permeability and is the same as μ_P of gaussian units.

We shall discuss magnetic-dipole moments in detail in Chap. 19. Now it is sufficient to assert that there are such induced moments in each atom or molecule, and we let $\mathfrak{M}_V$ be the induced magnetic moment per unit volume. Then, quite generally,

$$\mathbf{B} = \mathbf{H} + 4\pi\mathfrak{M}_V \qquad \mathbf{B} = \mu_0(\mathbf{H} + \mathfrak{M}_V) \qquad (17\text{-}20)$$

The magnetic volume susceptibility χ_V is rigorously defined (in an isotropic medium where $\mathfrak{M}_V$ and $\mathcal{H}$ are in the same direction) by

$$\chi_V = \frac{\mathfrak{M}_V}{H} \qquad (17\text{-}21)$$

so that

$$\mu_P = 1 + 4\pi\chi_V \qquad \mu_{\text{mks}} = \mu_{0,\text{mks}}(1 + \chi_V) \qquad (17\text{-}22)$$

The energy density in a magnetic field is

$$\frac{\mathbf{B}\cdot\mathbf{H}}{8\pi} \qquad \frac{\mathbf{B}\cdot\mathbf{H}}{2} \qquad (17\text{-}23)$$

For practical consideration of dimensions in Gaussian units, it is usually sufficient to recall that $\mathfrak{M}_V^2$, B^2, and H^2 have dimensions of energy per unit volume.

17-5. A Table of Conversion Factors. Some of the conversion factors between mks units and cgs-Gaussian units are presented in Table 17-1.*

17-6. Symbols. New difficulties regarding the choices of symbols and duplications of meanings for symbols arise when we take up electrical and magnetic problems. It is worthwhile to note specifically some of these duplications and to apologize for or justify the choices that have been made.

---

* There are similar tables in a number of texts, including W. K. H. Panofsky and M. Phillips, "Classical Electricity and Magnetism," p. 384, Addison-Wesley, Reading, Mass., 1955, and J. A. Stratton, "Electromagnetic Theory," p. 602, McGraw-Hill, New York, 1941.

The three common symbols for the dielectric constant are D, κ, and ϵ. This quantity occurs in our study of dielectrics and dipole moments (Chap. 18) and in the theory of ionic solutions (Chap. 21). It is inconvenient to use D, which is also the symbol for the electric-displacement vector. The symbol κ is traditionally used for an important reciprocal length which occurs in the Debye-Hückel theory of ionic solutions. We therefore use ϵ for the dielectric constant. We have used and shall also use ϵ for a molecular energy level. Fortunately, in our chapters on

Table 17-1. Conversion Factors

Multiply the number of mks units below	by	to obtain the number of Gaussian (cgs) units of
Volt	$\dfrac{1}{300}$*	statvolts
Volt m^{-1}	$\tfrac{1}{3} \times 10^{-4}$*	electric-field intensity E, statvolts cm^{-1}
Coulomb m^{-2}	$12\pi \times 10^{5}$*	electric displacement D, statvolts cm^{-1}
Coulomb	3×10^{9}*	charge, esu or statcoulombs
Ampere	3×10^{9}*	current, esu or statcoulombs sec^{-1}
Ohm = volt amp^{-1}	$\dfrac{10^{-11}}{9}$*	resistance, statvolt sec statcoulomb^{-1}
Farad = coulomb volt^{-1}	9×10^{11}*	capacitance, cm or statcoulombs statvolt^{-1}
Weber m^{-2}	10^{4}	magnetic induction B, gauss
Ampere-turns m^{-1}	$\dfrac{4\pi}{10^{3}}$	magnetic field H, oersteds
Joule	10^{7}	ergs
Newton	10^{5}	dynes

* In conversion factors marked by an asterisk, the number 3 is derived from the velocity of light. For highly accurate calculations, the number 2.99793 should be used.

electricity, the theory is almost entirely classical, and the only form of energy of any interest is potential energy, for which we can reasonably use the symbol u.

In the chapters on electricity, E is the electric-field vector, and the thermodynamic energy is denoted by U.

We use μ for both electric- and magnetic-dipole moments. There is little danger of confusion due to the use of the same symbol for the chemical potential.

In the chapter on magnetism, the g factor is of course denoted by g, thus conflicting with our notation for the degeneracy of a state. Confusion can be avoided by explicitly listing each individual quantum-mechanical state in the partition-function sum and other sums which occur.

18

Dielectric Phenomena

18-1. Electric-dipole Moments, Induced and Permanent.* Two equal and opposite charges $-q$ and $+q$ separated by the vector distance $\mathbf{r}$ (the vector points from the negative to the positive charge) comprise an electric dipole, with a moment

$$\boldsymbol{\mu} = q\mathbf{r} \tag{18-1}$$

More generally, for a cluster of discrete charges q_i which is electrically neutral ($\Sigma q_i = 0$), we choose an origin of coordinates at the center of mass of the system. Let the position vectors of the charges be $\mathbf{r}_i$; the dipole moment is then

$$\boldsymbol{\mu} = \Sigma q_i \mathbf{r}_i \tag{18-2}$$

(Actually, the dipole moment of a neutral molecule is independent of the choice of origin of coordinates; however, the above choice is convenient.)

The electrostatic potential and the field at a general field point, $\mathbf{r}$, due to this charge cluster, depend on all the q_i and $\mathbf{r}_i$; for example, the potential is given by

$$\phi(\mathbf{r}) = \sum_i \frac{q_i}{|\mathbf{r} - \mathbf{r}_i|} \tag{18-3a}$$

However, at field points reasonably far away from the cluster, i.e., for $|\mathbf{r}| > |\mathbf{r}_i|$ (for all i), the expression for the potential can be expanded in a power series in r_i/r. The leading term is

$$\phi(r) = \frac{\Sigma q_i \mathbf{r}_i \cdot \mathbf{r}}{r^3} = \frac{\boldsymbol{\mu} \cdot \mathbf{r}}{r^3} \tag{18-3b}$$

That is, at large distances from the charge cluster, the potential and field are determined, in first approximation, by the dipole moment $\boldsymbol{\mu}$.

---

* For general texts, see the following: P. Debye, "Polar Molecules," Chemical Catalog, New York, 1929 (also a Dover reprint); C. P. Smyth, "Dielectric Behavior and Structure," McGraw-Hill, New York, 1955; C. J. F. Böttcher, "Theory of Electric Polarization," Elsevier, Houston, Tex., 1952; J. T. Edsall and J. Wyman, "Biophysical Chemistry," chap. 6, Academic Press Inc., New York, 1958; H. Fröhlich, "Theory of Dielectrics," 2d ed., Oxford, New York, 1958.

Furthermore, the electrostatic energy of interaction of a rigid charge cluster with a *uniform* external field depends only on the dipole moment **μ**. Thus, for many purposes, the most important single parameter of a charge distribution of a neutral molecule is the electric-dipole moment.

The permanent dipole moment of a molecule is defined in a system of coordinates which is fixed with respect to the equilibrium position of the nuclei. Let ψ_j be the wave function of the molecule in the quantum state j. This wave function ψ_j, in a frame of coordinates fixed with respect to the equilibrium position of the nuclei, specifies the electronic and the vibrational state of the molecule but not its rotational state. The dipole moment of the molecule is then

$$\mathbf{\mu}_j = \langle \psi_j | \mathbf{\mu} | \psi_j \rangle \qquad (18\text{-}4)$$

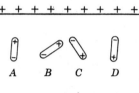

In the molecule HF, the fluorine is relatively negative, leaving the hydrogen relatively positive. The molecule has a permanent dipole moment pointing from the fluorine to the hydrogen. A molecule such as CHF_2Cl is also electrically unsymmetrical and has a permanent dipole moment, with a definite, albeit not necessarily simple, orientation with respect to a set of molecular axes. An atom such as argon and molecules such as SF_6 and CCl_4 are electrically symmetrical; they have no permanent dipole moment. However, an applied electric field will displace the average position of the negative charges in the atom or molecule with respect to the positive charges, resulting in polarization or an induced dipole moment in the direction of the field. The charge distribution of an unsymmetrical molecule is, of course, also polarized by an electric field; the total dipole moment is the vector sum of the induced and the permanent moments.

FIG. 18-1. A permanent dipole in the uniform electric field of a parallel-plate capacitor. Configuration A is one of unstable equilibrium; there is an orientating torque for configurations B and C tending to align the dipole as in D. D is the configuration of lowest energy.

In the discussion that follows, we take a simple classical model for the behavior of a molecule in an electric field and then apply classical statistical mechanics. The reader will appreciate that there is a corresponding quantum-mechanical theory; however, it is more difficult and is not needed for the problems that we consider.

In a uniform electric field, there is no net translational force on a permanent dipole, but there is an orienting torque which tends to align the dipole parallel to the external field. This situation is illustrated in Fig. 18-1. The orienting torque is $\mathbf{T} = \mathbf{\mu} \times \mathbf{E}$, or $|\mathbf{T}| = \mu E \sin \theta$, where θ is the angle between the field and the dipole. It then can be shown that the electrostatic energy u of the dipole in the electric field is

$$u = -\mathbf{\mu} \cdot \mathbf{E} = -\mu E \cos \theta \qquad (18\text{-}5)$$

In the classical statistical mechanics of a collection of independent permanent dipoles in an electric field, there is a competition between the orienting, ordering effects of the field and the randomizing effects of thermal agitation, which is quantitatively expressed by the Boltzmann factor $e^{\mu E \cos \theta / kT}$. The resulting partial orientation of the dipoles produces a net electric polarization of the medium which is called orientation polarization. This is the principal topic which we wish to study.

The quantitative energy relation is somewhat different for induced dipoles. It is illustrated by the following simple model: Let the molecule consist of two charges $+q$ and $-q$. Both charges are located at $x = 0$ in the absence of a field. (In a more realistic treatment of an atom, we would say that the center of charge of the electron cloud is located at the nucleus and that at zero field there is no dipole moment.) On application of an electric field, the two charges are displaced relative to each other so that their separation is x. There is a Hooke's law restoring force of fx, where f is the force constant. The electrostatic force separating the charges, which is qE, must balance the restoring elastic force fx; therefore $fx = qE$. The induced moment is then

$$m = qx = \frac{q^2 E}{f} = \alpha E \qquad (18\text{-}6)$$

(We use μ for a permanent moment and m for an induced moment.) We see that the induced moment is proportional to the applied field; the constant of proportionality, $\alpha = q^2/f$, is called the polarizability. The positive elastic energy is $\tfrac{1}{2}fx^2$; the negative electrostatic energy is $-qEx$ or $-mE$; the total energy u is

$$u = \tfrac{1}{2}fx^2 - qEx = -\frac{1}{2}\frac{q^2 E^2}{f} = -\tfrac{1}{2}\alpha E^2 = -\tfrac{1}{2}mE \qquad (18\text{-}7)$$

The model we have used in this derivation is oversimplified, but in general, for atoms or sufficient symmetrical molecules, it is a satisfactory approximation for the induced moment m to write

$$m = \alpha E \qquad (18\text{-}8)$$

from which it follows in general that

$$u = -\tfrac{1}{2}\alpha E^2 = -\tfrac{1}{2}mE \qquad (18\text{-}9)$$

For less symmetrical molecules (for example, a linear molecule such as CS_2), the polarizability is different in different directions, so that the energy of interaction of the molecule with the field is dependent on the orientation of the molecule. In many cases, the induced moment is much smaller than the permanent moment; and furthermore, the anisotropy of the polarizability is rather small. In an introductory treatment, therefore, it is usual to assume that the polarizability is isotropic and independent of the orientation of the molecule. The more general case is treated in Sec. 18-6. The polarizability α consists of two parts. The electronic polarizability α_e is due to the distortion of the electron cloud by the electric field. The atomic polarizability α_A is due to induced moments associated with changes in internuclear distances induced by the field. For example, if a CCl_4 molecule is placed in an electric field, the electron cloud will be displaced with respect to the nuclei, giving rise to an induced moment. This is electronic polarization. In addition, since we can think of CCl_4 as containing a slightly positive carbon and slightly negative chlorines, the electric field will cause the bond angles and distances to distort so that the negative chlorines are displaced with respect to the positive carbon atom. This is atomic polarization. These distortions of the bond angles and distances are closely related to the vibrations of the atoms responsible for the infrared vibration spectrum. In general, the electronic polarizability is much greater than the atomic polarizability. For CCl_4, they are 105×10^{-25} cc and 8×10^{-25} cc molecule^{-1}, respectively.

There are several additional facts about the field due to a dipole and the electrostatic interaction between two dipoles which should be mentioned, although we shall not use them explicitly.

Consider a dipole μ_1 (due to a charge distribution of infinitesimal extension) located at the origin. The electric field at a general position in space, r, due to μ_1 is

$$\mathbf{E} = -\nabla\phi(\mathbf{r}) = -\frac{\mu_1}{r^3} + \frac{3(\mu_1 \cdot \mathbf{u})\mathbf{u}}{r^3} \qquad (18\text{-}10)$$

where $r = |\mathbf{r}|$ and $\mathbf{u}$ is the unit vector in the direction of $\mathbf{r}$. Let there be a second dipole μ_2 at the point $\mathbf{r}_{12}$ and let the unit vector in this direction be $\mathbf{u}_{12}$. The electrostatic energy of interaction of the two dipoles is

$$u = -\mu_2 \cdot \mathbf{E}_{12} = \frac{\mu_1 \cdot \mu_2}{r_{12}^3} - \frac{3(\mu_1 \cdot \mathbf{u}_{12})(\mu_2 \cdot \mathbf{u}_{12})}{r_{12}^3} \qquad (18\text{-}11)$$

In qualitative thinking about dipoles, it is usually helpful and sufficient to picture two point charges $\pm q$, separated by a distance r. For q the charge on the electron and $r = 1$ Å $= 10^{-8}$ cm; the dipole moment is $\mu = 4.80 \times 10^{-18}$ esu. The unit 10^{-18} esu is called the debye; 10^{-18} esu = 1 debye.

Instead of thinking about or trying to remember the cgs units of a dipole moment, in terms of m, l, and t, it is usually better to recall that μ^2/r^3 and μE both have the dimensions of energy. The units of polarizability are l^3 or cubic centimeters.

The mks units of a dipole moment are coulomb meters; the energy of interaction of a dipole with an external field is still $-\boldsymbol{\mu} \cdot \mathbf{E}$; the field due to a dipole is

$$\mathbf{E} = \frac{1}{4\pi\epsilon_0}\left[-\frac{\boldsymbol{\mu}}{r^3} + \frac{3(\boldsymbol{\mu}\cdot\mathbf{u})\mathbf{u}}{r^3}\right] \tag{18-12}$$

18-2. Orientation Polarization. For a dilute gas, the effective electric field on a molecule is the applied external field E. Consider a nonvibrating diatomic molecule with a permanent dipole moment μ along the intermolecular axes, an isotropic polarizability α, and moment of inertia I. Let the orientation of the molecule with respect to the external field be specified by the usual polar angles θ and ϕ, with conjugate momenta p_θ and p_ϕ. The classical Hamiltonian is

$$H = \frac{p_\theta^2}{2I} + \frac{p_\phi^2}{2I\sin^2\theta} - \mu E\cos\theta - \tfrac{1}{2}\alpha E^2 \tag{18-13}$$

It should be recognized that this Hamiltonian contains terms for the kinetic energy of rotation, for the potential energy of orientation of the permanent dipole in the field, and for the potential energy due to induced polarization. The classical approximation for the partition function for the rotational degrees of freedom is then

$$q = \frac{1}{h^2}\int\cdots\int e^{-H(p_\theta, p_\phi, E, \theta, \phi)/kT}\,dp_\theta\,dp_\phi\,d\theta\,d\phi \tag{18-14}$$

The probability $f(\theta,\phi)\,d\theta\,d\phi$ of finding the system with orientation θ and ϕ, irrespective of the value of the kinetic energy of rotation, is

$$f(\theta,\phi)\,d\theta\,d\phi = \frac{\int_{-\infty}^{\infty}\int_{-\infty}^{\infty} e^{-H/kT}\,dp_\theta\,dp_\phi}{\int\cdots\int e^{-H/kT}\,dp_\theta\,dp_\phi\,d\theta\,d\phi} \tag{18-15}$$

The integral over the momenta can be performed in both the numerator and the denominator.

$$\int_{-\infty}^{+\infty}\int_{-\infty}^{+\infty} e^{-p_\theta^2/2IkT}e^{-p_\phi^2/2I\sin^2\theta kT}\,dp_\theta\,dp_\phi = 2\pi IkT\sin\theta$$

so that

$$f(\theta,\phi)\,d\theta\,d\phi = \frac{2\pi e^{\mu E\cos\theta/kT}e^{\alpha E^2/2kT}\sin\theta\,d\theta\,d\phi}{z} \tag{18-16}$$

where z is the configuration integral

$$z(E,T) = \iint e^{\mu E\cos\theta/kT}e^{\alpha E^2/2kT}\,2\pi\sin\theta\,d\theta\,d\phi \tag{18-17}$$

For orientation θ, ϕ, the component of the dipole moment along the field is $\mu\cos\theta + \alpha E$; the average value of the polarization in the field direction is then

$$\bar{\mu} = \iint (\mu \cos\theta + \alpha E) f(\theta,\phi)\, d\theta\, d\phi$$
$$= \frac{\iint (\mu \cos\theta + \alpha E) 2\pi e^{\mu E \cos\theta/kT} e^{\alpha E^2/2kT} \sin\theta\, d\theta\, d\phi}{\iint 2\pi e^{\mu E \cos\theta/kT} e^{\alpha E^2/2kT} \sin\theta\, d\theta\, d\phi} \quad (18\text{-}18)$$

We note in passing that we can write $\bar{\mu} = kT[\partial(\ln z)/\partial E]_T$; this relation will be discussed further for the analogous magnetic problem (Chap. 19). Without performing any integrations, the expression (18-18) reduces to

$$\bar{\mu} = \alpha E + \mu \overline{\cos\theta} \quad (18\text{-}19)$$

where
$$\overline{\cos\theta} = \frac{\int_0^\pi \cos\theta\, e^{\mu E \cos\theta/kT} \sin\theta\, d\theta}{\int e^{\mu E \cos\theta/kT} \sin\theta\, d\theta} \quad (18\text{-}20)$$

Expression (18-20) could have been written down directly on the basis of the elementary consideration that the energy for orientation θ is $-\mu E \cos\theta$ with the Boltzmann factor $e^{\mu E \cos\theta/kT}$ and the solid angle is $2\pi \sin\theta\, d\theta$. We have started from more fundamental expressions to emphasize the fact that, when there is a potential energy which is a function of the position coordinates only, the integrals over the momentum coordinates cancel from the significant answers. The same cancellation would occur for the terms giving the kinetic energy of rotation of a polyatomic molecule. If a molecule has an isotropic polarizability α and a permanent dipole moment μ, and if θ is the angle between the dipole and the applied field, expressions (18-19) and (18-20) give the average moment in the field direction. This is true for diatomic molecules and polyatomic molecules provided that the classical approximation for the partition function applies.

Introduce the dimensionless variable $y = \mu E/kT$. Equation (18-20) becomes

$$\overline{\cos\theta} = \frac{\int_0^\pi \cos\theta\, e^{(y \cos\theta)} \sin\theta\, d\theta}{\int_0^\pi e^{(y \cos\theta)} \sin\theta\, d\theta} = L(y) \quad (18\text{-}21)$$

The function $L(y)$ defined in (18-21) is known as the Langevin function. Apart from some numerical factors, the denominator of (18-21) is essentially a configuration integral; the numerator is the derivative of the denominator with respect to y. Then

$$\int_{\theta=0}^{\theta=\pi} e^{y \cos\theta}(-d\cos\theta) = \frac{e^y - e^{-y}}{y} \quad (18\text{-}22)$$

and
$$L(y) = \frac{d}{dy} \ln \frac{e^y - e^{-y}}{y} = \frac{e^y + e^{-y}}{e^y - e^{-y}} - \frac{1}{y} \quad (18\text{-}23)$$

Problem 18-1. Calculate the value of $y = \mu E/kT$ for $\mu = 10^{-18}$ esu $= 1$ debye, $E = 10^4$ practical volts cm^{-1}, and $T = 300°$K.

Problem 18-2. Show that for small values of y

$$L(y) \approx \frac{y}{3} - \frac{y^3}{45} \quad (18\text{-}24)$$

For the small values of y that occur in almost all practical situations, it is sufficient to take the leading term in (18-24); the mean moment per molecule is then

$$\bar{\mu} = \alpha E + \frac{\mu^2 E}{3kT} \qquad (18\text{-}25)$$

This is the desired result. If N is the number of molecules per unit volume, the polarization or net moment per unit volume is $P = N\bar{\mu}$. But, since $D = \epsilon E$, and $D = E + 4\pi P$, $P = (\epsilon - 1)E/4\pi$, where ϵ is the dielectric constant (which is the quantity that is experimentally measured),

$$\frac{(\epsilon - 1)E}{4\pi} = P = N\left(\alpha + \frac{\mu^2}{3kT}\right)E \qquad (18\text{-}26)$$

With M the molecular weight and ρ the density, the volume per mole is M/ρ; the polarization per mole is then

$$\frac{M}{\rho}P = \frac{M(\epsilon - 1)E}{\rho(4\pi)} = N_0\left(\alpha + \frac{\mu^2}{3kT}\right)E \qquad N_0 = \text{Avogadro's number} \qquad (18\text{-}27)$$

The quantity $N_0\bar{\mu}/E$ (or $MP/\rho E$) is clearly the moment per mole per unit field. However, it is customary to multiply by $4\pi/3$ and to call the quantity $4\pi N_0\bar{\mu}/3E$ the molar polarizability,

$$\frac{4\pi N_0 \bar{\mu}}{3E} = \frac{4\pi N_0}{3}\left(\alpha + \frac{\mu^2}{3kT}\right) = \frac{M}{\rho}\frac{\epsilon - 1}{3} \qquad (18\text{-}28)$$

Equation (18-28) is a useful working equation for the dielectric constants of dilute gases. The temperature dependence which it predicts is in fact observed. For molecules with a permanent dipole moment, the molar polarization $(M/\rho)(\epsilon - 1)/3$ decreases with increasing temperature. By plotting $(M/\rho)(\epsilon - 1)/3$ versus $1/T$, straight-line plots are obtained; the slope is related to the permanent moment μ; the intercept at $1/T = 0$ gives the polarizability α. Molecules with a center of symmetry have zero permanent moment and a molar polarizability which is independent of temperature.

We can rewrite (18-28) as

$$\frac{M}{\rho}\frac{\epsilon - 1}{3} = R_d + \frac{B}{T} \qquad (18\text{-}29a)$$

with
$$R_d = \frac{4\pi N_0}{3}\alpha \qquad B = \frac{4\pi N_0 \mu^2}{9k} \qquad (18\text{-}29b)$$

so that
$$\mu = 0.01281 B^{1/2} \qquad \text{debyes} \qquad (18\text{-}29c)$$
$$B = 6.10 \times 10^3 \mu^2 \qquad \mu \text{ in debyes} \qquad (18\text{-}29d)$$

The quantity R_d is known as the molar distortion polarization. There is a small contribution due to atomic polarization; otherwise R_d is equal to the molar refractivity R_e, which for a gas is given by

$$R_e = \frac{4\pi N_0}{3} \alpha_e = \frac{M}{\rho} \frac{n^2 - 1}{3} \quad (18\text{-}29e)$$

where n is the refractive index.

Furthermore, for an ideal gas, the molar volume M/ρ is given by $M/\rho = RT/P$, where R is the gas constant and P is the pressure. With P in atmospheres, use $R = 82.05$ cc atm mole^{-1} deg^{-1}. A useful working equation for an ideal gas is then

$$\frac{RT}{P} \frac{\epsilon - 1}{3} = R_d + \frac{B}{T} = \frac{4\pi N_0}{3} \left(\alpha + \frac{\mu^2}{3kT} \right) \quad (18\text{-}29f)$$

Representative values of R_d and of μ are given in Table 18-1.

Table 18-1. Dipole Moments of Some Gas Molecules*

Substance	CsCl	HF	HI	NH$_3$	CH$_3$Cl	CH$_3$NO$_2$
R_d, cc	....	2.0	13.5	5.3	11.7	18.6
$\mu(D)$	10.5	1.92	0.42	1.46	1.89	3.44

* A. A. Maryott and F. Buckley, "Table of Dielectric Constants and Electric Dipole Moments of Substances in the Gaseous State," *Natl. Bur. Standards Circ.*, 537, 1953. Another good compilation of dipole moments and polarizabilities is that by H. Stuart in A. Eucken and K. H. Hellwege (eds.), "Landolt-Börnstein Tabellen," part 3 (II), vol. I, pp. 386–517, Springer, Berlin, 1951.

The dielectric constant, of course, varies with the frequency of the electromagnetic radiation; the refractive index n for any frequency is related to the dielectric constant at that frequency by the formula $\epsilon = n^2$. For low frequencies, it is more convenient to measure ϵ; for high frequencies, such as visible light, it is only practical to measure n.

At frequencies greater than molecular-rotation frequencies (10^{10} to 10^{12} sec^{-1}), the permanent dipoles cannot rotate with sufficient rapidity to follow the oscillations of the electric-field vector; therefore at higher frequencies the orientation polarization does not contribute significantly to the total polarization.

The variation of the molar polarization (or the refractive index or the dielectric constant) with frequency is qualitatively depicted in Fig. 18-2.

As the frequency increases from zero, but for frequencies below the molecular-rotation frequencies, the polarization gradually increases. This is the usual behavior for the polarization as a function of frequency in a region where there are no significant absorption lines, and it is known as normal dispersion. The polarization goes through a set of maxima and minima in the neighborhood of the absorption frequencies for the pure rotation spectrum. This behavior is known as anomalous dispersion; a proper discussion of this region of the curve requires the use of quantum theory. At higher frequencies, the orientation polarization does not contribute appreciably to the total polarization, which is then due to atomic polarization and electronic polarization. There is another region of anomalous dispersion at the infrared absorption frequencies corresponding to the vibrational energy levels of the molecule. At still higher frequencies in the range for visible light, the atoms cannot vibrate with sufficient rapidity to follow the oscillations of the electric field, and the polarization is due solely to the electronic polarization. There is anomalous dispersion of the electronic polarization in the neighborhood

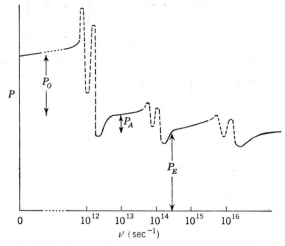

FIG. 18-2. A *very* schematic representation of the variation of the molar polarization with frequency. Below the molecular-rotation frequencies of 10^{10} to 10^{12} sec^{-1}, the orientation polarization P_0 contributes significantly to the total polarization. As the frequency goes through the rotational absorption lines, there is a series of maxima and minima of the polarization which is indicated very schematically by the dashed lines. Between about 10^{12} and 10^{14} sec^{-1}, the atomic polarization P_A and the electronic polarization P_E contribute. In the region of the infrared vibrational absorption bands, there is anomalous dispersion for P_A; above these frequencies, only P_E is significant. There is anomalous dispersion of the electronic polarization in the ultraviolet or visible electronic absorption bands.

of the wavelengths of the visible or ultraviolet electronic absorption bands of the substance. The atomic and electronic polarizabilities, as determined by refractive-index measurements with infrared and visible light, should be extrapolated to low frequencies if a precise comparison with the polarizability deduced from dielectric-constant measurements is desired.

The main subject of interest in the present chapter is the orientation polarization, since it is temperature-sensitive and requires statistical mechanical consideration. The atomic and electronic polarizabilities present no statistical mechanical problem, because they are almost the same for all the rotational and vibrational states of the ground electronic state of the molecule. They are therefore independent of temperature except at very high temperatures, where electronic excitation and dissociation occur.

The rotation of molecules in the gas phase is of course a quantal phenomenon. As an alternative to the above classical treatment of orientation polarization, we could inquire into the effect of an electric field on the rotational energy levels. It would be found that the energies of the different rotational states are shifted, either up or down, by an electric field (Stark effect). The net orientation polarization can be computed by an analysis of these shifts. Such an analysis is possible in closed form for diatomic molecules and symmetric tops. For asymmetric tops, numerical solutions can be obtained for any particular case. The Stark shifts of the energy levels are important for microwave spectroscopy; the classical treatment given here is adequate for computing the macroscopic dielectric constant.

We finally remark that, if the approximation $L(y) \approx y/3$ fails because $y = \mu E/kT$ becomes large, the polarization is no longer a linear function of the applied field. This is dielectric saturation. The electric fields required are so great that it is very difficult to achieve this condition in practice.

18-3. The Local-field Problem. The Clausius-Mosotti Relation. The electric field on a molecule in a condensed phase is in part due to the dipoles, permanent and induced, of the other molecules in the dielectric. The problem then is: What is the effective local field acting to orient any particular molecule? This is a problem of the statistical mechanics of condensed phases, and a rigorous statistical mechanical treatment is not yet available. A discussion of the more serious approximate statistical mechanical treatments is beyond the scope of our work; we shall, however, present several simple approximate treatments.

Consider the parallel-plate capacitor illustrated in Fig. 18-3. As already discussed in Chap. 17, if the potential is ϕ and the separation of the plates is L, the electric field in the dielectric is $E = \phi/L$ and the electric displacement is $D = \epsilon E$. The polarization vector in the medium is then $P = (D - E)/4\pi = (\epsilon - 1)E/4\pi$. There is a free surface charge σ on the plates such that $D = 4\pi\sigma$; there is a smaller induced charge of opposite sign on the surface of the dielectric, $\sigma' = P$. The electric field in the medium is $E = 4\pi(\sigma - \sigma')$; this relation is seen to be consistent with the relations $E = D - 4\pi P$, $D = 4\pi\sigma$, and $P = \sigma'$, already given.

Consider now a small conceptual cavity around the molecule of interest. The cavity is usually said to be somewhat larger than molecular dimensions, but small compared with the capacitor. We do not excavate the matter from this cavity, but we conceptually regard it as in a different category from the rest of the dielectric. There will be an induced surface charge σ_c on the walls of the cavity due to the polarization of the dielectric outside it. We then say that the orienting force **F** on the molecule of interest at the center of the cavity is composed of three parts,

$$\mathbf{F} = \mathbf{F}_1 + \mathbf{F}_2 + \mathbf{F}_3$$

Force $\mathbf{F}_1$ is the field due to the charges σ and $-\sigma'$ on the walls of the

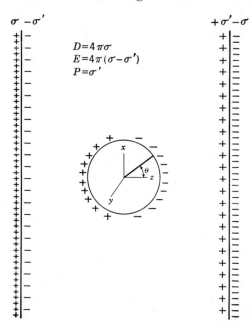

Fig. 18-3. The Clausius-Mosotti (or Lorentz-Lorenz) local field on a molecule in a dielectric. The molecule is assumed to be at the center of a small spherical cavity (which is rather large in the illustration). There is a free surface charge σ on the plates of the capacitor, an induced charge σ' on the surface of the dielectric, and a surface charge distribution σ_c on the walls of the cavity. It is assumed that the force on the molecule is the sum of the forces due to these three charge distributions.

capacitor; this is the macroscopic electric-field vector in the medium,

$$\mathbf{F}_1 = 4\pi(\sigma - \sigma') = \mathbf{E}$$

Force $\mathbf{F}_2$ is due to the surface charges σ_c on the walls of the cavity. We shall calculate this shortly. Force $\mathbf{F}_3$ is due to polarized material inside the cavity; this is assumed to be zero. It was shown by Lorentz[*] that, in a cubic crystal, $\mathbf{F}_3$ is indeed zero, and it is assumed that the same applies to gases or to liquids if there is no correlation between the orientation of one molecule and that of a neighbor.

[*] H. A. Lorentz, "Theory of Electrons," p. 306, note 55, Teubner Verlagsgesellschaft, Leipzig, 1909 (also a Dover reprint); this result is quoted in most texts and seldom proved.

We now calculate F_2. As shown in Fig. 18-3, take a system of polar coordinates with the z axis parallel to the electric field. We want to know the surface charge σ_c on the cavity due to the polarization of the external dielectric. It is a general theorem in electrostatics that the surface charge is the normal component of the electric polarization vector $\mathbf{P}$; thus

$$\sigma_c = -P \cos \theta \tag{18-30}$$

We now calculate the force $\mathbf{F}_2$ at the center of the cavity due to this surface-charge distribution σ_c. The component of force in the z direction due to an element of area dA on the surface of the cavity at angle θ is

$$-\frac{\sigma_c \, dA \, \cos \theta}{a^2}$$

where a is the radius of the sphere. But for a ring at angle θ,

$$dA = 2\pi a^2 \sin \theta \, d\theta$$

substitute for σ_c, and obtain

$$F_2 = \int_{\theta=0}^{\theta=\pi} 2\pi P \cos^2 \theta \sin \theta \, d\theta = \frac{4\pi P}{3} \tag{18-31}$$

The local field is then

$$F = F_1 + F_2 = E + \frac{4\pi P}{3} \tag{18-32}$$

This is usually called the Clausius-Mosotti or the Lorentz-Lorenz expression for the local field.

One comment should be added. Consider an ion immersed or dissolved in the dielectric. The force on the ion that causes its translational migration is E. This is because, as the ion moves, it carries whatever cavity it has with it. The force due to the induced charges on the inner surface of the cavity around the ion cannot influence its translational motion. In the same way, the translational force between two charges is $q_1 q_2 / \epsilon r^2$, provided that the charges are far enough apart so that there is a medium of dielectric constant ϵ between them. The local-field problem arises in connection with the polarizing force and orienting torque on a molecule, not in connection with the translational motion of a charged particle.

We now assert that the relations derived in the previous section between the polarization and the field still hold, but the field to be used is the local field. In particular, we can now rewrite expression (18-25) for the mean moment per molecule.

$$\bar{\mu} = \left(\alpha + \frac{\mu^2}{3kT} \right) F \tag{18-33a}$$

so that the polarization per unit volume is

$$P = N\bar{\mu} \quad \text{or} \quad P = N\left(\alpha + \frac{\mu^2}{3kT}\right)F \qquad (18\text{-}33b)$$

The macroscopic-field equation is

$$P = \frac{\epsilon - 1}{4\pi} E \qquad (18\text{-}34)$$

and the local-field equation is

$$F = E + \frac{4\pi P}{3} \qquad (18\text{-}32)$$

The object now is to eliminate F, P, and E from the three equations (18-33b), (18-34), and (18-32) and thus obtain a relation between the dielectric constant and the molecular parameters α and μ. From (18-32) and (18-34)

$$F = \frac{(\epsilon + 2)E}{3} \qquad (18\text{-}35)$$

Substitution of (18-34) and (18-35) into (18-33b) gives

$$\frac{\epsilon - 1}{\epsilon + 2} = \frac{4\pi}{3} N\left(\alpha + \frac{\mu^2}{3kT}\right) \qquad (18\text{-}36a)$$

Again, multiply by the molar volume M/ρ, and obtain an expression for the molar polarization,

$$\frac{M}{\rho} \frac{\epsilon - 1}{\epsilon + 2} = \frac{4\pi N_0}{3}\left(\alpha + \frac{\mu^2}{3kT}\right) \qquad (18\text{-}36b)$$

This is the desired result; it is the fundamental equation of the Debye theory of dielectrics. The corresponding equation for dilute gases was

$$\frac{M}{\rho} \frac{\epsilon - 1}{3} = \frac{4\pi N_0}{3}\left(\alpha + \frac{\mu^2}{3kT}\right) \qquad (18\text{-}28)$$

The equations are in practice identical for dilute gases for which ϵ is very closely equal to unity.

When applied to the electronic polarizability α_e of a molecule, Eq. (18-36) becomes

$$\frac{M}{\rho} \frac{n^2 - 1}{n^2 + 2} = \frac{4\pi N_0}{3} \alpha_e = R_e \qquad (18\text{-}37)$$

where n^2 is the refractive index and R_e is the molar refraction.

The electronic polarizability of a substance calculated by this equation from data on the liquid phase is almost always very closely the same as that calculated from gas-phase data—thus, to some extent, confirming the equation. The corresponding adaptation of Eq. (18-28) does not correlate liquid- and gas-phase refractive-index data.

For a dilute solution of a polar solute (component 2) in a nonpolar solvent (component 1), the obvious extension of (18-36a) is

$$\frac{\epsilon - 1}{\epsilon + 2} = \frac{4\pi}{3}\left[N_1\alpha_1 + N_2\left(\alpha_2 + \frac{\mu_2^2}{3kT}\right)\right] \quad (18\text{-}38)$$

where N_1 and N_2 are the number of molecules per unit volume of the two components. This equation is used for calculating the dipole moments of polar molecules from measurements of the dielectric constants of dilute solutions. Where comparison is possible, the dipole moments so calculated agree fairly well with gas-phase results.

The Debye equation [(18-36a) or (18-36b)] however, does not work at all for pure liquids composed of polar molecules. We study this subject in the next section.

Problem 18-3. The dipole moment of C_2H_5Br as measured in the gas phase is 2.0 debyes. The density of the liquid is 1.43 (the molecular weight is 109), and the electronic polarizability α_e is 1.0×10^{-23} cc molecule^{-1}. Use the Debye equation to calculate:

(a) The refractive index and the dielectric constant of the gas at 320°K and 1 atm pressure (neglect the contribution by the atom polarization).

(b) The refractive index and the dielectric constant of the liquid (the experimental values are 1.424 and 9.3).

18-4. Limitations of the Debye Equation. The Dielectric Constants of Polar Liquids.

The absurd result of the calculation of Prob. 18-3 illustrates the inapplicability of the Debye equation to highly polar liquids with large dielectric constants. Let us rewrite Eq. (18-36a), omitting the polarizability α, which is usually relatively small.

$$\frac{\epsilon - 1}{\epsilon + 2} = \frac{4\pi N}{9}\frac{\mu^2}{kT} \quad (18\text{-}36c)$$

The left-hand side of the equation is always less than unity, no matter how great ϵ. As the temperature is decreased, the right-hand side increases and there is a temperature at which it becomes greater than unity. We can define a critical temperature by

$$kT_c = \frac{4\pi N\mu^2}{9} \quad (18\text{-}39)$$

and rewrite (18-36c) as

$$\frac{\epsilon - 1}{\epsilon + 2} = \frac{T_c}{T}$$

For example, the values of T_c for liquid C_2H_5Br and for liquid H_2O, calculated using the dipole moments measured in the gas phase, are 320°K and 1140°K, respectively.

The singularity at $T = T_c$ is sometimes interpreted as meaning that the theory predicts that the material becomes ferroelectric and all the dipoles spontaneously align below T_c. This is not observed, and it is sufficient to say that the theory is simply inapplicable to polar liquids even at temperatures above T_c. Values of the dipole moment calculated from data for the pure liquid by using (18-36c) are always substantially lower than values obtained from gas phase or dilute solution in nonpolar solvent measurements. For example, the dielectric constant of liquid C_2H_5Br is 9.3; the value of μ calculated from Eq. (18-36c) is then 1.3 debyes, whereas the gas-phase value is 2.0 and the dilute-solution value is 1.80.*

The energy of interaction of two dipoles at a distance r is of the order of μ^2/r^3; the nearest intermolecular distance is of the order of $N^{-1/3}$; therefore, in a liquid, the nearest-neighbor dipole-dipole interaction energy is of the order of $N\mu^2$ per pair of dipoles. The critical temperature defined by Eq. (18-39) is therefore a temperature in the range where the energy of interaction of neighboring dipoles becomes comparable with kT. This supports our suspicion that the failure of the equation is related to an incorrect approximation for the effects of the interactions between neighboring dipoles.

The equations of the Debye theory are derived by assuming that the local field is $F = E + (4\pi P/3)$, or from Eq. (18-35)

$$F = \frac{\epsilon + 2}{3} E$$

For substances with a large dielectric constant, the Lorentz-Lorenz local field is very large; this would cause a large amount of orientation of the dipoles for a small applied field. This is the basic error in the theory.

The nature of the difficulty was clearly recognized by Onsager,† and it has been aptly stated by Kirkwood.‡ We recall the model used in Sec. 18-3 in calculating the internal field. There was a sphere of matter around the central dipole which contributed an electric field F_3. It was assumed that outside the sphere, the polarization was uniform and had the value of the macroscopic polarization vector in the medium.

For a large enough sphere, this assumption must be true. However, the precise size of the sphere did not affect the results. The Lorentz local field is obtained by assuming that $F_3 = 0$. This last assumption must then be false. However, rather than attempting to calculate F_3, we can shrink the cavity onto the surface of the molecule for which the local field is to be calculated. (We imagine that the molecule exists in

---

* W. M. Heston, Jr., E. J. Hennerly, and C. P. Smyth, *J. ACS*, **72**: 2071 (1950). There is a good compilation of the dielectric constants of pure liquids, A. A. Maryott and E. R. Smith, Table of Dielectric Constants of Pure Liquids, *Natl. Bur. Standards Circ.* 514, 1951.

† L. Onsager, *J. ACS*, **58**: 1486 (1936).

‡ J. G. Kirkwood, *J. Chem. Phys.*, **7**: 911 (1939); *Trans. Faraday Soc.*, **42A**: 7 (1946).

a roughly spherical cavity.) Now surely $F_3 = 0$, but we can no longer assume that the polarization of the surrounding medium is fixed at its average value. The polarization of the medium in the immediate neighborhood is in fact influenced by the orientation of the central dipole. In macroscopic electrostatics, a dipole in a spherical cavity in an isotropic dielectric produces a polarization of the medium, which in turn produces an electric field in the same direction as the dipole, as depicted in Fig. 18-4. This electric field is known as the "reaction field" **R**. Its magnitude is actually

$$R = \frac{2\mu}{a^3} \frac{\epsilon - 1}{2\epsilon + 1}$$

where a is the radius of the sphere. Since the reaction field always points in the direction of the dipole, it exerts no orienting torque and does not affect the average orientation of the dipole. But, to some extent at least, the reaction field associated with the mean favorable orientation of the dipole is included in the Lorentz internal field.

In estimating the local field on the dipole it would be better to start by assuming that the average orientation of the dipole is essentially random and has no effect on the polarization of the medium. The molecule can then be regarded as a sphere of low dielectric constant ϵ_1 (due to its electron polarizability) immersed in a medium of high dielectric constant ϵ. The lines of

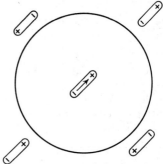

FIG. 18-4. The polarization of a dielectric by a dipole. The dipole is shown as an arrow and an ellipse within the spherical cavity. The polarization of the medium is indicated by the small ellipses outside the sphere. It should be apparent to the reader that the reaction field at the dipole due to its own polarization of the medium is in the direction of the dipole.

force in the dielectric tend to avoid the sphere of low dielectric constant; the surface polarization is therefore less than the uniform polarization assumed for the calculation of the Lorentz field. In fact, the solution of the electrostatic problem gives for the field inside the sphere*

$$F = \frac{3\epsilon}{2\epsilon + \epsilon_1} E \qquad (18\text{-}40)$$

When this is combined with the equations

$$P = \frac{\epsilon - 1}{4\pi} E$$

$$P = N\left(\alpha + \frac{\mu^2}{3kT}\right) F$$

---

* The calculation is clearly expounded by J. T. Edsall and J. Wyman, "Biophysical Chemistry," chap. 6, Academic Press Inc., New York, 1958.

we obtain

$$\frac{4\pi N}{3}\left(\alpha + \frac{\mu^2}{3kT}\right) = \frac{(2\epsilon + \epsilon_1)(\epsilon - 1)}{9\epsilon} \quad (18\text{-}41)$$

The limiting form for large ϵ and μ is

$$\frac{4\pi N}{3}\frac{\mu^2}{3kT} \approx \frac{2}{9}\epsilon \quad (18\text{-}42)$$

Equations (18-41) and (18-42) give better agreement with experiment for polar liquids than does (18-36).

The Onsager theory is more sophisticated but similar in general intent to the simple argument given above. The resulting equation is

$$\frac{4\pi N\mu^2}{9kT} = \frac{(\epsilon - \epsilon_1)(2\epsilon + \epsilon_1)}{\epsilon(\epsilon_1 + 2)^2} \quad (18\text{-}43)$$

(where, we repeat, ϵ_1 is the high-frequency dielectric constant or the square of the index of refraction). The Onsager equation is roughly similar to Eq. (18-41).

We shall not pursue these matters further here. It is to be noted that the subtleties and complexities introduced in the various improved theories are basically concerned with electrostatics but that the general statistical mechanical theories for systems of interacting particles are not explicitly used.

Kirkwood[*] has extended the Onsager theory. In this treatment, the correlation in the orientation of neighboring molecules is explicitly introduced, and the possibility that this correlation may be due to other interactions besides the dipolar interaction is admitted. Further progress would seem to depend on a general advance in our knowledge of the liquid state.

The method of diagonal sums mentioned in Sec. 19-14 for treating a system of interacting magnetic dipoles can also be applied to a set of interacting electric dipoles.[†] The computations are at present prohibitively difficult.

18-5. The Thermodynamics of the Electric Field. Consider an ideal parallel-plate capacitor. If the area of the plates is A and their separation is L, the capacitance *in vacuo* is $c_0 = A/4\pi L$ statfarads. If the potential applied to the capacitor is ϕ, the charge is $q = c_0\phi$. The work done to add an increment of charge dq is

$$dw = \phi\, dq = c_0\phi\, d\phi$$

The electrical energy stored in the capacitor is the work to increase the potential from zero to ϕ, or

$$\text{Energy stored} = \int_0^\phi c_0\phi\, d\phi = \tfrac{1}{2}c_0\phi^2 \quad (18\text{-}44)$$

---

[*] J. G. Kirkwood, *J. Chem. Phys.*, **7**: 911 (1939); *Trans. Faraday Soc.*, **42A**: 7 (1946).
[†] J. H. Van Vleck, *J. Chem. Phys.*, **5**: 556 (1937).

If we write $c_0 = A/4\pi L$ and, for the electric field, $E = \phi/L$, (18-44) becomes

$$\text{Energy stored} = \frac{AL}{8\pi} E^2 = \frac{VE^2}{8\pi} \qquad (18\text{-}45)$$

where V is the volume between the plates of the capacitor. Instead of regarding the stored energy as being due to the interaction between the charges on the plates of the capacitor, we can regard the energy as residing in the electric field. In view of (18-45), we then say that the energy density in an electric field *in vacuo* is $E^2/8\pi$. The result that the energy density in an electric field *in vacuo* is $E^2/8\pi$ can be derived for an arbitrary charge configuration and is not restricted to the field between parallel plates.

The capacitance is a constant *in vacuo*, independent of temperature, and there is no need to introduce thermodynamic considerations. However, if the capacitor is immersed in a medium with a dielectric constant which is a function of temperature and density, there may be heat and density changes during the charging process and a thermodynamic analysis is necessary.

There are some subtle and confusing points in the thermodynamics of a material medium in an electric field. In the interests of clarity and certainty we shall treat a very specific and simple system.

Consider the parallel-plate capacitor depicted in Fig. 18-5. The thermodynamic system consists of the plates of the ideal capacitor and the fluid in the volume $V = AL$ between the plates; i.e., the system is contained in the region defined by the solid lines and the dotted lines in the figure.

The system is an open system, and it is surrounded by additional fluid which is not in the system. There can be a transport of matter between the system (i.e., the region between the plates) and the surroundings (the outside fluid). We assume that the capacitor is ideal and that there are no electric fields in the outside fluid. Therefore the chemical potential μ of the outside fluid is determined by T and P (for pressure) only. The chemical potential inside must equal the chemical

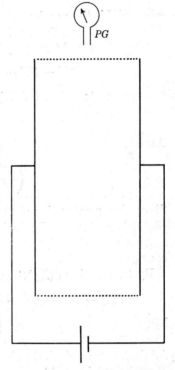

Fig. 18-5. A parallel-plate capacitor charged by a battery to a potential ϕ. The thermodynamic system consists of the plates of the capacitor and the fluid between the plates. This fluid is in equilibrium with fluid outside the capacitor; the pressure is measured in the outside fluid by the pressure gauge PG.

potential outside. We specify the pressure outside and evade difficult questions concerning electrical stresses and the meaning of the pressure inside.

The electrical work done in adding dq of charge to the capacitor is $\phi \, dq$.* The fundamental thermodynamic equation (with U, the thermodynamic energy, since E is now the electric field) is

$$dU = T \, dS + \phi \, dq + \mu \, dN \qquad (18\text{-}46)$$

where dN is the number of molecules of fluid which flow into the system. The system has a fixed volume, and there is no $P \, dV$ term.

With the usual transposition of differentials, we obtain

$$d(U - ST - q\phi) = -S \, dT - q \, d\phi + \mu \, dN \qquad (18\text{-}47)$$

But $q = c_0 \epsilon \phi$; so Eq. (18-47) becomes

$$d(U - ST - c_0 \epsilon \phi^2) = -S \, dT - c_0 \epsilon \phi \, d\phi + \mu \, dN \qquad (18\text{-}48)$$

A useful cross-differentiation relation is

$$\left(\frac{\partial S}{\partial \phi}\right)_{T,N} = c_0 \phi \left(\frac{\partial \epsilon}{\partial T}\right)_{\phi,N} \qquad (18\text{-}49a)$$

Since for most polar fluids (and certainly for an ideal polar gas) $(\partial \epsilon / \partial T)_{\phi,N} < 0$, it follows that $(\partial S / \partial \phi)_{T,N} < 0$. The entropy decreases and heat is released to the surroundings as the capacitor is charged at constant T and N.

By transposing $d(N\mu)$, Eq. (18-48) becomes

$$d(U - ST - c_0 \epsilon \phi^2 - N\mu) = -S \, dT - c_0 \epsilon \phi \, d\phi - N \, d\mu \qquad (18\text{-}50)$$

Thus
$$c_0 \phi \left(\frac{\partial \epsilon}{\partial \mu}\right)_{T,\phi} = \left(\frac{\partial N}{\partial \phi}\right)_{T,\mu} \qquad (18\text{-}51a)$$

As already remarked, the chemical potential μ is determined by the T and P of the outside fluid. In fact

$$\left(\frac{\partial \epsilon}{\partial \mu}\right)_{T,\phi} = \left(\frac{\partial \epsilon}{\partial P}\right)_{T,\phi} \left(\frac{\partial P}{\partial \mu}\right)_{T,N} = \frac{1}{\tilde{V}} \left(\frac{\partial \epsilon}{\partial P}\right)_{T,\phi}$$

where $\tilde{V}$ is the partial molecular volume of the outside fluid (that is, at

---

* It should be noted that the work term $\phi \, dq$ in Eq. (18-46) includes the energy of charging the capacitor *in vacuo*, i.e., the energy of interaction between the charges on the two plates of the capacitor, as well as the energy of interaction of these charges (or the field due to them) with the polarizable dielectric. Alternative definitions in which we exclude the energy of the capacitor *in vacuo* are possible. This point is mentioned because in our discussion of magnetism we do find it convenient to adopt the alternative definition of work and include only the energy of interaction of the magnetic material with the field. Either definition, consistently and correctly used, leads to the same final results.

zero field);

$$\tilde{V} = \left(\frac{\partial V}{\partial N}\right)_{T,P}$$

We thus have from (18-51a)

$$\frac{c_0 \phi}{\tilde{V}} \left(\frac{\partial \epsilon}{\partial P}\right)_{T,\phi} = \left(\frac{\partial N}{\partial \phi}\right)_{T,P} \qquad (18\text{-}51b)$$

Since $(\partial \epsilon/\partial P)_{T,\phi}$ is almost always positive, $(\partial N/\partial \phi)_{T,P} > 0$. The density of the fluid increases when it is placed in an electric field. This is electrostriction.

If we set $c_0 = A/4\pi L$ and $\phi = EL$ and eliminate c_0 and ϕ, Eq. (18-49a) becomes

$$\left(\frac{\partial S}{\partial E}\right)_{T,N} = \frac{VE}{4\pi} \left(\frac{\partial \epsilon}{\partial T}\right)_{E,N} \qquad (18\text{-}49b)$$

and Eq. (18-51b) becomes

$$\frac{E}{4\pi \tilde{V}} \left(\frac{\partial \epsilon}{\partial P}\right)_{T,E} = \left[\frac{\partial (N/V)}{\partial E}\right]_{T,P} \qquad (18\text{-}51c)$$

These are in fact generally applicable expressions for the entropy change on applying an electric field at constant T and N and the density change on applying an electric field at constant T and P, independent of the special configuration of the parallel-plate capacitor with which we started.

It is possible to define a suitable system partition function for the system in an electric field and obtain general statistical mechanical relations for the entropy, energy, and mean polarization as derivatives of the partition function. We shall not do this here, but the analogous magnetic problem is treated in the next chapter.

Problem 18-4. An ideal parallel-plate capacitor with plates 1 cm² in area and separation $L = 1$ cm contains a gas at a fixed concentration. The gas molecules have a polarizability α and a dipole moment μ_D.

(a) Derive an expression for the entropy change on isothermal charging to a potential ϕ.

(b) Give a qualitative discussion of the reason for the sign of the entropy change.

(c) Calculate the number of calories released to the surroundings if the apparatus contains C_2H_5Br gas at an initial pressure (zero field) of 1 atm at 320°K (cf. Prob. 18-3) and is then charged to a potential of 10^6 practical volts. Express the result as calories per mole of gas.

(d) Derive an expression for the pressure change of the gas on isothermal charging at constant concentration. Remember that the pressure of the gas is defined as the pressure of the outside gas such that there is no mass flow between inside and outside gas. Assume the ideal-gas law. Calculate the change in pressure for C_2H_5Br as in part (c) if the final potential is 10^6 volts.

Problem 18-5. In the same way, derive an expression for the density change in a gas if the capacitor is part of an open system at a constant pressure P. Calculate the fractional density change for C_2H_5Br at $320°K$ and $P = 1$ atm for $\phi = 10^6$ volts.

18-6. Electric Birefringence.

If a molecule has a permanent dipole moment and/or an anisotropic polarizability, it will be to some extent oriented in an electric field. The sample should then have different physical properties parallel and perpendicular to the electric field. As we have seen, typical values of the parameter $\mu E/kT$ are very small so that the degree of orientation is small and the anisotropy in whatever physical property is measured would be small. Very small differences in refractive index in different directions are readily measured. In the present context, this is electric birefringence or the Kerr electrooptical effect. The statistical mechanics of this problem is straightforward, after the necessary relations for the interaction of an anisotropic molecule with an electric field have been set forth.

Let $\mathbf{F}$ be the effective, local, electric field which acts to orient the molecules. We take cartesian axes 1, 2, 3 fixed in a molecule. For a given orientation of the molecule, an electric field F along the z axis in space can be resolved into components F_1, F_2, F_3 along the molecular axes. The induced moments m_1, m_2, and m_3 along the three molecular axes are

$$m_1 = \alpha_{11}F_1 + \alpha_{12}F_2 + \alpha_{13}F_3$$
$$m_2 = \alpha_{21}F_1 + \alpha_{22}F_2 + \alpha_{23}F_3 \qquad (18\text{-}52)$$
$$m_3 = \alpha_{31}F_1 + \alpha_{32}F_2 + \alpha_{33}F_3$$

We note that, according to this general equation, an electric field F_1 along axis 1 induces moments in the directions 2 and 3 as well as 1. The matrix of the α's is called the polarizability matrix, or tensor. It is possible to choose a set of axes in the molecule (the principal axes) for which the polarizability matrix is diagonal; that is, $\alpha_{ij} = 0$ $(i \neq j)$. We write α_1 for α_{11} in this coordinate system, etc., and have

$$m_1 = \alpha_1 F_1 \qquad m_2 = \alpha_2 F_2 \qquad m_3 = \alpha_3 F_3 \qquad (18\text{-}53)$$

Let μ_1, μ_2, and μ_3 be the components of the permanent dipole moment along the three axes. The energy of interaction of the molecule with the electric field is then

$$u = -\mu_1 F_1 - \mu_2 F_2 - \mu_3 F_3 - \tfrac{1}{2}(\alpha_1 F_1^2 + \alpha_2 F_2^2 + \alpha_3 F_3^2) \qquad (18\text{-}54)$$

Let $\cos z1$, $\cos z2$, $\cos x2$, $\cos y3$, etc., be the cosines of the angles between the respective molecular axes and the cartesian axes x, y, z fixed in space. Then, for an electric field along the z axis,

$$F_1 = F \cos z1 \qquad F_2 = F \cos z2 \qquad F_3 = F \cos z3 \qquad (18\text{-}55)$$

For a given orientation of the molecule we thus obtain the projection of the electric field along the three molecular axes. The moments along the three molecular axes are $\mu_1 + \alpha_1 F_1$, etc. For this orientation of the molecule, the moment along an axis in space is obtained by adding up the three projections of the molecular moments along this axis. Thus

$$\mu_x = (\mu_1 + \alpha_1 F \cos z1) \cos x1 + (\mu_2 + \alpha_2 F \cos z2) \cos x2$$
$$+ (\mu_3 + \alpha_3 F \cos z3) \cos x3$$
$$\mu_z = (\mu_1 + \alpha_1 F \cos z1) \cos z1 + (\mu_2 + \alpha_2 F \cos z2) \cos z2$$
$$+ (\mu_3 + \alpha_3 F \cos z3) \cos z3 \quad (18\text{-}56)$$

For the birefringence experiment, there is a static, or low-frequency, field F which we take along the z axis. There is then a high-frequency electric field (a light wave) which we denote by the vector **L**. Its components along the three directions in space are L_x, L_y, L_z. Suppose that we have light with its electric vector polarized along the z axis ($L_y = L_x = 0$).

The components of L_z along the three molecular axes are $L_z \cos z1$, $L_z \cos z2$, $L_z \cos z3$. In general, the polarizability of the molecule at optical frequencies may be different from its polarizability for low frequencies. We call the optical polarizabilities β, but we make the simplifying assumption that the directions for the principal axes of polarizability for the α matrix and the β matrix are the same. The rotations of the molecule are slow compared with optical frequencies, and the permanent dipoles do not contribute to the optical polarization. The induced moments due to the electric field of the light wave along the three molecular axes are then

$$m_1 = \beta_1 L \cos z1 \quad m_2 = \beta_2 L \cos z2 \quad m_3 = \beta_3 L \cos z3 \quad (18\text{-}57)$$

The projection of these moments along the z axis is then

$$m_z = m_1 \cos z1 + m_2 \cos z2 + m_3 \cos z3$$
$$= L(\beta_1 \cos^2 z1 + \beta_2 \cos^2 z2 + \beta_3 \cos^2 z3) \quad (18\text{-}58)$$

Because of the static electric field, different orientations have different probabilities. By taking the appropriate Boltzmann weighted orientation over all orientations, we shall calculate an average value of m_z, $\bar{m}_z$. If we accept the Lorentz local-field correction, the refractive index for light polarized in the z direction is

$$\frac{n_z^2 - 1}{n_z^2 + 2} = \frac{4\pi N}{3} \frac{\bar{m}_z}{L_z} \quad (18\text{-}59)$$

It may be noted that for an arbitrary orientation of the molecule there are induced moments m_x and m_y along the x and the y axes for a light vector along the z axis. This would indeed be true for an anisotropic crystal in an arbitrary orientation with respect to the light wave. However, for a molecule in a liquid oriented by an electric field, on averaging over all orientations, it will be found that $\bar{m}_x = \bar{m}_y = 0$.

Now, with the static field in the z direction, we can consider the light vector along the x axis. We can calculate the induced moment $\bar{m}_x$, and then refractive index n_x, for light vibrating along the x axis. The birefringence is then $n_z - n_x$.

The necessary calculations are somewhat long and tedious. The first thing to do is to express the various direction cosines ($\cos z1$, $\cos y3$, etc.) in terms of Eulerian angles. Then, after expanding the exponential $e^{-u/kT}$, the necessary integrations can be performed.

We shall illustrate this program for a somewhat simpler problem, that is, for an axially symmetrical molecule, with axis 3 the symmetry axis and

$$\mu_1 = \mu_2 = 0 \qquad \mu_3 \neq 0 \qquad \alpha_1 = \alpha_2 \neq \alpha_3 \qquad \beta_1 = \beta_2 \neq \beta_3$$

We use a set of polar coordinates θ, ϕ for the orientation of the symmetry axis with respect to the external axes. The component of the static field along the symmetry axis is $F_3 = F \cos \theta$. The component of the static field perpendicular to this axis is $F \sin \theta$. The problem is relatively simple because the polarizability of the molecule is α_1 for all directions perpendicular to the symmetry axis, and we need not resolve $F \sin \theta$ into components F_1 and F_2 along particular molecular axes. The energy then is

$$u(\theta) = -\mu_3 F \cos \theta - \tfrac{1}{2}\alpha_3 F^2 \cos^2 \theta - \tfrac{1}{2}\alpha_1 F^2 \sin^2 \theta \qquad (18\text{-}60)$$

If the electric vector of the light wave is along the z axis, the induced moments along and perpendicular to the symmetry axis are $\beta_3 L_z \cos \theta$ and $\beta_1 L_z \sin \theta$ and the induced moment along the z axis is

$$\frac{m_z}{L} = \beta_3 \cos^2 \theta + \beta_1 \sin^2 \theta \qquad (18\text{-}61)$$

The average value of the moment is therefore

$$\frac{\bar{m}_z}{L} = \beta_3 \overline{\cos^2 \theta} + \beta_1 \overline{\sin^2 \theta} \qquad (18\text{-}62)$$

$$\overline{\cos^2 \theta} = \frac{\iint \cos^2 \theta\, e^{-u(\theta)/kT} \sin \theta\, d\theta\, d\phi}{\iint e^{-u(\theta)/kT} \sin \theta\, d\theta\, d\phi} \qquad (18\text{-}63a)$$

and

$$\overline{\sin^2 \theta} = \frac{\iint \sin^2 \theta\, e^{-u(\theta)/kT} \sin \theta\, d\theta\, d\phi}{\iint e^{-u(\theta)/kT} \sin \theta\, d\theta\, d\phi} \qquad (18\text{-}63b)$$

The problem is now reduced to the evaluation of the integrals in (18-63). We shall return to this calculation shortly.

To calculate the refractive index for light polarized along the x axis, we could calculate the induced moment m_x (it is to be emphasized that the static field is still along the z axis, but the electric vector of the light is along the x axis). This calculation is more complicated than the calculation of m_z, and there is a functional dependence on ϕ as well as θ. Also, the necessary averaging is more tedious. The calculation is unnecessary, however. There is a general theorem that the sum of the optical polarizations in all three directions is a constant, independent of the existence of a static field. That is, if the polarized light vectors L_z, L_x, and L_y produce average polarizations $\bar{m}_z$, $\bar{m}_x$, and $\bar{m}_y$, respectively (for a single fixed direction of the static field),

$$\frac{\bar{m}_z}{L_z} + \frac{\bar{m}_y}{L_y} + \frac{\bar{m}_x}{L_x} = \beta_1 + \beta_2 + \beta_3 \tag{18-64}$$

(The theorem is true for an arbitrary anisotropic substance with $\beta_1 \neq \beta_2 \neq \beta_3$.) Since the only axis of preferred orientation in the problem is the z axis, the average polarizations in the x and y directions must be equal,

$$\frac{\bar{m}_y}{L_y} = \frac{\bar{m}_x}{L_x} \tag{18-65}$$

From Eqs. (18-65) and (18-64) it follows that

$$\frac{\bar{m}_x}{L_x} = \frac{\beta_1 + \beta_2 + \beta_3 - (\bar{m}_z/L_z)}{2} \tag{18-66}$$

The proof of the theorem (18-64) is as follows: Let L_ξ be a field along one of the axes in space ($\xi = x$, y, or z), so that $L_\xi \cos \xi 1$, $L_\xi \cos \xi 2$, and $L_\xi \cos \xi 3$ are the components of the field along the three molecular axes. The induced moment along the ξ axis is

$$\frac{m_\xi}{L_\xi} = \beta_1 \cos^2 \xi 1 + \beta_2 \cos^2 \xi 2 + \beta_3 \cos^2 \xi 3$$

The sum over x, y, z then gives

$$\sum_{\xi=x,y,z} \frac{m_\xi}{L_\xi} = \beta_1 \sum_{x,y,z} \cos^2 \xi 1 + \beta_2 \sum_{x,y,z} \cos^2 \xi 2 + \beta_3 \sum_{x,y,z} \cos^2 \xi 3 \tag{18-67}$$

But it is a general result that $\cos^2 x1 + \cos^2 y1 + \cos^2 z1 = 1$, so that Eq. (18-67) reduces to (18-64).

To evaluate the integrals in (18-63), with $u(\theta)$ given by (18-60), the term $e^{-u(\theta)/kT}$ is expanded in powers of F/kT. By consistently retaining terms in F^2, one obtains the desired final result. We shall not carry out the calculation, which is left as an exercise. The rash but industrious student* can separately calculate $\overline{\cos^2 \theta}$ and $\overline{\sin^2 \theta}$ in Eq. (18-63); however, it is more efficient to calculate one or the other and observe that $\overline{\cos^2 \theta} + \overline{\sin^2 \theta} = 1$.

* Like the author!

The final answer is

$$\frac{\bar{m}_z}{L_z} = \frac{\beta_3 + 2\beta_1}{3} + \frac{2F^2}{45kT}(\alpha_3' - \alpha_1)(\beta_3 - \beta_1) \qquad (18\text{-}68)$$

where
$$\alpha_3' = \alpha_3 + \frac{\mu_3^2}{kT}$$

and then
$$\frac{\bar{m}_x}{L_x} = \frac{\beta_3 + 2\beta_1}{3} - \frac{F^2}{45kT}(\alpha_3' - \alpha_1)(\beta_3 - \beta_1) \qquad (18\text{-}69)$$

In the absence of a field, the polarizabilities are equal and are the space average $(\beta_3 + 2\beta_1)/3$ [i.e., $(\beta_1 + \beta_2 + \beta_3)/3$].

For the orienting field F, one should use the estimated local field, either the Lorentz field $(\epsilon + 2)E/3$ (where E is the macroscopic electric field vector) or the field of Eq. (18-40) (approximately $[3\epsilon/(2\epsilon + 1)]E$, which is the same as the local orienting field assumed in the Onsager theory), as seems appropriate.

In the relations

$$\frac{n_z^2 - 1}{n_z^2 + 2} = \frac{4\pi N}{3}\frac{\bar{m}_z}{L_z} \qquad \frac{n_x^2 - 1}{n_x^2 + 2} = \frac{4\pi N}{3}\frac{\bar{m}_x}{L_x} \qquad (18\text{-}70)$$

the difference between $\bar{m}_z/L_z$ and $\bar{m}_x/L_x$ is small. Let n be the average refractive index; by subtraction, the equations become

$$\frac{2n}{n^2 + 2}(n_z - n_x) = \frac{4\pi N}{3}\left(\frac{\bar{m}_z}{L_z} - \frac{\bar{m}_x}{L_x}\right)$$
$$= \frac{4\pi N}{3}\frac{F^2}{15kT}(\alpha_3' - \alpha_1)(\beta_3 - \beta_1) \qquad (18\text{-}71)$$

This is the final expression for the electric birefringence. By suitable experiments with polarized light, it is possible to measure very small amounts of birefringence of the order of $n_z - n_x \approx 10^{-7}$.*

It is to be emphasized that the quantitative formulae (11-68), (11-69), and (11-71) are based on the assumption that $|u(\theta)/kT| \ll 1$, so that the degree of orientation is small. For some macromolecules with very large dipole moments, it is possible to obtain almost complete orientation in electric fields (saturation). The quantitative relations discussed above are then not appropriate. It may also be mentioned that, for polyelectrolyte macromolecules in ionizing solvents, the mechanism of

---

* For a further discussion, see P. Debye and H. Sack, in E. Marx (ed.), "Handbuch der Radiologie," vol. VI, pp. 768, 780, 1925; vol. VI/2, pp. 989 ff., 1934, Akademische Verlagsgesellschaft m.b.H., Leipzig. See also C. F. J. Böttcher, "Theory of Electric Polarization," pp. 286–291, and including footnote 65, Elesevier, Houston, Tex., 1952; these references give the formula for the general case. There is a modern review of the chemical applications of the Kerr effect by C. G. Le Fevre and R. J. W. Le Fevre, *Revs. Pure and Appl. Chem. (Australia)*, **5**(4) (1955).

orientation is somewhat different from that discussed here. For this case, the principal mechanism of polarization is the displacement of the charge cloud due to the counter ions in solution with respect to the fixed charges in the polyelectrolyte macromolecule.* For such systems, it is often possible to obtain almost complete orientation of the macromolecules.

PROBLEMS

18-6. Consider a diatomic molecule which is constrained to rotate in the xy plane and which has a permanent dipole moment μ. Calculate the average polarization in the direction of a field E along the x axis.

18-7. Calculate the value of the parameter $y = \mu E/kT$ and the value of $\bar{\mu}/\mu$ for a water molecule ($\mu = 1.85 \times 10^{-18}$ esu) in the field of an ion of charge $|e^-| = 4.8 \times 10^{-10}$ esu, at a distance of 2.5 Å at $T = 300°K$. The field is $|e|/\epsilon r^2$. Take for ϵ the two values 5 and 80. Repeat for a distance of 5 Å.

The purpose of the problem is to give some semiquantitative insight into the orientation of water molecules around an ion. The distance from ion to water molecule in the first coordination sphere is about 2.5 Å; and 5 Å is reasonable for the second shell of waters. Because of dielectric saturation and because there is no intervening solvent, the macroscopic dielectric constant of 80 is unreasonable for a molecule in the first coordination shell, and 5 is perhaps preferable. The macroscopic value of 80 is more reasonable for the second shell.

18-8. Apply the simple equation (18-42) and the Onsager equation (18-43) to calculate the dipole moment of C_2H_5Br, using the data on the dielectric constant and other properties of the liquid given in Prob. 18-3.

18-9. Complete the calculation for Eq. (18-68) from (18-60) and (18-63).

18-10. Dichroism is the variation in light absorption with the direction of the electric vector of a polarized light wave in an anisotropic sample. Consider an axially symmetric molecule as in Sec. 18-6, with $\mu_1 = \mu_2 = 0$, $\mu_3 \neq 0$, $\alpha_1 = \alpha_2 \neq \alpha_3$. Let the absorption coefficient per molecule for light vibrating along the symmetry axis be γ_3, and let the absorption coefficient perpendicular to the symmetry axis be zero. Let the static orienting field be along the z axis. If the electric vector of the light is polarized along the ξ axis ($\xi = x, y,$ or z), the electric vector is L_ξ and the intensity is proportional to L_ξ^2. The component along the symmetry axis of the molecule is $L_\xi \cos \xi 3$, and the probability of light absorption by this molecule is $\gamma_3 L_\xi^2 \cos^2 \xi 3$. The average absorption coefficients for light vibrating along the z and x directions are denoted by $\bar{\gamma}_z$ and $\bar{\gamma}_x$. Calculate the dichroic ratio

$$\frac{\bar{\gamma}_z - \bar{\gamma}_x}{(\bar{\gamma}_z + \bar{\gamma}_x)/2}$$

assuming as in Sec. (18-6) that $|u(\theta)/kT| \ll 1$.

18-11. The dielectric constant of a pure gas at one (1.00) atmosphere pressure is observed to be $\epsilon - 1 = 4.83 \times 10^{-3}$ at 300°K and $\epsilon - 1 = 3.06 \times 10^{-3}$ at 400°K. Calculate the dipole moment, the molar distortion polarization R_d, and the polarizability α for this substance.

---

* C. T. O'Konski and A. J. Haltner, *J. ACS*, **79**: 5634 (1957); C. T. O'Konski and B. H. Zimm, *Science*, **111**: 113 (1950); H. Benoit, *Ann. phys.*, **6**: 561 (1951).

19

Magnetic Phenomena*

19-1. Introduction, Magnetic Dipoles. Magnetic dipoles are created by closed current loops; in atomic physics, they are due to the orbital motion of an electron or to the spinning motion of an electron or nucleus.

A current I in a single-plane circular current loop of area A has a magnetic-dipole moment of $\mu = IA/c$ in cgs-Gaussian units with I in esu or $\mu = IA$ in mks units. The magnetic dipole is a vector perpendicular to the plane of the loop.

Suppose that a single electron moves in a circular orbit of radius r with velocity v; the frequency of rotation is $v/2\pi r$, and the average current (charge per second past a point) is $ev/2\pi r$. Note also that the angular momentum is $M = mvr$. Therefore, the magnetic moment is given by

$$\mu = \begin{cases} \dfrac{IA}{c} = \dfrac{ev\pi r^2}{2\pi rc} = \dfrac{evr}{2c} = \dfrac{eM}{2mc} & \text{Gaussian units} \\ \dfrac{evr}{2} = \dfrac{eM}{2m} & \text{mks units} \end{cases} \quad (19\text{-}1)$$

If the angular momentum is taken as $\hbar$ ($M \neq \hbar$), the magnetic moment is 1 Bohr magneton β,

$$\beta = \begin{cases} \dfrac{|e|\hbar}{2mc} = 0.927 \times 10^{-20} \text{ erg gauss}^{-1} & \text{Gaussian} \\ \dfrac{|e|\hbar}{2m} = 0.927 \times 10^{-23} \text{ amp m}^2 & \text{mks} \end{cases} \quad (19\text{-}2)$$

By our definition, β is a positive quantity. (In this chapter, we shall use β exclusively for the Bohr magneton, and never as a symbol for kT.)

For our present purpose, any magnetic dipole can be viewed as a little bar magnet—with the vector dipole pointing from the south to the north pole. When the dipole is placed in a *uniform* magnetic field, there is no

---

* A basic early reference for the entire field is J. H. Van Vleck, "The Theory of Electric and Magnetic Susceptibilities," Oxford, New York, 1932. Fowler and Guggenheim, chap. 14 [1], is also recommended. References for magnetic-resonance phenomena and other modern developments are given later.

Sec. 19-1] MAGNETIC PHENOMENA 429

net translational force on it but there is an orienting torque which tends to align it along the direction of the magnetic field. This is illustrated in Fig. 19-1.

It can be shown that the orientation energy ϵ of the dipole in a magnetic field is $\epsilon = -\boldsymbol{\mu} \cdot \mathbf{B} = -\mu B \cos \theta$, where θ is the angle between the vector magnetic field $\mathbf{B}$ and the vector magnetic dipole. The corresponding orienting torque is $\mathbf{T} = \boldsymbol{\mu} \times \mathbf{B}$, or $|\mathbf{T}| = \mu B \sin \theta$. For example, if in Fig. 19-1 the magnetic dipole $\boldsymbol{\mu}$ and the field $\mathbf{B}$ are both in the plane of the paper, the torque $\mathbf{T}$ is perpendicular to the plane of the paper. From the definition of torque in mechanics, the force is in the direction tending to orient the bar magnet along the field direction.

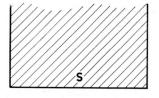

In quantum mechanics, there are only certain allowed orientations of the angular-momentum vector with respect to the field, with components $m\hbar$, m being an integer or a half integer. (In a very crude sense we can say that only certain values of θ as defined in the previous paragraph are allowed.) There is a corresponding set of allowed values for the energy of the magnetic dipole in the magnetic field. If there is no magnetic field, there is no difference in energy between the different orientations and each orientation is equally probable. In the presence of a field, the states of lower energy become more populated. Our principal business in the present chapter is to work out the statistical mechanical consequences of this situation.

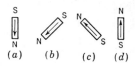

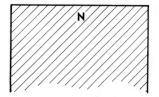

Fig. 19-1. A small bar magnet, or a magnetic dipole, in the uniform field of a large magnet; a, b, c, d are configurations of progressively increasing stability or lower energy. The orienting torque tends to pull the magnets into the configuration d. As indicated by the arrows, the magnetic dipole points from the south to the north pole.

A system of paired electrons in a closed shell or subshell has no net angular momentum or magnetic moment. Nevertheless, there is an energy of interaction of such an atom with a magnetic field. The simplest and most naïve picture is of two electrons, with their spins paired, circulating in opposite directions in circular orbits—with no net orbital or spin angular momentum. It can be shown that, if a magnetic field is applied to such a system, the orbital motions are affected in such a way (the Larmor precession) that there are a net induced circulating current and magnetic moment in the direction opposite to the applied field. Thus the energy of the system is increased in the field. This is diamagnetism. It is not very important for us, but a few quantitative relations are given in the next section.

19-2. Electronic Paramagnetism.

We first review the basic facts about the energy levels of an isolated atom in a magnetic field. There are a large number of cases which conform to the simple relations described below, although more complicated cases do occur (for example, the "Paschen-Back effect" for atoms in strong magnetic fields, as discussed in textbooks on atomic spectra).

Consider an atom in a particular internal quantum state, which we denote by k. There is a magnetic field B in the z direction. The energy of the atom in the field can be expressed as

$$\epsilon_k(B) = \epsilon_k(0) - \mu_k B + \frac{\alpha_k B^2}{2} \tag{19-3}$$

As we shall see, the total magnetic moment in the state k can be defined by

$$\mu_{tot,k} = -\frac{d\epsilon}{dB}$$

so that

$$\mu_{tot,k} = \mu_k - \alpha_k B \tag{19-4}$$

The μ_k term is the paramagnetic moment, whereas the second term, $\alpha_k B$, is the diamagnetic term. For atoms in spherically symmetrical states, the diamagnetic term is given by

$$\alpha_k = \frac{-e^2}{6mc^2} \left\langle \sum_i r_i^2 \right\rangle_k \tag{19-5}$$

In this formula, r_i is the radial coordinate of the ith electron, and $\langle \Sigma r_i^2 \rangle_k$ represents the quantum-mechanical expectation value of the sum in the state k. The diamagnetic contribution must be properly taken into account in direct measurements of magnetic susceptibility, but it can be omitted from an introductory discussion of the statistical mechanics of magnetic materials. In the first place, the quantity α_k is usually a constant which is the same for all the various low-lying quantum states k which are significantly populated. Therefore, diamagnetism contributes a constant term $e^{-\alpha B^2/2kT}$ to the partition function

$$q(B,T) = \sum_k e^{-\epsilon_k(B,T)/kT}$$

and the statistical mechanics are trivial. Furthermore, for the interesting cases, which are where μ_k is not zero, the diagmagnetic moment αB is usually much smaller than the paramagnetic term μ_k. We note in passing that the induced moment due to the diamagnetic term is in the opposite direction to the magnetic field.

Now, suppose that the atom or molecule is in a state with a total electronic angular-momentum quantum number of J (and with J integral or half integral, for an even or odd number of electrons in the atom or molecule). The mean-square angular momentum is $\langle M^2 \rangle = J(J+1)\hbar^2$. The component of angular momentum along the z axis is quantized and has the values $\langle M_z \rangle = m\hbar$, with m taking on the $2J+1$ values, $-J$, $-J+1$, $-J+2$, ..., $J-1$, J. The states of different m are the magnetic sublevels of the "state" J; for many practical cases, the set of general quantum states k mentioned in the preceding paragraphs are just these magnetic sublevels.

It can be shown that the expectation value of the square of the magnetic moment for each of the states m belonging to the state J is

$$\langle \mu^2 \rangle = g^2 \beta^2 J(J+1) \tag{19-6}$$

where g is a number which is independent of m and which we shall discuss shortly. The component of magnetic moment along the z axis in the state with magnetic quantum number m is given by

$$\langle \mu_z \rangle_m = \mu_m = -g\beta m \tag{19-7}$$

We shall use $\langle \mu_z \rangle_m$ so commonly that we abbreviate it as μ_m. In these expressions, β is the Bohr magneton [cf. Eq. (19-2)], and g is a number, the Lande g factor. In Russell-Saunders coupling, with orbital and spin angular-momentum quantum numbers L and S, it can be shown that

$$g = 1 + \frac{J(J+1) + S(S+1) - L(L+1)}{2J(J+1)} \tag{19-8}$$

For an atom with only spin angular momentum, $L = 0$, $J = S$, and the g factor is 2.00. For an atom with pure orbital angular momentum, $S = 0$, $J = L$, and the g factor is 1.00.

The relations given above are quite accurate for many gaseous atoms. We shall discuss the situation in condensed phases in more detail later, but it should be mentioned now that similar relations are often a very good approximation for ions or molecules (such as free radicals) in solution. For many such cases, it is found that we must in effect take $L = 0$, but $S \neq 0$. There are cases where $g = 2$ as predicted by (19-8); there are also cases where $g \neq 2$. Nevertheless, there are a set of $2S+1$ magnetic sublevels with $m = -S, -S+1, \ldots, +S$; and the magnetic moment is given by Eq. (19-7).* For ions or free radicals in crystals, the g factor may be different along different crystal axes; in fact, in general it is a tensor quantity.

In summary, then, we may say that there are a large number of cases for which the energy levels in the presence of a magnetic field are given by

$$\epsilon_m = -\mu_m B = g\beta m B \tag{19-9}$$

For gas-phase atoms with Russell-Saunders coupling, the g factor is

---

* See D. J. E. Ingram, "Spectroscopy at Radio and Microwave Frequencies," chap. 6, Butterworth, London, 1955.

given by Eq. (19-8). In other cases, it has a value which we can regard as given by experiment. The allowed values of m range from $-J$ to $+J$, where $2J+1$ is the multiplicity of this set of states. The maximum magnetic moment is

$$\mu_{\max} = g\beta J \qquad (19\text{-}10a)$$

and (19-9) is sometimes written as

$$\mu_m = -\frac{m}{J}\mu_{\max} \qquad \epsilon_m = \frac{m}{J}\mu_{\max}B \qquad (19\text{-}10b)$$

We have chosen to define β as a positive quantity [(19-2)]. Because the magnetic moment of an atom is due to the orbital motion or the spin of a negative charge, the direction of the magnetic-moment vector is opposite to that of the angular-momentum vector. Thus, the level with angular-momentum component $-m\hbar$ has a magnetic moment of $+g\beta m$, and $\epsilon_m = -\mathbf{\mu} \cdot \mathbf{B} = g\beta mB$. The sign of the gyromagnetic ratio is measurable by special experiments. However, all the phenomena considered in this chapter essentially depend on μ^2 and are independent of the sign of the magnetic moment.

19-3. Nuclear Magnetism. The formulae for nuclear magnetism are identical with those for electron paramagnetism. A nucleus with spin quantum number I can have components along the axis of quantization of $m\hbar$, with $m = -I, -I+1, \ldots, I-1, I$. The square of the nuclear magnetic moment is

$$\langle \mu^2 \rangle = g^2\beta_N^2 I(I+1) \qquad (19\text{-}11)$$

and $\qquad \langle \mu_z \rangle = g\beta_N m \quad \text{or} \quad \mu_m = g\beta_N m \qquad (19\text{-}12)$

where β_N is the nuclear magneton,

$$\beta_N = \frac{|e|\hbar}{2M_P c} = 0.5050 \times 10^{-23} \text{ erg gauss}^{-1} \qquad M_P = \text{proton mass} \qquad (19\text{-}13)$$

and g is the nuclear g factor. Because of the complexities of nuclear structure, g is in general not an integer; for example, for the proton, $g = 2.793$, and $I = \frac{1}{2}$; for the Hg^{201} nucleus, $g = -0.607$, and $I = \frac{3}{2}$. It is to be noted that nuclear magnetic moments are of the order of m_e/M_P, or 1/2,000 of the magnetic moments of atoms.

A number of textbooks and tabulations speak of the "magnetic moment of the nucleus" and mean $\mu_{\max}$, defined by $\mu_{\max} = g\beta_N I$, so that, in the state m, $\mu_m = (m/I)\mu_{\max}$.

In a static magnetic field B, the energy levels are given by

$$\epsilon_m = -g\beta_N mB = -\frac{m}{I}\mu_{\max}B \qquad (19\text{-}14)$$

From a naïve classical point of view, a proton is a spinning positive charge, and it would be expected that the magnetic-moment vector would have the same direction as the angular-momentum vector. This is in fact correct, and in (19-13) we have written $\mu_m = g\beta_N m$, with g a positive number. For other nuclei, because of the complexities of nuclear structure, the nuclear magnetic moment has the opposite sign to the nuclear angular momentum. This is indicated in nuclear theory by giving g a negative value.

Thus a positive g factor for a nucleus means that $\mu_m = g\beta_N m$ and $\epsilon_m = -g\beta_N mB$. On the other hand, a positive g factor for an atomic system, with the magnetism due to spinning and circulating electrons, means that $\mu_m = -g\beta m$ and $\epsilon_m = g\beta mB$. It will be obvious, however, that for all the problems treated here this difference in sign convention is unimportant.

19-4. The Statistical Mechanics of Noninteracting Magnets. Consider a set of N noninteracting atoms in the gas phase. The energy for any one atom is the sum of the translational energy and the electronic energy,

$$\epsilon = \epsilon_{tr} + \epsilon_{el}$$

and the partition function is

$$q = q_{tr} q_{el}$$

The Gibbs free energy is given by

$$F = -NkT \ln \frac{q}{N} = -NkT \left(\ln \frac{q_{tr}}{N} \right) - NkT \ln q_{el}$$

For the electronic contribution to the free energy, there is no distinction between F_{el} and A_{el}, and we have

$$F_{el} = A_{el} = -NkT \ln q_{el} \qquad (19\text{-}15)$$

The electronic energy is given by

$$E_{el} = NkT^2 \frac{\partial(\ln q_{el})}{\partial T} \qquad (19\text{-}16)$$

We first consider the case where all the excited states have energies which are quite large compared with kT and only the lowest electronic state is important. If this state has an angular-momentum quantum number of J and a multiplicity of $2J + 1$, the partition function in the absence of a field is

$$q_{el} = 2J + 1 \qquad (19\text{-}17)$$

In the presence of a magnetic field, the energy levels are given by Eq. (19-9),

$$\epsilon_m = g\beta mB \qquad m = -J, \ldots, +J$$

and the electronic-partition function is

$$q(B,T) = \sum_{m=-J}^{m=+J} e^{-g\beta mB/kT} = \sum_{m=-J}^{m=+J} e^{-my} \qquad (19\text{-}18)$$

where we have set $y = g\beta B/kT$. The probability of observing an atom in the state m is

$$p_m = \frac{e^{-\epsilon_m/kT}}{q} = \frac{e^{-g\beta Bm/kT}}{q} = \frac{e^{-my}}{q} \qquad (19\text{-}19)$$

The average magnetic moment per particle is

$$\bar{\mu} = \sum \mu_m p_m = \frac{\sum_m -g\beta m e^{-g\beta Bm/kT}}{q} = \frac{kT}{q}\left(\frac{\partial q}{\partial B}\right)_T \qquad (19\text{-}20a)$$

or, in dimensionless form,

$$\frac{\bar{\mu}}{g\beta} = \frac{\Sigma - me^{-my}}{\Sigma e^{-my}} = \frac{1}{q}\frac{dq}{dy} \qquad (19\text{-}20b)$$

(For the important special case that we consider here, the partition function and hence all the magnetic properties are functions of B/T. According to the first two relations in (19-20a), we could correctly calculate the magnetic moment from the formula $\bar{\mu} = -(g\beta kT^2/B)[\partial(\ln q)/\partial T]_B$. However, $kT[\partial(\ln q)/\partial B]_T$ is preferable because, as we shall see in the next section, it is more generally applicable.)

The partition function in (19-18) represents the contribution to the thermodynamic properties due to the splitting of a $(2J + 1)$-fold-degenerate state by the magnetic field. We have spoken of it as being applicable to a sample of paramagnetic gas atoms. It is obviously equally applicable if the gas atoms contain magnetic nuclei to which the formula $\epsilon_m = -g\beta_N mB$ applies, with $m = -I, \ldots, +I$. Of course, in this case, we must include the nuclear spin partition function in zero field, $q = 2I + 1$, in calculating the total partition function at zero field, contrary to our usual practice for nonmagnetic problems.

The partition function (19-18) and the formulae (19-15), (19-16), (19-19), and (19-20) are also applicable to a set of noninteracting magnetic particles, atoms or nuclei, at localized distinguishable lattice sites in a crystal. This statement is almost obvious, but we shall explicitly justify it later.

Thus the partition function (19-18) is quite generally applicable for noninteracting magnetic particles, in the gaseous, the liquid, or the crystalline state, if there is only one energy level at zero field which is $(2J + 1)$-fold-degenerate and if the energy levels in the presence of a magnetic field are given by $\epsilon_m = +g\beta m$. In general, we shall call it the single-particle magnetic partition function, although the term electronic partition function in a magnetic field is equally appropriate for paramagnetic atoms.

Similarly, the formula (19-20) for the mean magnetic moment is applicable to a collection of noninteracting magnets in general.

If there are several nondegenerate low-lying electronic levels in the absence of a field, each such level will be split by the magnetic field. The total electronic partition function in a magnetic field is then the appropriate sum of $e^{-\epsilon(B)/kT}$ over all the levels. (For an example, see Prob. 19-3.)

The magnetic contributions to the thermodynamic properties are calculated from the formulae

$$A = -NkT \ln q \qquad (19\text{-}15)$$

$$E = NkT^2 \left[\frac{\partial (\ln q)}{\partial T} \right]_B \qquad (19\text{-}16)$$

and
$$S = \frac{E - A}{T} \qquad (19\text{-}21)$$

By calculating $[\partial (\ln q)/\partial T]_B$ from (19-18) (or by using the equivalent statement that $E = N \Sigma p_m \epsilon_m$), we obtain

$$E = \frac{N}{q} \sum g\beta m B e^{-g\beta m B/kT} = -N\bar{\mu}B \qquad (19\text{-}22)$$

To evaluate the partition function, first observe that $y = g\beta B/kT$ is given by

$$y = \frac{6.72 \times 10^{-5} gB}{T} \qquad (19\text{-}23a)$$

for β, the Bohr magneton, and

$$y = \frac{3.66 \times 10^{-8} gB}{T} \qquad (19\text{-}23b)$$

for β, the nuclear magneton. Thus, except at very low temperatures and very large fields, the approximation $y \ll 1$ holds.

We first treat the case $y \ll 1$. Then

$$e^{my} \approx 1 + my + \frac{m^2 y^2}{2} + \cdots$$

Observe that, by symmetry,

$$\sum_{m=-J}^{m=+J} m = 0$$

Furthermore,
$$\sum_{m=-J}^{m=+J} m^2 = \frac{J(J+1)(2J+1)}{3} \qquad (19\text{-}24)$$

This theorem is true for J either an integer or a half integer and is readily proved by induction. Then the partition function (19-18) is given by

$$q = \sum_{m=-J}^{m=+J} e^{-my} \approx \sum_{-J}^{+J}\left(1 - my + \frac{m^2 y^2}{2}\right)$$

$$q = (2J + 1)\left[1 + \frac{J(J + 1)y^2}{6}\right] \qquad (19\text{-}25a)$$

$$q = (2J + 1)\left[1 + \frac{J(J + 1)g^2\beta^2 B^2}{6k^2 T^2}\right] \qquad (19\text{-}25b)$$

Then, using (19-6),

$$q = (2J + 1)\left[1 + \frac{\langle\mu^2\rangle B^2}{6k^2 T^2}\right] \qquad (19\text{-}25c)$$

Therefore
$$\bar{\mu} = \frac{g\beta}{q}\frac{dq}{dy} \approx \frac{g\beta J(J + 1)y}{3}$$

$$= \frac{g^2\beta^2 J(J + 1)B}{3kT} = \frac{\langle\mu^2\rangle B}{3kT} \qquad (19\text{-}26)$$

The molar susceptibility may be defined* as the moment per mole per unit field, or

$$\chi_M = \frac{N_0 \bar{\mu}}{B} = \frac{N_0\langle\mu^2\rangle}{3kT} = \frac{N_0 g^2\beta^2 J(J + 1)}{3kT} \qquad (19\text{-}27a)$$

If we add a constant term for the diamagnetic susceptibility, the resulting equation is

$$\chi_M = \frac{N_0\langle\mu^2\rangle}{3kT} - N_0\alpha \qquad (19\text{-}27b)$$

where α is the diamagnetic susceptibility per atom.

Equation (19-27b) or the statement that the paramagnetic contribution to the susceptibility is proportional to $1/T$ is known as Curie's law.

In the approximation that $y \ll 1$, the magnetic contributions to the thermodynamic quantities are

$$\frac{A}{NkT} = -\ln q = -\ln(2J + 1) - \ln\left(1 + \frac{\langle\mu^2\rangle B^2}{6k^2 T^2}\right)$$

$$= -\ln(2J + 1) - \frac{\langle\mu^2\rangle B^2}{6k^2 T^2} = -\ln(2J + 1) - \frac{g^2\beta^2 B^2 J(J + 1)}{6k^2 T^2}$$

$$= -\ln(2J + 1) - \frac{J(J + 1)y^2}{6} \qquad (19\text{-}28)$$

$$E = -N\bar{\mu}B = \frac{-N\langle\mu^2\rangle B^2}{3kT} \qquad (19\text{-}29)$$

* For a more exact definition, see Sec. 19-6.

$$\frac{S}{Nk} = \ln(2J+1) - \frac{\langle \mu^2 \rangle B^2}{6k^2T^2}$$
$$= \ln(2J+1) - \frac{J(J+1)y^2}{6} \qquad (19\text{-}30)$$

The entropy per atom is $k \ln(2J+1)$, and the free energy is given by $A/NkT = -\ln(2J+1)$ in the absence of a field, corresponding to the equal population of the $2J+1$ magnetic sublevels. When the field is applied, the entropy decreases because the distribution of the magnetic atoms among the various states is no longer random. The dimensionless entropy decrease $\Delta S/Nk$ is $-\langle \mu^2 \rangle B^2/6k^2T^2$; the dimensionless energy parameter E/NkT decreases by $-\langle \mu^2 \rangle B^2/3k^2T^2$; so the free energy $A/NkT = (E/NkT) - (S/Nk)$ decreases by $-\langle \mu^2 \rangle B^2/6k^2T^2$.

It is helpful to use the approximation $y = g\beta B/kT \ll 1$ when it is justified, but it is not necessary, and the magnetic partition function (19-18) can be summed exactly.

$$q = \sum_{m=-J}^{m=+J} e^{-my} = e^{-Jy}(e^{2Jy} + e^{(2J-1)y} + \cdots + e^y + 1)$$

The factor in parentheses is a geometrical series.

$$q = e^{-Jy} \frac{e^{(2J+1)y} - 1}{e^y - 1} = \frac{e^{(2J+1)y/2} - e^{-(2J+1)y/2}}{e^{y/2} - e^{-y/2}}$$
$$= \frac{\sinh[(2J+1)y/2]}{\sinh(y/2)} \qquad (19\text{-}31)$$

Then, by straightforward differentiation,

$$\frac{\bar{\mu}}{g\beta} = \frac{1}{q}\frac{dq}{dy}$$
$$= \frac{2J+1}{2} \frac{e^{(2J+1)y/2} + e^{-(2J+1)y/2}}{e^{(2J+1)y/2} - e^{-(2J+1)y/2}} - \frac{1}{2}\frac{e^{y/2} + e^{-y/2}}{e^{y/2} - e^{-y/2}}$$
$$= \frac{2J+1}{2} \coth \frac{(2J+1)y}{2} - \frac{1}{2}\coth\frac{y}{2} \qquad (19\text{-}32)$$

The other thermodynamic functions can be calculated from q and $\bar{\mu}$. In particular, the energy is given by $E = -N\bar{\mu}B$. A convenient formula for the entropy is useful for magnetic-cooling experiments.

$$\frac{S}{Nk} = \frac{E}{NkT} - \frac{A}{NkT} = \ln q - \frac{y\bar{\mu}}{g\beta} \qquad (19\text{-}33)$$

19-5. Magnetic Saturation. The general formulae (19-31) to (19-33) are needed in experiments with large magnetic fields at low temperatures where the assumption $y \ll 1$ does not hold. For example, for $g = 2$, $y = g\beta B/kT = 1.34 \times 10^{-4} B/T$; at $T = 1°K$ and $H = 50{,}000$ gauss (a field which can be achieved with powerful but conventional magnets), $y = 6.7$. From relations (19-32), (19-31), and (19-33) in the limit as $y \to \infty$,

$$\frac{\bar{\mu}}{g\beta} \approx J \qquad \bar{\mu} \approx g\beta J = \mu_{\max} \tag{19-34a}$$

$$q \approx e^{Jy} = e^{-\mu_{\max} B/kT} \tag{19-34b}$$

$$S = 0 \tag{19-34c}$$

When the magnetic energy $g\beta B$ is large compared with kT, all the particles are in the state $m = -J$, the magnetic moment reaches a saturation

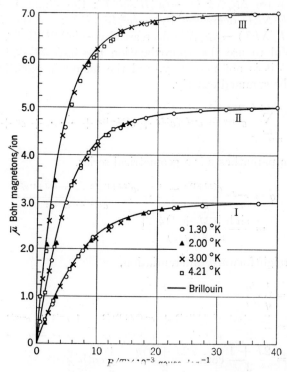

FIG. 19-2. Plot of the average magnetic moment per ion, $\bar{\mu}$ versus B/T. (I) Potassium chrome alum ($J = S = 3/2$). (II) Iron ammonium alum ($J = S = 5/2$). (III) Gadolinium sulfate octahydrate ($J = S = 7/2$). For all cases, $g = 2$, and in all cases the curves are normalized at the highest value of B/T. For $g = 2$, the value $B/T = 10^4$ gauss deg^{-1} corresponds to $y = 1.34$. [From W. Henry, Phys. Rev., **88**: 561 (1952).]

value of $g\beta J$, and the entropy approaches zero. The function that gives $\bar{\mu}/g\beta$ in Eq. (19-32) is known as Brillouin's function; it describes the approach to saturation as a function of B/T. Examples of this behavior are given in Fig. 19-2. In this and similar measurements, the agreement with experiment is very satisfactory.

Problem 19-1. As a collaborative class exercise, plot the Brillouin function

$$\frac{\bar{\mu}}{\mu_{max}} = \frac{2J+1}{2J} \coth \frac{(2J+1)y}{2} - \frac{1}{2J} \coth \frac{y}{2}$$

as a function of y for $J = \frac{1}{2}$ and $J = \frac{7}{2}$. Also plot the Langevin function of Chap. 18 [Eq. (18-23)] for the degree of orientation of a classical dipole

$$\frac{\bar{\mu}}{\mu_{max}} = L(y) = \frac{e^y + e^{-y}}{e^y - e^{-y}} - \frac{1}{y}$$

Under what conditions does the quantum-mechanical result approach the classical result? (Cf. Fowler and Guggenheim, p. 630 [1].)

19-6. Magnetic Susceptibility. Probably the two most common and important measurements of the magnetic properties of matter are (*a*) macroscopic magnetic susceptibility and (*b*) magnetic-resonance absorption. We shall consider the first of these two methods in this section.

A paramagnetic or a ferromagnetic material is attracted into a magnetic field; a diamagnetic material is repelled. This is essentially the basis for the measurement of magnetic susceptibility.

Suppose that a small sample of volume δV is suspended at a point $\mathbf{r}_1$ in a solenoid or between the pole faces of a magnet. Let $\mathbf{B}_0(\mathbf{r})$ be the magnetic-induction vector prior to the insertion of the sample. There is an induced magnetic moment $\mathfrak{M}_V$ per unit volume in the sample, so that its total magnetic moment is $\mathfrak{M} = \mathfrak{M}_V \delta V$. There are magnetic-field vectors $\mathbf{B}(\mathbf{r})$ and $\mathbf{H}(\mathbf{r})$ after inserting the sample. In general, $\mathbf{B}(\mathbf{r}) \neq \mathbf{B}_0(\mathbf{r})$, because of the magnetic field contributed by the sample. At any point inside the sample, $\mathbf{B} = \mathbf{H} + 4\pi\mathfrak{M}_V$ (in cgs-Gaussian units). If the sample is isotropic, $\mathbf{B}$, $\mathbf{H}$, and $\mathfrak{M}_V$ are all parallel.

The standard definition in electromagnetic theory of the volume susceptibility is

$$\chi_V = \frac{\mathfrak{M}_V}{H} \qquad (19\text{-}35a)$$

However for most experiments on paramagnetic or diamagnetic materials (as distinct from ferromagnetic materials) except at very low temperatures, $\mathfrak{M}_V \ll H$, and it is true that $B_0 \approx B \approx H$. Therefore, a satisfactory approximate definition of the susceptibility is

$$\chi_V = \frac{\mathfrak{M}}{B_0} \qquad (19\text{-}35b)$$

Now suppose that the small sample in the volume δV consists of N noninteracting magnetic atoms or nuclei. For any one particle, the energy levels are a function of the field B_0,

$$\epsilon_m = \epsilon_m(B_0) \qquad (19\text{-}36a)$$

In the particular case that we have considered so far,

$$\epsilon_m = g\beta m B_0 \tag{19-36b}$$

but we now consider the more general case where the energy levels are not necessarily a linear function of the field.

The partition function is

$$q(B,T) = \sum_m e^{-\epsilon_m(B_0)/kT} \tag{19-37}$$

and the probability of finding a particle in the state m is

$$p_m = \frac{e^{-\epsilon_m(B_0)/kT}}{q} \tag{19-38}$$

We now suppose that the external field is not uniform in space but varies in the x direction, with a derivative dB_0/dx. The force on a particle which is in the mth quantum state is

$$F_m = -\frac{d\epsilon_m(B_0)}{dx} = -\frac{d\epsilon_m(B_0)}{dB_0}\frac{dB_0}{dx} \tag{19-39}$$

We *define* the magnetic moment in the state m by the equation

$$\mu_m = -\frac{d\epsilon_m}{dB_0} \tag{19-40}$$

For the common and simple case that $\epsilon_m = g\beta m B_0$, this definition gives $\mu_m = -g\beta m$, in agreement with our previous treatment.

If a particle in the mth state moves from x to $x + dx$, the energy level changes according to the equation

$$d\epsilon_m = \frac{d\epsilon_m}{dB_0}\frac{dB_0}{dx}dx = -\mu_m \frac{dB_0}{dx}dx = -\mu_m\, dB_0 \tag{19-41}$$

The force on the sample at thermodynamic equilibrium is

$$\bar{F} = N\sum p_m F_m = N\sum p_m \left(-\frac{d\epsilon_m}{dB_0}\right)\frac{dB_0}{dx}$$
$$= N\left(\sum p_m \mu_m\right)\frac{dB_0}{dx} = N\bar{\mu}\frac{dB_0}{dx} \tag{19-42}$$

where the average moment $\bar{\mu}$ is defined by $\bar{\mu} = \Sigma p_m \mu_m$. The total moment $N\bar{\mu}$ is the same as the quantity $\mathfrak{M} = \mathfrak{M}_V\, \delta V$ defined previously, where $\mathfrak{M}_V$ is the magnetic moment per unit volume.

The work done by the sample on the surroundings in moving from x to $x + dx$ is

$$dw = \bar{F}\, dx = N\bar{\mu}\, dB_0 \tag{19-43a}$$

It should be noted that the work done can be written as

$$dw = \Sigma p_m \, d\epsilon_m \tag{19-43b}$$

in agreement with our previous general conclusion (Chap. 6) as to the expression for work in statistical mechanics.

The volume susceptibility of the sample is

$$\chi_V = \frac{\mathfrak{M}_V}{B_0} = \frac{N\bar{\mu}}{B_0 \, \delta V} \tag{19-44}$$

and according to (19-42) this can be measured as

$$\chi_V = \frac{\bar{F}}{B_0(dB_0/dx) \, \delta V} \tag{19-45}$$

where $\bar{F}$ is the force on an element of the sample where the field is B_0 and the gradient is dB_0/dx.

The molar susceptibility is given by

$$\chi_M = \chi_V V_M = \frac{N_0 \bar{\mu}}{B_0}$$

where V_M is the molar volume and N_0 is Avogadro's number.

For the case considered in Sec. 19-4, where $\epsilon_m = g\beta m B_0$, $\mu_m = -g\beta m$, in weak fields ($y \ll 1$),

$$\bar{\mu} = \frac{g^2 \beta^2 J(J+1) B}{3kT} = \frac{\langle \mu^2 \rangle B}{3kT} \tag{19-26}$$

and, as previously remarked [Eq. (19-27)], the molar susceptibility is

$$\chi_M = \frac{N_0 g^2 \beta^2 J(J+1)}{3kT} = \frac{N_0 \langle \mu^2 \rangle}{3kT} \tag{19-27}$$

For such a substance, the susceptibility is a constant.

A very common method of measuring the magnetic susceptibility of a sample is actually to measure the force on it in a nonuniform magnetic field and to apply Eq. (19-45).*

The mutual inductance of two coils, i.e., the induced emf in one coil when the current through the other coil is changed, is dependent on the magnetic susceptibility of the media in which the coils are immersed. Measurements of susceptibility at very low temperatures are commonly based on this general method.

In summary, there are three important points which have been developed in this section:

(a) The work done by a sample on its surroundings when it moves from a point where the field due to the surroundings is B_0 to a point where the field is $B_0 + dB_0$ is $N\bar{\mu} \, dB_0$, or $\mathfrak{M} \, dB_0$, where $\mathfrak{M}$ is the magnetic moment of the sample.

---

* Cf. P. W. Selwood, "Magnetochemistry," Interscience, New York, 1956.

(b) In cases where the energy of an atom in a field is not a linear function of the field (for example, in the Paschen-Back effect) we define μ_m as $-d\epsilon_m/dB_0$ and obtain the same basic statistical mechanical equations as before. In particular, the statement in paragraph (a) applies. In general, for such cases, the susceptibility will be a function of the magnetic field.

(c) For a set of noninteracting magnets which obey the usual equation $\epsilon_m = -g\beta Bm$, the molar susceptibility is given by Eq. (19-27). The susceptibility is then independent of the applied field and inversely proportional to the temperature.

19-7. General Statistical Mechanics of Magnetic Materials. It is not always a satisfactory approximation to assume that the energy levels of one paramagnetic atom are independent of the magnetic state of the neighboring atoms. This is most obviously true for a ferromagnetic material, where the spins of the paramagnetic atoms spontaneously align, even in the absence of an applied magnetic field. In an antiferromagnetic material, the interactions between the spins are such that the state of lowest energy is one in which all the spins are paired and the total spin of the system is zero. In general, in magnetically dilute materials such as chrome alum, $K_2Cr(SO_4)_2 \cdot 12H_2O$ (the Cr^{3+} ions being the only paramagnetic atoms), or in dilute solid solutions of Fe_2O_3 in Al_2O_3, the Curie law that $\chi_M = N_0 \langle \mu^2 \rangle / 3kT$ with a value of $\langle \mu^2 \rangle$ that is expected on the basis of the electron configuration of the ion usually holds over a wide range of temperature. As we shall see, the law $\chi \sim 1/T$ fails at some low temperature for any substance, no matter how dilute. In magnetically nondilute systems, such as CuO, MnO, or Fe_2O_3, the assumption of noninteracting magnets is not valid even at room temperature. The field at a vector distance $\mathbf{r}$ from a magnetic dipole is

$$\mathbf{B} = -\frac{\boldsymbol{\mu}}{r^3} + \frac{3(\boldsymbol{\mu} \cdot \mathbf{r})\mathbf{r}}{r^5} \qquad \text{cgs-Gaussian} \qquad (19\text{-}46a)$$

$$\mathbf{B} = -\frac{\mu_{0,\text{mks}}}{4\pi}\left[\frac{\boldsymbol{\mu}}{r^3} - \frac{3(\boldsymbol{\mu} \cdot \mathbf{r})\mathbf{r}}{r^5}\right] \qquad \text{mks} \qquad (19\text{-}46b)$$

Therefore, the potential energy of interaction of two magnetic dipoles *in vacuo* separated by the vector $\mathbf{r}_{12}$ is

$$U = \frac{\boldsymbol{\mu}_1 \cdot \boldsymbol{\mu}_2}{r_{12}^3} - 3\frac{(\boldsymbol{\mu}_1 \cdot \mathbf{r}_{12})(\boldsymbol{\mu}_2 \cdot \mathbf{r}_{12})}{r_{12}^5} \qquad \text{cgs} \qquad (19\text{-}47)$$

(In mks units, the expression should be multiplied by $\mu_{0,\text{mks}}/4\pi$.)

(Incidentally, this is an opportune occasion to remark that it is not worthwhile to remember the cgs units of magnetic moment or magnetic field; it is usually sufficient to recall that the units of μ/r^3 are gauss, that the units of μB and μ^2/r^3 are both energy, and that the units of B^2 are energy times volume^{-1}.)

Thus, for two neighboring magnetic dipoles, there is a direct magnetic-dipole interaction as given by (19-47). In addition, there are other quantum-mechanical interactions between the electron clouds which are dependent on the spin states of the interacting particles. Many of these interactions are lumped together under the general name of "exchange forces."

We shall not consider the difficult quantum-mechanical problem of calculating the energy levels of such a system. For statistical mechanical purposes, it is sufficient for us to assert that, for a system of N particles, there are a set of system energy levels which we can index with the general index i. For example, for a system containing N particles each of spin $\frac{1}{2}$, there are 2^N possible system states. For a system of N particles, each of spin J, there are $(2J + 1)^N$ system states.

The energy of any one system state will depend on the interactions between the particles in the particular configuration of spins specified by the index i plus the interactions of the particles with the external magnetic field B_0. Thus the energy can be written as $E_i(B_0)$, but we must understand that, even for $B_0 = 0$, the different system states have different energies because of the interactions between the particles. The magnetic moment $\mathfrak{M}_i$ of the system in the state i is defined by

$$\mathfrak{M}_i = -\frac{dE_i(B_0)}{dB_0} \tag{19-48}$$

The canonical partition function is of course

$$Q(B_0, T) = \sum_i e^{-E_i(B_0)/kT} \tag{19-49a}$$

The probability of finding a system in the state i is

$$p_i = \frac{e^{-E_i(B_0)/kT}}{Q(B_0, T)} \tag{19-49b}$$

The ensemble average magnetic moment of the system is

$$\overline{\mathfrak{M}} = \Sigma p_i \mathfrak{M}_i \tag{19-50}$$

Let each of the systems of the canonical ensemble be placed in an identical nonuniform field with a gradient dB_0/dx. The force on a system in the state i is

$$F_i = -\frac{dE_i}{dB_0}\frac{dB_0}{dx} = \mathfrak{M}_i \frac{dB_0}{dx} \tag{19-51a}$$

and the ensemble average force is

$$\bar{F} = \sum p_i F_i = \overline{\mathfrak{M}} \frac{dB_0}{dx} \tag{19-51b}$$

The average work done on the surroundings if the systems move from a point where the field is B_0 to a point where it is $B_0 + dB_0$ is

$$d\bar{w} = \bar{F}\,dx = \overline{\mathfrak{M}}\frac{dB_0}{dx}\,dx = \overline{\mathfrak{M}}\,dB_0 \qquad (19\text{-}52)$$

We have dealt in detail with the definition of thermodynamic work done by the system that we use in connection with the definition (19-48) of the magnetic moment of a system, because this is a topic for which some other authors use a different definition. In other respects, the statistical mechanics of a system of interacting magnets can be written out in a straightforward way. We restrict our considerations to a system at constant volume.

Thus, the magnetic free energy is given by

$$A(B_0, T) = -kT \ln Q(B_0, T) \qquad (19\text{-}53)$$

The ensemble average energy, or thermodynamic energy, is

$$E(B_0, T) = \frac{\Sigma E_i(B_0) e^{-E_i/kT}}{\Sigma e^{-E_i/kT}} = kT^2 \left\{\frac{\partial[\ln Q(B_0, T)]}{\partial T}\right\}_{B_0} \qquad (19\text{-}53a)$$

$$E(B_0, T) = -T^2 \left[\frac{\partial (A/T)}{\partial T}\right]_{B_0} \qquad (19\text{-}53b)$$

The entropy is then

$$S(B_0, T) = \frac{E - A}{T} = kT \left[\frac{\partial (\ln Q)}{\partial T}\right]_{B_0} + k \ln Q$$

$$= \left\{\frac{\partial[kT \ln Q(B_0, T)]}{\partial T}\right\}_{B_0} \qquad (19\text{-}54a)$$

$$S(B_0, T) = -\left[\frac{\partial A(B_0, T)}{\partial T}\right]_{B_0} \qquad (19\text{-}54b)$$

The magnetic moment may be written

$$\overline{\mathfrak{M}} = \frac{-\Sigma(dE_i/dB_0)e^{-E_i/kT}}{\Sigma e^{-E_i/kT}} = kT \left\{\frac{\partial[\ln Q(B_0, T)]}{\partial B_0}\right\}_T$$

$$= \left\{\frac{\partial[kT \ln Q(B_0, T)]}{\partial B_0}\right\}_T \qquad (19\text{-}55a)$$

or $\quad \overline{\mathfrak{M}} = -\left[\dfrac{\partial A(B_0, T)}{\partial B_0}\right]_T \qquad (19\text{-}55b)$

By comparison of (19-55a) and (19-54a), we derive the important cross-differentiation identity

$$\frac{\partial S(B_0, T)}{\partial B_0} = \frac{\partial^2 (kT \ln Q)}{\partial B_0\, \partial T} = \frac{\partial \overline{\mathfrak{M}}}{\partial T}$$

or $\quad \left(\dfrac{\partial S}{\partial B_0}\right)_T = \left(\dfrac{\partial \overline{\mathfrak{M}}}{\partial T}\right)_{B_0} \qquad (19\text{-}56)$

and similarly

$$\left[\frac{\partial(\overline{\mathfrak{M}}/T)}{\partial T}\right]_{B_0} = \left[\frac{\partial(E/T^2)}{\partial B_0}\right]_T \qquad (19\text{-}57)$$

The relations derived above are applicable in general and are based on the definition $\mathfrak{M}_i = -dE_i/dB_0$ of the magnetic moment of a system in a given quantum state. It is worth noting that, for the case of the simple model discussed below, the definition gives a reasonable and expected result.

Consider a system of N atoms, each with a spin of $\tfrac{1}{2}$. We regard the atoms as localized at definite lattice points: $1, 2, \ldots, N$. There are 2^N possible spin states of the system. A particular system state is specified by an array of N $+$'s or $-$'s, for example, $+, +, -, +, -, \ldots, -$, which gives the orientation of each of the spins. If in the state i there are n_i spins up and $N - n_i$ spins down (and let us assume that a plus spin has a magnetic moment of $-\mu$), then the sum of the magnetic moments of all the atoms is

$$\mathfrak{M}_i = n_i(-\mu) + (N - n_i)(+\mu) = (N - 2n_i)\mu \qquad (19\text{-}58)$$

The energy of interaction of an atom in the $+$ state with the external field is μB_0; the energy of interaction of an atom in the $-$ state is $-\mu B_0$. The energy of the system in the state i can be written as

$$E_i(B_0) = -(N - n_i)\mu B_0 + n_i \mu B_0 + E_i^\circ \qquad (19\text{-}59)$$

The first two terms represent the sum of the interactions with the external field, and the term E_i° is the energy of interaction of the atoms with each other when they are in the particular spin configuration, $+, +, -$, etc., of the ith quantum state. For this particular simple system, it is *assumed* that the energy of interaction is independent of the value of the external field. If (19-59) is accepted, our formal definition of a magnetic moment gives

$$\mathfrak{M}_i = -\frac{dE_i(B_0)}{dB_0} = (N - 2n_i)\mu \qquad (19\text{-}60)$$

in agreement with Eq. (19-58), which was derived in a more intuitive and direct way.

There are two comments that should be made about this simple illustration. It does clarify the reason why the magnetic moment is the negative derivative of the energy with respect to the field due to external sources. There are or may be magnetic fields at the position of one atom due to the neighboring atoms. But in a given quantum state, i.e., for a fixed relative configuration of the spins, this field does not change when the specimen changes its position in the external field, and it is only the field due to the external sources which exerts a force on the sample and does work on it.

In the second place, we should remark that the model used as an illustration is greatly oversimplified and unrealistic. Let us call an array of the N spins, $+$, $+$, $-$, ..., $-$, a simple configuration. In general, such a simple configuration is not an eigenfunction of the system. In some cases, the correct eigenfunctions are linear combinations of the simple configurations with the same value of n_i. In other cases, the eigenfunctions are more remotely related to the simple configurations. In all cases, however, the general definition $\mathfrak{M}_i = -dE_i/dB_0$ is applicable.

This is an appropriate point at which to comment on the logical status of the statistical mechanical treatment of a set of noninteracting identical magnetic atoms in a crystal. The atoms are identical, but the different lattice points are distinguishable. We start with the naïve point of view that we can identify the spin state of the atoms at the different lattice sites. We can say that the atom at site A is in the magnetic quantum state m_1 and the atom at site B is in the magnetic quantum state m_2, and this is a recognizably different state of the system from the one in which the atom at site A is in the state m_2 and the atom at site B is in the state m_1. There are no symmetry restrictions, because the atoms are at different sites and the spin orientations of any one atom are independent of the spin orientations of the others. We can then write

$$Q = Q_{\text{lattice}} q_{\text{mag}}^N$$

This equation is valid for a set of noninteracting magnetic atoms at localized distinguishable lattice sites in a crystal, provided that the lattice vibrations do not affect the magnetic-energy levels. The magnetic properties of such a set of localized particles and the magnetic properties of a set of nonlocalized gas atoms depend on the single-particle magnetic partition function in an identical way [cf. Eqs. (19-15) to (19-20)].

However, the naïve point of view that we can identify the spin states of identical atoms at different lattice sites is not really tenable. There are always weak interactions between magnetic atoms. Suppose that two neighboring atoms at equivalent lattice sites A and B are in the magnetic quantum states m_1 and m_2. There must be some sort of energy of interaction, ϵ, between them. This energy of interaction may be so small as to be negligible in calculating the partition function. However, quite generally atoms at A and at B can be said to be exchanging magnetic states at the frequency ϵ/h. So, for times longer than h/ϵ, one cannot say which atom is in the state m_1 and which is in the state m_2. This does not, however, invalidate the statement that $Q = q^N$, and we do not have to worry about Fermi-Dirac or Bose-Einstein restrictions. Consider the simplest possible example. Let the crystal consist of two protons, at the distinguishable (but equivalent) lattice sites A and B. Let α and β represent the spin-up and spin-down states for a single particle and ϵ_α and ϵ_β the respective energies in a magnetic field. The single-proton partition function is

$$q = e^{-\epsilon_\alpha/kT} + e^{-\epsilon_\beta/kT}$$

and we assert that the system partition function is $Q = q^2$. Because of spin exchange, we cannot actually say that A is in state α and B in state β, etc.; however, the wave functions and energy levels for the entire crystal are given by:

Wave function	Energy
$\alpha(A)\alpha(B)$	$2\epsilon_\alpha$
$\dfrac{1}{2^{1/2}}[\alpha(A)\beta(B) + \alpha(B)\beta(A)]$	$\epsilon_\alpha + \epsilon_\beta$
$\beta(A)\beta(B)$	$2\epsilon_\beta$
$\dfrac{1}{2^{1/2}}[\alpha(A)\beta(B) - \alpha(B)\beta(A)]$	$\epsilon_\alpha + \epsilon_\beta$

The system partition function is then

$$Q = e^{-2\epsilon_\alpha/kT} + e^{-(\epsilon_\alpha+\epsilon_\beta)/kT} + e^{-2\epsilon_\beta/kT} + e^{-(\epsilon_\alpha+\epsilon_\beta)/kT}$$
$$= (e^{-\epsilon_\alpha/kT} + e^{-\epsilon_\beta/kT})^2 = q^2$$

Thus exchange does not affect the partition functions. The point is that both the symmetrical wave function and the antisymmetrical wave function as regards spin exchange [$\alpha\beta + \beta\alpha$ and $\alpha\beta - \beta\alpha$, respectively] are allowed, because protons A and B are at distinguishable sites in the crystal. If the two protons were part of a hydrogen molecule in the gas phase, only one symmetry species would be allowed, as discussed in Chap. 9.

In general, then, for a crystal with N atoms at N distinguishable sites in the crystal, with weak interactions that do not appreciably affect the energy levels but do cause spin exchange, if there are $2J + 1$ different magnetic sublevels for one atom, there are $(2J + 1)^N$ different magnetic quantum states for the crystal, even though, because of exchange, we cannot say what is the state of a particular atom. The canonical partition function is then given by $Q = q^N$.

19-8. The Thermodynamics of Magnetic Materials. In the preceding section, we have already derived the basic equations for the thermodynamics of a magnetic material. It is convenient to reiterate the important formulae here.

Consider a magnetic field B_0 which is not necessarily uniform in space, created perhaps by a system of permanent magnets and solenoids. Now introduce a small specimen of matter of volume V. For simplicity, assume that the sample is incompressible and V is constant.

At a given position in the field, let the magnetic polarization per unit volume of the specimen be $\mathfrak{M}_V$ so that the total magnetic moment is $\mathfrak{M} = \mathfrak{M}_V V$. It can then be shown that the work done by the specimen in moving to a new position where the field due to the external sources has changed to $B_0 + dB_0$ is

$$dw = \mathfrak{M}\, dB_0 \qquad (19\text{-}61)$$

We can therefore write the first law of thermodynamics as

$$dE = T\, dS - \mathfrak{M}\, dB_0 \qquad (19\text{-}62)$$

For (19-61) and (19-62) to apply, the field must be uniform over the volume of the specimen. It is to be emphasized that B_0 in these equations is the field due to the external sources—the contribution to the field by the specimen itself is not included.

From (19-62)

$$dA = d(E - ST) = -S\, dT - \mathfrak{M}\, dB_0 \qquad (19\text{-}63)$$

so that

$$\mathfrak{M} = -\left(\frac{\partial A}{\partial B_0}\right)_T \qquad S = -\left(\frac{\partial A}{\partial T}\right)_{B_0} \qquad (19\text{-}64)$$

and

$$\left(\frac{\partial \mathfrak{M}}{\partial T}\right)_{B_0} = \left(\frac{\partial S}{\partial B_0}\right)_T \qquad (19\text{-}65)$$

We can also write

$$d\left(\frac{A}{T}\right) = -\frac{E}{T^2}\, dT - \frac{\mathfrak{M}}{T}\, dB_0 \qquad (19\text{-}66)$$

and

$$E = -T^2\left[\frac{\partial (A/T)}{\partial T}\right]_{B_0} \qquad (19\text{-}67)$$

$$\left[\frac{\partial (\mathfrak{M}/T)}{\partial T}\right]_{B_0} = \frac{1}{T^2}\left(\frac{\partial E}{\partial B_0}\right)_T \qquad (19\text{-}68)$$

These equations have already been derived by statistical mechanical arguments [cf. Eqs. (19-55) to (19-57)]. They are rederived here to emphasize that they have general thermodynamic validity and are true even when we cannot theoretically construct the canonical partition function for the system.

It should be remarked that the nature of the work term in the thermodynamics of magnetic materials is a confusing and ambiguous subject for which more than one answer is possible. In principle, it is a straightforward problem in electromagnetism to calculate the work done in establishing a certain configuration of solenoids or permanent magnets and magnetic materials. If we wish to regard one piece of material as the system and the solenoids or permanent magnets as surroundings, there is the problem of deciding how much of the total work was done on the system and how much was done on the surroundings. Thus it is permissible to assert that the work term is $-B_0\, d\mathfrak{M}$ instead of $+\mathfrak{M}\, dB_0$; the thermodynamic equation is then

$$dE = T\, dS + B_0\, d\mathfrak{M} \qquad (19\text{-}69)$$

It is easy to see that the thermodynamic consequences of this equation [for example, $(\partial S/\partial B_0)_T = (\partial \mathfrak{M}/\partial T)_{B_0}$] are identical with those previously derived. However, the thermodynamic energy of the system E in (19-69) has a different meaning from E in (19-62). Equation (19-62) is the appropriate one to use with the definitions of energy which have been used in our statistical treatment.*

19-9. Magnetic Resonance. If the ground state of an isolated magnetic atom or nucleus is $(2J + 1)$-fold-degenerate, this degeneracy will in general be removed by an external magnetic field and there will be $2J + 1$ energy levels which are functions of the applied field,

$$\epsilon_m(B_0) \qquad m = -J, -J+1, \ldots, +J$$

If there is a radiative transition between two adjacent levels with magnetic quantum numbers m and $m - 1$, the frequency of the light quantum absorbed is given by

$$h\nu = \epsilon_m(B_0) - \epsilon_{m-1}(B_0) \tag{19-70}$$

In the simplest possible case, the formula for the energy levels is

$$\epsilon_m = g\beta m B_0$$

Such a system of energy levels is displayed in Fig. 19-3. The spacing between any two adjacent levels is $\Delta\epsilon = \epsilon_m - \epsilon_{m-1} = g\beta B_0$; it is propor-

---

* We shall not endeavor to discuss this rather tricky subject in detail. It is discussed by Fowler and Guggenheim, pp. 673–674 [1], and Kittel, pp. 77–81 [12], and at length by E. A. Guggenheim, "Thermodynamics," 3d ed., chap. 13, North-Holland, Amsterdam, 1957, and especially app. C. In app. C, Guggenheim gives an argument to the effect that the work done by the field on a small sample in a solenoid when the field is increased from B_0 to $B_0 + dB_0$ is $B_0 \, d\mathfrak{M}$. According to this argument, if the sample consisted of a piece of a permanent magnet with a *fixed* magnetic moment along the axis of the solenoid, then there would be no work done on the sample (since $d\mathfrak{M} = 0$). But clearly the orientation energy of the permanent-magnet sample in the solenoid is $-MB_0$ and increases as the field increases. Thus, in Guggenheim's definition of the thermodynamic energy, the orientation energy is excluded; with our definition, it is included. The reader will observe that our molecular definitions of energy include the orientation energy $-\boldsymbol{\mu} \cdot \mathbf{B}$ also.

By including a $P \, dV$ term in the expression for the work and writing

$$dE = T \, dS - \mathfrak{M} \, dB_0 - P \, dV$$

one can in a straightforward way consider changes in the volume of the system. For example, the equation

$$\left(\frac{\partial V}{\partial B_0}\right)_{P,T} = -\left(\frac{\partial \mathfrak{M}}{\partial P}\right)_{B_0,T}$$

is an equation for one kind of magnetostriction.

The interesting practical applications of magnetostriction, however, deal principally with the change in shape of a material when a field is applied, with a relatively small change in volume. Detailed magnetostatic considerations are needed to deal with this situation.

tional to the applied field B_0 and is the same for any two adjacent levels, independent of the magnetic quantum number m. If a transition between two adjacent levels is caused by the absorption of a light quantum, the frequency of the light is

$$\nu = \frac{\Delta \epsilon}{h} = \frac{g\beta B_0}{h} = \frac{ge}{4\pi mc} B_0 \tag{19-71}$$

Such transitions are indeed possible; this is the field of magnetic-resonance spectroscopy. For example, the numerical factors for electron magnetic resonance (with β the Bohr magneton) are

$$\frac{\nu}{c} = \frac{gB}{21{,}418.4} \quad \text{cm}^{-1} \tag{19-72}$$

So for $g = 2$ and $B = 3{,}580$ gauss, $\nu/c \approx 0.33$ cm^{-1}, which is the so-called X band in microwave spectroscopy.

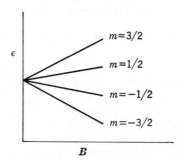

Fig. 19-3. Energy levels in a magnetic field for the case $J = \tfrac{3}{2}$.

For nuclear magnetic resonance, we use for β the nuclear magneton (0.505×10^{-23} erg gauss^{-1}). For protons, $g = 2.793$; and if $B_0 = 10{,}000$ gauss, $\nu = 4.26 \times 10^7$ cycles sec^{-1}, which is in the radiowave region.

In a typical magnetic-resonance experiment, there is a static field B_0 which produces the separation of the energy levels.

In addition, there must be electromagnetic radiation with the frequency ν given in the general case by Eq. (19-70) or for the special case when $\epsilon_m = g\beta m B_0$ by (19-71). By quantum-mechanical perturbation theory, it can be shown that the selection rule $\Delta m = \pm 1$ holds for most cases. The essential feature of the "light-quantum" or electromagnetic radiation at the frequency ν is that there should be an oscillating magnetic field $\mathbf{B}_1 \sin 2\pi\nu t$, with $\mathbf{B}_1$ perpendicular to the static field $\mathbf{B}_0$, and typically with the amplitude B_1 much less than B_0.

The classical picture of magnetic resonance is instructive. We have already seen that a magnetic dipole oriented at an angle θ with respect to a large static field B_0 experiences a torque $\mathbf{T} = \boldsymbol{\mu} \times \mathbf{B}$ which tends to orient the magnetic moment in the field direction. However, the magnetic dipole is like a top in a gravitational field in that it has angular momentum. The angular-momentum vector **M** is in the same direction as the magnetic-moment vector. The orienting torque therefore causes the top to precess about the magnetic field at the fixed angle θ.

The relation between the magnetic-moment vector and the angular-momentum vector is

$$\boldsymbol{\mu} = -\frac{g\beta}{\hbar}\mathbf{M} = -\frac{g|e|}{2mc}\mathbf{M} = \gamma\mathbf{M} \qquad (19\text{-}73)$$

where γ, the magnetogyric ratio, is implicitly defined by (19-73). An angular momentum subject to a torque obeys the equation

$$\frac{d\mathbf{M}}{dt} = \mathbf{T} = \boldsymbol{\mu} \times \mathbf{B}_0 = -\frac{g\beta}{\hbar}\mathbf{M} \times \mathbf{B}_0 \qquad (19\text{-}74)$$

The general solution of this differential equation is that the vector **M** precesses around $\mathbf{B}_0$ at a frequency of $g\beta B_0/2\pi\hbar$ or $g\beta B_0/h$, which is the same as the frequency quantum calculated mechanically for the transition with $\Delta m = \pm 1$ [Eq. (19-71)].

The precession is illustrated in Fig. 19-4. A circularly polarized oscillating magnetic field in a plane perpendicular to B_0 at the precession frequency will be in resonance with the precessing dipole and hence will either absorb energy from the dipole or give energy to it. This is the classical description of magnetic resonance.

If the energy levels are given by the simple equation $\epsilon_m = g\beta m B_0$, the magnetic-resonance spectrum consists of a single line of frequency $\nu = g\beta B_0/h$. In this case, the magnetic-resonance spectrum gives direct information about the g factor of the magnetic atom or nucleus.

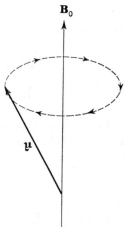

Fig. 19-4. The magnetic moment $\boldsymbol{\mu}$ with an associated angular momentum ($\boldsymbol{\mu} = -g\beta\mathbf{M}/\hbar$) precesses around the field direction (as indicated by the dashed line) with a frequency $g\beta B_0/h$.

In many cases, the magnetic-resonance spectrum is more complicated and consists of a number of absorption frequencies. This can be due to a variety of causes. There are interactions between neighboring paramagnetic atoms, between neighboring magnetic nuclei, and between magnetic atoms and magnetic nuclei. There are splittings due to the interaction of orbital angular momentum with electric fields in a crystal. For atoms or nuclei in crystals, the resonance spectrum may be different for different directions of the applied field B_0 and for different directions of the rf field B_1, with respect to the crystal axes.

It is not our purpose to expound on these interactions, but merely to point out that magnetic-resonance spectroscopy provides a very precise tool for the study of the energy levels of magnetic systems.*

19-10. Demagnetizing Factors. The Local Field. When a magnetic material is placed in a field, the field vectors at a particular point in the system differ from the field present before the insertion of the sample. This has an effect on the magnetic measurements which we briefly discuss here. The topic is not very important for understanding the basic principles of the magnetic properties of matter, but it is of practical significance in making and interpreting measurements.

We confine our attention to a small sample suspended either in a solenoid or between the pole faces of a permanent magnet. We are interested in the relation between the magnetic induction B_0 before the insertion of the sample and the effective fields in the sample. We assume that the source of B_0—the current through the solenoid or the magnetization of the permanent magnet—is unaffected by introducing the sample.

It can then be shown that there is a uniform magnetic moment per unit volume, $\mathfrak{M}_V$, in the sample; and that, for a spherical sample, the values of the field vectors B and H in the sample are

$$B_{\text{med}} = B_0 + \frac{8\pi\mathfrak{M}_V}{3} \quad \text{spherical sample} \quad (19\text{-}75a)$$

$$H_{\text{med}} = B_{\text{med}} - 4\pi\mathfrak{M}_V = B_0 - \frac{4\pi\mathfrak{M}_V}{3} \quad (19\text{-}75b)$$

For a long, thin cylinder with its axis transverse to the axis of the magnetic field (for example, a narrow test tube as in one of the typical setups for a nuclear magnetic-resonance experiment) the fields are

$$B_{\text{med}} = B_0 + 2\pi\mathfrak{M}_V \quad \text{(cylinder)} \quad (19\text{-}76a)$$
$$H_{\text{med}} = B_0 - 2\pi\mathfrak{M}_V \quad (19\text{-}76b)$$

In general, for any shape, $B_{\text{med}} = B_0 + \alpha\mathfrak{M}_V$, and α is known as the demagnetizing factor for this particular shape. It is to be emphasized that the calculation of α is a problem in macroscopic magnetostatics for which there is a definite and rigorous answer.

---

* For further information about electron and nuclear magnetic resonance, see J. D. Roberts, "Nuclear Magnetic Resonance," McGraw-Hill, New York, 1959; J. A. Pople, W. G. Schneider, and H. J. Bernstein, "High-resolution Nuclear Magnetic Resonance," McGraw-Hill, New York, 1959; E. R. Andrew, "Nuclear Magnetic Resonance," Cambridge, New York, 1955; A. K. Saha and T. P. Das, "Theory and Applications of Nuclear Induction," Saha Institute of Nuclear Physics, Calcutta, 1957; D. J. E. Ingram, "Spectroscopy at Radio and Microwave Frequencies," Butterworth, London, 1955; W. Gordy, W. V. Smith, and R. F. Trambarulo, "Microwave Spectroscopy," Wiley, New York, 1953; C. H. Townes and A. L. Schawlow, "Microwave Spectroscopy," McGraw-Hill, New York, 1955; W. Low, "Paramagnetic Resonance in Solids," Academic Press Inc., New York, 1960.

There is for any case a B_{med}, the macroscopic magnetic-induction vector in the medium. What then is the local field that acts on a little molecular magnetic dipole? There is an argument, explained in detail for the electrical case in Chap. 18, which indicates that the local field is the so-called Lorentz field, i.e., the magnetic induction at the position of the molecule due to all the medium except that contained in a small spherical cavity surrounding the molecule [Eq. (18-32)]. We have

$$B_{cavity} = B_{med} - \frac{8\pi \mathfrak{M}_V}{3} \tag{19-77}$$

For a spherical sample, $B_{med} = B_0 + (8\pi \mathfrak{M}_V/3)$; so

$$B_{cavity} = B_0 \tag{19-78}$$

whereas, for a transverse cylinder,

$$B_{cavity} = B_0 - \frac{2\pi}{3} \mathfrak{M}_V \tag{19-79}$$

(Remember, B_{cavity} is B in a spherical cavity, whatever the external shape of the sample.)

It is to be emphasized that the local field on a molecule is a molecular problem for which there is no known general, correct, and rigorous answer; the Lorentz field is an approximation or an assumption. Fortunately, in most cases, the magnetization $\mathfrak{M}_V$ is much less than B_0, and except for the very precise measurements of high-resolution nuclear magnetic resonance, the problem is not important. In ferromagnetic materials at low temperatures, $\mathfrak{M}_V$ may be comparable with B_0. Actually, in a ferromagnetic material, there are spin-dependent "exchange" forces which also affect the energy of interaction of neighboring spins. A knowledge of these forces as well as of the purely magnetostatic "local field" is needed for the statistical mechanical treatment of such substances.

The susceptibility is rigorously defined by the equation

$$\chi_V = \frac{\mathfrak{M}_V}{H_{med}} \tag{19-80a}$$

However, experimenters in the field of low-temperature physics very frequently use the definition

$$\chi_V = \frac{\mathfrak{M}_V}{B_0} \tag{19-80b}$$

The definition (19-80b) has the disadvantage that it depends on the shape of the sample; it has the advantage of being easily related to measurable quantities. If we believe that, for a spherical sample, the local field is actually B_0, the definition (19-80b) is especially convenient. On the other hand, χ as defined by (19-80a) is a property of the material, independent of sample shape.

For weakly paramagnetic materials and for diamagnetic materials, the susceptibility is close to zero, and the permeability $\mu_P = 1 + 4\pi\chi_V$ is close to unity, and both are independent of the field. For ferromagnetic substances or paramagnetic substances at low temperatures, χ and μ_P are not constant but are functions of the applied field. Furthermore, such systems frequently show the phenomenon of hysteresis, so that μ_P and χ depend on the past history of the sample.

19-11. More about Noninteracting Magnets. It is appropriate before proceeding further to provide a brief discussion of the magnetic properties of some typical substances that have been extensively studied and that illustrate the theory outlined in the preceding sections.

The theory for noninteracting magnets with an energy-level scheme that conforms to the equation $\epsilon_m = g\beta m B_0$ is widely applicable. For example, it is quite accurately applicable for experiments in nuclear magnetic resonance, at least when the nuclei are contained in a material which is electronically diamagnetic. It is applicable to dilute solutions of paramagnetic ions, such as the transition metal ions, or of paramagnetic free radicals. It is applicable to crystals containing paramagnetic ions, provided that the ions are sufficiently far apart so that the magnetic interaction between them is small compared with kT. Thus the theory works well for magnetically dilute crystals, such as the hydrated sulfates and alums, for example, $Gd_2(SO_4)_3 \cdot 8H_2O$ and $KCr(SO_4)_2 \cdot 12H_2O$. It does not work well for substances such as Cr_2O_3 or MnO where the interactions between the magnetic ions are large. The theory also works well for gaseous oxygen at moderate pressures; O_2 has two unpaired electrons and no orbital angular momentum, so that $J = S = 1$, and $g = 2.00$.*

Except at very low temperature and very high fields, the essential point is that the molar magnetic susceptibility fits the equation

$$\chi = \frac{A}{T} + B \tag{19-81}$$

where B is the diamagnetic term and, as previously shown,

$$A = \frac{N_0 \langle \mu^2 \rangle}{3k} = \frac{N_0 g^2 \beta^2 J(J+1)}{3k} \tag{19-82}$$

The result that, apart from a small diamagnetic correction, the susceptibility is proportional to the reciprocal of the temperature is known as Curie's law. The experimental data are often interpreted in terms of

---

* For further information about this and related topics, see J. H. Van Vleck, "The Theory of Electric and Magnetic Susceptibilities," Oxford, New York, 1932, and P. W. Selwood, "Magnetochemistry," Interscience, New York, 1956.

an effective magneton number μ_{eff}, which is determined by the equation

$$\langle \mu^2 \rangle = \mu_{\text{eff}}^2 \beta^2 \tag{19-83}$$

The trivalent rare-earth ions have the configuration (krypton shell) $4f^n\ 5s^2\ 5p^6$, with n, the number of $4f$ electrons, ranging from 0 to 14. The magnetism is due to the $4f$ electrons, which are shielded by the $5s$ and $5p$ electrons from the electrical forces of the environment, whether it be an aqueous solution or a crystal. The electronic states of the ion are therefore like the states of gaseous atoms, with both orbital and spin angular momentum. The g factors may be calculated from the Lande g-factor equation (18-8), although in some cases corrections are necessary, because there are several low-lying electronic terms (of different J) with energies above the ground states of the order of kT.

Table 19-1 is adapted from a now famous table of Van Vleck ("The Theory of Electric and Magnetic Susceptibilities," p. 243, Oxford, New York, 1932).

Table 19-1. Paramagnetism of the Rare-earth Ions

Ion	$4f^n$	Term	S	L	J	g	μ_{eff} Theory a	μ_{eff} Theory b	Exptl.
La^{3+}	0	$^1S_0$	0	0	0	...	0	0	0
Ce^{3+}	1	$^2F_{5/2}$	$1/2$	3	$5/2$	$6/7$	2.54	2.56	2.1–2.8
Pr^{3+}	2	$^3H_4$	1	5	4	$4/5$	3.58	3.62	3.4–3.5
Nd^{3+}	3	$^4I_{9/2}$	$3/2$	6	$9/2$	$8/11$	3.62	3.68	3.4–3.6
Sm^{3+}	5	$^6H_{5/2}$	$5/2$	5	$5/2$	$2/7$	0.84	1.6	1.3–1.6
Yb^{3+}	13	$^2F_{7/2}$	$1/2$	3	$7/2$	$8/7$	4.54	4.54	4.4–4.6

Prediction a is based on Eq. (19-8); prediction b considers the effect of other low-lying terms; for further details, see J. H. Van Vleck, "The Theory of Electric and Magnetic Susceptibilities," Oxford, New York, 1932.

For the first transition metal ions, the electron configuration is (argon shell) $3d^n$, with n varying from 0 to 10. The magnetic electrons are in the $3d$ shell; since this is an outer shell, the wave functions are strongly affected by the electrical influences from the environment. In solution, and in many crystals, the ions are present as hydrates, for example, $M(H_2O)_6^{2+\ \text{or}\ 3+}$. In this case, the principal effects of the environment on the ion are due to the octahedron or distorted octahedron of water molecules. The detailed discussion of these effects has come to be known as crystal-field theory or ligand-field theory. The electronic states of atoms or ions in the gas phase result from the spherical symmetry of the potential in which the electrons move. To a very good approximation, the atom has an orbital angular-momentum quantum number L, and there are $(2L + 1)$-degenerate states. Suppose, for simplicity of discussion, that there is no spin angular momentum. Because of the electrical fields due to the ligands or the crystal as a whole, the transition metal ions are exposed to a nonspherically symmetric potential in solution or in the crystal. If this ligand field is regarded as perturbation, we can say that the $(2L + 1)$-degenerate orbital states of the gas-phase ion are split by the ligand field. The precise results depend on the particular ion and the symmetry of the particular crystal; however, the usual final result when all the perturbations are considered is that the lowest state is orbitally nondegenerate. The above description is equally applicable when the ions in the gas phase have spin angular momentum as well as orbital angular momentum. The orientation of the electron spin has only a minor effect on the energy of interaction of the electrons with the electrical fields of the ligands. When the crystal-field perturbations are larger than the spin-orbit coupling (as is the case for the transition metal ions), the net result is that the lowest state is usually orbitally nondegenerate and the ion behaves as though it has "spin-only" magnetization, with the spin quantum number S equal to one-half the number of unpaired electrons and $\langle \mu^2 \rangle^{1/2} = g\beta[S(S + 1)]^{1/2}$, or $\mu_{\text{eff}} = g[S(S + 1)]^{1/2}$. The values of g are quite close to 2.00 when the "spin-only" approximation is good; they differ from 2.00 when the orbital contributions are not thoroughly "quenched."

The situation is actually more complicated than this. The magnetic susceptibilities in the crystal are different along different crystal axes, and in general there is a susceptibility tensor. This situation can be described by saying that the g factor is a tensor. For an ion in solution, one measures a space average g factor, due to the tumbling motion of the ion.

Thus, for example, the Cu^{++} ion with the electron configuration $3d^9$ has as its lowest state in the gas phase $^2D_{5/2}$, so that $S = 1/2$, $L = 2$, $J = 5/2$, $g = 1.20$, $\mu_{\text{eff}} = 3.57$. However, the experimental susceptibility in solution and in magnetically dilute crystals corresponds to $\mu_{\text{eff}} = 1.8 - 2.0$. If we take $S = 1/2$ and $\mu_{\text{eff}} = 1.9$, the calculated g factor is 2.19. Thus, the Cu^{++} ion behaves approximately, but not exactly, like an atom with one unpaired electron with $S = 1/2$ and $L = 0$. The hydrated Cu^{++} ion exhibits a moderate magnetic anisotropy. For example, in the crystal of Tutton's salt ($K_2CuSO_4 \cdot 6H_2O$), it is believed that the Cu^{++} ion is surrounded by a distorted octahedron of water molecules which has approximately tetragonal symmetry. There are four short $Cu—OH_2$ distances in one plane and two longer $Cu—OH_2$ distances in the perpendicular direction. The g factor is 2.43 when the magnitude field is parallel to the tetragonal axes and 2.05 when it is perpendicular to the tetragonal axis.* In some cases—including the hydrated ions Cr^{3+}, Fe^{3+},

* B. Bleaney, R. P. Penrose, and B. Plumpton, *Proc. Roy. Soc. (London)*, **A198**: 406 (1949).

Mn^{++}, Ni^{++}—the spin-only approximation works quite well; the g factors are close to 2 and reasonably isotropic. In other cases, including many Co^{++} salts, the g factors are quite different from 2.0 and highly anisotropic.*

The study of the electron-spin resonance of paramagnetic ions in solution and, more particularly, in crystals provides much more detailed information about the energy-level system than is available from susceptibility measurements.

This long digression has been necessary for the reader to begin to appreciate the factors which must be taken into consideration in applying the statistical formulae we have developed in practical situations.

The fact is, however, that the simple Curie law holds for magnetically dilute materials down to quite low temperatures. Results for potassium chrome alum between 14 and 290°K are displayed in Table 19-2.

Table 19-2. Magnetic Susceptibility of $KCr(SO_4)_2 \cdot 12H_2O$ †

$T°K$........	290.0	169.7	143.6	77.70	64.50	20.39	17.16	14.33
$\chi'T(\times 10^4)$ ‡	36.60	36.57	36.57	36.67	36.81	36.99	36.91	36.94

† Data from W. J. de Haas and C. J. Gorter, *Communs. Kamerlingh Onnes Lab. Univ. Leiden* 208C, 1940.

‡ The quantity χ' is the susceptibility per gram, with a small correction for diamagnetism and the so-called "demagnetization factor"; multiply by 499.4 to obtain the molar susceptibility; the experimental μ_{eff} is 3.60; the calculated spin-only value for $S = 3/2$, $g = 2$ is 3.88.

For some salts with anisotropic g factors, the susceptibilities along the different crystal axes separately obey the Curie law [for example, $K_2CuSO_4 \cdot 6H_2O$; J. C. Hupse, *Physica*, **9**: 633 (1942)]. In other cases, the average susceptibility as measured on a powdered sample satisfies the χT = constant relation, although the susceptibilities along the several axes do not.

As previously mentioned, there are cases where the system is magnetically dilute and the interactions are unimportant, but in which the simple equation for the energy levels, $\epsilon_m = g\beta mB$, does not hold. These are the so-called intermediate-field and Paschen-Back cases, as discussed in textbooks on spectroscopy. For example, in a weak field the $^2P_{1/2}$ and $^2P_{3/2}$ states (for the configuration $1s^22p$) of the lithium atom have g factors of $2/3$ and $4/3$, respectively. The separation of these two levels is 0.34 cm^{-1}; this small separation is a measure of the strength of the spin-orbit coupling in this light atom.

* For a general review, see D. J. E. Ingram, "Spectroscopy at Radio and Microwave Frequencies," Butterworth, London, 1955, and W. Low, "Paramagnetic Resonance in Solids," Academic Press Inc., New York, 1960.

Thus, in a weak field, the energy levels (we take the ground state $^2P_{1/2}$ in zero field as the zero of energy) are given by

$$^2P_{3/2}: \quad \epsilon = 0.34hc + \tfrac{4}{3}\beta Bm \quad m = -\tfrac{3}{2}, \ldots, \tfrac{3}{2}$$
$$^2P_{1/2}: \quad \epsilon = \tfrac{2}{3}\beta Bm \quad m = -\tfrac{1}{2}, \tfrac{1}{2} \quad (19\text{-}84)$$

In a very strong magnetic field, where the Zeeman splitting becomes greater than the spin-orbit separation of 0.34 cm^{-1}, the magnetic forces "uncouple" L and S, J ceases to have any meaning, and the orbital and spin angular momenta are independently quantized along the magnetic field.

In the strong-field limit, the energy levels are given by

$$\epsilon = 2\beta Bm_S + \beta Bm_L + 0.34hcm_Sm_L + 0.17hc \quad m_S = \pm\tfrac{1}{2};$$
$$m_L = \pm 1, 0 \quad (19\text{-}85)$$

where m_S and m_L are the components of spin angular momentum and orbital angular momentum along the axis of the magnetic field. In an intermediate field, the variation of ϵ with B is more complicated than for the weak-field [(19-84)] or strong-field [(19-85)] cases. In any event, the magnetic moment of a state is defined by $\mu_i = -d\epsilon_i/dB$, and the statistical calculations are straightforward.

There are similar strong-field and weak-field cases for the interaction between the electron magnetic moment and the nuclear moment of an atom; and, indeed, these usually occur at lower field than for the interaction between orbital and spin moments of atoms illustrated above.*

19-12. Magnetic Cooling. One of the most fascinating achievements that has resulted from an understanding of the ideas discussed in the preceding sections is that of magnetic cooling.†

We have already seen that the thermodynamic relation

$$\left(\frac{\partial S}{\partial B_0}\right)_{T,V} = \left(\frac{\partial \mathfrak{M}}{\partial T}\right)_{B_0,V} \quad (19\text{-}65)$$

holds for an arbitrary magnetic material (for experimental arrangements in which the magnetization is uniform throughout the sample).

For the situations that we have studied so far, $(\partial M/\partial T)_{B_0}$ is negative; therefore $(\partial S/\partial B_0)_T$ is negative. The entropy of a paramagnetic salt decreases as it is isothermally magnetized, because of the partial alignment of the randomly oriented magnetic atoms by the applied field.

---

* For a further discussion of these topics, see, for example, N. F. Ramsey, "Nuclear Moments," p. 15, Wiley, New York, 1953, and E. U. Condon and G. H. Shortley, "The Theory of Atomic Spectra," pp. 152–157, Cambridge, New York, 1957.

† W. F. Giauque, *J. ACS*, **49**: 1864 (1927); P. Debye, *Ann. Physik*, **81**: 1154 (1926). For further references and information about the topics of Secs. 19-12 to 19-15, see C. G. B. Garrett, "Magnetic Cooling," Harvard University Press, Cambridge, Mass., 1954; C. F. Squire, "Low Temperature Physics," McGraw-Hill, New York, 1953; K. Mendelssohn, "Cryophysics," Interscience, New York, 1960.

When the approximation $y = g\beta B/kT \ll 1$ holds, then, according to (19-30),

$$\frac{S}{Nk} = \ln(2J+1) - \frac{g^2\beta^2 J(J+1)B^2}{6k^2T^2} + \frac{S_{\text{lattice}}}{Nk}$$

$$= \ln(2J+1) - \frac{J(J+1)y^2}{6} + \frac{S_{\text{lattice}}}{Nk} \qquad (19\text{-}86)$$

(where $y = g\beta B/kT$).

In the expression above, we have included the vibrational entropy of the crystal, S_{lattice}, as well as the entropy of the spin system in the magnetic field.

When the assumption that $y \ll 1$ is not justified but we are still dealing with an ensemble of independent magnets which satisfy the relation $\epsilon_m = g\beta Bm$, the magnetic entropy is given by Eq. (19-33),

$$\frac{S}{Nk} = \ln q(y) - \frac{y\bar{\mu}(y)}{g\beta} \qquad (19\text{-}33)$$

and the total entropy is the magnetic entropy plus the vibrational entropy of the lattice. In (19-33), the magnetic entropy is a monotonically decreasing function of y or B/T which approaches zero as $y \to \infty$.

It might as well be noted right now that below 10°K the lattice entropy is quite small and in general almost negligible below 1°K. Thus, according to the Debye theory, at low temperatures

$$\frac{S_{\text{lattice}}}{Nk} \approx \frac{4\pi^4}{5}\left(\frac{T}{\theta_D}\right)^3 \qquad (19\text{-}87)$$

If $\theta_D \approx 10^2$, we have $S_{\text{lattice}}/Nk = 8 \times 10^{-5}T^3$.

For $J = \frac{1}{2}$ and $y = 1$ (which is easy to achieve at 1°K, since

$$y = 0.66 \times \frac{10^{-4}gB}{T}\bigg)$$

according to (19-33) the magnetic entropy S/Nk is 0.81. This is a decrease of 0.19 as compared with the field free value $\ln(2J+1) = 1.10$. Thus the lattice entropy ($S/Nk = 8 \times 10^{-5}$ at 1°K) is completely negligible compared with the entropy decrease due to magnetization of the sample. The comparison here is not quite fair. In the expression for the lattice entropy [(19-87)] N is approximately the total number of atoms in the sample; in the expressions for the magnetic entropy, N is the number of paramagnetic atoms. Since, as we shall see, magnetic-cooling experiments work best with highly dilute paramagnetic salts, the ratio of lattice entropy to magnetic entropy may be 10 to 100 times greater than calculated above. This is still pretty small below 1°K.

This suggests the following experiment: Let a sample of a paramagnetic salt A be suspended by a thread or other good heat-insulating support in a tube as in Fig. 19-5. The tube is immersed in a dewar vessel containing liquid helium. The latter can be cooled down to slightly below 1°K by pumping (the vapor pressure of liquid helium is 0.12 mm at 1.0°K and 0.011 mm at 0.80°K). The tube contains helium gas at a suitably low pressure (perhaps 10^{-3} mm), which provides heat contact between the outer bath and the crystal. The crystal may now be isothermally magnetized by applying a magnetic field via the electromagnet. Since the entropy of the salt decreases during this isothermal magnetization, heat is released, which simply causes some of the liquid helium to boil away. Now the thermal contact between the crystal A and the outer helium bath is removed by evacuating the tube. The magnetic field is now decreased. This is the adiabatic-demagnetization step. If the demagnetization proceeds reversibly, since it is adiabatic, it is isentropic. If B decreases at constant entropy, T must decrease. This is magnetic cooling. In fact, according to (19-33), when the lattice entropy is negligible, the entropy is a function of B/T. Therefore at constant entropy, if B is reduced to zero, T should be reduced to zero. We shall see shortly why this is not actually the case. However, it is possible and not too difficult to achieve temperatures of 0.005 to 0.01°K.

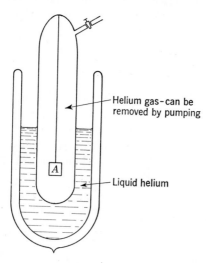

Fig. 19-5. Apparatus for magnetic cooling.

We shall not study the question of measuring the final temperature in great detail. If Curie's law holds, $\chi = A/T$; therefore, by measuring the susceptibility (using a weak field!), one can calculate a temperature which is called the magnetic temperature T^*. The constant A can be determined at higher temperatures where the temperature scale is well established.

As long as Curie's law holds, the thermodynamic temperature T and the magnetic temperature T^* will be equal.

However, as we shall see, it is a general rule that Curie's law does not hold at the lowest temperatures obtainable by isentropic demagnetization. One can then proceed as follows in order to establish an absolute temperature scale: The thermodynamic temperature can be rigorously defined by $T = (\partial E/\partial S)_{B_0, V}$. By a series of adiabatic demagnetizations, from different initial values of y one can determine the entropy as a function of the susceptibility or equivalently as a function of $T^* = A/\chi$. That is, by experiment we determine a curve $S_e(T^*)$ of the entropy at zero field as a function of T^*. One can now add an increment of energy δE to a cooled sample by γ-ray irradiation or a controlled heat leak of some sort. There is a resulting measured decrease in χ or increase in T^*, from which the increase in entropy δS can be determined using the experimental $S_e(T^*)$ curve. The thermodynamic temperature is then given by

$$T = \lim_{\delta E \to 0} \left(\frac{\delta E}{\delta S}\right)_{B_0}$$

The temperature-entropy diagram of Fig. 19-6 is one of the best ways of illustrating a magnetic-cooling experiment. Curve A is the entropy of the salt in the absence of a field. When the acoustical or lattice entropy is zero, the salt has a residual entropy of $\ln(2J+1)$ due to the random orientations of the spins. But we believe that, at absolute zero,

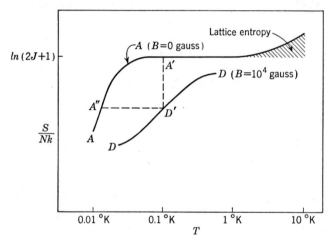

FIG. 19-6. Entropy-temperature diagram for a magnetic cooling experiment.

the substance should have zero entropy; therefore, at some low temperature, the spin system must somehow become ordered even at zero field. We shall consider in the next two sections how this can happen.

Let D be the entropy-temperature curve for a given field, say, 10^4 gauss. Then the isothermal magnetization is path $A'D'$ of the figure. The isentropic demagnetization is depicted by $D'A''$. The temperature at A'' is the final temperature achieved. It should be clear by inspection of the figure that, if a stronger field is used, the final temperature will be still lower; but, depending on the shape of the zero-field ST curve, the decrease in final temperature may not be very great.

19-13. Crystalline Stark-effect Splittings. The basic principles of this topic are simple and easy to describe, although in practice, the facts can be more complicated. Let us consider the chrome alums, $M_2Cr(SO_4)_2 \cdot 12H_2O$, where M is a unipositive alkali-metal ion. These are popular and useful salts for low-temperature magnetic experiments. The gas-phase ion Cr^{3+}, with the d^3 configuration, is in a ^4F state with $L = 3$

and $S = \frac{3}{2}$. In the crystal, there are four ions in the cubic unit cell, each surrounded by an octahedron of H_2O molecules. The electric fields of the crystal partially lift the orbital degeneracy; the lowest orbital state is nondegenerate. The other orbital states are about 20,000 cm^{-1} above the ground state and not important here.* The lowest state in the crystal is effectively in a spin-only $S = \frac{3}{2}$ state. There is however a slight distortion of the octahedral arrangement of the ligands along one of the threefold axes of the cubic crystal, so that the electric field at any one chromium ion has trigonal symmetry. The effect of the trigonal distortion is to split the fourfold-degenerate $S = \frac{3}{2}$ state into two substates, one with $S_z = \pm\frac{3}{2}$ and one with $S_z = \pm\frac{1}{2}$ (the z axis being the symmetry axis of the trigonal field). The separation between these two states is called δ and is about 0.12 cm^{-1} for the chrome alums, and, in this particular case, $|S_z| = \frac{3}{2}$ is the lower state.

Detailed information about these splittings is obtained by electron magnetic-resonance experiments.† The magnetic-resonance spectrum depends on the orientation of the magnetic field with respect to the trigonal axis of any particular ion. The energy-level diagram for the magnetic field along the trigonal axis is

$$\text{for } S_z = \pm\tfrac{3}{2}: \qquad \epsilon = 2\beta S_z B$$
$$\text{for } S_z = \pm\tfrac{1}{2}: \qquad \epsilon = \delta + 2\beta S_z B_0 \qquad (19\text{-}88)$$

The partition function is correspondingly

$$q(B,T) = (e^{+3\beta B/kT} + e^{-3\beta B/kT}) + e^{-\delta/kT}(e^{+\beta B/kT} + e^{-\beta B/kT}) \qquad (19\text{-}89)$$

Two particular situations may be noted. At high temperatures, with $\delta/kT \ll 1$ and $\beta B/kT \ll 1$, we can expand the exponentials and retain only the first nonzero terms in the variables δ/kT and $\beta B/kT$. The result is

$$q(B,T) \approx \left[4 + 10\left(\frac{\beta B}{kT}\right)^2\right] - \frac{\delta}{kT}\left[2 + \left(\frac{\beta B}{kT}\right)^2\right]$$
$$\approx \left[4 + 10\left(\frac{\beta B}{kT}\right)^2\right] - \frac{2\delta}{kT} \qquad (19\text{-}90)$$

By comparison with (19-25b), we see that the sum of terms enclosed in square brackets in the final expression is the weak-field partition function $q(B,T) = (2J + 1)[1 + J(J + 1)g^2\beta^2 B^2/6k^2T^2]$ for $J = \frac{3}{2}$ and $g = 2$ with no Stark splitting ($\delta = 0$). The magnetic properties all depend on $\partial(\ln q)/\partial B$, and to a first approximation the magnetic properties at high temperature are unaffected by the Stark splitting.

* B. Bleaney and W. K. Stevens, *Repts. Progr. Phys.*, **16**: 108 (1953); L. E. Orgel, *J. Chem. Phys.*, **23**: 1819 (1955); D. J. E. Ingram, "Spectroscopy at Radio and Microwave Frequencies," p. 156, Butterworth, London, 1955.

† D. M. S. Bagguley and J. H. E. Griffiths, *Proc. Roy. Soc. (London)*, **A204**: 188 (1950); B. Bleaney, *Proc. Roy. Soc. (London)*, **A204**: 203 (1950).

At zero field and low temperature, the partition function is given by

$$q(T) = 2(1 + e^{-\delta/kT}) \tag{19-91}$$

This is essentially the two-level problem analyzed in Prob. 8-2. The heat capacity shows a hump at $kT = \delta$, and the entropy decreases from $k \ln 4$ to $k \ln 2$, as the $\pm 3/2$ states are filled at the expense of the $\pm 1/2$ state.

According to this analysis, the entropy per chromic ion at zero field should decrease from $k \ln 4$ to $k \ln 2$ in the temperature range around 0.1°K. Below this temperature range, all the atoms are in the $\pm 3/2$ spin states. The additional interactions which are responsible for the loss of the residual entropy of $k \ln 2$ are discussed in the next section.

It should finally be remarked that the actual situation for rubidium, cesium, and methyl ammonium chrome alums conforms fairly well to the above description. However, for $K_2Cr(SO_4) \cdot 12H_2O$ the actual situation is more complicated. Above 160°K it behaves normally. Below this temperature, a phase transition occurs, and the magnetic-resonance pattern and the splitting pattern become more complicated [B. Bleaney, *Proc. Roy. Soc. (London)*, **A204**: 203 (1950)]. Unfortunately, it is this salt which has been most extensively studied from the point of view

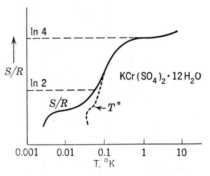

Fig. 19-7. Entropy vs. temperature for potassium chrome alum. [*Work of D. De Klerk, M. J. Steenland, and C. J. Gorter, Physica, **15**: 649 (1949); reproduced from C. Squire, "Low Temperature Physics," p. 161, McGraw-Hill, New York, 1953.*]

of magnetic cooling, low-temperature susceptibility, and the entropy-temperature relation. Figure 19-7 is a plot of the ST behavior of potassium chrome alum. It is seen that the entropy change around 0.1°K lowers the entropy somewhat below the theoretical value of $k \ln 2$.

19-14. The Interaction between Magnetic Moments. Antiferromagnetism. The energy of interaction of two magnetic dipoles is of the order of $\mu_1\mu_2/r^3$. In a magnetically dilute crystal, we should expect that when this interaction is of the order of kT the assumption of noninteracting magnets would fail. In chrome alum, the chromium atoms have a face-centered arrangement in the cubic lattice and the distance between two chromium atoms is $a_0/2^{1/2}$, where $a_0 = 12.2$ Å is the lattice constant of the cubic unit cell. If we take $\mu = 3\beta$ and set $kT = \mu^2/r^3$, we get $T = 0.008$°K. The spin-spin interaction should become very marked at temperatures in this range.

Figure 19-8 is a plot of the magnetic susceptibility in arbitrary units of potassium chrome alum as a function of temperature. There is a maximum in the susceptibility at 0.0037°, and below this temperature the susceptibility decreases. Thus the interaction of the spins results in the formation of an ordered state in which the spins are antiparallel. This is an example of the phenomenon of antiferromagnetism. We note, from Fig. 19-7, that the entropy is decreasing, presumably to zero apart from nuclear spin effects, in the same temperature range.

From the point of view of magnetic-cooling experiments, the interactions resulting in entropy decreases at finite temperatures present a formidable obstacle to the attainment of lower and lower temperatures.

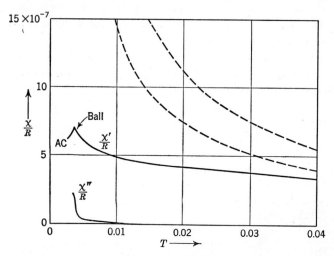

FIG. 19-8. The magnetic susceptibility of $KCr(SO_4)_2 \cdot 12H_2O$ as a function of the absolute temperature. The upper dashed curve is the Curie curve, based on the susceptibility at higher temperatures; the lower dashed curve is based on the theory of M. Hebb and E. Purcell [*J. Chem. Phys.*, **5**: 338 (1937)], which does endeavor to treat the mutual interaction of the magnetic dipoles.

The upper solid curve (χ'/R) is the experimental susceptibility of interest, measured by the "AC" (a-c bridge) and "Ball" (ballistic-galvanometer) methods. The other curve, χ''/R, concerns the imaginary component of the susceptibility, which is related to the absorption of energy on application of an a-c magnetic field. [*From D. De Klerk, M. J. Steenland, and C. J. Gorter, Physica*, **15**: 664 (1949).]

To some extent, the situation can be improved by using very powerful magnets (cf. Prob. 19-4). Progress can also be made by decreasing the magnitude of the interactions by using more dilute paramagnetic salts. This, of course, has the disadvantage that the amount of magnetic cooling per gram of salt is diminished.*

* We shall not treat the statistical mechanics of a system of magnets which interact by dipole-dipole forces in detail. We do wish to mention, however, that there is a rather general and powerful method of attacking such problems, known as the method of "diagonal sums" [I. Waller, *Z. Physik*, **104**: 132 (1936); J. H. Van Vleck, *J. Chem. Phys.*, **5**: 320 (1937), *Phys. Rev.*, **24**: 1168 (1948); M. Hebb and E. Purcell, *J. Chem. Phys.*, **5**: 338 (1937)].

We should also remark that, in systems that are not magnetically dilute, the phenomena of ferromagnetism and antiferromagnetism are common. These are ordered states in which the spins are all parallel or antiparallel, respectively. The susceptibility plots for MnO and MnS, two well-known antiferromagnetic materials, are displayed in Fig. 19-9.*

In general, for such substances, the dipole-dipole interaction between magnetic spins is not the only, or the principal, interaction. For magnetically concentrated materials, such as iron metal or MnO, the dominant interactions are the quantum-mechanical exchange interactions which lead to either ferromagnetism or antiferromagnetism and which are not understood in detail at present.

There is one unusual and interesting "interaction" between magnetic particles which can be mentioned at this point. The nucleus of He^3 has a spin of $\frac{1}{2}$ and a magnetic moment of 2.127 nuclear magnetons. The

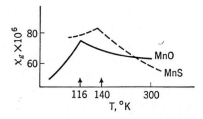

FIG. 19-9. Antiferromagnetism in manganous salts. Magnetic susceptibility per gram vs. temperature for MnO and MnS. (*From C. Squire, "Low Temperature Physics," p. 150, McGraw-Hill, New York, 1953*).

total canonical partition function of liquid He^3 is a product of the partition function of the atoms in the liquid and the nuclear spin partition function. We do not know the partition function of the liquid in detail, but this is not important. Except at low temperatures, the magnetic partition function is q_{mag}^N, where q_{mag} is the single-particle magnetic partition function of a He^3 nucleus. At a temperature of a few degrees Kelvin, the magnetic-dipole energy of interaction between two He^3 nuclei is negligible compared with kT. Nevertheless, the susceptibility falls well below the value predicted by Curie's law. The reason is that the He^3 atoms are fermions and have a low mass. At low temperatures, the atoms tend to occupy the lowest translational states available. For two particles to occupy the same translational state, their nuclear spins must be paired.†
This is entirely analogous to the magnetic properties of an "electron gas"

---

* H. Bizette, C. Squire, and B. Tsai, *Compt. rend.*, **207**: 449 (1938); C. Squire, *Phys. Rev.*, **56**: 922 (1939).

† See W. M. Fairbanks and G. K. Walters, Nuclear Resonance Experiments in He^3, in J. R. Dillinger (ed.), "Low Temperature Physics and Chemistry," p. 86, University of Wisconsin Press, Madison, Wis., 1958.

in a metal, where, at ordinary temperatures, most of the electrons are paired, in order to occupy the lowest available translational states.

19-15. Nuclear Cooling. For protons, the value of $y = g\beta_N B/kT$ is $1.01 \times 10^{-7} B/T$ (B in gauss). Therefore, for a sample of protons at 10^{-3}°K in a field of 10^4 gauss, $y = 1$; and there would be a very considerable degree of nuclear alignment. By adiabatic demagnetization of such a sample, one can expect to achieve temperatures of the order of 10^{-5}°K or less.

The procedure would be to establish a heat contact between the nuclear sample N (which is electronically diamagnetic) and a larger batch of a paramagnetic salt, P. N and P are now cooled to 0.01 to 0.001°K by isentropic demagnetization of P. N is now "isothermally magnetized," P being used as a heat sink (but care being taken to keep P out of the magnetic field!). The heat contact between N and P must now be broken. By adiabatic demagnetization, the nuclear sample N would then be cooled.

Preliminary experiments in this direction have been reported.* Temperatures of the order of 2×10^{-6} were achieved, as estimated by measuring the nuclear magnetic susceptibility.

Clearly, one can expect to cool the sample to the temperature where the energy of interaction of the nuclear spins is comparable with kT so that spontaneous ordering of the nuclear spins occurs.

If the nuclear spins in a substance, say CaF_2, became almost perfectly ordered owing to the spin-spin interaction at, say, 10^{-6}°K, and if there were no other quantum-mechanical degenerate states to be split by further interactions, then the system would truly be close to zero entropy with respect to all degrees of freedom. There might then be little point in cooling it to 10^{-8} or 10^{-10}°K, which would involve removing the last remnants of entropy.

It should be remarked that nuclear cooling in electronically diamagnetic substances as discussed above is not to be confused with nuclear alignment. The latter can be achieved, at least to some extent, in a paramagnetic material, by aligning the paramagnetic spins and relying on the strong interaction between an electron spin and the nuclear spin of the same atom to align the nuclear spin.†

PROBLEMS

19-2. (a) In $[Mn(H_2O)_6^{++}](SiF_6^=)$, the manganese cation has $S = 5/2$ and $g = 2$. The density is 1.9, and the formula weight is 305. Assuming the formula for ideal noninteracting magnets, calculate the volume susceptibility χ_V at 300°K and at 1°K.

(b) In $CuSO_4 \cdot 5H_2O$, the copper cation has $S = 1/2$ and $g = 2.2$. The density is 2.3, and the formula weight is 250. Again, assume the ideal formula, and calculate

---

* N. Kurti, F. N. H. Robinson, F. Simon, and D. A. Spohr, *Nature*, **178**: 450 (1956).
† For an introduction review, see N. Kurti, *Phys. Today*, **11**: 19 (March, 1958).

the volume susceptibility at 300 and 1°K. [*Note:* For both cases (*a*) and (*b*), the assumption of ideal noninteracting magnets is pretty good at 300°K but not good at 1°K. For references to the facts at low temperature, see D. J. E. Ingram, "Spectroscopy at Radio and Microwave Frequencies," pp. 166, 172, Butterworth, London, 1955.]

(c) Given a field of 10^4 gauss, with $\partial B/\partial x = 10^3$ gauss cm^{-1}. Calculate the forces on 1 mm^3 samples of the salts of parts (*a*) and (*b*) at 300 and 1°K.

19-3. The ground state of a fluorine atom is $^2P_{3/2}$, and the upper state of the spin-orbit multiplet is $^2P_{1/2}$ at 404.0 cm^{-1} above the ground state (Moore [30]). Other electronic states have very high excitation energies. Display a general formula for the electronic partition function, including the effects of a magnetic field (but assuming that $y \ll 1$), and for the molal susceptibility of a collection of fluorine atoms. Make a rough plot of the susceptibility as a function of temperature in the range from 100 to 1000°K. Is Curie's law obeyed by this system?

19-4. Assume that $K_2Cr(SO_4)_2 \cdot 12H_2O$ obeys the ideal formulae for $g = 2.00$ and $S = 3/2$ at 1°K. Calculate the entropy per atom at 1°K in fields of 7,450, 14,900, 29,800, and 37,250 gauss. From the data of De Klerk, Steenland, and Gorter [*Physica*, **15**, 649 (1949)] or from Fig. 19-7, estimate the temperature obtainable by adiabatic demagnetization. (*Suggestion:* Have each member of the class do the calculation for just one value of B.)

19-5. The heat of vaporization of liquid helium is 8×10^8 ergs mole^{-1} at 1°K (and 9×10^8 at 4°K; cf. Squire, "Low Temperature Physics," p. 67). How much liquid He will boil away for each of the isothermal magnetizations at 1°K described in the preceding problem?

19-6. Calculate the entropy loss per fluorine nucleus for the isothermal magnetization of CaF_2 at 0.01°K to a field of 50,000 gauss. The fluorine nucleus has a spin of $1/2$ and a magnetic moment of $g = 2.63$ nuclear magnetons.

19-7. How does the entropy of an antiferromagnetic material vary with applied field at a temperature just below the Néel temperature (temperature of maximum susceptibility—as measured in a weak field)? Can you explain this behavior qualitatively?

19-8. The equilibrium constant for the reaction of ferrous ion (Fe^{++}) with 1:10 phenanthroline (Ph) in aqueous solution to form the Fe(Ph)$_3^{++}$ complex is $10^{21.0}$ mole^{-3} liter^3

$$Fe^{++} + 3Ph \rightleftharpoons Fe(Ph)_3 \qquad K = 10^{21.0} \text{ moles}^{-3} \text{ liter}^3$$

Fe^{++} is paramagnetic, with a spin quantum number of $S = 2$ and an effective g factor of 2.05. Phenanthroline and the complex are diamagnetic. Estimate the change in log K at $T = 300°$K in a field of 30,000 gauss.

19-9. In an adiabatic-demagnetization experiment on a system of noninteracting magnets, assume that:

(a) The energy levels are given by the ideal formula $\epsilon_m = g\beta m B$.

(b) The lattice entropy is negligible.

(c) The value of B is not reduced to the point where T becomes so small that the assumption of noninteracting magnets fails.

How does the average magnetic moment of the sample vary with the field during the demagnetization?

19-10. Calculate the fractional difference in density for O_2 gas at, say, $P = 1$ atm pressure (measured outside the magnetic field) and $T = 100°$K between a region where there is a magnetic field, $B_0 = 5 \times 10^4$ gauss, and a region where there is no field. For O_2, $g = 2.00$, $S = 1$.

20

Distribution Functions and the Theory of Dense Fluids*

20-1. Introduction. The theory of the second virial coefficient (Chap. 15) is an approximation for calculating the properties of systems of interacting particles for systems which are sufficiently dilute so that the important interactions are due to only two particles at a time. In the present chapter, we derive exact statements that can be made about systems of interacting particles, whether the interactions be strong or weak. The material is inserted at this point because the results illuminate the Debye-Hückel theory, which is treated in the next chapter.

It was shown in Chap. 15 that the canonical partition function of a system of N atoms or molecules is

$$Q = (q_{\text{int}})^N \frac{Z}{N!\Lambda^{3N}} \qquad (20\text{-}1a)$$

where $\Lambda = (h^2/2\pi mkT)^{1/2}$ and the configuration integral Z is given by

$$Z = \int_V \cdots \int e^{-U(\mathbf{r}_1,\ldots,\mathbf{r}_N)/kT} \, d^3r_1 \cdots d^3\mathbf{r}_N \qquad (20\text{-}1b)$$

To derive these relations, it is assumed (a) that the system is sufficiently dilute so that corrected Boltzmann statistics apply; (b) that the internal energy levels of a molecule are unaffected by the interaction with neighboring molecules, and vice versa; (c) that the potential energy of interaction $U(\mathbf{r}_1, \ldots, \mathbf{r}_N)$ depends on the relative positions of the atoms or molecules, but not on their relative orientation; and (d) that classical statistical mechanics may be used for treating the interactions between the molecules. In view of assumption (b), the internal partition function q_{int} does not affect the P, V, T behavior of the gas and may be omitted in our subsequent derivations.

---

* The material herein is mainly taken from Hill, chap. 6, especially pp. 179–195 [9a]; see also J. De Boer, *Repts. Progr. Phys.*, **12**: 305 (1949).

20-2. Distribution Functions.

In our discussion, we can maintain the pretense that the particles are distinguishable, because the division by $N!$ in Eq. (20-1a) corrects for this error.

In general, in a canonical ensemble, the probability of observing the system in the system quantum state of energy E_i is $e^{-E_i/kT}/Q$. The probability of observing the system with particle 1 in the volume element $d^3\mathbf{r}_1$ at $\mathbf{r}_1$, particle j in the volume element $d^3\mathbf{r}_j$ at $\mathbf{r}_j$, etc., irrespective of the kinetic energy of the particles, is

$$p(\mathbf{r}_1, \ldots, \mathbf{r}_N) \, d^3\mathbf{r}_1 \cdots d^3\mathbf{r}_N = \frac{e^{-U(\mathbf{r}_1, \ldots, \mathbf{r}_N)/kT}}{Z} d^3\mathbf{r}_1 \cdots d^3\mathbf{r}_N \quad (20\text{-}2)$$

For conciseness, we shall henceforth speak of the probability of finding particle j at $\mathbf{r}_j$, instead of using the precise expression "the probability of finding particle j in the volume element $d^3\mathbf{r}_j$ at $\mathbf{r}_j$."

There are some more specialized distribution functions that interest us. Consider the subset of particles $1, 2, \ldots, n$, where $n < N$. The probability of finding particle 1 at $\mathbf{r}_1$, 2 at $\mathbf{r}_2$, and n at $\mathbf{r}_n$, averaged over all possible configurations of the remaining $N - n$ particles is

$$p^{(n)}(\mathbf{r}_1, \ldots, \mathbf{r}_n) = \frac{\int \cdots \int e^{-U(\mathbf{r}_1, \ldots, \mathbf{r}_N)/kT} \, d^3\mathbf{r}_{n+1} \cdots d^3\mathbf{r}_N}{Z} \quad (20\text{-}3)$$

Note that the integrand in the numerator of this equation is a function of all the coordinates $\mathbf{r}_1, \ldots, \mathbf{r}_N$; but we integrate only over the coordinates $\mathbf{r}_{n+1}, \ldots, \mathbf{r}_N$. The functions $p^{(n)}$ are called specific distribution functions. The generic distribution function $\rho^{(n)}(\mathbf{r}_1, \ldots, \mathbf{r}_n)$ is the probability, on making an observation of the configuration of the system, that a particle will be found at $\mathbf{r}_1$, a second at $\mathbf{r}_2$, etc., and an nth at $\mathbf{r}_n$. There are N ways of choosing the first particle, $N - 1$ of choosing the second, and $N - n + 1$ of choosing the nth; it is equally likely that any one of the particles should be found at $\mathbf{r}_1$, etc., and so

$$\rho^{(n)}(\mathbf{r}_1, \ldots, \mathbf{r}_n) = N(N-1) \cdots (N-n+1) p^{(n)}(\mathbf{r}_1, \ldots, \mathbf{r}_n)$$

$$= \frac{N!}{(N-n)!} p^n(\mathbf{r}_1, \ldots, \mathbf{r}_n) \quad (20\text{-}4)$$

The definitions of the general nth-order distribution functions are easy to understand, but the only functions that we shall seriously deal with are $\rho^{(2)}$ and $p^{(2)}$.

In a crystal, with the atoms at definite lattice positions, the probability of finding a particle at $\mathbf{r}_1$ is a periodic function of the variable position vector $\mathbf{r}_1$. For a liquid, however, except for points within a few atomic diameters of the walls, on the average, all positions $\mathbf{r}_1$ are equally probable, and we have

$$p^{(1)}(\mathbf{r}_1) = \frac{1}{V} \qquad \rho^{(1)}(\mathbf{r}_1) = \frac{N}{V} \quad (20\text{-}5)$$

For such a liquid, the quantity $\rho^{(2)}(\mathbf{r}_1,\mathbf{r}_2)$, which is the probability of finding one particle in the volume element $d^3\mathbf{r}_1$ at $\mathbf{r}_1$ and another simultaneously in $d^3\mathbf{r}_2$ at $\mathbf{r}_2$, should be a function only of the vector distance between the particles, $\mathbf{r}_{12} = \mathbf{r}_2 - \mathbf{r}_1$; in fact, for an isotropic liquid it should depend only on the scalar magnitude $r_{12} = |\mathbf{r}_{12}| = |\mathbf{r}_2 - \mathbf{r}_1|$. [This requirement plus Eq. (20-5) can be regarded as a microscopic definition of a fluid.] We shall interchangeably write $\rho^{(2)}(\mathbf{r}_1,\mathbf{r}_2)$, $\rho^{(2)}(\mathbf{r}_{12})$, and $\rho^{(2)}(r_{12})$ or even $\rho^{(2)}(r)$ as seems appropriate for emphasis in the particular context.

Suppose that there is no correlation whatsoever between the positions of the particles; then the probability of finding a particle at $\mathbf{r}_1$ is N/V; and the probability of finding a second particle at $\mathbf{r}_2$ (when we know there is a particle at $\mathbf{r}_1$) is $(N-1)/V$. Then

$$\rho^{(2)}(\mathbf{r}_1,\mathbf{r}_2) = \frac{N}{V}\frac{N-1}{V} \approx \frac{N^2}{V^2} \tag{20-6}$$

(We shall repeatedly make the approximation $N - 1 \approx N$ without explicit comment.) Now, because of the interaction between the particles, there should be some correlation between their positions; and the pair correlation function, or radial-distribution function, $g(r)$ is defined by

$$\rho^{(2)}(r_{12}) = \left(\frac{N}{V}\right)^2 g(r_{12}) \tag{20-7a}$$

or, more simply, $\rho^{(2)}(r) = (N/V)^2 g(r)$. An equivalent definition is

$$p^{(2)}(r) = \frac{1}{V^2} g(r) \tag{20-7b}$$

In a reasonably close-packed liquid, we expect the behavior for $g(r)$ indicated in the figure. There is zero probability of the two particles occupying the same space $[g(0) = 0]$; there is a high probability of the two particles being at a distance r_0 which is near the minimum of the potential curve describing their interaction. Because of the shell of molecules at r_0, there should be a second minimum farther out, and then a second maximum corresponding to the second coordination shell. Since there is no long-range order, $g(r) \to 1$ as $r \to \infty$.

The radial-distribution function can be observed experimentally by X-ray scattering in a liquid.*

Since by definition

$$\iint \rho^{(2)}(\mathbf{r}_1,\mathbf{r}_2)\, d^3\mathbf{r}_1\, d^3\mathbf{r}_2 = N(N-1) = N^2$$

then

$$\int_{\mathbf{r}_1} \int_{\mathbf{r}_2} g(\mathbf{r}_{12})\, d^3\mathbf{r}_1\, d^3\mathbf{r}_2 = V^2 \tag{20-8}$$

---

* A. Eisenstein and N. S. Gingrich, *Phys. Rev.*, **62**: 261 (1942); an extensive review of the data is given by N. S. Gingrich, *Revs. Modern Phys.*, **15**: 90 (1943).

Let the coordinates of particle 2 be measured with respect to the position of particle 1; since the value of $g(\mathbf{r}_{12})$ is independent of the position of the

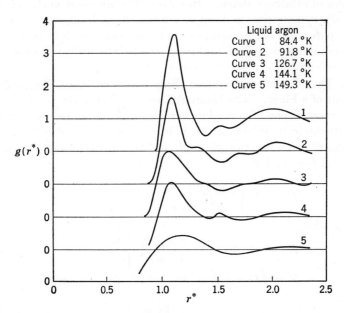

FIG. 20-1. The experimental radial-distribution function for liquid argon as a function of $r^* = r/\sigma$ ($\sigma = 3.42$ Å). [*The experiments are principally due to A. Eisenstein and N. S. Gingrich, Phys. Rev.*, **58**: 307 (1940), **62**: 601 (1942); *the curves are reproduced from J. DeBoer, Repts. Prog. Phys.*, **12**: 369 (1949).]

first particle (provided that it is not close to a wall), we can integrate over the coordinates of particle 1.

$$\int_{\mathbf{r}_1} g(\mathbf{r}_{12}) \, d^3\mathbf{r}_1 \, d^3\mathbf{r}_2 = V g(\mathbf{r}_{12}) \, d^3\mathbf{r}_{12} \quad (20\text{-}9)$$

and so

$$\int g(\mathbf{r}_{12}) \, d^3\mathbf{r}_{12} = V \quad (20\text{-}10)$$

Clearly, the probability of finding a molecule in the volume element $d^3\mathbf{r}_{12}$ at a distance $\mathbf{r}_{12}$ from another molecule is

$$\frac{N}{V} g(\mathbf{r}_{12}) \, d^3\mathbf{r}_{12} \quad (20\text{-}11a)$$

Another way of stating this is that $(N/V)g(r)$ is the probability per unit volume of finding a second particle at a distance r from a given particle. The probability of finding a molecule in the spherical shell at distance r is

$$\frac{N}{V} g(r) \, 4\pi r^2 \, dr \quad (20\text{-}11b)$$

Because of the isotropic nature of a liquid and the tumbling motions of the molecules, the considerations of the present section are applicable to molecules of arbitrary shape. However, they are most aptly suitable for spherically symmetrical molecules. When the molecules are not spherical and the potentials of interaction are angle-dependent, it is appropriate to introduce additional correlation functions for the probabilities of various relative orientations of the molecules.

20-3. The Radial-distribution Function and the Thermodynamic Functions of a Fluid. The Internal Energy. When the potential energy of the system of particles can be expressed as a sum of pair potentials, the thermodynamic functions of the fluid can be expressed in terms of the radial-distribution function $g(r)$. Recall that the thermodynamic energy function E is given by

$$E = kT^2 \left[\frac{\partial (\ln Q)}{\partial T}\right]_{N,V}$$

Then by (20-1)

$$E = \tfrac{3}{2}NkT + kT^2 \left[\frac{\partial (\ln Z)}{\partial T}\right]_V \qquad (20\text{-}12)$$

but

$$Z = \int_V \cdots \int e^{-U(\mathbf{r}_1,\ldots,\mathbf{r}_N)/kT} \, d^3\mathbf{r}_1 \cdots d^3\mathbf{r}_N$$

and, upon differentiating under the integral sign,

$$\frac{\partial Z}{\partial T} = +\frac{1}{kT^2} \int \cdots \int U(\mathbf{r}_1, \ldots, \mathbf{r}_N) \, e^{-U(\mathbf{r}_1,\ldots,\mathbf{r}_N)/kT} \, d^3\mathbf{r}_1 \cdots d^3\mathbf{r}_N \qquad (20\text{-}13)$$

Therefore

$$kT^2 \frac{\partial (\ln Z)}{\partial T} = \frac{\int \cdots \int U(\mathbf{r}_1, \ldots, \mathbf{r}_N) e^{-U(\mathbf{r}_1,\ldots,\mathbf{r}_N)/kT} \, d^3\mathbf{r}_1 \cdots d^3\mathbf{r}_N}{Z} \qquad (20\text{-}14a)$$

$$kT^2 \frac{\partial (\ln Z)}{\partial T} = \bar{U}(T,V,N) \qquad (20\text{-}14b)$$

The reader will recognize that $\bar{U}$ in (12-14b) is the ensemble average value of U, as explicitly displayed in (20-14a). Thus, from (20-12),

$$E = \tfrac{3}{2}NkT + \bar{U} \qquad (20\text{-}15)$$

The total energy is the ensemble average kinetic energy plus the ensemble average potential energy.

Now, introduce the assumption that the potential energy is a sum of pair potentials,

$$U(\mathbf{r}_1, \ldots, \mathbf{r}_N) = \sum_{1 \leq i < j \leq N} u(r_{ij}) \qquad (20\text{-}16)$$

The integral for $\bar{U}$ in (20-14a) is clearly then the sum of $N(N-1)/2$ identical terms; and

$$\bar{U} = \frac{[N(N-1)]}{2} \frac{\int_{\mathbf{r}_1}\int_{\mathbf{r}_2} u(r_{12}) \left[\int_{\mathbf{r}_3}\cdots\int_{\mathbf{r}_N} e^{-U(\mathbf{r}_1,\ldots,\mathbf{r}_N)/kT} d^3\mathbf{r}_3 \cdots d^3\mathbf{r}_N\right] d^3\mathbf{r}_1\, d^3\mathbf{r}_2}{Z} \quad (20\text{-}17)$$

The term in square brackets of (20-17) involving the integrations over $\mathbf{r}_3 \cdots \mathbf{r}_N$ gives $p^{(2)}(\mathbf{r}_1,\mathbf{r}_2)$; incorporating the $N(N-1)$ factor, we get

$$\bar{U} = \frac{1}{2} \iint u(r_{12}) \rho^{(2)}(\mathbf{r}_1,\mathbf{r}_2)\, d^3\mathbf{r}_1\, d^3\mathbf{r}_2 \quad (20\text{-}18)$$

Now set $\rho^{(2)}(\mathbf{r}_{12}) = (N/V)^2 g(\mathbf{r}_{12})$;

$$\bar{U} = \frac{1}{2}\left(\frac{N}{V}\right)^2 \int_{\mathbf{r}_1}\int_{\mathbf{r}_{12}} u(\mathbf{r}_{12}) g(\mathbf{r}_{12})\, d^3\mathbf{r}_1\, d^3\mathbf{r}_{12} \quad (20\text{-}19)$$

Integrate over the coordinates of particle 1, and we have

$$\frac{\bar{U}}{N} = \frac{1}{2}\left(\frac{N}{V}\right) \int u(r) g(r) 4\pi r^2\, dr \quad (20\text{-}20)$$

so that

$$\frac{E}{N} = \tfrac{3}{2}kT + \frac{1}{2}\frac{N}{V} \int u(r) g(r) 4\pi r^2\, dr \quad (20\text{-}21)$$

It is worth reemphasizing that, in (20-21), N/V is the average particle density and $(N/V)g(r)$ is the probability per unit volume of finding a second particle at a distance r from the first. Formula (20-21) is then eminently plausible.

20-4. The Equation of State and the Radial-distribution Function. We start from the relation

$$P = kT \frac{\partial (\ln Q)}{\partial V} = kT \frac{\partial (\ln Z)}{\partial V}$$

There is a neat way of carrying out the differentiation $\partial Z/\partial V$, due to H. S. Green.* Let the fluid be contained in a cubical box, so that each cartesian coordinate runs from 0 to $V^{1/3}$; then

$$Z = \int_0^{V^{1/3}} \cdots \int_0^{V^{1/3}} e^{-U/kT}\, dx_1\, dy_1\, dz_1 \cdots dx_N\, dy_N\, dz_N \quad (20\text{-}22)$$

To perform the differentiation, first transform to dimensionless scale variables,

$$x_i' = \frac{x_i}{V^{1/3}} \qquad dx_i = V^{1/3} dx_i'$$

---

* H. S. Green, "Molecular Theory of Fluids," North-Holland, Amsterdam, 1952; *Proc. Roy. Soc. (London)*, **A189**: 103 (1947).

The limits of integration, $0 \leq x_i \leq V^{1/3}$, become $0 \leq x'_i \leq 1$; the volume element is $dx_1\, dy_1 \cdots dy_N\, dz_N = V^N\, dx'_1 \cdots dy'_N\, dz'_N$.

Recall that U is a sum of terms

$$U(\mathbf{r}_1, \ldots, \mathbf{r}_N) = \sum_{1 \leq i < j \leq N} u(r_{ij})$$

where
$$r_{ij} = [(x_i - x_j)^2 + (y_i - y_j)^2 + (z_i - z_j)^2]^{1/2} = \left[\sum_{x,y,z}(x_i - x_j)^2\right]^{1/2}$$

$$= V^{1/3}\left[\sum_{x',y',z'}(x'_i - x'_j)^2\right]^{1/2}$$

so that
$$\left(\frac{\partial r_{ij}}{\partial V}\right)_{x'_i, y'_i, \ldots} = \tfrac{1}{3}V^{-2/3}\left[\sum_{x'y'z'}(x'_i - x'_j)^2\right]^{1/2} = \frac{1}{3}\frac{r_{ij}}{V} \qquad (20\text{-}23)$$

and
$$\left(\frac{\partial U}{\partial V}\right)_{x'\ldots} = \sum_{1 \leq i < j \leq N} \frac{du}{dr_{ij}}\frac{\partial r_{ij}}{\partial V} = \sum \frac{du}{dr_{ij}}\frac{1}{3}\frac{r_{ij}}{V} \qquad (20\text{-}24)$$

Now write the integral for Z in terms of the primed variables;

$$Z = V^N \int_0^1 \cdots \int_0^1 e^{-U/kT}\, dx'_1\, dy'_1 \cdots dz'_N \qquad (20\text{-}25)$$

$$\frac{\partial Z}{\partial V} = NV^{N-1}\int_0^1 \cdots \int_0^1 e^{-U/kT}\, dx'_1 \cdots dz'_N$$
$$- \frac{V^N}{kT}\int_0^1 \cdots \int_0^1 \left(\frac{\partial U}{\partial V}\right)_{x'\ldots} e^{-U/kT}\, dx'_1 \cdots dz'_N \qquad (20\text{-}26)$$

$$\frac{\partial Z}{\partial V} = \frac{N}{V}Z - \frac{V^N}{kT}\int_0^1 \cdots \int_0^1 \sum_{0 \leq i < j \leq N}\frac{1}{3}\frac{r_{ij}}{V}\frac{du(r_{ij})}{dr_{ij}}e^{-U/kT}\, dx'_1 \cdots dz'_N$$
$$(20\text{-}27)$$

Now, in the remaining integral, return to the unprimed variables; at the same time, recognize that the integral is the sum of $N(N-1)/2$ identical terms.

$$\frac{\partial Z}{\partial V} = \frac{N}{V}Z - \frac{N(N-1)}{6kTV}\int_0^{V^{1/3}} \cdots \int_0^{V^{1/3}} r_{12}\frac{du(r_{12})}{dr_{12}}e^{-U(\mathbf{r}_1,\ldots,\mathbf{r}_N)/kT}$$
$$dx_1 \cdots dz_N \qquad (20\text{-}28)$$

Now divide by Z, and recognize that the integral over the coordinates $\mathbf{r}_3, \ldots, \mathbf{r}_N$ gives $p^{(2)}(\mathbf{r}_1, \mathbf{r}_2)$;

$$\frac{\partial(\ln Z)}{\partial V} = \frac{N}{V} - \frac{N(N-1)}{6VkT}\int_{\mathbf{r}_1}\int_{\mathbf{r}_2} r_{12}\frac{du(r_{12})}{dr_{12}}p^{(2)}(r_{12})\, d^3\mathbf{r}_1\, d^3\mathbf{r}_2 \qquad (20\text{-}29)$$

Sec. 20-5] DISTRIBUTION FUNCTIONS 475

We substitute $p^{(2)}(r_{12}) = V^{-2}g(r_{12})$, and obtain a factor of V by integrating over $\mathbf{r}_1$ in the usual way;

$$\frac{\partial(\ln Z)}{\partial V} = \frac{N}{V} - \frac{N(N-1)}{6V^2kT} \int_{r_{12}} r_{12} \frac{du(r_{12})}{dr_{12}} g(r_{12}) 4\pi r_{12}^2 \, dr_{12} \quad (20\text{-}30)$$

Then

$$P = kT \frac{\partial(\ln Z)}{\partial V} = \frac{N}{V} kT - \frac{N^2}{6V^2} \int g(r) \frac{du}{dr} 4\pi r^3 \, dr$$

or

$$\frac{PV}{NkT} = 1 - \frac{N}{6VkT} \int g(r) \frac{du}{dr} 4\pi r^3 \, dr \quad (20\text{-}31)$$

This is the desired result. It is an exact relation which gives the equation of state in terms of the radial-distribution function, which in principle can be measured experimentally.

For a dilute gas, where only the term in N/V in the virial expansion

$$\frac{PV}{NkT} = 1 + \frac{B(T)N}{N_0 V} + C(T) \left(\frac{N}{VN_0}\right)^2 + \cdots$$

is important, we have therefore the approximate equation (N_0 is Avogadro's number)

$$B(T) = -\frac{N_0}{6kT} \int g(r) \frac{du}{dr} 4\pi r^3 \, dr \quad (20\text{-}32)$$

In Prob. 15-3, we saw that the second virial coefficient could be written as

$$B(T) = -\frac{N_0}{6kT} \int \frac{du}{dr} e^{-u(r)/kT} 4\pi r^3 \, dr \quad (20\text{-}33)$$

Comparison of (20-32) and (20-33) gives, for a dilute system,

$$g(r) = e^{-u(r)/kT} \quad (20\text{-}34)$$

When only the interactions of a single pair of molecules at a time are important, the radial-distribution function is simply the Boltzmann factor for the interaction between the pair. For more concentrated systems, this simple relation will not hold. The presence of a second molecule, and perhaps a third or a fourth molecule, for example, in contact with the first, influences the energies of the various positions that are available for the next molecule with respect to the first one. This is evident in our discussion of the qualitative behavior of the radial-distribution function and Fig. 20-1.

20-5. The Chemical Potential and the Radial-distribution Function. We have seen in the preceding sections that we can evaluate the thermodynamic internal energy and the equation of state if the radial-distribution function is known. If the radial-distribution function is known over a range of temperatures and densities, the free-energy change can be evaluated from the equation

$$\left[\frac{\partial(A/T)}{\partial(1/T)}\right]_V = E \qquad \left(\frac{\partial A}{\partial V}\right)_T = -P \qquad (20\text{-}35)$$

Thus the thermodynamic properties of the fluid can be calculated in terms of the pair-potential function and the radial-distribution function, which, in principle, can be measured experimentally. The statement is true only for those cases where the potential can be expressed as a sum of pair potentials. The latter is probably a reasonably accurate assumption for molecules where the main interaction is of the Lennard-Jones type. It should be remarked that, for practical computations of thermodynamic functions, the experimental accuracy that is needed in the measurement of the radial-distribution function by X-ray scattering is greater than has yet been obtained.

There is another way of calculating the chemical potential from the radial-distribution function, which is useful for theoretical purposes and which involves the idea of a coupling parameter. Suppose, for example, that the interaction between the particles is the ordinary coulomb interaction between two charges. For the interaction between the first and the jth particle, with charges $z_1 q$ and $z_j q$ and $u(r_{1j}) = z_1 z_j q^2 / r_{1j}$, where the valences of the ions are z_1 and z_j, respectively, q is the magnitude of the electron charge. Now imagine a charging process with the distance fixed at r_{1j} but the charge on particle 1 continuously changing from zero to $z_1 q$; at any point in the procedure, the charge on particle 1 is $z_1 q \xi_1$, where ξ_1 is the charging or coupling parameter for particle 1 and $0 \leq \xi_1 \leq 1$. When the charge on the first particle is $z_1 \xi_1 q$, the potential energy is $u(r_{1j}, \xi_1) = \xi_1 z_1 z_j q^2 / r_{1j}$.

In the same way, for an arbitrary interaction, we can write

$$u(r_{1j}, \xi_1) = \xi_1 u(r_{1j})$$

where $u(r_{1j})$ is the real interaction and $u(r_{1j}, \xi_1)$ is the hypothetical interaction when the coupling parameter for particle 1 is ξ_1. We can also introduce coupling parameters for both particles and write

$$u(r_{1j}, \xi_1, \xi_j) = \xi_1 \xi_j u(r_{1j})$$

but this procedure is not used for the present problem.

With a coupling parameter ξ_1 for the first particle only, the total potential energy of the fluid is

$$U(\mathbf{r}_1, \ldots, \mathbf{r}_N, \xi_1) = \sum_{j=2}^{N} \xi_1 u(r_{1j}) + \sum_{2 \leq i < k \leq N} u(r_{ik}) \qquad (20\text{-}36)$$

and

$$\frac{\partial U(\mathbf{r}_1, \ldots, \mathbf{r}_N, \xi_1)}{\partial \xi_1} = \sum_{j=2}^{N} u(r_{1j}) \qquad (20\text{-}37)$$

From Eq. (20-36), we see that, with $\xi_1 = 1$, $U(\mathbf{r},\xi_1)$ is the potential energy of the N particles; with $\xi_1 = 0$, $U(\mathbf{r},\xi_1)$ is the potential energy of $N - 1$ particles in the configuration $\mathbf{r}_2, \ldots, \mathbf{r}_N$.

The configuration integral then is

$$Z_N(\xi_1) = \int \cdots \int e^{-U(\mathbf{r}_1,\ldots,\mathbf{r}_N,\xi_1)/kT}\, d^3\mathbf{r}_1 \cdots d^3\mathbf{r}_N \qquad (20\text{-}38)$$

where the subscript N emphasizes that this is the configuration integral for N particles in the system. Now $Z_N(\xi_1 = 1) = Z_N$, the real value of the configuration integral for N particles. Furthermore,

$$Z_N(\xi_1 = 0) = V Z_{N-1} \qquad (20\text{-}39)$$

where Z_{N-1} is the real value of the configuration integral for $N - 1$ particles. Statement (20-39) should be obvious, since for $\xi_1 = 0$ the potential function $U(\mathbf{r}, \xi_1 = 0)$ is the potential for $N - 1$ particles and is not a function of $\mathbf{r}_1$; the integration over $\mathbf{r}_1$ in (20-38) contributes a factor of V.

The free energy of the system of N particles is, from (20-1a),

$$-\frac{A_N}{kT} = \ln Q_N(T,V) = \ln Z_N(T,V) - \ln N! - 3N \ln \Lambda \qquad (20\text{-}40)$$

The chemical potential is

$$\mu = \left(\frac{\partial A}{\partial N}\right)_{T,V} = A_N(T,V) - A_{N-1}(T,V)$$

so that

$$-\frac{\mu}{kT} = \ln \frac{Z_N}{Z_{N-1}} - \ln N - 3 \ln \Lambda \qquad (20\text{-}41)$$

But $Z_{N-1} = (1/V) Z_N(\xi_1 = 0)$; so we can write

$$\ln \frac{Z_N}{Z_{N-1}} = \ln \frac{V Z_N(\xi_1 = 1)}{Z_N(\xi_1 = 0)} = \ln V + \ln \frac{Z_N(\xi_1 = 1)}{Z_N(\xi_1 = 0)}$$

$$= \ln V + \int_0^1 \frac{\partial [\ln Z_N(\xi_1)]}{\partial \xi_1}\, d\xi_1 \qquad (20\text{-}42)$$

Now differentiate Eq. (20-38),

$$\frac{\partial Z_N(\xi_1)}{\partial \xi_1} = -\frac{1}{kT} \int \cdots \int \frac{\partial U(\mathbf{r}_1,\ldots,\mathbf{r}_N,\xi_1)}{\partial \xi_1} e^{-U(\mathbf{r}_1,\ldots,\mathbf{r}_N,\xi_1)/kT}\, d^3\mathbf{r}_1 \cdots d^3\mathbf{r}_N \qquad (20\text{-}43)$$

Substitute Eq. (20-37) for $\partial U/\partial \xi_1$,

$$\frac{\partial Z_N(\xi_1)}{\partial \xi_1} = -\frac{1}{kT} \int \cdots \int \left[\sum_2^N u(r_{1j}) e^{-U(\mathbf{r}_1,\ldots,\mathbf{r}_N,\xi_1)/kT}\right] d^3\mathbf{r}_1 \cdots d^3\mathbf{r}_N \qquad (20\text{-}44)$$

The sum in Eq. (20-44) contributes $N - 1$ identical terms; divide by $Z_N(\xi_1)$, and obtain

$$\frac{\partial[\ln Z_N(\xi_1)]}{\partial \xi_1} = -\frac{N-1}{kT} \frac{\int_{\mathbf{r}_1}\int_{\mathbf{r}_2} u(r_{12}) \left[\int_{\mathbf{r}_3} \cdots \int_{\mathbf{r}_N} e^{-U(\mathbf{r}_1,\ldots,\mathbf{r}_N,\xi_1)/kT} d^3\mathbf{r}_3 \cdots d^3\mathbf{r}_N\right] d^3\mathbf{r}_1 d^3\mathbf{r}_2}{Z}$$

(20-45)

The term in square brackets is clearly $p^{(2)}(\mathbf{r}_1,\mathbf{r}_2,\xi_1)$ or $V^{-2}g(r_{12},\xi_1)$, where $g(r_{12},\xi_1)$ is the radial-distribution function when the coupling parameter for particle 1 has the value ξ_1. Then, integrate as usual over $\mathbf{r}_1$, and

$$\frac{\partial[\ln Z_N(\xi_1)]}{\partial \xi_1} = -\frac{N}{VkT} \int u(r_{12}) g(r_{12},\xi_1)\, d^3\mathbf{r}_{12} \qquad (20\text{-}46)$$

Now use Eqs. (20-41) and (20-42),

$$\frac{\mu}{kT} = \ln \Lambda^3 \frac{N}{V} + \frac{N}{VkT} \int_{r_{12}}\int_{\xi_1} u(r_{12}) g(r_{12},\xi_1) 4\pi r_{12}^2\, dr_{12}\, d\xi_1 \qquad (20\text{-}47)$$

If we have a theory that predicts $g(r,\xi)$ as a function of the coupling parameter, the chemical potential can be calculated from (20-47).

The absolute activity λ, which occurs in the grand partition function, is given by $\mu/kT = \ln \lambda$; so

$$\lambda = \frac{N\Lambda^3}{V} e^{(N/VkT)[\int\int u(r)g(r,\xi) 4\pi r^2\, dr\, d\xi]} \qquad (20\text{-}48)$$

If there is no interaction $[u(r) = 0]$, the system behaves ideally and

$$\lambda_{\text{ideal}} = \frac{N\Lambda^3}{V} \qquad (20\text{-}49)$$

The activity coefficient γ may be defined by

$$\lambda = \gamma \lambda_{\text{ideal}} \qquad (20\text{-}50)$$

and so

$$\ln \gamma = \frac{N}{VkT} \int_0^\infty \int_0^1 u(r) g(r,\xi) 4\pi r^2\, dr\, d\xi \qquad (20\text{-}51)$$

For an ideal gas,

$$P = \frac{NkT}{V}$$

and thus

$$\frac{\mu_{\text{ideal}}}{kT} = \ln \frac{P}{kT} \Lambda^3 = \ln \frac{\Lambda^3}{kT} + \ln P \qquad (20\text{-}52)$$

The fugacity f may be defined for the general case by

$$\frac{\mu}{kT} = \ln \frac{\Lambda^3}{kT} + \ln f \qquad (20\text{-}53)$$

Therefore, $\ln f = \ln \dfrac{NkT}{V} + \dfrac{N}{VkT} \displaystyle\int_0^\infty \int_0^1 u(r)g(r,\xi)4\pi r^2\, dr\, d\xi \qquad (20\text{-}54)$

20-6. Radial-distribution Functions and Chemical Potentials in a Two-component System. This material is needed in the next chapter. Suppose that there are N_a particles of type a, whose coordinates we specify by $\mathbf{r}_i$, and N_b particles of type b, with coordinates the position vectors $\mathbf{s}_k$.

With the usual assumptions about classical mechanics and corrected Boltzmann statistics, the partition function is

$$Q = \frac{Z}{(\Lambda_a)^{3N_a}(\Lambda_b)^{3N_b}\, N_a!N_b!} \qquad (20\text{-}55)$$

where

$$Z = \int \cdots \int e^{-U(\mathbf{r}_1,\ldots,\mathbf{r}_{N_a};\mathbf{s}_1,\ldots,\mathbf{s}_{N_b})/kT}\, d^3\mathbf{r}_1 \cdots d^3\mathbf{r}_{N_a}\, d^3\mathbf{s}_1 \cdots d^3\mathbf{s}_{N_b} \qquad (20\text{-}56)$$

We use $\mathbf{r}_{ij}$ and r_{ij} for the vector $\mathbf{r}_j - \mathbf{r}_i$ and its scalar magnitude; for brevity we use $\mathbf{t}_{ij}$ and t_{ij} for the vector $\mathbf{s}_j - \mathbf{r}_i$ and $|\mathbf{s}_j - \mathbf{r}_i|$.

The total potential energy is a sum of pair potentials

$$U(\mathbf{r}_1,\ldots,\mathbf{r}_N;\mathbf{s}_1,\ldots,\mathbf{s}_N) = \sum_{1\le i<j\le N_a} u_{aa}(r_{ij}) + \sum_{1\le i<j\le N_b} u_{bb}(s_{ij}) + \sum_{\substack{1\le i\le N_a \\ 1\le j\le N_b}} u_{ab}(t_{ij}) \qquad (20\text{-}57)$$

There are $N_a(N_a - 1)/2$ of the u_{aa} terms, $N_b(N_b - 1)/2$ of the u_{bb} terms, and $N_a N_b$ of the u_{ab} terms.

The several specific pair distribution functions $p_{aa}^{(2)}(\mathbf{r}_1,\mathbf{r}_2)$ or $p_{aa}^{(2)}(r_{12})$, $p_{bb}^{(2)}(s_{12})$, and $p_{ab}^{(2)}(\mathbf{r}_1,\mathbf{s}_1) = p_{ab}^{(2)}(t_{11})$ are calculated exactly as before by integrating the phase integral over all other coordinates. The generic pair distribution functions are given by

$$\begin{aligned}\rho_{aa}^{(2)}(\mathbf{r}_1,\mathbf{r}_2) &= N_a(N_a - 1)p_{aa}^{(2)}(\mathbf{r}_1,\mathbf{r}_2) = N_A^2 p_{aa}^{(2)}(\mathbf{r}_1,\mathbf{r}_2) \\ \rho_{bb}^{(2)}(\mathbf{s}_1,\mathbf{s}_2) &= N_b^2 p_{bb}^{(2)}(\mathbf{s}_1,\mathbf{s}_2) \\ \rho_{ab}^{(2)}(\mathbf{r}_1,\mathbf{s}_1) &= \rho_{ab}^{(2)}(t_{11}) = N_a N_b p_{ab}^{(2)}(t_{11})\end{aligned} \qquad (20\text{-}58a)$$

and

The radial-distribution functions are given by

$$\begin{aligned}g_{aa}(r_{12}) &= \frac{V^2}{N_a(N_a - 1)}\rho_{aa}^{(2)}(r_{12}) = V^2 p_{aa}^{(2)}(r_{12}) \\ g_{bb}(s_{12}) &= \frac{V^2}{N_b(N_b - 1)}\rho_{bb}^{(2)}(s_{12}) = V^2 p_{bb}^{(2)}(r_{12}) \\ g_{ab}(t_{11}) &= \frac{V^2}{N_a N_b}\rho_{ab}^{(2)}(t_{12}) = V^2 p_{ab}^{(2)}(t_{12})\end{aligned} \qquad (20\text{-}58b)$$

Now let us repeat the calculation of the chemical potential. By removing one atom of type a from the system we obtain

$$-\frac{\mu_a}{kT} = \ln \frac{Z_{N_a, N_b}}{Z_{N_a-1, N_b}} - \ln N_a - \ln \Lambda_a^3 \qquad (20\text{-}59)$$

We introduce a coupling parameter ξ_{1a} for the first particle of type a; but for brevity we call it simply ξ. Then, just as before,

$$Z_{N_a, N_b}(\xi = 0) = V Z_{N_a-1, N_b} \qquad (20\text{-}60)$$

The chemical potential is then given by

$$\frac{\mu_a}{kT} = \ln \Lambda^3 \frac{N_a}{V} - \int_0^1 \frac{\partial [\ln Z(\xi)]}{\partial \xi} d\xi \qquad (20\text{-}61)$$

and $\quad U(\mathbf{r}_1, \ldots, \mathbf{s}_N, \xi) = \sum_{j=2}^{N_a} \xi u_{aa}(r_{1j}) + \sum_{k=1}^{N_b} \xi u_{ab}(t_{1k})$

$$+ \sum \text{ (other terms not involving particle 1 of type } a) \qquad (20\text{-}62)$$

Therefore $\quad \dfrac{\partial U}{\partial \xi} = \sum\limits_{j=2}^{N_a} u_{aa}(r_{1j}) + \sum\limits_{i=1}^{N_b} u_{ab}(t_{1i}) \qquad (20\text{-}63)$

Now $\quad Z_{N_a N_b}(\xi) = \int \cdots \int e^{-U(\mathbf{r}_1, \ldots, \mathbf{s}_{N_b}, \xi)/kT} d^3\mathbf{r}_1 \cdots d^3\mathbf{s}_N$

and $\quad \dfrac{\partial Z_{N_a N_b}(\xi)}{\partial \xi} = -\dfrac{1}{kT} \int \cdots \int \dfrac{\partial U}{\partial \xi} e^{-U(\mathbf{r}_1, \ldots, \mathbf{s}_{N_b}, \xi)/kT} d^3\mathbf{r}_1 \cdots d^3\mathbf{s}_N$

$$(20\text{-}64)$$

When we substitute for $\partial U/\partial \xi$ from Eq. (20-63), there will be $N_a - 1$ identical integrals for u_{aa} terms and N_b identical integrals for u_{ab} terms. We write

$$\frac{\partial Z}{\partial \xi} = -\frac{N_a}{kT} \int \cdots \int u_{aa}(r_{12}) e^{-U(\mathbf{r}_1, \ldots, \mathbf{s}_{N_b}, \xi)/kT} d^3\mathbf{r}_1 \cdots d^3\mathbf{s}_{N_b}$$

$$-\frac{N_b}{kT} \int \cdots \int u_{ab}(t_{11}) e^{-U(\mathbf{r}, \ldots, \mathbf{s}_{N_b}, \xi)/kT} d^3\mathbf{r}_1 \cdots d^3\mathbf{s}_{N_b} \qquad (20\text{-}65)$$

For the first term, the integral over $\mathbf{r}_3, \ldots, \mathbf{r}_{N_a}; \mathbf{s}_1, \ldots, \mathbf{s}_{N_b}$ does not involve $u_{aa}(r_{12})$. We recall that

$$p_{aa}^{(2)}(r_{12}, \xi) = \frac{\int_{\mathbf{r}_3} \cdots \int_{\mathbf{s}_{N_b}} e^{-U(\mathbf{r}_1, \ldots, \mathbf{s}_{N_b}, \xi)/kT} d^3\mathbf{r}_3 \cdots d^3\mathbf{s}_{N_b}}{Z} \qquad (20\text{-}66)$$

Similarly, for the second term in Eq. (20-65), the integral over $\mathbf{r}_2, \ldots, \mathbf{r}_N; \mathbf{s}_2, \ldots, \mathbf{s}_N$ essentially gives $p_{ab}^{(2)}(t_{11},\xi)$. Therefore, dividing (20-65) by Z,

$$\frac{\partial(\ln Z_{N_a,N_b})}{\partial \xi} = -\frac{N_a}{kT} \int_{\mathbf{r}_1} \int_{\mathbf{r}_2} u_{aa}(r_{12}) p_{aa}^{(2)}(r_{12}) \, d^3\mathbf{r}_1 \, d^3\mathbf{r}_2$$

$$-\frac{N_b}{kT} \int_{\mathbf{r}_1} \int_{\mathbf{s}_1} u_{ab}(t_{11}) p_{ab}^{(2)}(t_{11}) \, d^3\mathbf{r}_1 \, d^3\mathbf{s}_1 \quad (20\text{-}67)$$

Now substitute the radial-distribution functions $g(r)$ for the $p^{(2)}$ functions, using (20-58b); at the same time integrate over $d^3\mathbf{r}_1$, and get a factor of V,

$$\frac{\partial[\ln Z_{N_aN_b}(\xi)]}{\partial \xi} = -\frac{N_a}{VkT} \int u_{aa}(r_{12}) g_{aa}(r_{12},\xi) 4\pi r_{12}^2 \, dr_{12}$$

$$-\frac{N_b}{VkT} \int u_{ab}(t_{11}) g_{ab}(t_{11},\xi) 4\pi t_{11}^2 \, dt_{11} \quad (20\text{-}68)$$

$$\frac{\mu_a}{kT} = \ln \Lambda^3 \frac{N_a}{V} - \int_0^1 \frac{\partial[\ln Z(\xi)]}{\partial \xi} \, d\xi$$

$$= \ln \Lambda^3 \frac{N_a}{V} + \frac{N_a}{VkT} \int_{r=0}^{r=\infty} \int_{\xi=0}^{\xi=1} u_{aa}(r) g_{aa}(r,\xi) 4\pi r^2 \, dr \, d\xi$$

$$+ \frac{N_b}{VkT} \int_{r=0}^{r=\infty} \int_{\xi=0}^{\xi=1} u_{ab}(r) g_{ab}(r,\xi) 4\pi r^2 \, dr \, d\xi \quad (20\text{-}69)$$

This is the desired result. It is an almost obvious generalization of (20-47).

20-7. The Potential of Mean Force. We saw that for a dilute gas, in which only the second virial coefficient is important,

$$g(r_{12}) = V^2 p^{(2)}(r_{12}) = e^{-u(r_{12})/kT} \quad (20\text{-}34)$$

where $u(r_{12})$ is the actual potential of interaction between particles 1 and 2. The relation is not true for more concentrated systems where the simultaneous interaction of three or more particles is important. The pair potential of mean force, $w^{(2)}(r_{12})$, is defined as that potential which, in general, bears the same relation to $g(r)$ as $u(r_{12})$ does for the dilute case; i.e.,

$$g(r_{12}) = V^2 p^{(2)}(r_{12}) = e^{-w^{(2)}(r_{12})/kT} \quad (20\text{-}70)$$

From the definition of $p^{(2)}$, it follows that

$$e^{-w^{(2)}(r_{12})/kT} = V^2 \int \cdots \int \frac{e^{-U(\mathbf{r}_1,\ldots,\mathbf{r}_N)/kT}}{Z} \, d^3\mathbf{r}_3 \cdots d^3\mathbf{r}_N \quad (20\text{-}71)$$

In mechanics, if the potential energy of a system of particles is $U(\mathbf{r}_1, \ldots, \mathbf{r}_N)$, the force on the first particle when the other particles are at $\mathbf{r}_2, \ldots, \mathbf{r}_N$ is $-\nabla_{(1)} U(\mathbf{r}_1, \ldots, \mathbf{r}_N)$, where $\nabla_{(1)}$ means the gradient taken with respect to the coordinates of the first particle. Then take the gradient of both sides of Eq. (20-71), with respect to the coordinates of particle 1,

$$-\frac{1}{kT} e^{-w^{(2)}(r_{12})/kT} \nabla_{(1)} w^{(2)}(r_{12})$$
$$= -\frac{(\nabla^2/kT) \int \cdots \int \nabla_{(1)} U(\mathbf{r}_1, \ldots, \mathbf{r}_N) e^{-U(\mathbf{r}_1, \ldots, \mathbf{r}_N)/kT} d^3\mathbf{r}_3 \cdots d^3\mathbf{r}_N}{Z}$$

(20-72)

Dividing Eq. (72) by (71),

$$\nabla_{(1)} w^{(2)}(r_{12}) = \frac{\int \cdots \int \nabla_{(1)} U(\mathbf{r}_1, \ldots, \mathbf{r}_N) e^{-U(\mathbf{r}_1, \ldots, \mathbf{r}_N)/kT} d^3\mathbf{r}_3 \cdots d^3\mathbf{r}_N}{\int \cdots \int e^{-U(\mathbf{r}_1, \ldots, \mathbf{r}_N)/kT} d^3\mathbf{r}_3 \cdots d^3\mathbf{r}_N}$$

(20-73)

The right-hand side of Eq. (20-73) is the ensemble average force on particle 1 at $\mathbf{r}_1$, when particle 2 is at $\mathbf{r}_2$, but averaged over all configurations of the particles 3, . . . , N; the potential $w^{(2)}(r_{12})$, which is defined so that $g(r_{12})$ satisfies a simple Boltzmann relation (20-70), has the property that its gradients give the mean forces on particles 1 and 2.

The potentials of mean force and the distribution functions are used extensively in advanced statistical mechanical theories of condensed phases (cf. Hill, pp. 193–285 [9a]). We shall have some small occasion to use them in our study of the Debye-Hückel theory.

20-8. Approximate Theories of Dense Gases and Liquids. If the classical approximation is satisfactory, the thermodynamic properties of a fluid can be deduced if the configuration integral can be evaluated. For a typical liquid, a cube of edge 1,000 Å contains about 10^7 molecules. The straightforward evaluation of the configuration integral for such a system requires an integration in a (3×10^7)-dimensional space; this is not practicable at present for a realistic intermolecular potential or even for a rigid-sphere potential.

It has therefore been necessary to make very drastic simplifying assumptions in order to make approximate calculations about actual fluids. These calculations are not very successful, and we shall only mention them here.* Many of the theories are semiempirical in the sense, for example, that some of the parameters in the theoretical equation of state are chosen to fit the properties of the liquid at one temperature and pressure.

There are a number of "cell" and "hole" theories of liquids. In the cell theories, the volume V occupied by the N molecules of the liquid is divided up into a lattice of N cells, each of volume V/N. Each molecule

---

* They are described in some detail in a number of treatises, including Hill, chap. 16 [9b], and Hirschfelder, Curtiss, and Bird, chap. 4 [15]. See also J. S. Rowlinson, "Liquids and Liquid Mixtures," Butterworth, London, 1959; I. Prigogine, "The Molecular Theory of Solutions," North-Holland, Amsterdam, 1957.

is confined in a single cell. The potential field in which the molecule moves is usually calculated on the assumption that the other molecules are at their equilibrium positions at the centers of their own cells. In effect it is assumed that the motion of a molecule is independent of the motion of its neighbors.

The ideal that each molecule is confined in a cell and cannot exchange places with its neighbors leads to a certain decrease in the calculated entropy of the system (cf. Prob. 20-2 and the discussion of communal entropy). The cell picture is obviously most appropriate for a liquid at a high pressure and density and not appropriate for a gas of intermediate density.

In the hole theories, the possibility is recognized that some of the cells may be empty, thus allowing an additional degree of randomness in the fluid.

A particularly promising variation of these treatments is the theory of "significant structures" due to Eyring, Ree, and their collaborators.* The liquid is viewed as a mixture of microcrystals, which have the partition function for a solid, and gaslike regions. Rather successful semiempirical equations of state have been derived for a variety of liquids.

Another general approach to the calculation of liquid properties is based on an approximate solution of a set of integral equations which relate the distribution function $\rho^{(n)}(\mathbf{r}_1, \ldots, \mathbf{r}_n, \xi_1)$ to the pair potential and the distribution function of next higher order† (cf. Prob. 20-3). Kirkwood and his collaborators have made extensive studies of this treatment.‡

Finally, there are numerical calculations of the partition function of a system containing several hundred particles. Even for this small system, it is not practical with present-day computers§ to make a rigorous evaluation of the partition function. In one treatment, a representative sample of configurations in phase space is constructed by the so-called Monte Carlo procedure.‖ An alternative procedure is to calculate the actual trajectories for a system of particles with given initial conditions and to take the configurations encountered in a representative sample of such trajectories as representative of the important configurations of the liquid.¶

---

* H. Eyring, T. Ree, N. Hirai, C. M. Carlson, E. J. Fuller, and T. R. Thompson, *Proc. Natl. Acad. Sci.*, **44**: 683 (1958), **45**: 1594 (1959), **46**: 333, 336, 639 (1960).

† J. G. Kirkwood, *J. Chem. Phys.*, **3**: 300 (1935); Hill, pp. 198–203 [9a]; HCB, pp. 325–326 [15].

‡ J. G. Kirkwood, E. Maun, and B. Alder, *J. Chem. Phys.*, **18**: 1040 (1950).

§ 1960!

‖ N. Metropolis, A. W. Rosenbluth, M. N. Rosenbluth, A. H. Teller, and E. Teller, *J. Chem. Phys.*, **21**: 1087 (1953), **22**: 881 (1954); W. W. Wood, F. R. Parker, and J. D. Jacobson, *J. Chem. Phys.*, **27**: 720, 1207 (1957).

¶ B. J. Alder and T. E. Wainwright, *J. Chem. Phys.*, **27**: 1208 (1957), **31**: 459 (1959), **33**: 1439 (1960).

There are difficult computational problems in the application of the numerical method. There are some technical questions about the representative character of the configurations or trajectories considered. There are more fundamental and as yet unanswered questions about the difference in properties between a small sample and a real macroscopic sample and about the effect of the shape of the container on the partition function for a very small sample. Nevertheless, straightforward numerical calculation of the partition function and other properties of a fluid may prove to be an especially powerful method of attack in this difficult field.

PROBLEMS

20-1. For a dilute gas, $g(r,\xi) = e^{-\xi u(r)/kT}$ [(20-34)]. Then use the relation between the fugacity and the radial-distribution function [(20-54)] to derive a formula for fugacity in terms of the potential $u(r)$. Using this result, find a relation between the fugacity and the second virial coefficient.

20-2. Calculate the canonical partition function and the entropy of a crystal consisting of N atoms, with each atom confined in a cell of volume v. The cells are distinguishable from each other because they are at different lattice points in the crystal. Assume that the potential energy of an atom inside a cell is zero, except at the cell boundaries, where the potential is infinite. Compare this result with the canonical partition function and the entropy of a perfect gas of N atoms in a volume $V = Nv$.

The difference between these two results for the entropy is the so-called communal entropy, which, in a sense, results from the possibility of exchange between molecules in different cells. One of the difficult problems in cell theories of liquids (cf. Sec. 20-8) is concerned with how much of the communal entropy to include for the liquid.

20-3. For a shorthand notation, write $U(\mathbf{r},\xi)$ for $U(\mathbf{r}_1, \ldots, \mathbf{r}_N, \xi_1)$, and

$$U(\mathbf{r},\xi) = U(\mathbf{r}_1, \ldots, \mathbf{r}_N, \xi) = \sum_{j=2}^{N} \xi u(r_{1j}) + \sum_{2 \leq i < k \leq N} u(r_{ik})$$

Recall the definitions of the specific distribution functions,

$$p^{(2)}(r_{12},\xi) = \frac{\int \cdots \int e^{-U(\mathbf{r},\xi)/kT} \, d^3\mathbf{r}_3 \cdots d^3\mathbf{r}_N}{Z(\xi)}$$

and

$$p^{(3)}(r_{12},r_{13},r_{23},\xi) = \frac{\int \cdots \int e^{-U(\mathbf{r},\xi)/kT} \, d^3\mathbf{r}_4 \cdots d^3\mathbf{r}_N}{Z(\xi)}$$

(a) What are the units of $p^{(2)}$ and $p^{(3)}$?
(b) Show that

$$-kT \frac{\partial [\ln Z(\xi)]}{\partial \xi} = NV \int \cdots \int u(r_{13}) p^{(2)}(r_{13},\xi) \, d^3\mathbf{r}_{13}$$

(c) Show that

$$-kT \frac{\partial [\ln p^{(2)}(r_{12},\xi)]}{\partial \xi} = u(r_{12}) + N \int u(r_{13}) \left[\frac{p^{(3)}(\mathbf{r}_{12},\mathbf{r}_{13},\mathbf{r}_{23},\xi)}{p^{(2)}(r_{12})} - V p^{(2)}(r_{13},\xi) \right] d^3\mathbf{r}_{13}$$

Integration of the equation in part (c) with respect to ξ gives rise to the Kirkwood integral equation, which is the basis of one calculational approach to the study of liquid structure (cf. the references cited in Sec. 20-8).

21

Solutions of Electrolytes. The Debye-Hückel Theory

21-1. Introduction. Dimensional Considerations. For a 1:1 electrolyte in water, the deviations from the laws of ideal dilute solution are of the order of 20 per cent for a 0.01-molal solution where the mole fraction of solute is $\sim 4 \times 10^{-4}$, whereas most solutions of small neutral molecules in ordinary solvents behave almost ideally at these low concentrations. It is a valid generalization of experience that for solutions of small molecules in ordinary solvents the deviations from ideal behavior become important at a very much lower concentration for ionic solutes than for neutral molecular solutes. One naturally suspects that this is due to the long-range nature of coulombic forces as compared with intermolecular forces of the van der Waals type.

As is well known, the Debye-Hückel theory does not accurately predict the equilibrium properties for a typical concentrated electrolyte solution; however, it is remarkably and admirably effective for calculating the limiting laws for the deviations from ideal behavior as the solute concentration goes to zero. With a certain amount of judicious semiempirical adjustment of parameters, the theory gives excellent agreement with activity-coefficient measurements for 1:1 electrolytes for concentrations up to 0.1 molal, and it is useful even for 1-molal solutions. For electrolytes containing polyvalent ions, it works only in somewhat more dilute solutions.

We have not studied the general theory of solutions, including the effects of solvent-solvent, solvent-solute, and solute-solute interactions. In the present treatment, we focus our attention exclusively on the electrostatic solute-solute interactions. The solute-solvent interactions (which are briefly discussed in Sec. 21-15) and the solvent-solvent interactions will affect the chemical potentials of the ionic species at infinite dilution. However, for moderately dilute solutions, these interactions do not contribute to the deviations from ideal behavior with changing ion concentration, because the interactions are not significantly altered by the concentration changes.

The variables which will affect the electrostatic energy of interaction of the ions in an electrolyte solution can be recognized by a dimensional analysis. Consider a solution containing a 1:1 electrolyte A^zB^{-z}, which is completely dissociated in solution to give ions of charge $\pm zq$ (q being the magnitude of the fundamental charge). Let N be the number of AB "molecules" dissolved in volume V. There are two important distances. For a random distribution of ions, the average distance between two ions is of the order of $(V/N)^{1/3}$. If ϵ is the dielectric constant of the solvent, the potential energy of interaction between the two ions at a distance of $(V/N)^{1/3}$ is $z^2q^2/\epsilon(V/N)^{1/3}$. The corresponding interesting dimensionless quantity for statistical mechanics is $z^2q^2/kT\epsilon(V/N)^{1/3}$. If a is the radius of an ion, an independent dimensionless quantity is $z^2q^2/\epsilon akT$. We expect that the dimensionless intensive free-energy quantities F/NkT or μ/kT (where μ is the chemical potential of one of the components) will be functions of the above dimensionless parameters, the quantity $z^2q^2/kT\epsilon(V/N)^{1/3}$ being more important in dilute solution, where the ions, on the average, will be relatively far apart.

21-2. More Electrostatics. Before taking up the Debye-Hückel theory, it is worthwhile to review some theorems in electrostatics. We are particularly concerned with the calculation of the electrostatic potential and the electrostatic energy of interaction of a set of charges.

In a medium of dielectric constant ϵ, the electrostatic potential $\phi(\mathbf{r})$ at a general field point $\mathbf{r}$ due to a set of point charges q_i at the points $\mathbf{r}'_i$ is

$$\phi(\mathbf{r}) = \sum_i \frac{q_i}{\epsilon|\mathbf{r} - \mathbf{r}'_i|} \tag{21-1}$$

If, instead of a set of point charges, there is a continuous charge distribution $\rho(\mathbf{r}')$, where $\mathbf{r}'$ is a variable position vector, then the expression for the potential at the field point $\mathbf{r}$, analogous to (21-1), is

$$\phi(\mathbf{r}) = \int_V \frac{\rho(\mathbf{r}')\,d^3\mathbf{r}'}{\epsilon|\mathbf{r} - \mathbf{r}'|} \tag{21-2}$$

Note that $\mathbf{r}'$ is a variable of integration, but $\mathbf{r}$ is a free variable in (21-2). Take the Laplacian derivative with respect to $\mathbf{r}$ of both sides of (21-2);

$$\nabla^2 \phi(\mathbf{r}) = \nabla^2 \int_V \frac{\rho(\mathbf{r}')\,d^3\mathbf{r}'}{\epsilon|\mathbf{r} - \mathbf{r}'|}$$

It can be shown by differentiation under the integral sign that

$$\nabla^2 \int_V \frac{\rho(\mathbf{r}')\,d^3\mathbf{r}'}{\epsilon|\mathbf{r} - \mathbf{r}'|} = -\frac{4\pi}{\epsilon}\rho(\mathbf{r}) \tag{21-3a}$$

so that

$$\nabla^2 \phi(\mathbf{r}) = -\frac{4\pi\rho(\mathbf{r})}{\epsilon} \tag{21-3b}$$

which is Poisson's equation. This may be taken as the fundamental equation of electrostatics. It can then be shown that Eq. (21-2) is the general solution of this differential equation.

Given a charge distribution, one can calculate a potential by solving Eq. (21-3b) subject to appropriate boundary conditions. In applying electrostatic theory to ionic solutions, the solvent molecules are regarded as contributing a continuous dielectric constant ϵ. We disregard, for example, the violent fluctuations in potential as we go through one molecule, such as a water molecule with its complicated charge distribution.

In the Debye-Hückel theory, we shall be concerned with a continuous-charge density function around a given ion due to the ion atmosphere. The most important point is that this continuous-charge distribution arises by averaging over all possible configurations of the ions. If we regard the ions as *point* charges, a particular configuration, specified by the position vectors $\mathbf{r}_1, \mathbf{r}_2, \ldots, \mathbf{r}_N$ of the centers of the ions, corresponds to a charge distribution which is discontinuous at the actual positions of the ions. However, we can smear out this charge distribution by saying that at $\mathbf{r}_1$ there is the charge q_1 somewhere within the volume element d^3r_1. We thus can construct a continuous-charge distribution even for a configuration of point charges. If we wish, we can also use the fact that each ion has a finite radius and is itself a continuous charge distribution and thus a particular configuration of the ions corresponds to a continuous-charge distribution.

Gauss' law is a general theorem in electrostatics which asserts that the surface integral of the normal component of the electric-displacement vector over any closed surface is 4π times the total amount of charge enclosed by the surface. For example, for a spherically symmetrical ion of radius $a/2$ immersed in a dielectric medium with no other charges, the potential outside the ion ($r > a/2$) is

$$\phi(r) = \frac{zq}{\epsilon r} \tag{21-4a}$$

The electric-displacement vector is directed radially outward, and its magnitude is

$$D = -\epsilon \frac{d\phi}{dr} = \frac{zq}{r^2} \tag{21-4b}$$

The surface integral over the sphere of radius r is

$$4\pi r^2 D = 4\pi zq \tag{21-4c}$$

in agreement with Gauss' theorem.

Let $\rho_1(\mathbf{r})$ be a charge distribution which gives rise to the potential $\phi_1(\mathbf{r})$ such that $\nabla^2 \phi_1 = -4\pi \rho_1/\epsilon$, and let $\rho_2(\mathbf{r})$ be a second charge distribution which gives rise to $\phi_2(\mathbf{r})$; then, for a charge distribution which is the sum $\rho_1(\mathbf{r}) + \rho_2(\mathbf{r})$ the potential is the sum of potentials $\phi_1(\mathbf{r}) +$

$\phi_2(\mathbf{r})$. This is the superposition principle. Clearly, the sum of the potentials satisfies the Poisson equation $\nabla^2(\phi_1 + \phi_2) = -4\pi(\rho_1 + \rho_2)/\epsilon$.

In what follows we shall take an origin of coordinates at the center of one particular ion. We shall call the total electrostatic potential ϕ and the electrostatic potential due to all the other ions ψ. According to the superposition principle, the total potential due to the central (chosen) ion and the external charge distribution is the sum of the potentials of the separate charge distributions.

Suppose now that there is an ion of radius $a/2$ and charge zq surrounded by a medium of dielectric constant ϵ. Suppose that between the spheres of radius $r = a$ and $r = a/2$ the charge density $\rho(r)$ is zero but that for $r > a$ there may be a finite charge density. We wish to calculate the contribution to the potential and the field due to the external charge distribution in the region $a/2 < r < a$. In this region, where $\rho(r) = 0$, the potential satisfies the equation $\nabla^2\phi(r) = 0$, or for spherical symmetry

$$\frac{1}{r^2}\frac{d}{dr}\left(r^2\frac{d\phi}{dr}\right) = 0 \qquad \frac{a}{2} < r < a \qquad (21\text{-}5a)$$

to which the general solution is

$$\phi(r) = G + \frac{F}{r} \qquad \frac{a}{2} < r < a \qquad (21\text{-}5b)$$

By Gauss' theorem, since the only charge inside the sphere $r = a$ is that of the central ion, we have for the region $a/2 < r < a$

$$-4\pi r^2 \epsilon \frac{\partial \phi}{\partial r} = 4\pi\epsilon F = 4\pi zq$$

so that $F = zq/\epsilon$, and

$$\phi(r) = G + \frac{zq}{\epsilon r} \qquad \frac{a}{2} < r < a \qquad (21\text{-}5c)$$

The term $zq/\epsilon r$ is the potential due to the ion itself. The important result of this argument is that an external spherically symmetrical charge distribution contributes only a constant (G) to the potential.

It is worthwhile to review some calculations of the electrostatic energy of a charge configuration. For a set of point charges $z_i q$ at positions $\mathbf{r}_i$, the potential energy of interaction is

$$U(\mathbf{r}_1, \ldots, \mathbf{r}_N) = \sum_{1 \leq i < j \leq N} \frac{z_i z_j q^2}{\epsilon r_{ij}} = \frac{1}{2}\sum_{i \neq j} \frac{z_i z_j q^2}{\epsilon r_{ij}} = \frac{1}{2}\sum_{i=1}^{N} z_i q \sum_{j \neq i} \frac{z_j q}{\epsilon r_{ij}} \qquad (21\text{-}6)$$

The electrostatic potential ψ_i at $\mathbf{r}_i$ *due to the other charges* is

$$\psi_i(\mathbf{r}_i) = \sum_{j \neq i} \frac{z_j q}{\epsilon r_{ij}} \qquad (21\text{-}7)$$

We then recognize from the final expression for U in (21-6) that it may also be written as

$$U(\mathbf{r}_1, \ldots, \mathbf{r}_N) = \frac{1}{2} \sum z_i q \psi_i \tag{21-8}$$

The potential energy U is the work required to bring all the charges from infinity to their positions at $\mathbf{r}_i$, $\mathbf{r}_j$, etc., with the dielectric constant held constant. However, there is another way that is useful of calculating the electrostatic work of assembling the charge configuration. The ions are at fixed positions, but the charges on the ions are treated as variables which are simultaneously increased from zero to the final values. Let ξ be a charging parameter which varies from 0 to 1 so that at any stage in the charging process the charge on the ions is $z_i q \xi$ ($i = 1, \ldots, N$). The potential due to the other ions at the position of the jth ion is

$$\psi_j = \sum_{i \neq j} \frac{z_i q \xi}{\epsilon r_{ij}} \tag{21-9}$$

The work to increase the charge on the jth ion by $z_j q\, d\xi$ is $\psi_j q z_j\, d\xi$. The total work in increasing the charge on all ions is

$$dU = \sum_{j=1}^{N} \psi_j(z_1, \ldots, z_N) q z_j\, d\xi \tag{21-10}$$

It is to be emphasized that we are taking into account only the energy of mutual interaction of the ions, but not the electrostatic self-energy of each ion. The potential ψ_j is the potential due to the other ions, and the work term $\psi_j\, d(z_j q \xi)$ is the work done against the forces due to the other ions. The repulsive force of the charge $z_j q \xi$ of the jth ion itself on the increment in charge of the jth ion, $z_j q\, d\xi$, is excluded from the calculation.

From (21-9) and (21-10) we get

$$dU = \sum_{j=1}^{N} \sum_{i \neq j} \frac{z_i z_j q^2}{\epsilon r_{ij}} \xi\, d\xi$$

If the distances r_{ij} are held fixed during the charging process and if ϵ is a constant, integration of the above expression from $\xi = 0$ to $\xi = 1$ gives Eq. (21-6) again.

The charges are immersed in a medium of dielectric constant ϵ. We have not recognized here the molecular nature of the medium, but we know that the dielectric constant is a function of temperature. Since, as already emphasized, the charging process above is at constant ϵ, it is at constant temperature. The energy, therefore, is work at constant temperature and has the nature of a free energy.

We now consider the free energy of immersing a spherical ion in a dielectric. This discussion traditionally begins with the calculation of the electrostatic self-energy of a solid spherical conductor. In electrostatics, the charge on a solid spherical conductor of radius b is all on the surface of the sphere. The potential outside the sphere is $zq/\epsilon r$, where zq is the total charge and ϵ is the dielectric constant. The potential at the surface of the conductor is $zq/\epsilon b$, and it is constant at this value inside the conductor.

To calculate the self-energy of the charged sphere, we proceed as follows: Let zq be the final charge and ξ a charging parameter so that the variable charge on the sphere is $zq\xi$ and ξ is increased continuously from 0 to 1. At any stage the potential at the surface is $zq\xi/\epsilon b$, and the work to increase the charge by $zq\,d\xi$ is $(z^2q^2/\epsilon b)\xi\,d\xi$.

The total work to charge the sphere from zero to zq is therefore

$$w = \int_0^1 \frac{z^2q^2}{\epsilon b}\,\xi\,d\xi = \frac{1}{2}\frac{z^2q^2}{\epsilon b} \tag{21-11a}$$

This is the electrostatic self-energy of a charged conducting sphere in a medium of dielectric constant ϵ. The self-energy of the charged sphere can also be regarded as the energy to establish the electric field in the medium. The field E at a point outside the sphere is $zq/\epsilon r^2$; the electric displacement D is $\epsilon E = zq/r^2$. In electrostatics, the energy per unit volume in an electric field is $DE/8\pi$. In the present instance, this gives for the energy stored in the electric field

$$\int_b^\infty \frac{DE}{8\pi} 4\pi r^2\,dr = \int_b^\infty \frac{z^2q^2}{2\epsilon r^4}\,r^2\,dr = \int_b^\infty \frac{z^2q^2}{2\epsilon r^2}\,dr = \frac{z^2q^2}{2\epsilon b}$$

which is the same as the answer obtained above.

If such a sphere is transferred from a vacuum ($\epsilon = 1$) to a medium of dielectric constant ϵ, work is done by the system and the free-energy change is

$$\Delta F = -\frac{1}{2}\frac{z^2q^2}{b}\left(1 - \frac{1}{\epsilon}\right) \tag{21-11b}$$

This equation is often taken as an approximate calculation of the free energy of solvation of an ion.

There are several comments about Eq. (21-11b) which should be made here. An ion is not a spherical conductor; its charge distribution and self-energy are determined by the solution of the Schrödinger equation and not by elementary electrostatics. However, the external field of any spherical charge distribution confined within a sphere $r \leq b$ is $zq/\epsilon r$ for $r \geq b$. If the electron distribution and the quantized energy level of the ion were not affected by the solvent, the electrostatic energy gained in transferring the ion from a vacuum to a medium of dielectric constant ϵ would still be given by Eq. (21-11b).

Actually, it is now known that the charge distribution and the internal electronic energy of an ion in a prescribed quantum state are changed if the ion is immersed in a solvent. This subject is discussed in the ligand field theory of the transition-metal ions. By taking into account these factors plus the electrostatic contribution contained in Eq. (21-11b) it is possible to give a fairly good quantitative discussion of the free energies of solution of the transition-metal ions.*

Our principal concern in the Debye-Hückel theory will be to calculate the free energy due to the interaction between the ions and to avoid including the energy of interaction with the solvent. We shall briefly consider Eq. (21-11b) and its implications for the entropy of solvation in Sec. 21-15.

21-3. The Poisson-Boltzmann Equation. In the next few sections we present the derivation of the Poisson-Boltzmann equation, and then its approximate solution, more or less in the standard form in which these topics are found in almost all textbook treatments.† It will be clear that the formulation of the fundamental equation is based on a plausible hypothesis and does not proceed rigorously from the fundamental principles of statistical mechanics. To some extent, at least, we shall expose and examine the assumptions that have been made in following sections; however, a presentation of the rigorous derivations of the Debye-Hückel law that are now known is slightly beyond the scope of our treatment.

Consider, for definiteness, a single binary electrolyte $A_{\nu_+}^{z_+}B_{\nu_-}^{z_-}$ which dissociates to give ν_+ cations of type A and ionic valence z_+ and ν_- anions of type B with valence z_-. There are N_+ cations and N_- anions in the volume V; and to conserve electrical neutrality,

$$\nu_+ z_+ + \nu_- z_- = 0 \qquad (21\text{-}12a)$$
$$N_+ z_+ + N_- z_- = 0 \qquad (21\text{-}12b)$$

In fact, if N is the number of "molecules" of the formula $A_{\nu_+}B_{\nu_-}$,

$$N_+ = \nu_+ N \qquad N_- = \nu_- N \qquad (21\text{-}12c)$$

---

* O. G. Holmes and D. S. McClure, *J. Chem. Phys.*, **26**: 1686 (1957); L. E. Orgel, *J. Chem. Soc.*, 4756 (1952).

† We shall not give specific references to most of the derivations that follow. There is an admirable general treatment in Fowler and Guggenheim, chap. IX, pp. 377–419 [1]. The original Debye-Hückel papers [P. Debye and E. Hückel, *Physik. Z.*, **24**: 185 (1923)] have been reprinted in English translation in P. Debye, "Collected Papers," Interscience, New York, 1954. Both these sources give references to the earlier history of the subject, including the work of S. R. Milner, *Phil. Mag.*, **23**: 441 (1912), who correctly recognized many of the important features of the problem. Valuable recent general references are H. S. Harned and B. B. Owen, "The Physical Chemistry of Electrolytic Solutions," 3d ed., Reinhold, New York, 1958, and R. A. Robinson and R. H. Stokes, "Electrolyte Solutions," 2d ed., Butterworth, London, 1959. There are some interesting discussions in W. J. Hamer (ed.), "The Structure of Electrolytic Solutions," Wiley, New York, 1959.

Let α and β be variables which range over the different kinds of ions; for a single binary electrolyte, $\alpha = +$ or $-$, $\beta = +$ or $-$. The above relations can then be written

$$\sum_{\alpha=+,-} \nu_\alpha z_\alpha = 0 \qquad \sum_{\alpha=+,-} N_\alpha z_\alpha = 0 \qquad N_\alpha = \nu_\alpha N \qquad (\alpha = +\text{ or }-)$$
(21-12d)

Problem 21-1. Show that, for a binary electrolyte,

$$\Sigma \nu_\alpha z_\alpha^2 = \nu_+ z_+^2 + \nu_- z_-^2 = (\nu_+ + \nu_-)|z_+ z_-| = \nu|z_+ z_-| \qquad (21\text{-}13a)$$

where
$$\nu = \nu_+ + \nu_- \qquad (21\text{-}13b)$$

The function $\Sigma \nu_\alpha z_\alpha^2$ occurs repeatedly in the Debye-Hückel theory.

Take a system of coordinates with origin at the center of one of the ions which is of type β ($\beta = +$ or $-$). This is the chosen ion. For any given charge distribution of the other ions, the total electrostatic potential ϕ_β (which includes the contribution to the potential from the chosen ion) satisfies the equation

$$\nabla^2 \phi_\beta(\mathbf{r}) = -\frac{4\pi \rho_\beta(\mathbf{r})}{\epsilon}$$

where the subscript emphasizes the origin of coordinates that has been chosen. This instantaneous charge distribution and potential are not necessarily spherically symmetrical. By the standard statistical mechanical procedures for a canonical ensemble, we could calculate the probability of any given configuration and the ensemble average potential $\overline{\phi_\beta(r)}$ around the chosen ion. The rigorous calculation is of course not practical, but it is obvious that the resulting average potential and charge distribution are spherically symmetrical.

If, for a given configuration, ion j of charge $z_j q$ is at point $\mathbf{r}$, its potential energy of interaction with the other ions is not actually $z_j q \phi(\mathbf{r})$, because part of the potential ϕ is due to the ion j itself. However, when this contribution is small, the potential energy of the jth ion is

$$z_j q \phi(\mathbf{r})$$

The fundamental simplifying assumption in the Debye-Hückel theory is that the probability of finding an ion of type α ($\alpha = +$ or $-$) at a place where the average potential is $\overline{\phi_\beta(r)}$ is the Boltzmann factor, $e^{-z_\alpha q \overline{\phi_\beta(r)}/kT}$. Since the average number density of ions of type α throughout the solution is N_α/V, the average number density of ions at radius r is

$$\frac{N_\alpha}{V} e^{-z_\alpha q \overline{\phi_\beta(r)}/kT} \qquad (21\text{-}14a)$$

The average charge density due to ions of type α is therefore

$$\overline{\rho_\alpha(r)} = \frac{N_\alpha}{V} z_\alpha q e^{-z_\alpha q \overline{\phi_\beta(r)}/kT} \qquad (21\text{-}15a)$$

The total average charge density at r is

$$\overline{\rho(r)} = \sum_{\alpha=+,-} \overline{\rho_\alpha(r)} = \sum_{\alpha=+,-} \frac{z_\alpha q N_\alpha}{V} e^{-z_\alpha q \overline{\phi_\beta(r)}/kT}$$

$$= \frac{z_+ q N_+}{V} e^{-z_+ q \overline{\phi_\beta(r)}/kT} + \frac{z_- q N_-}{V} e^{-z_- q \overline{\phi_\beta(r)}/kT} \quad (21\text{-}15b)$$

We note in passing that, since the radial-distribution function (with respect to the chosen ion) of the ions of type α is just the excess probability over a random distribution of finding the ion at r, our fundamental assumption then can be restated as

$$g_{\beta\alpha}(r) = e^{-z_\alpha q \overline{\phi_\beta(r)}/kT} \quad (21\text{-}14b)$$

The average potential $\overline{\phi_\beta(r)}$ and the average charge density $\overline{\rho_\beta(r)}$ must satisfy the Poisson equation:

$$\nabla^2 \overline{\phi_\beta(r)} = -\frac{4\pi \overline{\rho_\beta(r)}}{\epsilon} = -\frac{4\pi}{\epsilon V} \sum_{\alpha=+,-} N_\alpha z_\alpha q e^{-z_\alpha q \overline{\phi_\beta(r)}/kT} \quad (21\text{-}16a)$$

This is called the Poisson-Boltzmann equation; it is the basic equation of the Debye-Hückel theory. We rewrite it, recognizing the spherical symmetry,

$$\frac{1}{r^2}\frac{d}{dr}\left(r^2 \frac{d\phi_\beta}{dr}\right) = -\frac{4\pi}{\epsilon V} \sum_{\alpha=+,-} N_\alpha z_\alpha q e^{-z_\alpha q \phi_\beta(r)/kT} \quad (21\text{-}16b)$$

In (21-16b) we have omitted the bar over the ϕ_β since it will be clear throughout that we are dealing with the mean potential. We shall frequently omit the subscript β, except when it is desirable to emphasize that the chosen ion is of type β.

The equation cannot be solved in closed form. To obtain a first approximate solution, assume that the exponent of the expotential is small, and expand

$$e^{-z_\alpha q \phi_\beta(r)/kT} \approx 1 - \frac{z_\alpha q \phi_\beta(r)}{kT} \quad (21\text{-}17)$$

The right-hand side of (21-16b) then becomes

$$-\frac{4\pi q}{\epsilon V} \sum_\alpha N_\alpha z_\alpha e^{-z_\alpha q \phi_\beta(r)/kT} \approx -\frac{4\pi q}{\epsilon V} \sum_{\alpha=+,-} N_\alpha z_\alpha$$

$$+ \frac{4\pi q^2}{\epsilon V kT} \sum_\alpha N_\alpha z_\alpha^2 \phi_\beta \quad (21\text{-}18)$$

The term $\Sigma N_\alpha z_\alpha$ is set equal to zero; it actually equals $-z_\beta$ since it is a sum over all the ions other than the central ion, but it can be shown that this term is small enough to be neglected except for a very highly charged central ion. The Poisson-Boltzmann equation thus becomes

$$\frac{1}{r^2}\frac{d}{dr}\left(r^2\frac{d\phi}{dr}\right) = \frac{4\pi q^2}{\epsilon V k T}(\Sigma N_\alpha z_\alpha^2)\phi \qquad (21\text{-}19)$$

This is known as the approximate, or linearized, Poisson-Boltzmann equation.

21-4. Introduction of Dimensionless Variables. Further calculations are simplified by the introduction of explicit symbols for several of the characteristic lengths which occur in the theory and the introduction of dimensionless energy variables.

Call κ^2 the coefficient of ϕ on the right-hand side of (21-19); its dimensions are (length)$^{-2}$,

$$\kappa^2 = \frac{4\pi q^2}{\epsilon k T}\sum_{\alpha=+,-}\frac{N_\alpha}{V}z_\alpha^2 \qquad (21\text{-}20a)$$

and it depends on the concentrations of the various ions. The quantity κ^2 is proportional to the "ionic strength" of the solution. The numerical value of the proportionality factor is given in Sec. 21-10. Let

$$s = \frac{q^2}{\epsilon k T} \qquad (21\text{-}20b)$$

where s is the length at which the electrostatic interaction between two unit charges is kT. Note that

$$\kappa^2 = 4\pi s \left(\sum_\alpha \frac{N_\alpha}{V}z_\alpha^2\right) \qquad (21\text{-}20c)$$

If ϕ is the electrostatic potential, $q\phi$ is an energy; we set

$$\Phi = \frac{q\phi}{kT} \qquad (21\text{-}20d)$$

Similarly, we are going to call ψ_β the potential around β due to all the other ions, and its dimensionless form is

$$\Psi_\beta = \frac{q\psi_\beta}{kT} \qquad (21\text{-}20e)$$

21-5. Solution of the Approximate Poisson-Boltzmann Equation. Equation (21-19) now becomes

$$\frac{1}{r^2}\frac{d}{dr}\left(r^2\frac{d\phi}{dr}\right) = \kappa^2\phi \qquad (21\text{-}21a)$$

or

$$\frac{1}{r^2}\frac{d}{dr}\left(r^2\frac{d\Phi}{dr}\right) = \kappa^2\Phi \qquad (21\text{-}21b)$$

The general solution to (21-21) is

$$\Phi = \frac{Ae^{-\kappa r}}{r} + \frac{Be^{+\kappa r}}{r}$$

but we require the potential to vanish at infinity; so $B = 0$, and

$$\Phi(r) = \frac{Ae^{-\kappa r}}{r} \qquad \text{for } r > a \tag{21-22}$$

We assume that the radius of each ion is $a/2$ and that the closest distance of approach is a. Inside the sphere, $r = a$, but outside the ion itself $(a/2 < r < a)$, the dielectric constant is ϵ. The assumption is usually made that, in this spherical shell $(a/2 < r < a)$ the charge density due to the ion atmosphere is zero. This assumption is not entirely reasonable for if a second ion were in contact with the chosen ion, part of the charge of this ion would be within the sphere $r = a$. However, as we shall see, under the conditions where the Debye-Hückel theory is a good approximation, there is not much likelihood of two ions being in contact, and the assumption $\rho(r) = 0$ for $a/2 < r < a$ is a good approximation. It should also be observed that it is not really necessary that all the ions have the same radius $a/2$. If the cations and anions have different radii $a_+/2$ and $a_-/2$ we would take $a = (a_+ + a_-)/2$. This is because the important contacts are between anions and cations; anion-anion and cation-cation contacts are improbable because of energy considerations. In any case, as we shall see, the parameter a, the distance of closest approach is one of the less important and least satisfactory features of the Debye-Hückel theory. It does not enter into the limiting laws, but only into the calculations of activity coefficients at finite concentrations. The exact value of a is difficult to determine and possibly not too meaningful.

If, in the sphere $a/2 < r < a$, $\rho(r) = 0$, then according to our previous discussion culminating in Eq. (21-5c) the potential is the sum of two terms; the contribution of the central ion is $z_\beta q/\epsilon r$, and the contribution of the external ion atmosphere is a constant (G).

$$\phi_\beta(r) = \frac{z_\beta q}{\epsilon r} + G \quad \text{or} \quad \Phi_\beta(r) = \frac{z_\beta s}{r} + G' \qquad \frac{a}{2} < r < a \tag{21-23}$$

However, outside the sphere $r = a$, where there is an ion atmosphere, the approximate solution of the Poisson-Boltzmann equation is [(21-22)]

$$\Phi_\beta(r) = \frac{Ae^{-\kappa r}}{r}$$

The standard boundary conditions in electrostatics for joining up these two solutions at $r = a$ are (1) Φ is continuous and (2) the normal component of the electric displacement $D = -\epsilon(\partial \phi/\partial r)$ is continuous.

$$\frac{A e^{-\kappa a}}{a} = z_\beta \frac{s}{a} + G' \qquad \text{condition 1} \qquad (21\text{-}24a)$$

$$\frac{\epsilon A e^{-\kappa a}}{a^2} + \frac{\epsilon A \kappa e^{-\kappa a}}{a} = \frac{\epsilon z_\beta s}{a^2} \qquad \text{condition 2} \qquad (21\text{-}24b)$$

so that
$$A = \frac{z_\beta s e^{\kappa a}}{1 + \kappa a} \qquad G' = \frac{-z_\beta s \kappa}{1 + \kappa a} \qquad (21\text{-}25)$$

The potential at any point for $r > a$ is

$$\Phi_\beta(r) = \frac{z_\beta e^{\kappa(a-r)}}{1 + \kappa a} \frac{s}{r} \qquad (21\text{-}26)$$

The potential $\psi_\beta(r)$ due to all the ions, other than the chosen ion, is

$$\psi_\beta(r) = \phi(r) - \frac{z_\beta q}{\epsilon r} \qquad (21\text{-}27a)$$

or
$$\Psi_\beta(r) = \Phi_\beta(r) - \frac{z_\beta q^2}{\epsilon r k T} = \Phi_\beta(r) - \frac{z_\beta s}{r} \qquad (21\text{-}27b)$$

We are particularly interested in the value of Ψ at $r = a$;

$$\Psi_\beta(a) = \frac{z_\beta s}{a}\left(\frac{1}{1 + \kappa a} - 1\right) = -z_\beta \frac{s \kappa}{1 + \kappa a} \qquad (21\text{-}28)$$

Indeed, $\Psi_\beta(a)$ is just the constant G' of Eqs. (21-23) and (21-25).

Since we have both $\nabla^2 \phi = -4\pi\rho/\epsilon$ and $\nabla^2 \phi = \kappa^2 \phi$, the total charge density at distance r from the central ion β is

$$\rho_\beta(r) = -\frac{\epsilon \kappa^2 \phi_\beta(r)}{4\pi} \qquad (21\text{-}29a)$$

or, in dimensionless form,

$$\frac{\rho_\beta(r)}{q} = -\frac{\kappa^2 \Phi_\beta(r)}{4\pi s} \qquad (21\text{-}29b)$$

The separate charge density around the chosen ion of type β due only to ions of type α is

$$\rho_{\alpha\beta}(r) \approx z_\alpha q \frac{N_\alpha}{V} e^{-z_\alpha \Phi_\beta(r)} \qquad (21\text{-}30a)$$

$$\rho_{\alpha\beta}(r) \approx z_\alpha q \frac{N_\alpha}{V}[1 - z_\alpha \Phi_\beta(r)] \qquad (21\text{-}30b)$$

Equations (21-30) are written as approximate equalities to emphasize that they are based on the solution of the linearized Poisson-Boltzmann equation derived from the approximation expansion of the potential. Given this initial approximation, it is a somewhat controversial question whether the "best" equation for the charge density is obtained by substituting the approximate potential in the "exact" equation (21-30a) or in the self-consistent approximation (21-30b). We shall use the latter in our further developments.

The radial-distribution function as defined in Sec. 20-2 for ions of type α around an ion of type β is the ratio of the actual to the average number density of the ions

$$g_{\alpha\beta}(r) = \frac{\rho_{\alpha\beta}(r)/z_\alpha q}{N_\alpha/V} = 1 - z_\alpha \Phi_\beta \qquad (21\text{-}31)$$

For example, around a central positive ion, ϕ is positive, g_{++} is everywhere less than unity, and g_{-+} is everywhere greater than unity.

The total amount of charge in the spherical shell r, $r + dr$ around a β ion is $\rho_\beta(r)4\pi r^2\, dr$ or, from (21-26) and (21-29b),

$$\rho_\beta(r)4\pi r^2\, dr = -\frac{z_\beta q \kappa^2 e^{\kappa(a-r)}}{1+\kappa a} r\, dr \qquad (21\text{-}32)$$

The maximum of the charge distribution occurs at

$$\frac{d}{dr}(re^{-\kappa r}) = 0 \quad \text{or} \quad r_{\max} = \frac{1}{\kappa} \qquad (21\text{-}33)$$

So $1/\kappa$ should properly be called the most probable thickness of the ion atmosphere, but it is commonly called the mean thickness (cf. Prob. 21-12).

The total amount of charge outside the central ion is $\int_a^\infty \rho(r)4\pi r^2\, dr$, or

$$-\int_a^\infty \frac{z_\beta q \kappa^2}{1+\kappa a} e^{\kappa(a-r)} r\, dr = -z_\beta q \qquad (21\text{-}34)$$

Thus, our approximate solution is self-consistent, at least to the extent that the total charge outside the ion is the negative of the charge of the ion.

21-6. Activity Coefficients. The Güntelberg Charging Process. There are two methods that have been used for calculating the electrical contribution to the free energy and chemical potential of the ions.

In the Güntelberg charging process, the charge on the chosen ion is continuously increased from 0 to $z_\beta q$. Let the charge at any stage be $z_\beta q \xi$, with ξ increasing continuously from 0 to 1. However, all the other ions have their full charge; so κ^2 is a constant during the charging process. As the central ion becomes charged, the ion atmosphere around it is reversibly established. The potential at the surface of the ion, due to the ion atmosphere at any stage, is [from (21-28)]

$$\psi_\beta(a,\xi) = \frac{-\xi z_\beta q}{\epsilon}\frac{\kappa}{1+\kappa a} \qquad \Psi_\beta(a,\xi) = \frac{-\xi z_\beta s \kappa}{1+\kappa a} \qquad (21\text{-}35)$$

We recall at this point that the potential due to the other ions is constant at $\psi(a,\xi)$ at all points inside the sphere $r = a$, up to the surface of the chosen ion.

The electrical work dw_e in increasing the charge by $d(z_\beta q\xi)$ is $\psi_\beta(a,\xi)\, d(z_\beta q\xi)$, or

$$\frac{dw_e}{kT} = \Psi_\beta(a,\xi) z_\beta \, d\xi \tag{21-36}$$

The total electrical work in charging up this single ion is taken as the electrical contribution to the chemical potential of this ion. Thus,

$$\frac{\mu_{\beta,\text{el}}}{kT} = \int_0^1 - \frac{z_\beta^2 s\kappa}{1 + \kappa a} \xi \, d\xi = -\frac{1}{2} \frac{z_\beta^2 s\kappa}{1 + \kappa a} \tag{21-37}$$

This is the desired final result.

It may not be obvious to the reader (as it was not to the author) that this plausible calculation is a correct statistical mechanical procedure. However, it is easy to show that the calculation is rigorous in the sense that, when the linearized Poisson-Boltzmann equation gives a satisfactory approximation for the charge distribution, the Güntelberg charging process is an example of the charging process for a two-component system which was considered in Sec. 20-6 and which led to Eq. (20-69) for the chemical potential. This demonstration is left to the reader as a (reasonably difficult) problem.

Problem 21-2. Apply Eq. (20-69) for the calculation of the electrical contribution to the chemical potential of a chosen ion of type β, with charge $z_\beta q\xi$, using the expressions previously derived from the linearized Poisson-Boltzmann equation for $g_{+\beta}(r,\xi)$ and $g_{-\beta}(r,\xi)$. (*Hint:* Try to calculate separately the contribution of the positive ions and the negative ions to the chemical potential of the chosen ion. If this fails, try to combine the terms.)

We have now calculated the chemical potentials of the ions of type $\beta(+$ or $-)$. The expression for the chemical potential of a binary salt is given in Sec. 21-8.

21-7. The Electrical Free Energy. The Debye Charging Process. In this procedure all the ions are charged proportionately and simultaneously, allowing the ion atmosphere to establish itself at each stage of the process. The total work is identified with the electrical free energy of the system of ions.

The charge of an ion of type α is $z_\alpha q\xi$. The dimensionless potential at the surface of this ion due to the other ions is

$$\Psi_\alpha(\xi) = -\xi s z_\alpha \frac{\kappa(\xi)}{1 + \kappa(\xi)a} \tag{21-38}$$

The quantity $\kappa(\xi)$ is a function of the charging parameter ξ,

$$\kappa(\xi) = \left(\frac{4\pi s}{V}\right)^{1/2} (\Sigma N_\alpha z_\alpha^2)^{1/2} \xi = \kappa\xi \tag{21-39}$$

where it is understood that the symbol κ means $\kappa(1)$ or the value of κ for $\xi = 1$.

The total electrical work to increase each of the charges by an amount $z_\alpha q\, d\xi$ (for $\alpha = +$ and $\alpha = -$) is dA_{el}, and by analogy with (21-36)

$$\frac{dA_{\text{el}}}{kT} = N_+\Psi_+(\xi)z_+\, d\xi + N_-\Psi_-(\xi)z_-\, d\xi = -s\left(\sum N_\alpha z_\alpha^2\right)\frac{\kappa\xi^2\, d\xi}{1 + \kappa a\xi} \tag{21-40}$$

or

$$\frac{A_{\text{el}}}{kT} = -\sum_{\alpha=+,-} N_\alpha z_\alpha^2 s \int_0^1 \frac{\kappa\xi^2}{1 + \kappa a\xi}\, d\xi \tag{21-41}$$

The integration gives

$$\frac{A_{\text{el}}}{kT} = -\frac{(\Sigma N_\alpha z_\alpha^2)s\kappa}{3}\tau(\kappa a) \tag{21-42}$$

where

$$\tau(\kappa a) = \frac{3}{\kappa^3 a^3}\left[\log(1 + \kappa a) - \kappa a + \frac{\kappa^2 a^2}{2}\right] \tag{21-43}$$

By expanding (21-43) for the case $\kappa a < 1$, we obtain

$$\tau(\kappa a) \approx 1 - \frac{3\kappa a}{4} + \frac{3\kappa^2 a^2}{5} + \cdots \tag{21-44}$$

The general expression (21-42) can be simplified by recalling that $\Sigma N_\alpha z_\alpha^2 = \kappa^2 V/4\pi s$; then

$$\frac{A_{\text{el}}}{kT} = -\frac{V}{4\pi}\frac{\kappa^3}{3}\tau(\kappa a) \tag{21-45}$$

In dilute solution, where $\kappa a \ll 1$, this becomes

$$\frac{A_{\text{el}}}{kT} = -\frac{V\kappa^3}{12\pi} \tag{21-46}$$

The electrical contribution to the chemical potential of component β is $\mu_{\beta,\text{el}} = \partial A_{\text{el}}/\partial N_\beta$. It is easy to show, by differentiation of (21-45), that this calculation leads to the expression (21-37) for μ_β derived by the Güntelberg charging process.

Problem 21-3. Show that $\partial A_{\text{el}}/\partial N_\beta$ gives $\mu_{\beta,\text{el}}$ as stated above.

Since we alleged that we carried out the charging process described above at constant volume, it would seem that the calculated electrical work should be identified with the Helmholtz free energy A. If the process were carried out at constant pressure, the electrical work would give the Gibbs' free energy F. Since there would be some electrostriction of the dielectric as the ions were charged, the Helmholtz free-energy change for a constant-pressure process should include a $P\,\Delta V$ term; however, a fine distinction between A and F is not warranted in view of

the various approximations that are made in carrying out the calculation.*
For example, it is assumed that the dielectric constant is constant
during the charging process. But because of dielectric saturation (cf.
Prob. 18-7), the dielectric constant would decrease during the charging
process. In any case, we shall ignore the distinction between A and F
and shall take (21-42), (21-45), and (21-46) as expressions for F_{el}, the
electrical contribution to the Gibbs' free energy.

21-8. Some Thermodynamics. The Chemical Potential of the Solvent. We consider a binary salt $A_{\nu_+}B_{\nu_-}$ and, in calculating the mole
fractions, assume that it is fully dissociated. Let N_+, N_-, and N_1 be the
moles of each species of ion and of the solvent in the volume V. Let N_2
be the number of "moles" or formula weights of the salt $A_{\nu_+}B_{\nu_-}$;

$$N_+ = \nu_+ N_2 \qquad N_- = \nu_- N_2 \qquad (21\text{-}47)$$

The mole fractions of each species of ion and of the solvent are

$$x_\alpha = \frac{N_\alpha}{N_+ + N_- + N_1} \qquad (\alpha = +\text{ or } -) \qquad x_1 = \frac{N_1}{N_+ + N_- + N_1}$$

$$x_+ + x_- + x_1 = 1 \qquad (21\text{-}48)$$

The general relation between the Gibbs' free energy and the chemical
potentials of solute (μ_2) and solvent (μ_1) is

$$F = N_1 \mu_1 + N_2 \mu_2 \qquad (21\text{-}49)$$

For a change in composition at constant T and P, the Gibbs-Duhem
relation is

$$N_1 \, d\mu_1 = -N_2 \, d\mu_2 \qquad (21\text{-}50)$$

For the salt $A_{\nu_+}B_{\nu_-}$, we write

$$F = N_+ \mu_+ + N_- \mu_- + N_1 \mu_1 = N_2(\nu_+ \mu_+ + \nu_- \mu_-) + N_1 \mu_1 \qquad (21\text{-}51)$$

Then the chemical potential of the salt is given by

$$\mu_2 = \frac{\partial F}{\partial N_2} = \nu_+ \mu_+ + \nu_- \mu_- \qquad (21\text{-}52)$$

This is the fundamental relation between the chemical potential of the
salt and the chemical potentials of the separate ions.

Let us now recall the situation for ideal dilute solutions. The chemical
potentials of the components of an ideal dilute solution of a nondissociat-

---

* Fowler and Guggenheim, sec. 917 [1], assert that it is F, not A, which is calculated.
I do not understand their argument.

ing solute (component 2) in a solvent are given by

$$\mu_{1,\text{id}} = \mu_1^\circ(T,P) + kT \ln \frac{N_1}{N_1 + N_2}$$
$$= \mu_1^\circ + kT \ln \left(1 - \frac{N_2}{N_1 + N_2}\right) \quad (21\text{-}53a)$$

$$\mu_{2,\text{id}} = \mu_2^\circ(T,P) + kT \ln \frac{N_2}{N_1 + N_2} \quad (21\text{-}53b)$$

If the solution is truly ideal, Eqs. (21-53) hold over the entire composition range $0 \leq N_2 \leq 1$, and the μ_i° are the chemical potentials of the pure components. If the formulae hold only for a dilute solution ($N_2 \ll 1$), μ_1° is the chemical potential of pure solvent but μ_2° is calculated from measurements in dilute solution and is not the chemical potential of the pure solute. One can readily show that the Gibbs-Duhem relation (21-50) holds for Eqs. (21-53); in fact (21-53b) can be deduced from (21-53a) plus (21-50).

If the ionic solutions were ideal, the corresponding formulae for the chemical potentials would be

$$\mu_{\alpha,\text{id}} = \mu_\alpha^\circ + kT \ln \frac{N_\alpha}{N_+ + N_- + N_1} \quad \alpha = + \text{ or } -$$
$$\mu_{1,\text{id}} = \mu_1^\circ + kT \ln \frac{N_1}{N_+ + N_- + N_1} \quad (21\text{-}54)$$
$$= \mu_1^\circ + kT \ln \left(1 - \frac{N_+ + N_-}{N_+ + N_- + N_1}\right)$$

For the cases of interest to us, the solutions are sufficiently dilute so that $N_+, N_- \ll N_1$, and Eqs. (21-54) can be written

$$\mu_{\alpha,\text{id}} \approx \mu_\alpha^\circ + kT \ln \frac{N_\alpha}{N_1}$$
$$\mu_{1,\text{id}} \approx \mu_1^\circ + kT \ln \left[1 - \frac{(\nu_+ + \nu_-)N_2}{N_1}\right] \approx \mu_1^\circ - kT \frac{(\nu_+ + \nu_-)N_2}{N_1} \quad (21\text{-}55)$$

The ideal chemical potential of the salt is then given by

$$\mu_{2,\text{id}} = \sum \nu_\alpha \mu_\alpha = \mu_2^\circ + kT \ln (\nu_+^{\nu_+} \nu_-^{\nu_-}) + kT \ln \left(\frac{N_2}{N_1}\right)^{\nu_+ + \nu_-} \quad (21\text{-}56)$$

Problem 21-4. Show that the ideal chemical potentials of the ionic solutions satisfy the Gibbs-Duhem equations

$$\Sigma N_\alpha \, d\mu_{\alpha,\text{id}} + N_1 \, d\mu_{1,\text{id}} = 0$$
$$N_2 \, d\mu_{2,\text{id}} + N_1 \, d\mu_{1,\text{id}} = 0$$

We can now return to the interesting problem, namely, the study of the electrical contributions to the free energy and the chemical potentials.

We attribute all the deviations from ideality to the electrical interactions. It is then true that

$$F_{el} = N_+\mu_{+,el} + N_-\mu_{-,el} + N_1\mu_{1,el} \quad (21\text{-}57a)$$
$$F_{el} = N_2\mu_{2,el} + N_1\mu_{1,el} \quad (21\text{-}57b)$$
where
$$\mu_{2,el} = \nu_+\mu_{+,el} + \nu_-\mu_{-,el} \quad (21\text{-}57c)$$

and it is also true that

$$N_2\, d\mu_{2,el} + N_1\, d\mu_{1,el} = 0 \quad (21\text{-}57d)$$

In the discussions that follow, by the Debye-Hückel limiting-law case we mean the equations for the chemical potentials, free energy, and charge distributions that are obtained when $\kappa a \ll 1$ (that is, as the solutions become sufficiently dilute) in equations such as (21-28), (21-37), and (21-45).

We propose to calculate the chemical potential of the solvent $\mu_{1,el}$ for the Debye-Hückel limiting-law case and then to observe a very simple relation between $\mu_{1,el}$, $\mu_{2,el}$, and F_{el}.

We start with the formulae (21-37) for the chemical potentials of the ions,

$$\frac{\mu_{\beta,el}}{kT} = -\frac{z_\beta^2 s \kappa}{2(1+\kappa a)}$$

The limiting-law equation is

$$\frac{\mu_{\beta,el}}{kT} = -\tfrac{1}{2} z_\beta^2 s \kappa \quad (21\text{-}58a)$$

so that
$$\frac{\mu_{2,el}}{kT} = -\frac{1}{2}\left(\sum \nu_\alpha z_\alpha^2\right) s\kappa \quad (21\text{-}58b)$$

But
$$\kappa^2 = \frac{4\pi s}{V}\left(\sum N_\alpha z_\alpha^2\right) = \frac{4\pi s}{V}\left(\sum \nu_\alpha z_\alpha^2\right) N_2 \quad (21\text{-}59)$$

and so
$$\frac{\mu_{2,el}}{kT} = -\frac{1}{2}\left(\frac{4\pi}{V}\right)^{1/2}\left(s\sum \nu_\alpha z_\alpha^2\right)^{3/2} N_2^{1/2} \quad (21\text{-}58c)$$

or
$$\mu_{2,el} = aN_2^{1/2} \quad (21\text{-}58d)$$

where a is a constant, provided that V (and therefore N_1) is constant.

We can now calculate $\mu_{1,el}$ from the relation $d\mu_1 = -(N_2/N_1)\, d\mu_2$. From (21-58d),

$$d\mu_{2,el} = \tfrac{1}{2} a N_2^{-1/2}\, dN_2 \quad (21\text{-}59a)$$

and so
$$d\mu_{1,el} = -\frac{aN_2^{1/2}}{2N_1}\, dN_2 \quad (21\text{-}59b)$$

Now integrate (21-59b) at constant N_1, recalling that $\mu_{1,el} = 0$ at infinite dilution ($N_2 = 0$);

$$\mu_{1,\text{el}} = -\frac{1}{3}\frac{aN_2^{3/2}}{N_1} \tag{21-59c}$$

or
$$\mu_{1,\text{el}} = -\frac{N_2}{N_1}\frac{1}{3}aN_2^{1/2} = -\frac{N_2}{N_1}\frac{\mu_{2,\text{el}}}{3} \tag{21-59d}$$

Then
$$F_{\text{el}} = N_1\mu_{1,\text{el}} + N_2\mu_{2,\text{el}} = \tfrac{2}{3}N_2\mu_{2,\text{el}} \tag{21-60a}$$

or
$$\frac{F_{\text{el}}}{N_2} = \tfrac{2}{3}\mu_{2,\text{el}} \tag{21-60b}$$

Thus, at high dilutions, when the limiting-law expression (21-58c) for $\mu_{2,\text{el}}$ holds, the electrical free energy of the solution per molecule of salt is two-thirds of $\mu_{2,\text{el}}$. There is a simple formula for $\mu_{1,\text{el}}$ also.

Problem 21-5. Show that Eq. (21-46) for A_{el} (or F_{el}) in dilute solution is identical with (21-60a).

To calculate the chemical potential of the solvent, $\mu_{1,\text{el}}$, according to the Debye-Hückel theory at somewhat higher concentrations, the simplest procedure is to use

$$F_{\text{el}} = N_1\mu_{1,\text{el}} + N_2\mu_{2,\text{el}}$$

with (21-45) for F_{el} and (21-37) for $\mu_{2,\text{el}}$.

21-9. More Thermodynamics. Activity Coefficients. Osmotic Coefficient of the Solvent. We again consider the fully dissociated salt $A_{\nu_+}B_{\nu_-}$. The mole fractions of the several components were calculated in the preceding section.

The weight molalities of the ions (moles per kilogram of solvent) are

$$m_\alpha = \frac{N_\alpha}{N_1}\frac{1{,}000}{M_1} \tag{21-61a}$$

where M_1 is the gram-molecular weight of the solvent.

The volume molalities of the ions (moles per liter) are

$$c_\alpha = \frac{N_\alpha}{N_0}\frac{1{,}000}{V} \tag{21-61b}$$

We shall, whenever convenient, deal with the chemical potentials of the individual ions, although, in thermodynamics, we can measure only the chemical potential of the solute, $\mu_2 = \nu_+\mu_+ + \nu_-\mu_-$. For an ideal solution, the chemical potential of a solute component is $\mu_{\alpha,\text{id}} = \mu_\alpha^\circ + kT\ln x_\alpha$. The rational activity coefficient γ is defined for a real solution by

$$\mu_\alpha = \mu_\alpha^\circ + kT\ln \gamma_\alpha x_\alpha \tag{21-62}$$

For the binary salt $A_{\nu_+}B_{\nu_-}$, the mean activity coefficient is defined by

$$\gamma_\pm = (\gamma_+^{\nu_+}\gamma_-^{\nu_-})^{1/\nu} \tag{21-63}$$

The chemical potential per solute molecule is then given by

$$\mu_2 = \sum_{\alpha=+,-} \nu_\alpha(\mu_\alpha^\circ + kT \ln \gamma_\alpha x_\alpha) = \sum_\alpha \nu_\alpha[(\mu_\alpha^\circ + kT \ln x_\alpha) + kT \ln \gamma_\alpha]$$
$$= \mu_2^\circ + kT \ln (x_+^{\nu_+} x_-^{\nu_-}) + kT \ln (\gamma_+^{\nu_+} \gamma_-^{\nu_-})$$
$$= \mu_2^\circ + kT \ln (x_+^{\nu_+} x_-^{\nu_-}) + kT \ln (\gamma_\pm)^\nu \quad (21\text{-}64)$$

where $\mu_2^\circ = \nu_+ \mu_+^\circ + \nu_- \mu_-^\circ$.

For a real solution, the quantity μ_2° must be determined by extrapolation to infinite dilute, so that $\ln \gamma_\pm \to 0$; that is, μ_2° is operationally defined by

$$\mu_2^\circ = \lim_{x_\alpha \to 0} [\mu_2 - kT \ln (x_+^{\nu_+} x_-^{\nu_-})] \quad (21\text{-}65a)$$

The weight-molal and volume-molal activity coefficients are defined by similar relations with concentrations expressed in these units; for example,

$$\mu_\alpha = \mu_\alpha^\circ + kT \ln \gamma_\alpha m_\alpha \quad (21\text{-}66)$$

Although we use the same symbol μ_α° in (21-62) and (21-66), the value of μ_α° depends on the concentration units being used. This is easy to grasp, because for an ideal solution μ° is the chemical potential at unit concentration, which is certainly different for unit mole fraction and for unit weight molality. The value of μ_2° for weight-molal units, for example, is defined by

$$\mu_2^\circ = \lim_{m_\alpha \to 0} [\mu_2 - kT \ln (m_+^{\nu_+} m_-^{\nu_-})] \quad (21\text{-}65b)$$

For dilute solutions for which the mole fraction x_α, the weight molality m_α, and the volume molality c_α of an ion are proportional to each other, the rational activity coefficient, the weight-molal activity coefficient, and the volume-molal activity coefficient are the same. This follows directly from the definitions (21-62) and (21-66). Even for 1-molal solutions, the differences are small.*

In the Debye-Hückel theory, the only important cause of the deviations from ideality is the electrical interactions and

$$\frac{\mu_{\alpha,\text{el}}}{kT} = \ln \gamma_\alpha = \frac{-z_\alpha^2 \kappa s}{2(1 + \kappa a)} \quad (21\text{-}67a)$$

so that
$$\frac{\mu_{2,\text{el}}}{kT} = \nu \ln \gamma_\pm = \frac{-(\Sigma \nu_\alpha z_\alpha^2) \kappa s}{2(1 + \kappa a)} \quad (21\text{-}67b)$$

The limiting law is

$$\frac{\mu_{2,\text{el}}}{kT} = \nu \ln \gamma_\pm = -\frac{1}{2}\left(\sum \nu_\alpha z_\alpha^2\right) \kappa s \quad (21\text{-}67c)$$

We now turn to the calculation of the chemical potential of the solvent.

---

*See, for example, H. S. Harned and B. B. Owen, "The Physical Chemistry of Electrolytic Solutions," 3d ed., pp. 10–12, Reinhold, New York, 1958.

For an ideal solution, the chemical potential of the solvent is given by

$$\mu_{1,\text{id}} = \mu_1^\circ + kT \ln x_1$$

The deviations from ideality are conveniently represented by the osmotic coefficient g; for a real solution

$$\begin{aligned}\mu_1 &= \mu_1^\circ + gkT \ln x_1 \\ &= \mu_1^\circ + gkT \ln(1 - \Sigma x_\alpha)\end{aligned} \qquad (21\text{-}68)$$

$$\mu_1 \approx \mu_1^\circ - gkT \Sigma x_\alpha \qquad \text{dilute solutions} \qquad (21\text{-}69)$$

The Gibbs-Duhem relation is

$$x_1 d\mu_1 + \Sigma x_\alpha d\mu_\alpha = 0 \qquad T, P \text{ constant}$$

or, by calculating $d\mu_1$ from (21-69) and $d\mu_\alpha$ from (21-62),

$$-x_1 d(g\Sigma x_\alpha) + \Sigma x_\alpha d(\ln \gamma_\alpha) + \Sigma dx_\alpha = 0$$

But $x_1 = 1 - \Sigma x_\alpha \approx 1$, and, by combining the first and third terms,

$$d[(1 - g)(\Sigma x_\alpha)] = -\Sigma x_\alpha d(\ln \gamma_\alpha) \qquad (21\text{-}70)$$

Since in dilute solutions $x_\alpha \sim m_\alpha$, the above equation can be rewritten as

$$d[(1 - g)(\Sigma m_\alpha)] = -\Sigma m_\alpha d(\ln \gamma_\alpha)$$

But $m_\alpha = \nu_\alpha m_2$, where m_2 is the molality of the salt; therefore,

$$d[(1 - g)m_2] = -m_2 d(\ln \gamma_\pm) \qquad (21\text{-}71)$$

These relations can be used with experimental data over a range of concentrations which approach infinite dilution to calculate $\gamma_\pm$ from measurements of g, or vice versa.

To calculate g in the Debye-Hückel theory, we start with

$$\mu_{1,\text{el}} = \mu_1 - \mu_{1,\text{id}} = (1 - g)kT \sum x_\alpha = (1 - g)kT \frac{\nu N_2}{N_1} \qquad (21\text{-}72)$$

But for the limiting-law case ($\kappa a \ll 1$), we have already deduced [Eq. (21-59d)]

$$\mu_{1,\text{el}} = -\frac{N_2}{N_1} \frac{\mu_{2,\text{el}}}{3}$$

so that we have the pleasingly simple final result

$$g - 1 = \frac{1}{3} \frac{\mu_{2,\text{el}}}{\nu kT} = \frac{1}{3} \ln \gamma_\pm \qquad (21\text{-}73a)$$

or

$$g - 1 = -\frac{1}{6} \frac{(\Sigma \nu_\alpha z_\alpha^2)\kappa s}{\nu} \qquad (21\text{-}73b)$$

As already remarked, the calculation of the chemical potential of the solvent in the Debye-Hückel theory when it is not assumed that $\kappa a \ll 1$

can be made by straightforward substitution of the formulae for F_{el} and $\mu_{2,el}$ in the equation

$$F_{el} = N_1\mu_{1,el} + N_2\mu_{2,el}$$

Problem 21-6. The Debye-Hückel expression for g when it is not assumed that $\kappa a \ll 1$ is usually given by the equation

$$1 - g = \frac{\kappa s}{6}\frac{\Sigma x_\alpha z_\alpha^2}{\Sigma x_\alpha}\sigma(\kappa a) \tag{21-74}$$

Find the function $\sigma(\kappa a)$ and its expansion in powers of κa, and show that (21-74) reduces to (21-73) as $\kappa a \to 0$ (cf. Fowler and Guggenheim, p. 399 [1]; H. S. Harned and B. B. Owen, "The Physical Chemistry of Electrolytic Solutions," 3d ed., p. 176, Reinhold, New York, 1958).

21-10. Practical Formulae.* For the parameter s

$$s = \frac{q^2}{\epsilon kT} = \frac{1.67 \times 10^{-3}}{\epsilon T} \quad \text{cm} \tag{21-75a}$$

In water at 25°C ($\epsilon = 78.5$)

$$s = 7.135 \text{ Å} \tag{21-75b}$$

The volume-molar ionic strength is defined by

$$I_v = \frac{1}{2}\frac{1{,}000}{N_0}\sum_\alpha \frac{N_\alpha z_\alpha^2}{V} = \frac{1}{2}\sum_\alpha c_\alpha z_\alpha^2 \tag{21-76}$$

where c_α is the concentration in moles per liter of ions of type α. (The "ional strength" Γ is $2I_v$.) The weight-molal ionic strength I_w is defined in terms of the molalities m_α (moles per kilogram of solvent) of the ions

$$I_w = \frac{1}{2}\sum m_\alpha z_\alpha^2 \tag{21-77}$$

For a dilute solution $I_v = \rho_s I_w$ (ρ_s = density of solvent).

Weight-molal concentrations are preferred by workers in this area for exact thermodynamic work, because these units are invariant under pressure and temperature changes. Volume-molar units are more convenient and more commonly used for less exact work by investigators whose interest is focused on chemical aspects of the problem.

The Debye-Hückel parameter κ is defined by

$$\kappa^2 = 4\pi s \sum_\alpha \left(\frac{N_\alpha}{V}\right)z_\alpha^2$$

---

* The numerical constants quoted are mostly from H. S. Harned and B. B. Owen, "The Physical Chemistry of Electrolytic Solutions," 3d ed., chaps. 3 and 5, Reinhold, New York, 1958, and R. A. Robinson and R. H. Stokes, "Electrolyte Solutions," 2d ed., app. 7.1, p. 468, Butterworth, London, 1959.

and so κ^2 is proportional to the ionic strength I_v and

$$\kappa^2 = \frac{2N_0}{1{,}000} 4\pi s I_v = \frac{2N_0}{1{,}000} 4\pi s \rho_s I_w \qquad (21\text{-}78a)$$

In general,

$$\kappa = \frac{50.29}{(\epsilon T)^{1/2}} I_v^{1/2} \quad \text{Å}^{-1} \qquad (21\text{-}78b)$$

In H$_2$O at 25°C

$$\kappa = 0.329 I_v^{1/2} \quad \text{Å}^{-1} \qquad (21\text{-}78c)$$

The formula for the activity coefficient of an ion of valence z_α is

$$\ln \gamma_\alpha = \frac{-z_\alpha^2 \kappa s}{2(1 + \kappa a)}$$

so that with the ionic diameter a in angstroms,

$$\log \gamma_\alpha = \frac{-z_\alpha^2 \times 1.8246 \times 10^6 (\epsilon T)^{-3/2} I_v^{1/2}}{1 + 50.29 a (\epsilon T)^{-1/2} I_v^{1/2}} \qquad (21\text{-}79)$$

or, for H$_2$O at 25°C,

$$\log \gamma_\alpha = \frac{-0.509 z_\alpha^2 I_v^{1/2}}{1 + 0.33 a I^{1/2}} \qquad (21\text{-}79a)$$

For a binary salt $A_{\nu_+} B_{\nu_-}$,

$$\log \gamma_\pm = \frac{1}{\nu} \log (\gamma_+^{\nu_+} \gamma_-^{\nu_-}) = -\frac{\Sigma \nu_\alpha z_\alpha^2}{2\nu} \frac{\kappa s}{1 + \kappa a} = \frac{-|z_+ z_-|}{2} \frac{\kappa s}{1 + \kappa a}$$

so that

$$\log \gamma_\pm = \frac{-|z_+ z_-| 1.8246 \times 10^6 (\epsilon T)^{-3/2} I_v^{1/2}}{1 + 50.29 a (\epsilon T)^{-1/2} I_v^{1/2}} \qquad (21\text{-}80a)$$

and for H$_2$O at 25°C

$$\log \gamma_\pm = \frac{-0.509 |z_+ z_-| I_v^{1/2}}{1 + 0.33 a I_v^{1/2}} \qquad (21\text{-}80b)$$

21-11. The Bjerrum Theory of Ion Pairs. The Debye-Hückel theory is least accurate when the interactions between the ions are large compared with kT. The expansion $e^{-z_\alpha q \phi_\beta(r)/kT} \approx 1 - [z_\alpha q \phi_\beta(r)/kT]$ was made to obtain the linearized Poisson-Boltzmann equation, which has been used for all our derivations. Close to the surface of the chosen ion, the field is mainly due to this ion so that $\phi_\beta \approx z_\beta q/\epsilon r$; therefore,

$$\frac{z_\alpha q \phi_\beta}{kT} \approx \frac{z_\alpha z_\beta q^2}{\epsilon r kT} = \frac{z_\alpha z_\beta s}{r}$$

For H$_2$O at 25°C this quantity is $z_\alpha z_\beta (7.14/r)$; and for $r = 5$ Å, it is about 1.4 for a 1:1 electrolyte and 5.6 for a 2:2 electrolyte. The smallest value allowed for r is the ionic diameter a; typical values of a are 3.7 Å. The linearized approximation is clearly poor close to the surface of the ion. The errors due to this approximation increase as the concentration increases because of the corresponding increase in the probability of finding two ions close to each other. Thus, we expect the theory as developed in the preceding sections to be least applicable to concentrated solutions of polyvalent ions and to solvents of low dielectric constant; this assessment of the situation is, of course, confirmed by experiment.

One approximate treatment for avoiding the difficulty is the theory of ion-pair formation as proposed by Bjerrum.* We say that any two ions of opposite charge that are sufficiently close such that the electrostatic energy of interaction is greater than $2kT$, that is, $-z_+z_-q^2/\epsilon r > 2kT$, will be regarded not as independent ions but as a molecule or ion pair, with a charge $z_+ + z_-$, which is, of course, less in absolute magnitude than the greater of $|z_+|$ and $|z_-|$.

We assume that the energy of interaction between the two ions of opposite charge that can form the ion pair is given by

$$u_{AB}(r) = \infty \qquad r \leq a \qquad (21\text{-}81a)$$

$$\frac{u_{AB}(r)}{kT} = \frac{z_+z_-q^2}{\epsilon r kT} = \frac{z_+z_-s}{r} \qquad r > a \qquad (21\text{-}81b)$$

That is, the ions are rigid spheres of radius sum a but otherwise interact according to a Coulomb potential.

The equilibrium constant for the reaction

$$A + B \rightleftharpoons AB$$

for spherical particles with no internal degrees of freedom (or for spherical particles with internal degrees of freedom that are not affected by the association) is given by Eq. (15-122) without the symmetry factor of 2;

$$K_{Bj}(\text{atom}^{-1}\text{ cc}) = 4\pi \int_0^{r_{\text{mol}}} e^{-u_{AB}(r)/kT} r^2\, dr \qquad (21\text{-}82)$$

where r_{mol} is the greatest separation for which we are willing to call to AB pair a molecule. The potential is given by (21-81), and r_{mol} is defined by the condition

$$\left|\frac{u_{AB}(r)}{kT}\right| = \frac{-z_+z_-q^2}{\epsilon r_{\text{mol}} kT} = \frac{-z_+z_-s}{r_{\text{mol}}} = 2 \qquad (21\text{-}83)$$

Thus, the association constant is given by

$$K_{Bj}(\text{atom}^{-1}\text{ cc}) = 4\pi \int_a^{-z_+z_-s/2} e^{(-z_+z_-s/r)} r^2\, dr \qquad (21\text{-}84a)$$

* N. Bjerrum, *Kgl. Danske Videnskab Selskab Mat.-fys. Medd.*, **7** (9) (1926); see also the general references.

Set $y = -z_+z_-s/r$. The upper limit of integration corresponds to $y = 2$; the lower limit is $y_a = -z_+z_-s/a$ (note that, since z_- is a negative number, y is positive). Thus,

$$K_{Bj}(\text{atom}^{-1}\text{ cc}) = 4\pi|(z_+z_-s)^3|\int_2^{y_a} e^y y^{-4}\,dy = 4\pi|(z_+z_-s)^3|Q(y_a) \quad (21\text{-}84b)$$

Values of the function $Q(y_a)$ are tabulated;* typical values are

$$y_a = 3 \quad Q = 0.33 \qquad y_a = 6 \quad Q = 1.04 \qquad y_a = 10 \quad Q = 4.6$$

We now consider the effect of the ion-pair association on the predicted equilibrium properties of the solution and on the calculated activity coefficients of the ions for the binary salt $A_{\nu_+}^{z_+}B_{\nu_-}^{z_-}$. Take unit volume of solvent, and let the "true" concentrations of the several species be denoted by N_A, N_B, and N_{AB}. The "formal" concentrations, on the assumption the salt is fully dissociated but that there is no ion-pair formation, are denoted by N_A' and N_B', and of course $N_A' = \nu_+ N_2$, $N_B' = \nu_- N_2$, where N_2 is the formal concentration of $A_{\nu_+}B_{\nu_-}$. The stoichiometric relations are

$$\begin{aligned} N_A' &= N_A + N_{AB} \\ N_B' &= N_B + N_{AB} \end{aligned} \quad (21\text{-}85)$$

The constant K_{Bj} of Eq. (21-84) is a thermodynamic-equilibrium constant, and the equilibrium equation for finite concentrations is

$$\frac{N_{AB}}{N_A N_B}\frac{\gamma_{AB}}{\gamma_A \gamma_B} = K_{Bj} \quad (21\text{-}86)$$

Since the ion-pair treatment is intended to remove the principal cause of deviations from the linearized Debye-Hückel theory, we assume that the latter theory may be used to calculate the activity coefficients of the ions A^{z+}, B^{z-}, and $AB^{(z_++z_-)}$. In using the theory, we take r_{mol} for the distance of closest-approach parameter. This is certainly reasonable for the approach of an A^{z+} to a B^{z-} ion, since, by definition, an AB pair at a closer distance than this is a different ion, $AB^{(z_++z_-)}$. The distance r_{mol} is a reasonable sort of diameter for the AB ion pair also. Therefore the activity coefficients of the three kinds of ions are given by

$$\ln \gamma_\alpha = \frac{-z_\alpha^2 \kappa s}{2(1 + \kappa r_{\text{mol}})} \qquad \alpha = A, B, \text{ and } AB \quad (21\text{-}87)$$

---

* N. Bjerrum, *Kgl. Danske Videnskab. Selskab Mat.-fys. Medd.*, **7**(9) (1926); H. S. Harned and B. B. Owen, "The Physical Chemistry of Electrolytic Solutions," 3d ed., p. 171, Reinhold, New York, 1958; E. A. Guggenheim, *Discussions Faraday Soc.*, **24**: 59 (1957); R. A. Robinson and R. H. Stokes, "Electrolyte Solutions," 2d ed., p. 549, Butterworth, London, 1959.

By combining (21-86) and (21-87) we get*

$$\frac{N_{AB}}{N_A N_B} = K_{Bj}\frac{\gamma_A \gamma_B}{\gamma_{AB}} = K_{Bj}e^{z_+ z_- \kappa s/(1+\kappa r_{\text{mol}})} \qquad (21\text{-}88)$$

The value of κ is given by

$$\kappa^2 = 4\pi s[N_A z_A^2 + N_B z_-^2 + N_{AB}(z_+ + z_-)^2] \qquad (21\text{-}89)$$

(we have taken unit volume). The three equations (21-88), (21-89), and (21-85) must be solved as simultaneous equations to calculate the composition of the solution. This complicated situation arises because the equilibrium degree of association depends on the ionic strength, and vice versa.

Having calculated the "true" concentrations of the several species and the ionic strength, we can now calculate the chemical potentials of the solute components and the solvent. The general condition for chemical equilibrium [Eq. (4-53)] in the present instance is

$$\mu_{AB} = \mu_A + \mu_B \qquad (21\text{-}90)$$

The free energy of the solution is given by

$$F = N_1\mu_1 + N_A\mu_A + N_B\mu_B + N_{AB}\mu_{AB} \qquad (21\text{-}91a)$$

which, in view of (21-90) and (21-85), becomes

$$F = N_1\mu_1 + N'_A\mu_A + N'_B\mu_B \qquad (21\text{-}91b)$$

We note that, even if we had ignored the possibility of association, we would still have written the free-energy equation (21-91b). Thus the assumption of an association equilibrium does not affect the value of the chemical potential of a component as determined by experiment. However, to calculate the free energy and chemical potentials according to the Debye-Hückel theory, we must use Eq. (21-91a) with true concentrations and calculate the chemical potentials μ_A, μ_B, and μ_{AB} according to the theoretical equations.

The "true" activity coefficient of ion A is given by

$$\mu_A = \mu_A^\circ + kT \ln \gamma_A m_A \qquad (21\text{-}92a)$$

(when we use weight-molal units), and, by assumption,

$$\ln \gamma_A = \frac{-z_+^2 \kappa s}{1 + \kappa r_{\text{mol}}}$$

* We use the relation $(z_+ + z_-)^2 - z_+^2 - z_-^2 = 2z_+ z_-$. We note in passing that the combination $e^{z_+ z_- \kappa s/(1+\kappa a)}$ occurs in any problem involving the effect of ionic strength on an ionic association reaction, including the transition-state theory of the effect of ionic strength on the rates of ionic reactions. S. Glasstone, K. J. Laidler, and H. Eyring, "The Theory of Rate Processes," pp. 424–427, McGraw-Hill, New York, 1941.

If we had ignored ionic association, the "formal" activity coefficient of ion A, γ'_A, would be calculated from

$$\mu_A = \mu_A^\circ + kT \ln \gamma'_A m'_A \tag{21-92b}$$

At infinite dilution, the degree of association approaches zero, and μ_A° in (21-92a) and (21-92b) is the same quantity. The formal activity coefficient is then given by

$$\gamma'_A = \gamma_A \frac{m_A}{m'_A} = \gamma_A \frac{N_A}{N_A + N_{AB}} \tag{21-93}$$

The chemical potential of the solvent can be calculated in a straightforward way by using the Gibbs-Duhem relationship.

Problem 21-7. (a) Derive an expression for the limiting-law behavior of a 1:1 electrolyte, $A^z B^{-z}$, assuming a given value of K_{Bj}, and using as a variable $N_2 = N'_A = N'_B$.
(b) Derive an expression for the limiting-law value of the osmotic coefficient of the solvent for the same case.

On the face of it, the most unsatisfactory feature of the theory of ion pairs appears to be the arbitrary nature of the definition of a molecule as a pair at such a distance that their energy of interaction is $2kT$. This criticism is only in part justified; the results are rather insensitive to the choice of $2kT$, instead of, say, $3kT$ or $1.5kT$. This problem has been discussed in a more rigorous and detailed way by Fuoss;* the main point can be developed by a simple argument.

Neglecting activity coefficients, the equilibrium law can be written as $N_{AB}/N_B = K_{Bj} N_A$, which, in effect, says that the probability of finding an A ion within a certain distance r_m (short for r_{mol}) of a B molecule is $K_{Bj} N_A$. But

$$K_{Bj} = 4\pi \int_a^{r_m} e^{-z+z-s/r} r^2 \, dr \tag{21-94}$$

If we change r_m, the degree of association N_{AB}/N_B changes. Assume that the degree of association $\alpha = N_{AB}/N_B$ is small so that N_A is essentially constant at N'_A. Then

$$\frac{d\alpha}{dr_m} = N_A \frac{dK_{Bj}}{dr_m} \tag{21-95}$$

But by differentiating the expression for K_{Bj},

$$\frac{dK_{Bj}}{dr_m} = 4\pi r_m^2 e^{-z+z-(s/r_m)} \tag{21-96}$$

An equivalent statement to Eqs. (21-95) and (21-96) is that the probability of finding an ion (A) within a spherical shell between r and $r + dr$

---

* R. M. Fuoss, *Trans. Faraday Soc.*, **30**: 967 (1934); *Chem. Rev.*, **17**: 27 (1935); see also Fowler and Guggenheim, pp. 413–415 [1].

of a B ion is $N_A 4\pi r^2 g(r)\, dr$ and

$$4\pi r^2 g(r) = 4\pi r^2 e^{-z_+ z_- s/r} \tag{21-97}$$

The function $4\pi r^2 g(r)$ is displayed in Fig. 21-1. The minimum occurs at

$$\frac{d}{dr}(r^2 e^{-z_+ z_- s/r}) = 0 \quad \text{or} \quad r = \frac{-z_+ z_- s}{2}$$

which is just the distance selected for r_m in the Bjerrum theory. The existence of this minimum makes the degree of ion-pair formation

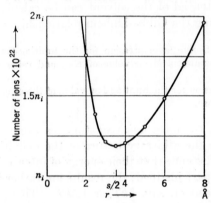

Fig. 21-1. The number of uninegative ions within a spherical shell 0.1 Å thick around a central unipositive ion for a concentration of n_i ions of each type per cubic centimeter, calculated from the expression

$$4\pi n_i r^2 e^{+q^2/\epsilon k t r} = 4\pi n_i r^2 e^{s/r}$$

The minimum occurs at $r = s/2$. (From R. A. Robinson and R. H. Stokes, "Electrolyte Solutions," 2d ed., p. 395, Butterworth, London, 1959.)

relatively insensitive to variations in r_m. Furthermore, if a larger value of r_m is chosen, the number of free ions decreases and the corresponding decrease in ionic strength increases the calculated activity coefficients. This compensating factor also makes the calculated activity coefficients relatively less sensitive to the choice of r_m.

We shall refer briefly to the experimental data bearing on ion pairs in a later section.

Finally, it should be noted that, for the study of ionic solutes in solvents of low dielectric constant, it has proved fruitful to extend the treatment to include clusters of triple ions and quadruple ions.*

21-12. Heats of Dilution and Other Aspects of the Debye-Hückel Theory. Other properties of electrolyte solutions can be calculated from the theoretical expressions for the free energy or chemical potential.

* R. M. Fuoss and C. A. Kraus, *J. ACS*, **55**: 2317 (1933), **57**: 1 (1935).

Sec. 21-12] ELECTROLYTES 513

We shall principally consider the partial molar enthalpy and the heat of dilution.

We first review the elementary thermodynamics of the situation. For a two-component system, the relation between the total enthalpy and the partial molecular enthalpies is

$$H = \tilde{H}_1 N_1 + \tilde{H}_2 N_2$$
$$\tilde{H}_i = \left(\frac{\partial H}{\partial N_i}\right)_{T,P} \quad i = 1, 2 \quad (21\text{-}98)$$

Operationally, $\tilde{H}_2$ is the enthalpy change per molecule if a small amount of component 2 is added to a large amount of solution of composition N_2/N_1. The values at infinite dilution ($N_2/N_1 \to 0$) are denoted by $\tilde{H}_1^\circ$ and $\tilde{H}_2^\circ$; $\tilde{H}_1^\circ$ is the enthalpy per molecule of pure solvent, and $\tilde{H}_2^\circ$ is the enthalpy increase upon dissolving one molecule of solute in a large amount of pure solvent.

The relative partial molecular enthalpies are defined by

$$\tilde{L}_i = \tilde{H}_i - \tilde{H}_i^\circ \quad (21\text{-}99)$$

The enthalpy of infinite dilution, ΔH_D, is the enthalpy change per molecule of solute when a solution at a finite concentration is diluted with an infinite amount of solvent (in an isobaric isothermal process). It can be shown that the enthalpy of infinite dilution is given by

$$N_2 \Delta H_D = -N_1 \tilde{L}_1 - N_2 \tilde{L}_2 \quad (21\text{-}100)$$

Problem 21-8. Prove relation (21-100).

The temperature coefficient of the electrical free energy is related to the electrical contribution to the heat of dilution of an ionic solution. We start with

$$d\left(\frac{F}{T}\right) = -\frac{H}{T^2} dT + \frac{V}{T} dP + \sum \frac{\mu_i}{T} dN_i$$

so that $\quad \tilde{H}_2 = \left(\frac{\partial H}{\partial N_2}\right)_{T,P,N_1} = -T^2 \left[\frac{\partial(\mu_2/T)}{\partial T}\right]_{P,N_1,N_2} \quad (21\text{-}101)$

The chemical potential μ_2 can be written as

$$\frac{\mu_2}{T} = \frac{\mu_2^\circ}{T} + k \ln (N_2/N_1) + \frac{\mu_{2,\text{el}}}{T} \quad (21\text{-}102)$$

if we attribute all the nonideality to electrical interactions. Then, at infinite dilution,

$$\tilde{H}_2^\circ = -T^2 \frac{\partial(\mu_2^\circ/T)}{\partial T} \quad (21\text{-}103)$$

The quantities $\tilde{H}_2^\circ$ and μ_2° relate to the interaction of the solute with the solvent and not to the electrical interactions between solute molecules.

From (21-101) to (21-103) we obtain

$$\tilde{L}_2 = \tilde{H}_2 - \tilde{H}_2^\circ = -T^2 \frac{\partial(\mu_{2,\text{el}}/T)}{\partial T} \tag{21-104}$$

Thus, the relative partial molecular enthalpy is related to the electrical interactions between the solute ions.

We calculate $\tilde{L}_2$ for the limiting-law case. For the salt $A_{\nu_+}B_{\nu_-}$, the limiting law gives

$$\frac{\mu_{2,\text{el}}}{kT} = -\frac{\Sigma \nu_\alpha z_\alpha^2}{2} \kappa s = -\frac{(4\pi)^{1/2}}{2} \left(\frac{N_2}{V}\right)^{1/2} \left(\sum \nu_\alpha z_\alpha^2\right)^{3/2} s^{3/2}$$

We neglect terms in $\partial V/\partial T$; then the temperature dependence of μ_{el}/T is the temperature dependence of $s = q^2/\epsilon kT$. By differentiation and then rearranging, we get

$$\tilde{L}_2 = -\tfrac{3}{2} T^2 \left(\frac{\mu_{2,\text{el}}}{T}\right) \frac{\partial(\ln s)}{\partial T} \tag{21-105}$$

and
$$\frac{\partial(\ln s)}{\partial T} = -\frac{\partial(\ln \epsilon)}{\partial T} - \frac{1}{T} \tag{21-106}$$

It has been shown that, in the limiting-law range, $N_1 \mu_{1,\text{el}} = -\tfrac{1}{3} N_2 \mu_{2,\text{el}}$; it then follows from (21-104) that

$$N_1 \tilde{L}_1 = -\tfrac{1}{3} N_2 \tilde{L}_2 \tag{21-107a}$$
and
$$\Delta H_D = (-\tfrac{2}{3}) \tilde{L}_2 \tag{21-107b}$$

For H_2O at 25°C the quantity $\tilde{L}_2$ is positive since the dominant term of $\partial(\ln s)/\partial T$ is $-\partial(\ln \epsilon)/\partial T$, which is positive. Thus, in the region where the limiting law applies, the enthalpy of dilution $\Delta H_D = -\tfrac{2}{3}\tilde{L}_2$ is negative or heat is liberated as an ionic solution is diluted, even though work is done to separate the ions from their oppositely charged ion atmosphere. The nonequality of the heat and work terms is due to the nonzero and negative value of $\partial \epsilon/\partial T$; this emphasizes the aptness of the remark in Sec. 21-2 that the electrical work at constant ϵ is of the nature of a free energy.

According to (21-105) the quantity $\tilde{L}_2$ is proportional to the square root of the ionic strength. We can write

$$\tilde{L}_2 = A_H \frac{\Sigma \nu_\alpha z_\alpha^2}{2} I^{1/2} \tag{21-108a}$$

and for aqueous solutions at 25°C

$$A_H = 718 \text{ cal mole}^{-3/2} \text{ liter}^{1/2} \tag{21-108b}$$

A similar discussion of the effect of the electrical interactions on the relative partial molar volume of the solute is possible, starting with $\tilde{V}_2 = \partial \mu_2/\partial P$. The quantity of significance here is $\partial \epsilon/\partial P$.

We conclude this section by mentioning that the theory of the effect of concentration on conductivity of ionic solutions is based on the Debye-Hückel theory of the ionic atmosphere, plus additional consideration regarding the drag between two ions moving in opposite directions and the drag of the oppositely charged ion atmosphere on a moving ion.*

21-13. Critique of the Debye-Hückel Theory. Having outlined the basic features of the theory, we now consider the nature and validity of some of its assumptions and approximations.

One of the most drastic oversimplifications is the assumption that the solvent is a continuous medium, with a constant dielectric constant, rather than a system of molecules. Water, for example, has a unique, open, quasi-crystalline molecular structure with a tetrahedral arrangement of the oxygens around each other. There is a characteristic organization of water molecules (a hydration sphere) around a solute ion. The structure of the hydration sphere is different from the structure of pure water. In between the hydration sphere around the ion and the main body of the solvent, there may be a transition zone which does not have either structure. (For an introductory discussion of the structure of aqueous solutions, see R. A. Robinson and R. H. Stokes, "Electrolyte Solutions," 2d ed., chap. 1, Butterworth, London, 1959.) The only point which we need emphasize here is that the interaction between solute ions and solvent is a complicated molecular process which can be only approximately described by the concept of a continuous medium.

Since the Debye-Hückel theory is concerned exclusively with the changes in properties with changing solute concentration, the complexities of the interactions alluded to above are not directly pertinent except in so far as they change with concentration. It is probable, however, that attempts to extend the theory of interionic attraction to higher concentrations without considering the molecular structure of the solvent can have only limited success.

The solvent structure must have a direct effect on the validity of the assumption of a constant dielectric constant. A water dipole in the first coordination sphere of a cation is almost completely oriented (cf. Prob. 18-7) and cannot be properly regarded as part of a medium of constant dielectric constant. From a macroscopic point of view, this is dielectric saturation. It may, however, be a fairly good approximation to treat the second hydration layer as a medium with the macroscopic dielectric constant. Suppose that two ions of opposite charge come so close that there is no water molecule between them. What then is the dielectric constant of the medium between them?

* See H. S. Harned and B. B. Owen, "The Physical Chemistry of Electrolytic Solutions," 3d ed., chaps. 4, 6, Reinhold, New York, 1958; R. A. Robinson and R. H. Stokes, "Electrolyte Solutions," 2d ed., chap. 7, Butterworth, London, 1959.

These difficulties, like most of the other difficulties with the Debye-Hückel theory, are most serious when the ions approach each other closely; i.e., they are most serious in more concentrated solutions.

We now consider several of the technical statistical mechanical questions associated with the Poisson-Boltzmann equation, granting the model of a continuous structureless medium with no dielectric saturation.

We first remark that Poisson's equation $\epsilon \nabla^2 \phi(\mathbf{r}) = -4\pi \rho(\mathbf{r})$, between the electrostatic potential and the charge density of any point $\mathbf{r}$ in the solution, does apply to the ensemble average values of the potential $\overline{\phi(\mathbf{r})}$ and charge density $\overline{\rho(\mathbf{r})}$; that is, $\epsilon \nabla^2 \overline{\phi(\mathbf{r})} = -4\pi \overline{\rho(\mathbf{r})}$. To prove this, we number the ions individually from 1 to N ($N = N_+ + N_-$). For a particular configuration of the ions $\mathbf{r}_1, \mathbf{r}_2, \ldots, \mathbf{r}_N$ there is a charge-density function $\rho(\mathbf{r}, \mathbf{r}_1, \ldots, \mathbf{r}_N)$ and a corresponding electrostatic potential function $\phi(\mathbf{r}, \mathbf{r}_1, \ldots, \mathbf{r}_N)$, where $\mathbf{r}$ is a general position vector in the solution. For this particular configuration, the Poisson equation is satisfied; $\nabla^2 \phi = -4\pi \rho/\epsilon$, where the differentiation ∇^2 is with respect to the variable $\mathbf{r}$. There are several possible ways of expressing $\rho(\mathbf{r}, \mathbf{r}_1, \ldots, \mathbf{r}_N)$: we can regard each ion as corresponding to a continuous charge density, or we can smear out the position vectors $\mathbf{r}_i$ of each ion over the volume element $d^3\mathbf{r}_i$, or we can express ρ as a sum of delta functions centered on $\mathbf{r}_1, \mathbf{r}_2, \ldots, \mathbf{r}_N$. The particular method of expressing ρ is immaterial since any one of them will give rise to the equation $\nabla^2 \phi(\mathbf{r}) = -4\pi \rho(\mathbf{r})/\epsilon$. The ensemble average functions are

$$\overline{\rho(\mathbf{r})} = \frac{\int \cdots \int \rho(\mathbf{r}, \mathbf{r}_1, \ldots, \mathbf{r}_N) e^{-U(\mathbf{r}_1, \ldots, \mathbf{r}_N)/kT} d^3\mathbf{r}_1 \cdots d^3\mathbf{r}_N}{Z} \quad (21\text{-}109a)$$

$$\overline{\phi(\mathbf{r})} = \frac{\int \cdots \int \phi(\mathbf{r}, \mathbf{r}_1, \ldots, \mathbf{r}_N) e^{-U(\mathbf{r}_1, \ldots, \mathbf{r}_N)/kT} d^3\mathbf{r}_1 \cdots d^3\mathbf{r}_N}{Z} \quad (21\text{-}109b)$$

Now, differentiate (21-109b) with respect to $\mathbf{r}$ under the integral sign,

$$\nabla^2 \overline{\phi(r)} = \frac{\int \cdots \int \nabla^2 \phi(\mathbf{r}, \mathbf{r}_1, \ldots, \mathbf{r}_N) e^{-U(\mathbf{r}_1, \ldots, \mathbf{r}_N)/kT} d^3\mathbf{r}_1 \cdots d^3\mathbf{r}_N}{Z}$$

$$= -\frac{4\pi}{\epsilon} \frac{\int \cdots \int \rho(\mathbf{r}, \mathbf{r}_1, \ldots, \mathbf{r}_N) e^{-U(\mathbf{r}_1, \ldots, \mathbf{r}_N)/kT} d^3\mathbf{r}_1 \cdots d^3\mathbf{r}_N}{Z}$$

$$= -\frac{4\pi}{\epsilon} \overline{\rho(\mathbf{r})} \quad (21\text{-}110)$$

which was to be proved.

The fundamental assumption in setting up the Poisson-Boltzmann equation is that the average number density of ions of type α ($\alpha = +$ or $-$) at a distance r from the chosen ion, which is of type β ($+$ or $-$), is $(N_\alpha/V) e^{-z_\alpha q \overline{\phi_\beta(r)}/kT}$ [(21-14a)] or $g_{\beta\alpha}(r) = e^{-z_\alpha q \overline{\phi_\beta(r)}/kT}$ [(21-14b)].

In general, in statistical mechanics, the probability of finding an ion of type α at r with respect to β is $g_{\beta\alpha}(r)$. The average charge density around ion β due to the other ions is

$$\overline{\rho_\beta(r)} = \sum_{\alpha=+,-} \frac{N_\alpha z_\alpha q}{V} g_{\beta\alpha}(r) \qquad (21\text{-}111)$$

The potential of mean force $w^{(2)}_{\alpha\beta}(r)$ was defined by (20-70) and (20-71).

$$g_{\beta\alpha}(r_{12}) = e^{-w_{\beta\alpha}{}^{(2)}(r_{12})/kT} = \frac{V^2 \int \cdots \int e^{-U(\mathbf{r}_1,\ldots,\mathbf{r}_N)/kT} d^3\mathbf{r}_3 \cdots d^3\mathbf{r}_N}{\int \cdots \int e^{-U(\mathbf{r}_1,\ldots,\mathbf{r}_N)/kT} d^3\mathbf{r}_1 \cdots d^3\mathbf{r}_N}$$

where we let the chosen ion of type β have coordinates $\mathbf{r}_1$ and the particular α ion have coordinates $\mathbf{r}_2$.

The rigorous form of the Poisson-Boltzmann equation is

$$\nabla^2 \overline{\phi_\beta(r)} = -\frac{4\pi \overline{\rho_\beta(r)}}{\epsilon} = -\sum_{\alpha=+,-} \frac{N_\alpha z_\alpha q}{V} e^{-w_{\beta\alpha}{}^{(2)}(r)/kT} \qquad (21\text{-}112)$$

The Debye-Hückel assumption, therefore, is that

$$z_\alpha q \overline{\phi_\beta(r)} = w^{(2)}_{\beta\alpha}(r) \qquad (21\text{-}113)$$

This assumption is plausible, but it cannot be said to be obviously true. Kirkwood and Poirier[*] have shown, by a rigorous statistical mechanical treatment, that Eq. (21-113) does hold in the limit of high dilution or weak electrostatic interactions between the ions. When the interactions become larger, there are higher terms in a series expansion of the potential of mean force that become important and so (21-113) and the conventional Poisson-Boltzmann equation are no longer exact.

There are two interesting and useful consistency checks that can be applied to the Debye-Hückel theory. In the first place, by definition, $w^{(2)}_{\beta\alpha}(r) = w^{(2)}_{\alpha\beta}(r)$; therefore, from (21-113),

$$z_\alpha \overline{\phi_\beta(r)} = z_\beta \overline{\phi_\alpha(r)}$$

or
$$\frac{\overline{\phi_\beta(r)}}{z_\beta} = \frac{\overline{\phi_\alpha(r)}}{z_\alpha} \qquad (21\text{-}114)$$

For the second condition, we refer to the section on the Debye charging process and now let the positive ions and negative ions be charged independently. It is notationally convenient to regard z_+ and z_- as variables, rather than using coupling parameters ξ_+ and ξ_-. The equation analogous to (21-40) for the increment in the electrical free energy is (we are again discussing exclusively ensemble average potentials and write ψ instead of $\bar{\psi}$)

$$dA_{el} = N_+ \psi_+(z_+,z_-,a)\, d(z_+q) + N_- \psi_-(z_+,z_-,a)\, d(z_-q) \qquad (21\text{-}115)$$

so that
$$\frac{\partial \psi_+(a)}{\partial z_-} = \frac{\partial \psi_-(a)}{\partial z_+}$$

[*] J. G. Kirkwood and J. C. Poirier, *J. Phys. Chem.*, **58**: 591 (1954).

or, to emphasize the comparison with (21-114),

$$\frac{\partial \psi_\beta(a)}{\partial z_\alpha} = \frac{\partial \psi_\alpha(a)}{\partial z_\beta} \qquad (21\text{-}116)$$

Equations (21-114) and (21-116) are consistency tests that can be applied to the approximate solutions in the Debye-Hückel theory. The solutions of the linearized Poisson-Boltzmann equation satisfy these consistency requirements. Inspection of (21-26) shows that (21-114) is satisfied. The potential $\Psi_+(a)$ is given by (21-28),

$$\Psi_+(a) = \frac{-z_+ s\kappa}{1 + \kappa a} \qquad \text{with } \kappa^2 = 4\pi s \sum \frac{N_\alpha}{V} z_\alpha^2$$

The reader can verify (Prob. 21-14) that

$$\frac{\partial \Psi_+(a)}{\partial z_-} = \frac{\partial \Psi_-(a)}{\partial z_+} \qquad (21\text{-}117)$$

The linearized equation $\nabla^2 \phi = \kappa^2 \phi$ is obtained by expanding the Poisson-Boltzmann equation,

$$\nabla^2 \Phi_\beta = -4\pi s \sum_{\alpha=+,-} \frac{N_\alpha}{V} z_\alpha e^{-z_\alpha \Phi_\beta(r)} \approx 4\pi s \frac{\Sigma N_\alpha z_\alpha^2}{V} \Phi_\beta(r) \qquad (21\text{-}118)$$

Neglecting the ion atmosphere, $\Phi_\beta(r) \approx z_\beta s/r$. For H_2O at 25°

$$z_\alpha \Phi_\beta(r) = z_\alpha z_\beta \frac{7.14}{r}$$

Thus, for $z_\alpha = z_\beta = 1$ and for the ionic diameter $a = 4$ Å, $z_\alpha \Phi_\beta(a) = 1.8$, and the expansion in (21-118) is certainly not accurate at $r = a$. However, detailed calculations show that for a 1:1 electrolyte with $a = 4$ Å, with $1/\kappa = 8$ Å (which corresponds to a 0.16-molal solution in water), the solution of the linearized equation for the potential is close to the solution of the exact equation.* However, for 2:2 electrolytes with $a = 4$ Å $[z_\alpha \Phi_\beta(a) = 7.14]$, the linearized approximation for the potential is poor even for a 0.001-molal solution.

Several more accurate solutions of the Poisson-Boltzmann equation (21-118) have been obtained by numerical integration or a suitable series-expansion method. The corrections to the linearized solution are the so-called "extended terms" of the Debye-Hückel theory.†

* E. Guggenheim, *Discussions Faraday Soc.*, **24**: 53 (1957).

† H. Muller, *Physik. Z.*, **28**: 324 (1927); T. H. Gronwald, V. K. La Mer, and K. Sandred, *Physik. Z.*, **29**: 358 (1928); E. Guggenheim, *Discussions Faraday Soc.*, **24**: 53 (1957), *Trans. Faraday Soc.*, **55**: 1714 (1959); H. S. Harned and B. B. Owen, "The Physical Chemistry of Electrolytic Solutions," 3d ed., pp. 66–69, Reinhold, New York, 1958.

The exact Poisson-Boltzmann equation (21-118) is not linear in ϕ; it can be shown without much difficulty that its solutions do not satisfy the consistency requirements (21-114) and (21-116). The arguments of Kirkwood and Poirier indicate that there are errors in the fundamental assumption of the Poisson-Boltzmann equation, $w^{(2)}_{\alpha\beta}(r) = z_\alpha \phi_\beta(r)$, which are of the same order of magnitude as the errors due to the use of the linearized equation. The more exact solutions of the Poisson-Boltzmann equation appear to be corrections in the right direction. Nevertheless, for the reasons indicated, the accuracy and usefulness of the solutions so obtained are in doubt.

It may be noted that, for a symmetrical electrolyte $z_+ = -z_- = z$,

$$\Sigma N_\alpha z_\alpha = \Sigma N_\alpha z_\alpha^3 = \Sigma N_\alpha z_\alpha^{2n+1} = 0 \qquad (21\text{-}119a)$$

so that in the expansion of

$$\sum N_\alpha z_\alpha e^{-z_\alpha \phi \beta(r)} = \sum N_\alpha z_\alpha \left(1 - z_\alpha \phi_\beta + \frac{z_\alpha^2 \phi_\beta^2}{2} - \frac{z_\alpha^3 \phi_\beta^3}{6} \cdots \right) \qquad (21\text{-}119b)$$

the first, third, and subsequent odd terms all vanish. Including the next term after the linearized approximation gives

$$\nabla^2 \Phi(r) = -4\pi s \left(\sum \frac{N_\alpha}{V} z_\alpha^2 \right) \Phi - 4\pi s \left(\sum \frac{N_\alpha z_\alpha^4}{6V} \right) \Phi^3 \qquad (21\text{-}120)$$

For unsymmetrical electrolytes, there is also a term in Φ^2; it is therefore generally believed that the linearized equation is a better approximation for symmetrical electrolytes than for unsymmetrical ones.

One general effect of the more exact solutions is to increase the probability of finding an ion of opposite charge around the chosen ion.

Mayer has treated the problem of the interaction of a system of ions by an adaptation of the cluster theory of imperfect gases. The obvious difficulty that integrals of the virial-coefficient type

$$\int [e^{-u_{\alpha\beta}(r)/kT} - 1] r^2 \, dr$$

diverge at ∞ for $u_{\alpha\beta}(r) = z_\alpha z_\beta / \epsilon r$ is avoided by recognizing that the ion atmosphere essentially introduces a screening factor $e^{-\kappa r}$ into the Coulomb potential. This treatment justifies the limiting law at high dilutions and gives a useful procedure for calculating higher terms that are important at moderate but finite concentrations.*

Fowler and Guggenheim assert that the Bjerrum ion-pair theory is one of the best ways to correct for the deficiencies of the linearized Debye-

---

* J. E. Mayer, *J. Chem. Phys.*, **18**: 1426 (1950); J. C. Poirier, *J. Chem. Phys.*, **21**: 965, 922 (1953); H. S. Harned and B. B. Owen, "The Physical Chemistry of Electrolytic Solutions," 3d ed., pp. 69–70, Reinhold, New York, 1958; H. Friedman, *Mol. Phys.*, **2**: 23 (1959).

Hückel theory. The advantages are that the treatment is simple and direct and is clearly a correction in the right sense.

We finally remark that the effect of any of the improved theories is to place greater emphasis on the particular interaction between highly charged ions of opposite sign. In the Debye-Hückel theory, the only significant parameter is the ionic strength, and it is predicted, for example, that 0.012-molal NaCl and 0.002-molal LaCl$_3$, which have the same ionic strength, would have the same effect on the activity coefficient of a polyvalent negative ion, such as the Fe(CN)$_6^{-4}$ ion. In the improved theories ion-pair formation or the especially strong interaction between La^{3+} and Fe(CN)$_6^{-4}$ would cause the LaCl$_3$ solution to have a greater effect than the NaCl solution. There is abundant experimental evidence that this is the case.

21-14. Comparison with Experiment. It would be possible to write either a single paragraph or a book about the comparison between the theory of electrolyte solutions and experiment. The paragraph states that the Debye-Hückel theory brilliantly explains the properties of solutions of electrolytes in solvents of high dielectric constant at high dilution. However, in more concentrated solutions, there are discrepancies between theory and experiments. These discrepancies increase with increasing concentration; at a given concentration, they are more marked the greater the charges on the ions.

There are a diversity of experiments and a diversity of theoretical treatments of the deviations from the simple Debye-Hückel theory. The theories generally contain one or more empirical parameters that can be adjusted to fit the data. Different investigators are not in agreement as to the relative merits of these various theories.

We shall try to convey, in a few pages, a general feeling for the concentration range in which the theory works and for the order of magnitude of the corrections. The examples cited are chosen almost at random from the large body of data available. The reader should consult specialized treatises for a more critical and extensive discussion.* Although data on heat content and other properties are pertinent, we shall principally emphasize the measurements of activity coefficients of solute and solvent.

The Debye-Hückel equation for the mean activity coefficient of a salt is

$$\log \gamma_\pm = \frac{-\nu|z_+z_-|Ad_0 I_w^{1/2}}{1 + A'd_0 a I_w^{1/2}} \qquad (21\text{-}121a)$$

where A and A' are known constants and concentrations are now expressed in weight-molality units; d_0 is the density of the solvent.

* H. S. Harned and B. B. Owen, "The Physical Chemistry of Electrolytic Solutions," 3d ed., Reinhold, New York, 1958; R. A. Robinson and R. H. Stokes, "Electrolyte Solutions," 2d ed., Butterworth, London, 1959.

The limiting law at low concentrations is

$$\log \gamma_\pm = -\nu |z_+ z_-| A d_0 I_w^{1/2} \tag{21-121b}$$

The limiting-law equation (21-121b) works very well for quite dilute solutions. The effect of the denominator in the more complete equation (21-121a) is to make the log γ versus $I_w^{1/2}$ plot gradually curve up toward

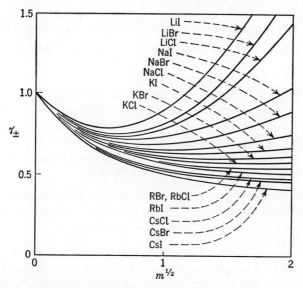

FIG. 21-2. The mean activity coefficients of the alkali halide salts at 25°. (*From H. S. Harned and B. B. Owen, "The Physical Chemistry of Electrolytic Solutions," 3d ed., p. 513, Reinhold, New York, 1958.*)

zero slope with increasing concentration. This is qualitatively in agreement with experiment for most salts for a limited concentration range. However, it is usually found that Eq. (21-121a) does not give a satisfactory quantitative fit over a very wide concentration range. Furthermore, the distance of closest approach, a, which occurs in this equation is essentially an arbitrary adjustable parameter; it cannot be predicted by using known interatomic distances and other structural data.

Figure 21-2 displays the concentration dependence of the activity coefficients of aqueous solutions of the alkali halides. The order of magnitude of the variation in activity coefficients can be seen in the plot, as well as the typical upward curvature. For solutions of concentration greater than about 1 molal, the plots of log $\gamma_\pm$ versus $I_w^{1/2}$ (or $m^{1/2}$) in the figure have a positive slope for some of the salts. This is a qualitative contradiction to the theoretical Eq. (21-121a). Even in somewhat more dilute solutions, Eq. (21-121a) does not give a good quantitative fit to the data.

One common empirical modification of the theoretical equation is to add a linear term in concentration.

$$\log \gamma_\pm = \frac{-\nu|z_+z_-|A d_0(I_w)^{1/2}}{1 + A' d_0 a(I_w)^{1/2}} + Bm \qquad (21\text{-}122)$$

The parameters a and B for aqueous solutions of several salts are displayed in Table 21-1. The equations fit the activity-coefficient data to ± 0.002 for concentrations up to 0.1 molal.

Table 21-1. Parameters of Eq. (21-122) for Several Salts at 25°C†

	a, Å	B, moles^{-1} kg
NaCl	4.12	0.031
KCl	3.97	-0.007_5
CaCl$_2$	4.57_5	0.203

Note: From Sec. 21-10, $A = -0.509$, $A' = 0.33$.

† From H. S. Harned and B. B. Owen, "The Physical Chemistry of Electrolytic Solutions," 3d ed., p. 490, Reinhold, New York, 1958.

From the values of B, it is seen that the Debye-Hückel equation (21-121a) without the Bm term fits the data for NaCl and KCl to within ± 0.003 in $\log \gamma_\pm$ for 0.1-molal solutions; the corrections are much larger for the CaCl$_2$ solutions.

Problem 21-9. With the value of the parameter a given above for KCl, and neglecting the Bm term, what is the difference in the calculated values of $\log \gamma_\pm$ for 0.1-molal KCl between the limiting-law equation (21-121b) and the more complete equation (21-121a)?

Another modified equation recommended by Guggenheim* is

$$\log \gamma_\pm = \frac{-\nu|z_+z_-|A d_0(I_w)^{1/2}}{1 + (I_w)^{1/2}} + \frac{4\nu_+\nu_-}{\nu_+ + \nu_-} B'm \qquad (21\text{-}123)$$

This is equivalent to taking $a = 3.0$ Å for all ions; B' is now the only adjustable parameter. One advantage of the correction terms Bm and $B'm$ of Eqs. (21-122) and (21-123) is that it is easy to generalize them to include specific interaction between ions in solutions containing several salts.

Guggenheim ("Thermodynamics," 3d ed., pp. 355–361, Interscience, New York, 1957) gives values for the parameter $4\nu_+\nu_-B'/(\nu_+ + \nu_-)$ of Eq. (21-123); for 2:1 electrolytes they are usually ~ 1. The corresponding error in $\log \gamma$ if this term were omitted would be 0.01 for a 0.01-molal solution.

* E. A. Guggenheim, "Thermodynamics," 3d ed., pp. 350–356, Interscience, New York, 1957.

Activity coefficients for the rare-earth trihalides* have been reported. In this case, the Debye-Hückel equation (21-120) with $a = 5$ to 7 Å depending on the salt is said to fit the data up to 0.03 molal with deviations in log $\gamma_\pm$ of less than ± 0.002.

As indicated previously, the Poisson-Boltzmann equation can be solved to any desired degree of accuracy without making the linear approximation of the Debye-Hückel theory. This procedure should give a correction in the right direction, although the Poisson-Boltzmann equation is not itself a rigorous statistical mechanical equation. La Mer and coworkers have provided examples of the application of the "extended-term" solutions of the Poisson-Boltzmann equation. For example, it is claimed that this treatment works well for $ZnSO_4$ up to 0.05 molal, although the simple Debye-Hückel theory is not satisfactory for the same concentration range.†,‡

Poirier§ has made calculations using the Mayer cluster-sum theory for the activity coefficients of several electrolytes, which fit the data fairly well up to 0.3 molal for NaCl solutions, and up to 0.05 molal for $ZnSO_4$, for example.

There is an extensive literature on the use of the Bjerrum ion-pair theory to extend the concentration range over which activity-coefficient data can be correlated. Association constants for ion-pair formation can also be estimated from conductance data, absorption spectra, and reaction kinetics.‖

For example, Brown and Prue¶ have measured the freezing points of a

---

* F. H. Spedding and I. S. Yaffe, *J. ACS*, **74**: 4751 (1952); F. H. Spedding and J. L. Dye, *J. ACS*, **76**: 881 (1952).

† I. A. Cowperthwaite and V. K. La Mer, *J. ACS*, **53**: 4333 (1931).

‡ R. M. Fuoss and L. Onsager [*Proc. Nat. Acad. Sci. U.S.*, **47**: 818 (1961)] have proposed a method of integrating the nonlinearized Poisson-Boltzmann equation with different boundary conditions than those used by Debye and Hückel [cf. Eqs. (21-24)]. Instead of the usual result for a symmetrical electrolyte, $\ln \gamma_\pm = -z^2\kappa s/2(1 + \kappa a)$, they obtain for the leading term in their equation for the activity coefficient $\ln \gamma_\pm = -z^2\kappa s/2[1 + (z^2\kappa s/2)]$. This replaces the ion diameter a with the Bjerrum radius r_{mol}. It is claimed that the agreement with experiment is improved.

§ J. C. Poirier, *J. Chem. Phys.*, **21**: 972 (1953); H. S. Harned and B. B. Owen, "The Physical Chemistry of Electrolytic Solutions," 3d ed., p. 544, Reinhold, New York, 1958; calculations have also been made by H. Friedman, *J. Chem. Phys.*, **32**: 1351 (1960).

‖ For a summary, see R. A. Robinson and R. H. Stokes, "Electrolyte Solutions," 2d ed., chap. 14, Butterworth, London, 1959. Association constants for particular ion pairs are given in the compilation by J. Bjerrum, G. Schwarzenbach, and L. Sillen, Stability Constants of Metal-ion Complexes. Part I. Organic Ligands, Part II. Inorganic Ligands, *Chem. Soc. (London) Spec. Publ.* 6 and 7, 1957 and 1958. Special mention should be made of the many contributions by C. B. Monk and by C. A. Davies, which can be found in the above references.

¶ P. G. M. Brown and J. E. Prue, *Proc. Roy. Soc. (London)*, **232**: 320 (1955); see also E. A. Guggenheim, *Discussions Faraday Soc.*, **24**: 54 (1957).

number of divalent sulfates up to concentrations of 0.1 molal. The data could be correlated by the systematic procedure outlined in Sec. 21-11, with values of the association constant of the order of 300 moles^{-1} kg, which correspond to values of the distance of closest approach, a [Eq. (21-84a)], in the neighborhood of 3.5 Å. (Incidentally, these authors give examples of the relatively small effect on the calculated results of changes in the parameter r_{mol} from the Bjerrum value of $z_+z_-q^2/2\epsilon kT$.)

It should be remarked that the distinction between ion-pair association due to purely electrostatic interaction and complex-ion formation due to covalent binding is not always sharp and clear. Thus it is plausible in terms of our general chemical knowledge to regard the complex $CaSO_4$ (in solution) between Ca^{++} ions and $SO_4^=$ ions as essentially an ion pair. On the other hand, the complex ion $HgCl^+$, which forms in solutions containing Hg^{++} ions and Cl^- ions, is held together largely by covalent binding. For other cases however, especially with the transition-metal ions, the situation is less clear.

Even for the "ion-pair" complex $CaSO_4$, our knowledge is very incomplete. Thus, presumably calcium ion in solution occurs as a hydrated ion, perhaps $Ca(H_2O)_6^{++}$. The $SO_4^=$ ion is weakly hydrated also. We do not know how many $CaSO_4$ pairs there are in solution with no intervening water molecules, Ca—O—SO_3, and how many pairs there are with at least one intervening water molecule,

$$\begin{array}{c} \text{H} \\ | \\ \text{Ca—O—H—O—SO}_3 \end{array}$$

At least in the former case, the notion of a coulombic interaction in a medium of dielectric constant 80 is obviously absurd.

It can also be mentioned that in favorable cases, where the hydration sphere around a cation is not rapidly displaced, it is possible to distinguish experimentally between "inner-sphere" and "outer-sphere" complexes. For example, it can be shown that there are distinct species, $Co(NH_3)_5SO_4^+$ and $\{[Co(NH_3)_5(H_2O)]^3(SO_4^=)\}^+$. In the former case, the sulfate anion is directly bonded to the cobaltic cation; the latter complex is an "outer-sphere" complex which is presumably held together principally by electrostatic interactions.*

Some of the most convincing evidence for the ion-pair theory comes from the work of Fuoss, Kraus, and their collaborators on salts of the type $[NR_4^+]A^-$, where R is an alkyl group (C_5H_{11}, for example) and A^- is a large organic anion. These substances are soluble in organic solvents with low dielectric constants, but there is no reasonable possibility of covalent binding between cation and anion. By considering the formation of ion pairs, ion triplets, and ion quadrupoles, it is possible to correlate a

---

* F. A. Posey and H. A. Taube. *J. ACS*, **78**: 15 (1956).

large number of experimental data on conductivities and thermodynamic properties.*

Finally, we reemphasize that all the theories discussed for extending the Debye-Hückel theory to higher concentrations have the effect of increasing the importance of the specific interactions between ions, especially polyvalent ions of opposite charge, whereas in the Debye-Hückel theory the sole parameter which affects the activity coefficients of the ions is the ionic strength.†

21-15. The Entropy of Solvation. We are reluctant to leave the subject of ionic solutions without commenting on one important property which can be understood qualitatively, although not quantitatively, on the basis of a simple electrostatic model. We are concerned here with the interaction between an ion and solvent, rather than with ion-ion interactions.

According to (21-11) the free energy of transferring an ion of radius b from vacuum to a medium of dielectric constant ϵ is

$$\Delta F_{\alpha,\text{el}} = \frac{-z_\alpha^2 q^2}{2b}\left(1 - \frac{1}{\epsilon}\right) \qquad (21\text{-}11)$$

The electrostatic contribution to the entropy of solution is

$$\Delta S_{\alpha,\text{el}} = -\frac{\partial F_{\alpha,\text{el}}}{\partial T} = \frac{z_\alpha^2 q^2}{2\epsilon b}\frac{\partial(\ln \epsilon)}{\partial T} = \frac{kTz_\alpha^2}{2}\frac{s}{b}\frac{\partial(\ln \epsilon)}{\partial T} \qquad (21\text{-}124)$$

Since, in general, $\partial(\ln \epsilon)/\partial T < 0$, the electrical contribution to the entropy of solution of an ion is negative. For aqueous solutions, with $b \approx 2.0$ A, at 25°C Eq. (21-124) predicts $\Delta S_{\alpha,\text{el}} = -5z_\alpha^2$ entropy units mole^{-1}. The entropy decrease is due to the polarization of the dielectric or, from a molecular point of view, to the orientation of polar solvent molecules around the solute ion. The absolute entropies of monatomic ions in the gas phase can, of course, be calculated by elementary statistical mechanics. The partial molal entropy of transferring a positive ion plus a negative ion from the gas phase to solution is experimentally accessible, but the entropies of hydration of the individual ions cannot be rigorously determined. Reasonable estimates, however, can be made.‡

---

* For summaries, see C. A. Kraus, *J. Phys. Chem.*, **60**: 129 (1956); H. S. Harned and B. B. Owen, "The Physical Chemistry of Electrolytic Solutions," 3d ed., chap. 7, Reinhold, New York, 1958.

† For example, in connection with a recent controversy, see V. K. La Mer and R. W. Fessenden, *J. ACS*, **54**: 2351 (1932); G. Scatchard, Electrochemical Constants, *Natl. Bur. Standards Circ.* 524, p. 185, 1953; P. Wyatt and C. W. Davies, *Trans. Faraday Soc.*, **45**: 774 (1949); A. R. Olson and T. R. Simonson, *J. Chem. Phys.*, **17**: 1167 (1949).

‡ R. E. Powell and W. M. Latimer, *J. Chem. Phys.*, **19**: 1139 (1951); R. E. Powell, *J. Phys. Chem.*, **58**: 528 (1954); R. E. Connick and R. E. Powell, *J. Chem. Phys.*, **21**: 2206 (1953); there is a good simple discussion in R. A. Robinson and R. H. Stokes, "Electrolyte Solutions," 2d ed., pp. 14–18, Butterworth, London, 1959.

The qualitative situation is very clear. There is a very marked negative entropy of solvation for ions, and this is greater the greater the charge and the smaller the size of the ion. For the reasons previously discussed relating to the molecular structure of water and to the effects of dielectric saturation, the simple equation (21-124) cannot be expected to be accurate.*

According to (21-124) in an ionic-association reaction

$$A^{z+} + B^{z-} \rightleftharpoons (AB)^{(z_+ + z_-)}$$

there is an increase in the electrostatic entropy which is given by

$$\Delta S = \frac{z_+ z_- q^2}{b\epsilon} \frac{\partial (\ln \epsilon)}{\partial T}$$

This factor may predominate over the loss of translational entropy on association. Thus, the association reactions of neutral molecules in the gas phase or in nonpolar solvents are invariably accompanied by an entropy decrease. For ions, the entropy change is much less negative and frequently positive. This result has profound significance for both the equilibrium and reaction kinetic properties of ionic solutions.†

PROBLEMS

21-10. In Sec. 21-1, several dimensionless parameters are derived, and it is stated that it is these parameters which will occur in the theory of electrolyte solutions. Is this statement true for the actual theoretical results of the Debye-Hückel theory?

21-11. The ionization potential of argon is 15.756 ev.

$$A \rightarrow A^+(^2P_{3/2}) + e^- \qquad \Delta \epsilon_0 = 15.756 \text{ ev}$$

(a) Assuming that all species obey the perfect-gas law, calculate the partial pressures of A^+ and e^- and the degree of ionization of the gas at 12,000°K when the partial pressure of argon is 1.00 atm. Neglect any excited electronic states of argon and A^+, except for $^2P_{1/2}$ for A^+, where $^2P_{1/2} - ^2P_{3/2} = 1,432$ cm^{-1} ($^2P_{3/2}$ is the ground state of A^+). ($k = 8.617 \times 10^{-5}$ ev molecule^{-1} deg^{-1}, and 1.380×10^{-16} erg molecule^{-1} deg^{-1}; 1 ev = 8,068 cm^{-1}.)

---

* For a recent contribution with references, see E. L. King, *J. Phys. Chem.*, **63**: 1070 (1959).

† S. Glasstone, K. J. Laidler, and H. Eyring, "The Theory of Rate Processes," pp. 433–439, McGraw-Hill, New York, 1941.

In the course of the chapter, a number of suggestions and references for further study have already been given. In addition, we wish to mention two related topics which involve the Debye-Hückel theory. The structure of the ionic double layer around a charged colloid particle or a charged macromolecule is discussed in a number of texts, including H. R. Kruyt (ed.), "Colloid Science," Elsevier, Houston, Tex., 1949. For an introduction and further references to the theory of the titration curves and other properties of polyelectrolytes, see S. Lifson and A. Katchalsky, *J. Polymer Sci.*, **13**: 43 (1954); A. Katchalsky, N. Shavit, and H. Eisenberg, *J. Polymer Sci.*, **13**: 69 (1954); F. Wall, J. A. Cote, and S. M. Rucker, *J. Chem. Phys.*, **6**: 1640 (1959).

(b) Use the Debye-Hückel theory to estimate the "activity-coefficient" corrections in the above situation, and calculate the degree of dissociation, taking into account this correction.

21-12. With reference to the remark following Eq. (21-33) that $1/\kappa$ is the most probable thickness of the ion atmosphere, calculate the average values $\langle r \rangle$ and $\langle 1/r \rangle$ for the ion atmosphere around a given ion.

21-13. As shown in Eq. (21-34), the charge-density function (21-32) does give the correct result for the total charge of the ion atmosphere around a chosen ion.

(a) Obtain a general expression for the fractional amount of charge within a sphere of radius r around the chosen ion.

(b) For a 0.01-volume-molal solution of a 1:2 electrolyte, what is the radius of the sphere containing $1 - (1/e)$ of the ion atmosphere? What is the total number of positive ions and of negative ions within this radius around a given positive ion? [Cf. H. S. Frank and P. T. Thompson, in W. J. Hamer ed., "The Structure of Electrolytic Solutions," chap. 8, Wiley, New York, 1959.]

21-14. Verify the statement of Eq. (21-117).

21-15. Calculate the mean activity coefficient of a 0.01-volume-molal $CaSO_4$ solution at 25°:

(a) Using the Debye-Hückel limiting law.

(b) Using the complete Debye-Hückel equation with $a = 4$ Å.

(c) Using the Bjerrum theory of ion-pair formation. [Calculate the degree of association and the mean "formal" activity coefficient, as defined by Eq. (21-93).]

General References

Brief comments about the special features and virtues of several of the books are included. There is no implication that books not singled out for comment are less valuable or interesting.

Statistical mechanics

1. Fowler, R. H., and E. A. Guggenheim: "Statistical Thermodynamics," Cambridge, New York, 1939. This is the classical advanced treatise on the application of statistical mechanics. It is admirable in scope and treatment. Fundamentals are developed via the "method of steepest descents."
2. Tolman, R. C.: "The Principles of Statistical Mechanics," Oxford, New York, 1938. A clear, extensive discussion of the fundamentals of statistical mechanics.
3. Fowler, R. H.: "Statistical Mechanics," Cambridge, New York, 1929.
4. Wilson, A. H.: "Thermodynamics and Statistical Mechanics," Cambridge, New York, 1957.
5. Mayer, J. E., and M. G. Mayer: "Statistical Mechanics," Wiley, New York, 1940.
6. ter Haar, D.: "Elements of Statistical Mechanics," Rinehart, New York, 1954.
7. Rushbrooke, G. S.: "Introduction to Statistical Mechanics," Oxford, New York, 1949.
8. Schrödinger, E.: "Statistical Thermodynamics," Cambridge, New York, 1948. A delightful, short discussion of the fundamentals of statistical mechanics.
9a. Hill, T. L.: "Statistical Mechanics," McGraw-Hill, New York, 1956. An excellent advanced book.
9b. Hill, T. L.: "Statistical Thermodynamics," Addison-Wesley, Reading, Mass., 1960. This admirable, intermediate-level text describes a variety of modern applications, especially for systems of interacting particles. It appeared just as this volume was being finished, and hence is not often referred to.
10. Pitzer, K. S.: "Quantum Chemistry," Prentice-Hall, Englewood Cliffs, N.J., 1953. Elementary statistical mechanics and quantum mechanics; a number of important applications are explained at an intermediate level in a practical and an instructive way.
11. Aston, J. G., and J. J. Fritz: "Thermodynamics and Statistical Thermodynamics," Wiley, New York, 1959.
12. Kittel, C.: "Elementary Statistical Physics," Wiley, New York, 1958.
13. Munster, A.: "Statistiche Thermodynamik, Springer, Berlin, 1956.
14. Landau, L. D., and E. M. Lifschitz: "Statistical Physics," translated from the Russian by E. Pierls and R. F. Pierls, Pergamon, London, 1958.
15. Hirschfelder, J. O., C. F. Curtiss, and R. B. Bird: "Molecular Theory of Gases and Liquids," Wiley, New York, 1954. Intermolecular forces, virial coefficients, theory of liquids, etc.

Other books (spectroscopy, quantum mechanics, etc.)

16. Herzberg, G.: "Atomic Spectra and Atomic Structure," Prentice-Hall, Englewood Cliffs, N.J., 1937; also a Dover reprint.
17. Herzberg, G.: "Molecular Spectra and Molecular Structure, I. Spectra of Diatomic Molecules," 2d ed., Van Nostrand, Princeton, N.J., 1950.

18. Herzberg, G.: "Infrared and Raman Spectra of Polyatomic Molecules," Van Nostrand, Princeton, N.J., 1945.
19. Landau, L. D., and E. M. Lifschitz: "Quantum Mechanics: Non-relativistic Theory," translated from the Russian by J. B. Sykes and J. S. Bell, Pergamon, London, 1958.
20. Eyring, H., J. Walter, and G. E. Kimball: "Quantum Chemistry," Wiley, New York, 1944.
21. Pauling, L., and E. B. Wilson, Jr.: "Introduction to Quantum Mechanics," McGraw-Hill, New York, 1935.
22. Leighton, R. B.: "Principles of Modern Physics," McGraw-Hill, New York, 1959.
23. Margenau, H., and G. M. Murphy: "The Mathematics of Physics and Chemistry," 2d ed., Van Nostrand, Princeton, N.J., 1956.

Tabulations of thermodynamic and spectroscopic properties*

24. Rossini, F. D., et al.: Selected Values of Chemical Thermodynamic Properties, *Natl. Bur. Standards Circ.* 500, 1952. Series I gives values of entropies, heat capacities, heats, and free energies of formation at 25°C; and heats of formation at 0°K for all substances for which the data are available. It does not give such quantities as $(F° - F_0°)/T$ or $H° - H_0°$. Of the common substances, the only major error in the table(s) is the heat of dissociation of nitrogen, which is now known to be $\Delta H_0° = 225.04$ kcal mole^{-1}, rather than 170.2 kcal mole^{-1}.
25. Rossini, F. D., et al.: "Selected Values of Physical and Thermodynamic Properties of Hydrocarbons and Related Compounds," American Petroleum Institute Research Project 44, Carnegie Press, Pittsburgh, 1953. Thermodynamic data, including free-energy functions, etc., for the hydrocarbons and a number of related simple molecules (O_2, OH, CO, CO_2, NO, etc.) over a wide range of temperatures.
26. Hilsenrath, J., et al.: Tables of Thermal Properties of Gases, *Natl. Bur. Standards Circ.* 564, 1955. Statistical thermodymamic functions and other properties for a number of the simple gases.
27. Selected Values of Chemical Thermodynamic Properties, Series III, National Bureau of Standards, 1954. A loose-leaf compilation of thermodynamic functions over a wide range of temperatures for a selection of simple substances. These extremely valuable tables have never been officially published and are not available on the open market; copies can be found in a large number of libraries, however.
28. Brewer, L., L. Bromley, P. W. Gilles, and N. Lofgren: In L. Quill, "Chemistry and Metallurgy of Miscellaneous Materials: Thermodynamics," papers 3–7, National Nuclear Energy Series IV-19B, McGraw-Hill, New York, 1950. Miscellaneous thermodynamic data.
29. Stull, D. R., and G. C. Sinke: "Thermodynamic Properties of the Elements," American Chemical Society, Washington, D.C., 1956. Heat capacities, enthalpies, entropies, and free-energy functions for the elements over a wide temperature range.
30. Moore, C. E.: "Atomic Energy States, *Natl. Bur. Standards Circ.* 467, vol. I, ^{1}H–^{23}V, 1949; vol. II, ^{24}Cr–^{41}Nb, 1952; vol. III, ^{42}Mo–^{57}La, ^{72}Hf–^{89}Ac, 1958.

* *Note added in proof:* There are tables of thermodynamic properties as a function of temperature for a number of substances in G. N. Lewis and M. Randall, "Thermodynamics," 2d. ed., revised by K. S. Pitzer and L. Brewer, pp. 669–686, McGraw-Hill, New York, 1961.

Note: Spectroscopic data for diatomic molecules are given in Herzberg [17]; data for polyatomic molecules are given in a less systematic way in Herzberg [18].

Appendix 1
Physical Constants and Conversion Factors

The values given here are taken from E. R. Cohen, J. W. M. DuMond, T. W. Layton, and J. S. Rollett, *Revs. Modern Phys.*, **27**: 363 (1955). The conversion factor between physical and chemical scales is taken as 1.000272. It should be realized that the "best" values of the fundamental physical constants have undergone periodic revisions since 1940, and different tabulations of data have been calculated, using different physical constants. Some of the important constants have changed by as much as 1/5,000. The changes are not significant for most thermodynamic calculations, but it is clear that a thoroughly consistent tabulation of statistical thermodynamic properties of substances will require recalculation of many of the present data by specialists.

Avogadro's number

Physical scale ($O^{16} = 16.0000$):
$N_0' = (0.602486 \pm 0.000016) \times 10^{24}$ atoms (g mole)$^{-1}$
Chemical scale ($N_0 = N_0'/1.000272$):
$N_0 = (0.602322 \pm 0.000016) \times 10^{24}$ atoms (g mole)$^{-1}$

Gas constant per mole (chemical scale)

$R = (8.31470 \pm 0.00034) \times 10^7$ ergs (g mole)$^{-1}$ deg^{-1}
$= 1.98726 \pm 0.00008$ cal (g mole)$^{-1}$ deg^{-1}
$= 82.054 \pm 0.004$ cc atm (g mole)$^{-1}$ deg^{-1}

Planck constant

$h = (6.62517 \pm 0.00023) \times 10^{-27}$ erg sec
$\hbar = (1.05443 \pm 0.00004) \times 10^{-27}$ erg sec

Boltzmann constant

$k = (1.38044 \pm 0.00007) \times 10^{-16}$ erg deg^{-1}

Velocity of light

$c = (299793.0 \pm 0.3) \times 10^5$ cm sec^{-1}

Second radiation constant

$hc/k = 1.43880 \pm 0.00007$ cm deg

PHYSICAL CONSTANTS AND CONVERSION FACTORS

Wien-displacement law constant

$\lambda_{max}T = (hc/k)(1/4.96511) = 0.289782 \pm 0.000013$ cm deg

Stefan-Boltzmann constant

$\sigma = 2\pi^5 k^4/15h^3c^2 = (5.6687 \pm 0.0010) \times 10^{-5}$ erg cm^{-2} deg^{-4} sec^{-1}

Electronic charge

$|e| = (4.80286 \pm 0.00009) \times 10^{-10}$ esu
$|e'| = |10e/c| = (1.60206 \pm 0.00003) \times 10^{-19}$ coulomb

Electron rest mass

$m = (9.1083 \pm 0.0003) \times 10^{-28}$ g

Proton rest mass

$m = (1.67237 \pm 0.00004) \times 10^{-24}$ g

Ratio proton mass to electron mass

$m_p/m_e = 1{,}836.12 \pm 0.02$

Schrödinger constant for electron

$2m/\hbar^2 = (1.63836 \pm 0.00007) \times 10^{27}$ ergs^{-1} cm^{-2}

Bohr magneton

$\beta = \hbar e/2mc = (0.92731 \pm 0.00002) \times 10^{-20}$ erg gauss^{-1}

Nuclear magneton

$\beta_n = (0.505038 \pm 0.000018) \times 10^{-23}$ erg gauss^{-1}

$\pi = 3.14159265358979$*

Energy-conversion Factors

Ergs molecule^{-1}	Ev	Cm^{-1}	Cal mole^{-1} (chemical scale)
1	6.2420×10^{11}	5.0348×10^{15}	1.4396×10^{16}
1.6021×10^{-12}	1	8,066.04	23,063
1.9862×10^{-16}	1.2398×10^{-4}	1	2.8593
6.9465×10^{-17}	4.3360×10^{-5}	0.34973	1

These numbers have been rounded off to 1 part in 10,000 or the next place thereafter and the probable errors omitted. Electrical units are absolute units. The defined calorie is 4.1840 absolute joules or 4.1840×10^7 ergs.

* The value of π is readily deduced from the mnemonic due, I believe, to Prof. R. C. Tolman:

> Yes, I need a drink, alcoholic of course, after
> the heavy sessions involving quantum mechanics.

Appendix 2
Thermodynamic Functions for the Harmonic Oscillator

These functions are defined in Eqs. (8-25) and (8-26). They are an abridgment of the tables in Pitzer, pp. 457–467 [10], which are in turn an abridgment of tables contained in an Office of Naval Research report by H. L. Johnston, L. Savedoff, and L. Belzer (Washington, 1949). There are also tables taken from Torkington, *J. Chem. Phys.*, **18**: 1373 (1956), and various other sources.

The harmonic-oscillator thermodynamic functions depend on the variable $u = h\nu/kT = 1.4388\omega/T$.

u	$\dfrac{C}{R}$	$\dfrac{H - H_0}{RT}$	$\ln q = -\dfrac{F - F_0}{RT}$	$\dfrac{S}{R}$
0.00	1.0000	1.0000	∞	∞
0.05	0.9998	0.9752	3.0206	3.9958
0.10	0.9992	0.9508	2.3522	3.3030
0.15	0.9981	0.9269	1.9712	2.8981
0.20	0.9967	0.9033	1.7078	2.6111
0.25	0.9948	0.8802	1.5087	2.3889
0.30	0.9925	0.8575	1.3502	2.2077
0.35	0.9898	0.8352	1.2197	2.0549
0.40	0.9868	0.8133	1.1096	1.9229
0.45	0.9833	0.7918	1.0151	1.8069
0.50	0.9794	0.7708	0.9328	1.7036
0.55	0.9752	0.7501	0.8603	1.6104
0.60	0.9705	0.7298	0.7959	1.5257
0.65	0.9655	0.7100	0.7382	1.4482
0.70	0.9602	0.6905	0.6863	1.3768
0.75	0.9544	0.6714	0.6394	1.3108
0.80	0.9483	0.6528	0.5966	1.2494
0.85	0.9419	0.6345	0.5576	1.1921
0.90	0.9352	0.6166	0.5218	1.1384
0.95	0.9281	0.5991	0.4890	1.0881
1.00	0.9207	0.5820	0.4587	1.0407
1.10	0.9050	0.5489	0.4048	0.9537
1.20	0.8882	0.5172	0.3584	0.8756

HARMONIC OSCILLATOR

u	$\dfrac{C}{R}$	$\dfrac{H - H_0}{RT}$	$\ln q = -\dfrac{F - F_0}{RT}$	$\dfrac{S}{R}$
1.30	0.8703	0.4870	0.3182	0.8052
1.40	0.8515	0.4582	0.2832	0.7414
1.50	0.8318	0.4308	0.2525	0.6833
1.60	0.8114	0.4048	0.2255	0.6303
1.70	0.7904	0.3800	0.2017	0.5817
1.80	0.7687	0.3565	0.1807	0.5372
1.90	0.7466	0.3342	0.1620	0.4962
2.00	0.7241	0.3130	0.1454	0.4584
2.20	0.6783	0.2741	0.1174	0.3915
2.40	0.6320	0.2394	0.0951	0.3345
2.60	0.5859	0.2086	0.0772	0.2858
2.80	0.5405	0.1813	0.0627	0.2440
3.00	0.4963	0.1572	0.0511	0.2083
3.20	0.4536	0.1360	0.0416	0.1776
3.40	0.4129	0.1174	0.0340	0.1514
3.60	0.3743	0.1011	0.0277	0.1288
3.80	0.3380	0.0870	0.0226	0.1096
4.00	0.3041	0.0746	0.0185	0.0931
4.25	0.2652	0.0615	0.0144	0.0759
4.50	0.2300	0.0506	0.0112	0.0618
4.75	0.1986	0.0414	0.0087	0.0501
5.00	0.1707	0.0339	0.0068	0.0407
5.50	0.1264	0.0226	0.0041	0.0267
6.00	0.08968	0.01491	0.00248	0.01739
6.50	0.06371	0.00979	0.00151	0.01130
7.00	0.04476	0.00639	0.00091	0.00730
7.50	0.03115	0.00415	0.00055	0.00470
8.00	0.02148	0.00269	0.00034	0.00303
9.00	0.01000	0.00111	0.00012	0.00123

Index

Activity coefficient and radial distribution function, 478
Activity coefficients in electrolyte solutions, 497–498, 503–507, 522–525
  and ion-pair formation, 510–512
Angular momentum (see Spin angular momentum)
Angular momentum operators, 25
Anharmonic oscillator effects, for diatomic molecules, 116–119
  for polyatomic molecules, 194
Antiferromagnetism, 463–466
Antisymmetric wave functions, 36–39
Arrhenius equation, 167

Bjerrum theory of ion-pair formation, 507–512, 523–525
Black-body radiation law, 215–217
Bohr magneton, 428
Boltzmann distribution, 80–81, 91, 150, 155–156
Boltzmann statistics, 77–78
  applicability of, conditions for, 94–95
  corrected, 78–79
Boltzons, corrected, 78–79
  defined, 36
Bose-Einstein statistics, 77, 80, 92–94
Bosons defined, 36
Boyle point, 325
Brownian motion, 285–286
  in galvanometer, 301–302

Capacitor, parallel-plate, 397–398, 418–421

Chemical equilibrium, 54–55
  in perfect gas, 58–60
  and statistical mechanics, 97–104
Chemical potentials in electrolyte solutions, 497–512
  Debye charging process, 498–500
  Güntelberg charging process, 497–498
  and ion-pair formation, 507–512
Classical statistical mechanics, 149–151
Clausius-Mosotti equation, 411–415
Collisions, molecular, 161–167
Combinatorial problems, 65–67
Compressibility factor for a real gas, 310–311
Configuration integral, 207, 314, 468
Conversion factors, for electric and magnetic units, 401
  for energy units, 431
Correlation functions for fluctuations, 292–295
Corresponding states, principle of, 333–337
Curie law for paramagnetism 436, 454

Debye theory, of heat capacities of crystals, 356–362
  of molar polarization, 406–415
Debye-Hückel theory of electrolyte solutions, 485–527
Demagnetizing factors, 452–454
Detailed balancing, 223, 230–235, 237
Diamagnetism, 429–430
Diameter, molecular, 163

Dielectric constants, of gases, 408–409
　of nonpolar liquids, 414–415
　of polar liquids, 415–418
Diffusion, 284–285
Diffusion coefficients and electrical
　　mobilities, 286–288
Dipole moments, electric, defined, 402–403
　magnetic, 428–432
　transition, 226–227
Dispersion, anomalous and normal, 410
Distribution, 40–42
　defined, 75–76
　most probable, 80–81, 242–245, 250–251
　probability of, 307
Distribution functions, 68–71
　for a fluid, 469
　　integral equation for, 483–484
　　(Prob. 20-3)
Dulong and Petit heat-capacity law,
　　351

Effusion, molecular, 158–161
Einstein coefficients of light emission
　　and absorption, 221–226
Einstein theory of heat capacity of
　　crystals, 353–355
Electric birefringence, 422–427
Electric field, thermodynamics of, 418–421
Electrical noise, 295–298
Electrolyte solutions, 485–527
　heat of dilution in, 512–514
　osmotic coefficient of solvent in, 505–506
　(See also Activity coefficients of electrolyte solutions; Chemical potentials in electrolyte solutions)
Electronic energy levels, of atoms, 108–110
　of diatomic molecules, 119–123
　and magnetic susceptibility, 431–434, 440
Electrostriction, 421
Energy, equipartition of, 151–155
Energy conversion factors, 431

Ensemble, canonical, 241–242
　perfect gas in, 245–247
　grand canonical, 248–249
　perfect gas in, 252
Entropy, 46
　of carbon monoxide, 372
　communal, 483–484 (Prob. 20-2)
　and disorder in crystals, 371–378
　of ice, 373–375
　of ion-pair formation, 526
　of mixing, 60–61
　of solvation of ions, 525–526
　statistical-mechanical definition of,
　　87–89, 247, 253–254
Equilibrium, chemical, 54–55, 97–104
　criteria of, 53–54
　between phases, 55
Equipartition of energy, 151–155
Euler's theorem, 51

f numbers, 226–227
Fermi-Dirac statistics, 76, 80, 92–94
Fermions, 36
Fluctuations, in concentration, and
　　density in a two-component
　　system, 272–280
　and light scattering, 272–280
　in density, 267–268
　in energy, in canonical ensemble,
　　266–267
　in grand ensemble, 268–270
　for perfect gas, 267
　in radioactive disintegrations, 298
　spectral density of, 292–295
　Taylor-expansion, Gaussian-approximation method for, 270–280
Fluids, 468–484
　cell theory for, 482–484
　communal entropy, 483–484
　　(Prob. 20-2)
Fourier analysis of random function,
　　290–295
Fourier series and Fourier transforms,
　　288–290
Franck-Condon principle, 236
Fugacity, 331–332
　and radial distribution function, 479

INDEX

g factor, 431–432
Gibbs-Duhem relations, 51–53
Grüneisen rule, 352

Hamiltonian operator, 25–27
Hamilton's equations of motion, 11–17
Harmonic oscillator, 30–31, 113–116
  table of thermodynamic functions for, 532–533
Heat, 43–44
  of dilution in electrolyte solutions, 512–514
  statistical mechanical definition, 86–87, 253–254
Heat capacity, of crystals, 351–371
  at constant pressure and constant volume, 351–352, 371
  Debye theory, 356–362
  Einstein theory, 353–355
  and elastic constants, 360–361
  of metals, 361–362
  of solid orthohydrogen, 375–378
Helix-coil transition in polypeptides, 385–393
Hohlraum, 212–214

Identical particles, wave functions for, 35–39
Integrals, table of, 73
Intermolecular potential, Lennard-Jones, 322–326
  between unlike molecules, 329
  modified Buckingham, 322
  and molecular beams, 341–343
  and second virial coefficient, 315–320
Internal rotation, 194–202
Ion-pair formation, 507–512, 523–526
  activity coefficients and, 510–512
  Bjerrum theory of, 507–512, 523–525
Ions, solvation of, entropy of, 525–526
  free energy of, 490, 525
Ising model for order-disorder transitions in one-dimensional systems, 378–393
Isotope effects in chemical equilibria, 141–144, 203–206

Isotopic molecules, normal vibrations and product rule, 202–203
  thermodynamic functions for, 125–126

Jacobian, 18
Joule-Thomson coefficient, 325

Kerr effect, 422–427
Kirchhoff's law for reflection and emission, 212

Lagrange multipliers, 67–68
Lagrange's equations of motion, 7–15
Lagrangian function, 5, 7–17
Lambda point transitions, 375–378
Lennard-Jones potential, 322–326, 342
Light-absorption coefficients, Einstein and conventional, 220–222
  and emission coefficients, 222–225
  relation to transition moments and f numbers, 226–227
Light emission, induced and spontaneous, 222–226
Light scattering and fluctuations, 272–280
London dispersion forces, 321
Lorentz-Lorenz equation, for local electric field, 412–413
  for local magnetic field, 453

Magnetic cooling, 458–461
  nuclear, 466
Magnetic dipoles, 428–429
Magnetic materials, thermodynamics of, 447–449
Magnetic moment, 430–433, 442–445, 451
  effective, 455
Magnetic resonance, 449–452
Magnetic saturation, 437–439
Magnetic susceptibility, 436, 439–442
  and crystalline Stark-effect splittings, 461–463

Mass action, law of, 97–101
Matrix method for one-dimensional order-disorder transitions, 378–393
Maxwell-Boltzmann velocity distribution, 155–158
Mean-square deviation, 69–70, 266
Microscopic reversibility, 223, 230–235
  and radiative recombination, 235–238
Molecular collisions, 161–167
Molecular diameter, 163
Molecular effusion, 158–161
Molecular weight, number average, 280
  weight average, 280
Moments of inertia of polyatomic molecules, 170–171

Newton's equations of motion, 8
Noise, electrical, 295–298
  shot, 299–301
Normal coordinates, 181–182
Normal vibrations, of a crystal, 355–356, 368–371
  of a one-dimensional crystal, 362–367
  of a polyatomic molecule, 178–194
  degeneracy, 191–192
Nuclear magnetic cooling, 466
Nuclear magnetism, 432–433
Nuclear spin degeneracy, 133, 136–138

Onsager theory for dielectric constants, 418
Operators, quantum-mechanical, 23–25
Order-disorder transitions, in crystals, 375–378, 463–466
  in one-dimensional systems, 378–393
Ortho hydrogen, 130–136
Ortho spin states for diatomic molecules, 133, 136–137
Osmotic coefficient of solvent in electrolyte solutions, 505–506

Para hydrogen, 130–136
Para spin states for diatomic molecules, 133, 136–137

Parallel-plate capacitor, 397–398, 418–421
Paramagnetism, 430–437, 439
  Curie law for, 436, 454
  of rare earth ions, 455
  of transition metal ions, 455–458
Particle in a box, energy levels and forces, 39–40
  quantum-mechanical, 27–29
Partition function, for anharmonic nonrigid rotor, 116–119
  canonical, 244, 252
  for a fluid, 468
  classical, 149
  and configuration integral, 207, 314
  for a polyatomic molecule, 206–210
  electronic, of atoms, 107–110
  grand canonical, 250–252
  for harmonic oscillator, 115–116
  for independent particles, 81
  for interacting magnetic dipoles, 443–446
  and magnetic and crystalline Stark-effect splittings, 462–463
  molecular, 81
  for noninteracting magnetic dipoles, 434–437
  for one-dimensional system, 382–393
  for quantum-mechanical perfect gas, 84–85
  for rigid diatomic rotor, 112–113
  for rotation, of asymmetric top molecules, 177–178
  of symmetric top molecules, 172–173
  for vibration, of a crystal, 355–356
  of a polyatomic molecule, 193
Paschen-Back effect and paramagnetism, 457–458
Perfect gas, chemical equilibrium in, 58–60
  chemical potential for, 56–60
  quantum-mechanical, 39–42, 84–86
  partition function for, 84–85
  volume in phase space for, 304–306

INDEX

Perturbation theory, 33–34
Phase space, 17–20, 146–149
   and number of quantum states, 146–149
   volume in, for a perfect gas, 304–306
Physical constants, 530–531
Poisson-Boltzmann equation, 491–494, 517
   solution of, 494–497, 518–519
Polarizability, electric, 404–405
   molar, 408
Polarization, electric, 406–411
   molar, Debye theory of, 406–415
Potential of mean force, 481–482, 517
Product rule, 202–204

Quantum-mechanical operators, 23–25
Quantum-mechanical particle in a box, 27–29
Quantum-mechanical perfect gas, 39–42, 84–86

Radial-distribution function, 470–472
   and chemical potential, 475–479
      in two-component system, 479–481
   and equation of state, 473–475
   in an ionic solution, 493, 497, 512, 517
   and potential energy, 472–473
Radiation, 211–239
   maximum emission, frequency of, 219
      wavelength of, 219
   by semitransparent gas, 228–230
   statistical mechanics of, 211–239
   thermodynamics of, 219–220
Radiative recombination, 235–238
Random walk, 283–284
   and diffusion, 284–285
Rayleigh-Jeans formula for black-body radiation, 215
Reversible and irreversible changes, 44–46
Rotation, of a diatomic molecule, 29–30, 111–113, 116–119, 125

Rotation, internal, for polyatomic molecule, 194–202
   of molecules in crystals, 375–378
   and nuclear spin, 129–138, 144–145
   of a polyatomic molecule, 125, 170–178

Schrödinger equation, 26–27
Secular equation, 180, 184, 188, 383–384, 389
Shot noise, 299–301
Spin angular momentum, 34–35
   electronic, of atoms, 108–110
   and magnetic moment, 430–432
Spin-orbit coupling, 108–109
Standard deviation, 69
Standard states, 56–59, 62–63, 102–103, 124, 330
Stefan-Boltzmann constant, 218
Stirling's approximation, 71–72
Symmetric top molecule, rotational energy levels of, 172–173
Symmetric wave functions, 36–39
Symmetry coordinates, 186–187
Symmetry number, 112, 138–139, 173–176
   and internal rotation, 196–200

Teller-Redlich product rule, 202–203
Thermodynamic functions, 48–51
   for bosons, 92–93
   in canonical ensemble, 247–248, 261
   for corrected boltzons, 89–91
   for crystals, 356–357
   for diatomic and polyatomic molecules, 123–126, 178, 193
   for dielectrics, 419–421
   for dilute systems of noninteracting particles, 89–91
   for fermions, 92–93
   in grand canonical ensemble, 252–256, 262–263
   for harmonic oscillator, tables of, 532–533
   for isotopic molecules, 125–126

Thermodynamic functions, for magnetic materials, 435–437, 444, 448
  for nuclear spin degeneracy, 138
  for radiation field, 219–220
  for real gases, 329–333
  tables of, 61–64
Thermodynamics, 43–64
  criteria of equilibrium in, 53–54
  derived from statistical mechanics, 89
  of electric field, 418–421
  first law, 44
  of magnetic materials, 447–449
  of radiation, 219–220
  second law, 46
  third law, 47–48
Tunneling and internal rotation, 198

Uncertainty principle, 27, 31, 42
  and volume in phase space, 149
Units, electrical and magnetic, 394–401

van der Waals forces, 312
Vibration, of diatomic molecules, 113–119
  of polyatomic molecules, 178–194
Virial coefficient, second, and chemical potential, 330
  and fugacity, 331
  and intermolecular potential, 315–320
  for Lennard-Jones 6–12 potential, 323–326
  of mixtures, 326–329
  and pair formation, 337–341
  and radial distribution function, 475
  for rigid spheres, 320–321
Virial coefficients, 313
  general derivation for, 345–349

Wiener-Khintchine theorem, 290–295
Work, 43–45
  in statistical mechanics, 86, 253